ILSI Human Nutrition Reviews

Series Editor: Ian Macdonald

Already published:

Sweetness
Edited by John Dobbing

Forthcoming titles in the Series:

Sucrose: Nutritional and Safety Aspects
Edited by Gaston Vettorazzi and Ian Macdonald

Zinc in Human Biology
Edited by C. F. Mills

B. E. C. Nordin (Ed.)

Calcium in Human Biology

With 110 Figures

Springer-Verlag
London Berlin Heidelberg New York
Paris Tokyo

B. E. C. Nordin
Institute of Medical and Veterinary Science, Adelaide,
South Australia 5000, Australia

Cover design permission. The calcium triangle within the calcium hexagons
showing the position of the phosphate groups (not to scale and a projection on a flat
surface of ions arranged three-dimensionally) is reproduced by permission from
Jenkins G.N. The Physiology and Biochemistry of the Mouth 4th ed (1978),
Blackwell Scientific Publications, Oxford

ISBN 3–540–17475–3 Springer-Verlag Berlin Heidelberg New York
ISBN 0–387–17475–3 Springer-Verlag New York Berlin Heidelberg

British Library Cataloguing in Publication Data
Nordin, B.E.C. (Borje Edgar Christopher)
Calcium in human biology. — (ILSI human nutrition reviews).
1. Man. Calcium & phosphorous. Metabolism
I. Title II. Series
612'.3924
ISBN 3–540–17475–3

Library of Congress Cataloging-in-Publication Data
Calcium in human biology / B. E. C. Nordin (ed.).
    p.   cm. — (ILSI human nutrition reviews)
  Includes bibliographies and index.
  ISBN 0–387–17475–3 (U.S.)
  1. Calcium—Physiological effect.   2. Calcium—Metabolism.   I. Nordin,
B. E. C. (Borje Edgar Christopher)   II. Series. [DNLM: 1. Calcium—metabolism.
QV 276 C1436]   QP535.C2C2635 1988   612'.3924—dc 19   DNLM/DLC
for Library of Congress   487–37662 CIP

The use of names, trademarks, etc. in this publication does not imply,
even in the absence of a specific statement, that such names are exempt from the
relevant protective laws and regulations and therefore free for general use.

Product Liability: The publisher can give no guarantee for information about drug
dosage and application thereof contained in this book. In every individual case the
respective user must check its accuracy by consulting other pharmaceutical
literature.

Filmset by Wilmaset, Birkenhead, Wirral
Printed at the University Printing House Oxford

2128/3916–543210

Foreword

The present volume is one of a series concerned with topics considered to be of growing interest to those whose ultimate aim is the understanding of the nutrition of man. One volume, *Sweetness*, has already been published, and another, *Zinc in Human Biology*, is in preparation.

Written for workers in the nutritional and allied sciences rather than for the specialist, the series aims to fill the gap between the textbook on the one hand and the many publications addressed to the expert on the other. The target readership spans medicine, nutrition and the biological sciences generally and includes those in the food, chemical and allied industries who need to take account of advances in these fields relevant to their products.

The reason for choosing calcium as the subject of this monograph is the growing awareness of the importance of this mineral, especially in medicine. Its role in human physiology and pathology is becoming increasingly apparent and this book brings together metabolic and biochemical aspects concerned in understanding the part played by calcium in health and disease.

The International Life Sciences Institute (ILSI) is a non-profit-making scientific foundation established to encourage and support research and educational programmes in nutrition, toxicology and food safety and to foster cooperation in these programmes between scientists from universities, industry and government in order to facilitate the resolution of health and safety issues worldwide.

London
January 1988

Ian Macdonald
Series Editor

Preface

It is a sign of the times that a work designed to review the role of calcium in human biology should have devoted so much of its space to cellular calcium. Only five years ago the internal balance of a work such as this would have been quite different. However, with a realization that calcium is actively involved in virtually every biological process has come the need for all of us to understand or at least to respect its role as a messenger. No doubt it is this ubiquitous role that has necessitated its almost uniquely precise extracellular regulation, which in turn relies on the virtually unlimited stores in the skeleton to cope with intra- and inter-individual variation in calcium intake and absorption. The fact that the skeleton's humble role as a source of calcium sooner or later comes into conflict with its more obvious mechanical role is of little interest to the evolutionary process, since the final skeletal failure, when it occurs, occurs too late in life to influence the evolutionary pattern.

No one author could be expected to command the expertise which this volume required, but neither can 17 be brought together to contribute the individual components without unevenness and overlap occurring between the chapters. At least, or so it seems to the Editor, there are more repetitions than gaps, and if the repetitions serve to reinforce important messages they will have served their turn. They could hardly have been avoided without injury to the smooth flow of the argument in one or other chapter and so they have been left untouched, even to the extent of the same figure appearing in almost identical guise in two places.

The Editor is not responsible for all the opinions or statements in this volume; in fact, there are some with which he sharply disagrees. But editorship is not the same as censorship and views honestly held and based on respectable data must be respected, even if some would interpret the data differently from others.

I am grateful to the International Life Sciences Institute for asking me to edit this volume and hope that it will prove useful to a

wide readership. It is not to be equated with the proceedings of a
scientific symposium. Although the material is up to date it is not,
we hope, ephemeral but rests on a solid foundation of well-
established knowledge. At least, this was the remit to the
contributors which I hope they have discharged.

Adelaide   B. E. C. Nordin
September 1987

Contents

Contributors

Prof. P. H. Adams
Department of Medicine, The Royal Infirmary, Manchester
M13 9WL, UK

Prof. G. D. Aurbach
Department of Health and Human Services, National Institutes of
Health, Bethesda, Maryland 20205, USA

Prof. M. P. Blaustein
Department of Physiology, University of Maryland, School of
Medicine, 655 West Baltimore Street, Baltimore, Maryland 21201,
USA

Dr. A. L. Boskey
Ultrastructural Biochemistry Laboratory, The Hospital for Special
Surgery, Cornell University Medical College, 535 East 70th Street,
New York, New York 10021, USA

Dr. F. Bronner
School of Dental Medicine, University of Connecticut Health
Center, Farmington, Connecticut 06032, USA

Dr. A. K. Campbell
Department of Medical Biochemistry, University of Wales
College of Medicine, Heath Park, Cardiff CF4 4XN, UK

Dr. J. Dequeker
Arthritis and Metabolic Bone Disease Research Unit,
Universitaire Ziekenhuizen, Weligerveld 1,3041 Pellenberg,
Belgium

Prof. S. Ebashi
National Institute for Physiological Sciences, Myodaiji, Okazaki
444, Japan

Prof. D. R. Fraser
Department of Animal Husbandry, University of Sydney, Sydney
2006, New South Wales, Australia

Prof. T. Fujita
Department of Medicine, Kobe University School of Medicine,
5-1 Kusunoki-cho 7-Chome, Chuo-ku, Kobe, Japan

Prof. W. J. Malaisse
Laboratory of Experimental Medicine, Boulevard de Waterloo
115, B-1000 Brussels, Belgium

Dr. D. H. Marshall
MRC Unit for Hearing Research, University of Nottingham,
Nottingham, UK

Dr. F. Melsen
University Institute of Pathology, Aarhus Amtssygehus, DK-8000
Aarhus C, Denmark

Prof. L. Mosekilde
University Department of Medicine, Aarhus Amtssygehus,
DK-800 Aarhus C, Denmark

Dr. Y. Nakao
Department of Medicine, Kobe University School of Medicine,
5-1 Kusunoki-cho 7-Chome, Chuo-ku, Kobe, Japan

Prof. B. E. C. Nordin
Institute of Medical and Veterinary Science, Box 14 Rundle Mall
Post Office, Adelaide 5000, South Australia, Australia

Dr. M. Peacock
Department of Medicine, Emerson Hall–421, 545 Barnhill Drive,
Indiana University Medical Center, Indianapolis, Indiana 46223,
USA

Dr. W. G. Robertson
King Faisal Specialist Hospital, Research Centre, Biological and
Medical Research, PO Box 3354, 11211 Riyadh, Kingdom of
Saudi Arabia

Dr. G. Schaafsma
Department of Nutrition, CIVO-Institutes TNO, PO Box 360,
3700 AJ Zeist, The Netherlands

Dr. R. H. Wasserman
Department of Physiology, 717 Veterinary Research Tower, New
York State College of Veterinary Medicine, Cornell University,
Ithaca, New York, 14853, USA

Chapter 1

Chemistry and Biochemistry of Calcium

W. G. Robertson

Introduction

Calcium is the fifth most abundant element in the human body and the most common of the mineral ions (Table 1.1). It is also the most important structural element, occurring not only in combination with phosphate in bone and teeth but also with phospholipids and proteins in cell membranes where it plays a vital role in the maintenance of membrane integrity and in controlling the permeability of the membrane to many ions including calcium itself (Table 1.2). It is widely involved in many physiological and biochemical processes through-out the body including the coagulation of blood, the coupling of muscle excitation and contraction, the regulation of nerve excitability, the motility of spermatozoa, the fertilization of ova, cell reproduction, the control of many enzyme reactions and (in its role as a "second or third messenger") the transmission of many hormone actions at the appropriate receptor site on the cell membrane. [For reviews see Bianchi (1968) and Scarpa and Carafoli (1978).]

Table 1.1. The elemental composition of the human body (after Lehninger 1975)

Element	% Total no. of atoms	Weight (%)
H	63.0	10.0
O	25.5	64.5
C	9.5	18.0
N	1.4	3.1
Ca	0.31	1.96
P	0.22	1.08
Cl	0.08	0.45
K	0.06	0.37
S	0.05	0.25
Na	0.03	0.11
Mg	0.01	0.04

Table 1.2. Distribution of calcium in a 70-kg man (after Nordin 1976)

Site	Calcium (g)	Weight (%)
Skeleton	1355	98.90
Teeth	7	0.51
Soft tissues	7	0.51
Plasma	0.35	0.026
Extravascular fluid	0.70	0.052

Since calcium is necessary for both the reproduction and growth of humans, it is an essential nutrient. Because of its importance, many mechanisms have evolved to preserve the body stores of the ion and to ensure a sufficient supply to the organism so that it can maintain relatively constant concentrations of both intra- and extracellular calcium. It is so vital to the body's normal functioning that if the plasma level of ionized calcium falls below 0.6 to 0.7 mmol/l (normal range 1.10 to 1.30 mmol/l) then the neuromuscular system ceases to function normally and bone fails to mineralize properly. Abnormally high concentrations of ionized calcium (>1.6 mmol/l), on the other hand, are toxic to many enzyme systems so that the level must also be kept below this critical upper limit to ensure the continuance of normal cellular function. Thus a finely tuned mechanism for calcium homeostasis has evolved to maintain a constant extracellular fluid (ECF) concentration of the cation (Nordin 1976).

In practice, ECF calcium homeostasis is achieved by the steady-state control of calcium fluxes into and out of the ECF by a number of hormones, mainly parathyroid hormone, and the active metabolites of vitamin D (but also, in the view of some workers, calcitonin). These act on the main target organs for calcium, namely, kidney, intestine and bone, of which the kidney is by far the most important regulatory organ for calcium homeostasis (Nordin and Peacock 1969). Deviations from the normal ECF level of calcium occur in certain disease states, particularly those involving alterations in the circulating concentrations of the hormones mentioned above.

In order to fulfil its various functions, calcium must often be transferred from one body compartment to another or from one cellular compartment to another. The cells involved in the translocation of calcium must be able to protect themselves against a surfeit of the ion, which, although necessary for some intracellular activities, is toxic to many others. To achieve both objectives, highly specific transport and buffering mechanisms for calcium have had to be developed within these cells.

The universality of calcium within the various organs and compartments of the body and the large variety of roles which this cation seems to play lead to the intriguing question – why is it calcium that plays these roles and not one of the other biological cations?

Evolutionary Aspects of Calcium

Animal life arose in the sea where calcium is also the fifth most abundant element, paralleling its relative frequency of occurrence in the Earth's crust

(Table 1.3). The total concentration of the ion in present-day oceans is relatively high (about 10 mmol/l) compared with that in the ECF (2.5 mmol/l). Although the calcium level of primaeval seas may not have been as high as 10 mmol/l, evolution of organisms from salt water, firstly to fresh water (where the calcium concentration may range from 0.1 to 3 mmol/l) and then to dry land, required adaptation from a high to a low calcium environment. To ensure a sufficient supply of calcium during periods of low availability of the ion, it was necessary for mammalian vertebrates to develop the ability not only to accumulate and store calcium but also to release it again when required. These transfers had to be made while at the same time maintaining constant concentrations of ionized calcium within both intracellular and extracellular fluids.

Table 1.3. The main elemental composition of the Earth's crust and of sea water (after Lehninger 1975 and Weast and Astle 1980)

Element	Earth's crust		Sea water	
	% No. of atoms	Weight (%)	Element	Weight (%)
O	47.0	31.6	Cl	58.2
Si	28.0	33.1	Na	32.4
Al	7.9	9.0	Mg	3.90
Fe	4.5	10.6	S	2.71
Ca	3.5	5.9	Ca	1.23
Na	2.5	2.4	K	1.17
K	2.5	4.1	Br	0.20
Mg	2.2	2.3	C (inorganic)	0.09
Ti	0.46	0.9	Sr	0.04
H	0.22	<0.01	Si	0.02
C	0.19	0.1	B	0.01

An evolutionary basis for the development of specific transport systems has been proposed by Rasmussen (1972). He suggested that the first simple biochemical reactions probably occurred in a primordial sea rich in K^+ and Mg^{2+} ions (these being the cations derived from the most soluble rock salts of the time). The first primitive cells probably possessed membranes that lacked the various ion pumps which became characteristic of later animal cells.

Following various geological changes, the composition of the sea changed, becoming richer in Na^+ and Ca^{2+} ions (Table 1.3), and the primitive cells had to adapt to this new environment. The evolutionary response was for cells to develop semipermeable membranes which incorporated ion-selective pumps to limit the accumulation of Na^+ and Ca^{2+} inside, but at the same time retain K^+ and Mg^{2+} ions. This enabled the original "primordial sea" to be preserved as the normal intracellular ionic environment and the younger Na^+-and Ca^{2+}-rich sea to be adopted as the new extracellular environment. The relative roles of these four cations in the various biochemical and physiological processes of the body stem partly from the chronological order of their evolutionary incorporation into living matter, partly from their prevailing availability at a particular point in the evolutionary time scale and partly from their respective physical and chemical properties.

Chemical Properties

Electrostatic Properties

Calcium belongs to the divalent metals in group IIa of the periodic table. This group, which includes beryllium (Be), magnesium (Mg), calcium (Ca), strontium (Sr), barium (Ba) and radium (Ra), has two valency electrons which are lost when the metals ionize. Both calcium and magnesium have highly negative electrode potentials (Table 1.4) indicating that they have a very strong tendency to ionize in aqueous solution, as do their corresponding alkali metal ions in group I, potassium and sodium. In the ionic form, Na^+ and Mg^{2+} have the same electronic configuration with full 2s and 2p shells. K^+ and Ca^{2+} exhibit the equivalent correspondence with complete 3s and 3p shells. Mg^{2+} and Ca^{2+} ions, however, have a much smaller crystal ionic radius than their group I counterparts because the doubly positive charge on the nucleus pulls the outer electronic shells into a more tightly bound configuration (Table 1.4). Overall, the ionic radius follows the order: $Mg^{2+} < Na^+ < Ca^{2+} < K^+$. In turn, the lower ionic radii of the divalent cations produce a much higher charge density (or ionic potential) on the divalent compared with the monovalent ions. This is defined by Ze/r, where Ze is the electronic charge and r is the ionic radius of the ion. The order of increasing charge density is now the reverse of that for ionic radii such that $K^+ < Na^+ < Ca^{2+} < Mg^{2+}$ (Table 1.4). As we shall see, it is this sequence which to a large extent defines the properties of these ions in aqueous solution and their relative activities in various biological processes.

Table 1.4. Some physical and chemical properties of cations (after Weast and Astle 1980)

Cation	Charge	Ionic radius (Å)	Ionic potential (Ze/r)	Electrode potential (E^0 volts)
Li	+1	0.68	1.47	−3.045
Na	+1	0.95	1.05	−2.714
K	+1	1.33	0.75	−2.925
Mg	+2	0.65	3.08	−2.37
Ca	+2	0.99	2.02	−2.87
Ba	+2	1.35	1.48	−2.90

Levine and Williams (1982a, b) have produced the following general expression to describe the particular biological activity of Ca^{2+}:

$$\text{Biological activity} \propto [Ca^{2+}] \times K_{aq} \times P \times (\text{Structural factors}) \qquad (1)$$

where $[Ca^{2+}]$ is the concentration of ionized (or free) calcium in the medium under consideration (more correctly this should be $\{Ca^{2+}\}$, the chemical activity of the ion), K_{aq} is the stability constant of Ca^{2+} with any free aqueous binding agent (or ligand), and P is the partition coefficient which modifies K_{aq} to define the stability of binding in a non-aqueous phase (for example, a membrane). The product $[Ca^{2+}] \times K_{aq} \times P$ defines the binding to a ligand but does not allow for the kinetic activity of the ion. This requires the inclusion of a "structural factors" term, the relevance of which will be explained in more detail on p. 11.

Hydration of Ions

In aqueous solution ions do not exist with the small ionic radii shown in Table 1.4. The high electrical field associated with them causes them to take up water molecules to shield their charge. This water of hydration, as it is known, can be considered to be located in two zones. The first zone is a tight inner one consisting of highly oriented water molecules, where the amount of water taken up depends partly on the charge density of the ion and partly on the coordination number associated with it. Mg^{2+}, for example, has a fixed coordination number of 6, so that the inner zone of water molecules is distributed in a tetrahedral manner around the ion. Ca^{2+}, on the other hand, in common with Na^+ and K^+, has a much more flexible coordination number than Mg^{2+}, which can be anywhere between 6 and 12. Beyond the inner zone of hydration there is a second shell of more loosely orientated water molecules.

Table 1.5. Physical chemical data on hydrated cations (after Dack 1975; Burgess 1978 and Conway 1981)

Cation	Radius of inner hydration sphere (Å)	Radius of outer hydration sphere (Å)	Number of water molecules	Hydration energy (kcal/cal)
Li	3.82	6	3.5– 7	−122
Na	3.58	4	2 – 4	−98
K	3.31	3	1 – 3	−81
Mg	4.28	8	10.5–13	−456
Ca	4.12	6	7.5–10	−381
Ba	4.04	5	5 – 9	−315

It is difficult to determine the size of the hydrated ions as this depends on a number of physical factors in the surrounding solution. However, a summary of the best estimates is shown in Table 1.5 which indicates how much larger are the hydrated ions than the original ionic radii defined from their dehydrated forms in crystals of their salts (Table 1.4). Table 1.5 also contains current estimates of the number of water molecules thought to be associated with each ion. With their loosely bound water molecules the sizes of the ions increase from around 1 Å to between 3 and 8 Å. Furthermore, the cation with the smallest ionic radius (Mg^{2+}) is now the largest of the group. Indeed, by including data from other cations, Tables 1.4 and 1.5 show that the effective size of the solvated ions, as judged by either their inner or outer zone radii, is directly proportional to their ionic potentials. This, in turn, affects many of the solution properties of the ions concerned, including their chemical activities, their electrophoretic mobilities, their diffusion coefficients and their abilities to traverse semipermeable membranes (Table 1.6).

Table 1.6. Kinetic properties of cations (after Robinson and Stokes 1959 and Burgess 1978)

Cation	Equivalent conductivity (ohms/cm^2/equiv)	Electrophoretic mobility (cm^2/volt/s $\times$ 10^4)	Diffusion coefficient (cm^2/s $\times$ 10^5)
Li	40.0	4.20	1.37
Na	50.9	5.20	1.61
K	74.5	7.62	2.00
Mg	53.1	5.50	1.25
Ca	60.0	6.16	1.34
Ba	65.0	6.59	1.40

Ion Activity

Another factor which influences the chemical activity of ions in solution is the prevailing electrical field due to other "bystander" ions in the same solution. According to Debye and Huckel (1923), the effective concentration of an ion (M) in solution (expressed as its chemical activity) differs from its actual concentration by a factor dependent on the electrical field strength. The latter is defined by the ionic strength (I) of the solution and is given by the expression:

$$I = 0.5 \Sigma (c_i \times Z_i^2) \tag{2}$$

In simple terms the correction allows for the non-specific "crowding effect" of other ions in the same solution on the true chemical activity of the ion under consideration. Table 1.7 shows how varying the ionic strength affects the interionic distance in solutions of increasing concentration of sodium chloride. At a physiological ionic strength of $I = 0.15$, the distance is reduced to 19 Å. Since the effective radius of the fully hydrated Ca^{2+} ion is about 6 Å, it is easy to see how its activity is reduced in such an environment.

Table 1.7. Average interionic distances in various solutions of sodium chloride (after Lehninger 1975)

Ionic strength	Interionic distance (Å)
0.001	94
0.01	44
0.10	20
0.15	19
1.00	9.4

In practice, the relationship between ion activity ($\{M\}$) and ion concentration ($[M]$) is given by:

$$\{M\} = [M] \, \gamma_M \tag{3}$$

where γ_M is the activity coefficient of ion M and is defined from:

$$\log \gamma = A \, Z_M^2 \, [\sqrt{I}/(1 + \sqrt{I}) - 0.3 \, I] \tag{4}$$

which is the Davies (1962) modification of the original Debye–Huckel equation. In this expression A is a constant and Z_M is the valency of ion M.

Activity corrections are extremely important in physiological solutions, particularly for multivalent ions and in media where I varies to a large degree, such as in urine (Robertson 1969). One of the main problems, however, is in obtaining an estimate of I in most biological fluids. Since it is not directly measurable, it must be determined by combining all the individual terms in Eq. (2). This necessitates the determination of the concentrations of all the charged ionic species in the solution concerned. Fortunately, most biological fluids (with the exception of urine) have values of I within a fairly narrow range of 0.145 to 0.165. Within this range, γ_1 and γ_2, the activity coefficients of monovalent and divalent ions respectively, are relatively constant and no appreciable error arises from the omission of small terms in the summation term in Eq. (2).

For urine, there are several computer programs available that calculate the prevailing ionic strength from a knowledge of the concentrations of all the ionic species which it contains (Robertson 1969; Finlayson 1977). Within the cell, however, the values of ionic strength, even in the cytosol, are unknown and difficult to estimate. There are also questions concerning what exactly is meant by free ion concentrations and activities in the finite microenvironment of intracellular organelles or in the narrow channels down which Ca^{2+} ions are transported into the cell. When the space in which an individual Ca^{2+} ion is located is of the same order of magnitude as its hydrated size, what is meant by the ionic strength of the milieu and how does one assess the activity of an ion in such an environment? These questions have yet to be addressed by biophysicists.

Ion-Pairing, Complexation and Chelation

Another property of the divalent cations (M^{2+}), which is important for an understanding of their relative biological roles, is their ability to interact with specific anions (particularly divalent anions, A^{2-}) to form soluble ion pairs or complexes (MA) according to the equilibrium:

$$M^{2+} + A^{2-} \rightleftharpoons MA \tag{5}$$

which is governed by the association or stability constant (K_{MA}) where

$$\begin{aligned} K_{MA} &= \{MA\}/\{M^{2+}\} \{A^{2-}\} \\ &= [MA]\gamma_{MA}/[M^{2+}] [A^{2-}]\gamma_M\gamma_A \end{aligned} \tag{6}$$

This effectively reduces the "free" or "ionized" concentrations of the two ions concerned while at the same time preserving the mass balance of each total (T_M and T_A) according to the equations:

$$T_M = [M^{2+}] + [MA] \tag{7}$$

$$T_A = [A^{2-}] + [MA] \tag{8}$$

In most physiological media, a large number of such ion-pair interactions may be encountered. Unfortunately, it is impossible to measure all the resultant ion activities directly although a few free ion concentrations are amenable to quantitation. Ionized calcium is one such ion which can be measured in a number of biological fluids using colorimetry or ion-selective electrode procedures (Robertson and Marshall 1981; Thomas 1982). It is also possible to measure calcium ion activity within the cell using chemoluminescence or microcolorimetric techniques (Shimomura and Johnson 1973; Tsien 1980). However, no electrodes have yet been developed which will measure satisfactorily the free ion concentrations of any of the principal anions, such as phosphate or oxalate, either within the cell or in the ECF.

Most of the binding properties between calcium and various anionic ligands can be explained on the basis of mutual electrostatic attraction. Williams (1976), in his analysis of this phenomenon, has concluded that it is the ratio of cation/anion radii which dominates the strength of the association. As an approximate generalization, anions and cations of similar size tend to associate more strongly than ions of different sizes. Thus with small ions, such as OH^-, the smaller Mg^{2+} ion forms a more stable complex than does the larger Ca^{2+} ion (Table 1.8). In fact, it is the size of the anionic ligand relative to the number of water molecules displaced in the process of forming the complex which is one of the critical factors in determining the resultant stability of a given complex. Strong acid anions, such as sulphate, cannot displace sufficient inner sphere water molecules from around Mg^{2+} ions but, instead, tend to form weak outer sphere complexes. In contrast, Ca^{2+} allows the sulphate anion to enter its inner coordination sphere where it displaces many more water molecules than from Mg^{2+} and forms a more stable complex. This difference is largely due to the relative amounts of energy required to dehydrate the ions which is in the order: $Mg^{2+} > Ca^{2+} >> Na^+ > K^+$, as shown in Table 1.5.

Table 1.8. Stability constants of some calcium and magnesium complexes at 37°C

Ligand (L)	Log K_{MgL}	Log K_{CaL}
HPO_4^{2-}	3.00	2.83
SO_4^{2-}	2.27	2.36
OH^-	2.58	1.36
F^-	1.50	1.00
$C_2O_4^{2-}$	3.60	3.43
$P_2O_7^{4-}$	7.20	5.75
AMP	1.95	1.76
ADP	3.15	2.82
ATP	4.75	3.77
EDTA	8.7	10.6
EGTA	5.2	11.0

With weakly acidic electron donors [for example RCO_2^- and $RO-(PO_3)^{2-}$] both Mg^{2+} and Ca^{2+} form inner sphere complexes. When these complexing agents are multidentate molecules, thermodynamic and steric factors play a major part in determining the strength of the stability and solubility constants. Since Ca^{2+}, but not Mg^{2+}, has a more flexible coordination number, it can more readily form multipoint links or chelates with the extra donor groups possessed by these molecules (Levine and Williams 1982a, b).

Thermodynamically the net number of water molecules released or taken up during complex formation has an effect on the stability of that complex through the change produced in the entropy of the system. (Entropy may be thought of as the energy which is unavailable for chemical use due to the degree of randomness present among the ions and molecules in a given system.) In practice, the complexing reaction described in Eq. (5) also involves a change in the degree of hydration of the ions concerned. Thus the complete reaction should really be written in the form:

$$M^{2+}(H_2O)_w + A^{2-}(H_2O)_x \rightleftharpoons MA(H_2O)_y + (w + x - y)H_2O \qquad (9)$$

The greater the number of freely moving particles on the right-hand side of the reaction, the greater the resultant degree of randomness in the solution. In this situation, the entropy of the system increases and the value of ΔS (i.e. the change in entropy of the system) is positive. It is clear that the number of freely moving particles in the system is largely dependent on the value of the expression $(w + x - y)$.

From thermodynamic considerations, the stability constant of the equilibrium is dependent on the change in free energy of the reversible reaction $(-\Delta G)$ and is given by:

$$-\Delta G = RT \ln K_{MA} \qquad (10)$$

The higher the value of $-\Delta G$, the more stable is the resulting complex (MA). Since $-\Delta G$ is related to the change in internal energy $(-\Delta H)$ and to the change in entropy (ΔS) of the system by the equation:

$$-\Delta G = -\Delta H + T\Delta S \qquad (11)$$

then by combining Eqs. (10) and (11) we can see that:

$$\ln K_{MA} = -\Delta H/RT + \Delta S/R \qquad (12)$$

Thus, for comparable values of $-\Delta H$, the higher the value of ΔS, the stronger the stability of the complex formed (Moeller and Horwitz 1960).

This conclusion has particular relevance when one is comparing the binding of various cations to large multidentate anions, such as ethylenediaminetetra-acetic acid (EDTA) and ethyleneglycol (bis β-amino)-N,N,N',N'-tetra-acetic acid (EGTA). For these ligands, the stability constants for the Ca^{2+}–EDTA and Ca^{2+}–EGTA complexes exceed those of the corresponding Mg^{2+} complexes severalfold, at least in part because of the greater increase in entropy generated during the binding reaction,

$$Ca(H_2O)_A{}^{2+} + EDTA(H_2O)_B \rightleftharpoons CaEDTA(H_2O)_C +$$
$$(A + B - C)H_2O \qquad (13)$$

than during the corresponding reaction involving Mg^{2+}:

$$Mg(H_2O)_a{}^{2+} + EDTA(H_2O)_B \rightleftharpoons MgEDTA(H_2O)_c +$$
$$(a + B - c)H_2O \qquad (14)$$

This means that $(A + B - C)$ must be greater than $(a + B - c)$, i.e. more water molecules are displaced from around the hydrated Ca^{2+} ion than from around the Mg^{2+} ion when chelated by EDTA. The main reason for this is the ability of the Ca^{2+} ion to adapt to a greater coordination number than the rigid value of 6 held by the Mg^{2+} ion, as was mentioned above. Thus the chelating anion can form many more links with an "enmeshed" Ca^{2+} ion than it can with Mg^{2+}. This is not a factor when one is considering the complexing of these cations with small unidentate anions. It is clear that the multidentate nature of

EDTA-like molecules, with their four –COOH groups, reverses the normal order of stability of calcium and magnesium complexes (Table 1.8). If, in addition, the chelating anion has weak acidic groups and either a hydroxyl or an ether–oxygen donor centre, then the degree of selectivity for calcium over magnesium is greatly accentuated. The incorporation of two neutral O-atoms, as in EGTA, for example, slightly increases the stability of the Ca^{2+} chelate but markedly reduces that associated with the more structurally rigid Mg^{2+} ion (Table 1.8). As will be seen below, a very similar grouping of donor centres is present in the calcium-binding macromolecules troponin C and calmodulin and it is this which provides these biologically important molecules with their high degree of selectivity for Ca^{2+} over other cations.

The irregular coordination sphere of Ca^{2+} is also important for the cross-linking of organic macromolecules and for the rapid binding and release processes necessary for the ion to act as a messenger for various hormone actions at cell membranes.

Calmodulin and Other Calcium-Binding Proteins

Calmodulin is a ubiquitous protein of molecular weight 16 500 containing 148 amino acids and has been found in every eukaryotic cell so far examined. It is clearly of fundamental biological importance since its structure appears to have been conserved with only six or fewer changes in its amino acid content between species separated by millions of years of evolution. The molecule possesses four calcium-binding domains, each of which contains the alternating acidic amino acid residues necessary for the chelation of calcium (Cheung 1970, 1980, 1982; Klee et al. 1980; Means and Dedman 1980). These domains contain a high proportion of glutamic and aspartic acids, and to a lesser extent, serine, tyrosine and asparagine. The stability constants for the four binding sites with calcium are of the order of 10^6 (Table 1.9). Crystallographic studies have shown that in three dimensions the molecule consists of two globular lobes connected by a long exposed α-helix (Babu et al. 1985). Each lobe binds two Ca^{2+} ions through helix–loop–helix domains similar to those observed in other calcium-binding, so-called "trigger" proteins. These include troponin C (Potter and Johnson 1982), calcineurin (Klee et al. 1979) and parvalbumin (Wnuk et al. 1982). Homology with intestinal calcium-binding protein (Wasserman and Fullmer 1982), however, is restricted to the calcium-binding regions alone (Babu et al. 1985).

Table 1.9.   Stability constants for cations at four sites in calmodulin (after Demaille 1982)

Cation	Site 1	Site 2	Site 3	Site 4
K^+	270	94	115	667
Mg^{2+}	1.4×10^4	3.7×10^3	1.0×10^4	1.1×10^4
Ca^{2+}	1.5×10^7	5.9×10^6	1.7×10^6	1.1×10^6

Other calcium-binding proteins in the body include the Gla proteins found in bone matrix (Hauschka et al. 1975; Price et al. 1976) although their role is not yet clear. These have a high proportion of γ-carboxyglutamic acid (Gla), a constituent of vitamin K-dependent blood-clotting protein. (For a review see Furie et al. 1982.) Teeth also possess a number of proline-rich calcium-binding molecules, including statherin which is involved in the controlled growth of enamel crystals (Moreno and Varughese 1981). Another protein containing an unusual sequence of phosphorylated serine and acidic amino acids has been found in dentine (Cookson et al. 1980).

A feature of all these calcium-binding proteins is that they bind with stability constants in the range 10^5 to 10^8 mol/l and that this occurs very rapidly with rates close to that of the diffusion control limit of around 10^{10}/s. This implies that the site of binding, which involves three or four –COOH groups plus two neutral oxygen donors on a continuous segment of 12 amino acids, has considerable mobility to allow entry of Ca^{2+} ions into the binding domain. This flexibility, however, does not extend to allowing the Mg^{2+} ion to be bound, because of its rigid tetrahedral coordination requirements. The discrimination between Ca^{2+} and Mg^{2+} is of the order of 10^2 to 10^3 (Table 1.9).

Levine and Williams (1982a, b) have drawn attention to the fact that the binding constant of calcium with large loosely structured molecules may not necessarily be "constant". Under mechanical, electrical or chemical "stress" the conformation of the chelating anion may change. The actual stability constant (K^1_{Ca}) may be defined from:

$$\log K^1_{Ca} = \log K^0_{Ca} - f(\Delta S/RT) \tag{15}$$

where K^0_{Ca} is the standard stability constant and ΔS is the change in entropy at the binding site produced by the applied constraint. This means that a continuous range of binding constants can be produced and explains the necessity for including the "structural factors" term in Eq. (1). This phenomenon has important implications for the allosteric switching from a relaxed to a tense structure (the so-called $R \rightleftharpoons T$ switch), since it is possible for this to take place progressively rather than suddenly in response to a conformational stress at the binding site. The phenomenon is similar to that described in the next section which accounts for the variable solubility product of bone mineral.

Solubility Considerations

In parallel with the relative order of stability of the Ca^{2+} and Mg^{2+} complexes with simple anions is the solubility of their respective salts (Table 1.10). Thus Ca^{2+} forms more insoluble salts than Mg^{2+} with large strongly acidic anions, such as sulphate, whereas the order of solubilities is reversed in the case of precipitation with small anions such as OH^-. Precipitation appears to be controlled by the same factors as complexation but, in addition, there is the cooperative energy of the crystal lattice being created. In some instances, however (for example, in the formation of the oxalate salts of calcium and magnesium), the order of solubilities is the opposite of that expected from the simple size rule. Oxalate might have been anticipated to associate more tightly with Mg^{2+} than with Ca^{2+} [as it does in the formation of their soluble complexes (Table 1.8)]. However, the difficulty of packing Mg^{2+} ions into magnesium oxalate monohydrate, while at the same time preserving the rigid octahedral packing structure and the fixed Mg–0 bond distance of 2.05 Å, leads to a lattice of lower stability than that for the corresponding calcium oxalate monohydrate (Table 1.10). This has important implications for the risk of forming urinary calculi consisting of this salt and explains why magnesium oxalate is never found in these concretions.

Table 1.10. Solubility products of some calcium and magnesium salts

Calcium salt	$-\log_{10}K_{sp}$	Magnesium salt	$-\log_{10}K_{sp}$
$Ca(OH)_2$	5.15	$Mg(OH)_2$	11.15
CaF_2	10.57	MgF_2	8.19
$CaSO_4$	4.60	$MgSO_4$	<1.0
$CaCO_3$	8.42	$MgCO_3$	7.46
$CaHPO_4$	7.00	$MgHPO_4$	6.40
CaC_2O_4	8.77	MgC_2O_4	5.00

The irregularity of coordination number associated with the Ca^{2+} ion has a major influence on several of its chemical and biological properties relative to those of Mg^{2+}. For example, it is this versatility which accounts for the tendency of Ca^{2+} to form "amorphous" rather than highly organized crystal lattices with many anions. Amorphous solids are those in which lattice order is maintained over only a few metal ion to metal ion distances and from which no x-ray diffraction pattern can be obtained. Thus the so-called "amorphous calcium phosphate" of young bone (Termine and Posner 1967) is analogous to the amorphous salts of building materials used in bricks, cement, mortar and plaster in which calcium can occupy many different lattice sites. The amorphous calcium phosphate laid down initially may not be the most thermodynamically stable form of the salt but slow dissolution and recrystallization into the more crystalline and stable hydroxyapatite eventually takes place. Amorphous materials have an advantage over more crystalline salts in that fracture along set planes is avoided. Thus amorphous materials, in general, have greater mechanical strength than their crystalline counterparts. This partly accounts for the enormous strength of bone, although other factors are also important such as the tubular structure of cortical bone, the interlocking of the building blocks of bone (osteons), and the elasticity imparted to bone by the incorporation of an organic matrix.

The classical solubility product (K_{sp}) which is given by:

$$K_{sp} = \{M^{2+}\}\,\{A^{2-}\} \qquad (16)$$

does not take into account the possible role of physical, electrical or chemical stress on the salt crystals in equilibrium with their bathing solution. One simple example of such a stress is the situation which arises when crystals of the salt concerned are small (i.e. $< 1\ \mu m$) or charged. These conditions impose an additional surface energy constraint on the crystal, which should be taken into account in assessing its solubility (Best 1959). For the small crystals of $CaCO_3$ in shells and of $Ca_{10}(OH)_2(PO_4)_6$ in bone, K_{sp} can vary as much as tenfold (Neuman and Neuman 1958) depending on the size of the crystals concerned. Even for a given size of bone crystal the apparent solubility product may be affected by changes in the mechanical, electrical or chemical field surrounding the crystal. For example, pressure on the bone during weight-bearing produces a lower solubility than during prolonged bed-rest or during the weightlessness of space flight (Whedon et al. 1976). There is a continuous, rather than an all-or-none, response in solubility to such stimuli depending on the magnitude of the stress involved.

Kinetic Properties of Cations

One factor which influences the rate of transport of the simple cations in bulk solution is their mobility through aqueous media. Table 1.6 contains the electrophoretic mobilities and diffusion coefficients of various cations. When comparing these rates with the corresponding crystal ionic radii (Table 1.4) it would appear that the *largest* cations travel fastest (i.e. $Li^+ < Na^+ < K^+$ and $Mg^{2+} < Ca^{2+} < Ba^{2+}$) and that for ions of approximately the same ionic radius, the one with the higher charge has more restricted mobility (i.e. $Mg^{2+} < Li^+$ and $Ca^{2+} < Na^+$). However, after allowance is made for the size of the primary hydration sphere, then, for a given charge, the observed order of electrophoretic mobility is inversely proportional to the size of the hydrated ions (Tables 1.5 and 1.6) and for a given size, the divalent cations travel about twice as fast as the corresponding monovalent cations, as might be anticipated. Simple diffusion of the groups I and II cations through aqueous solution, not under electrical stimulation, is also inversely proportional to the size of the corresponding hydrated ion.

Ion movement within a polar matrix of charged groups or within a lipid environment, such as is encountered in a cell membrane, will not necessarily follow the same rules as for free ions in bulk aqueous solution. As mentioned above, it will depend more on the facilitated or, in some cases, active transport of the ion via some binding/release association with a chelating ligand. These rates are related to the stability constant of the chelate concerned (K_{MA}) and to the rate of dissociation of water molecules (k_w) from the cation, since the latter will determine the rate of association of the cation with the chelate.

Water molecules around the cations can, in fact, exchange rapidly with those in the bulk solution. Table 1.11 shows the rate constants for water substitution in the inner coordination spheres of various cations. These are extremely fast for $Na^+(aq)$, $K^+(aq)$, $Ca^{2+}(aq)$ and $La^{3+}(aq)$ but much slower for the more hydrated $Mg^{2+}(aq)$ ion. This has implications for the relative kinetics of binding and release of these ions to and from ionophores or transport proteins involved in the movement of calcium across cell membranes. Also important in determining the rate of binding/release reactions is the flexibility of the chelating

Table 1.11. Rate constants for water substitution in the inner coordination sphere of various cations (after Levine and Williams 1982a)

Cation	Rate of substitution (s^{-1})
Na^+	ca 5×10^8
K^+	ca 2×10^9
Mg^{2+}	ca 1×10^5
Ca^{2+}	ca 5×10^8
La^{3+}	ca 5×10^7

anion. If Ca^{2+} is to be bound selectively (and therefore strongly) to a multidentate ligand, and at the same time the binding/release reactions have to be fast, then the chelating anion must be flexible (Levine and Williams 1982a, b). This property, in association with the flexibility in coordination number of Ca^{2+}, largely determines the ability of the latter to be transported across cell membranes and between intracellular compartments with great rapidity.

The rate constant (k_{off}) for the process of releasing the bound Ca^{2+} from the chelate is given by:

$$\log k_{off} = \log (k_w/K_{MA}) \tag{17}$$

Binding rate constants for Ca^{2+} have been shown to be of the order of 10^9/s. With a stability constant of about 10^6, this means that the dissociation rates for Ca^{2+} ions are still of the order of 10^3/s and that Ca^{2+} can diffuse rapidly even through a matrix of chelating agents with a high affinity for it. Since Mg^{2+} binding rates are slower than those of Ca^{2+} by a factor of approximately 10^4, then Ca^{2+} can still diffuse as rapidly as Mg^{2+} in spite of the fact that it binds to the transport chelator up to 10^4 times more strongly.

This has particular relevance to the development of specific Ca^{2+} channels which are not blocked by Mg^{2+} ions and which do not allow the leakage of Mg^{2+} ions through them. The entrances to such channels have to be controlled and the channels themselves must have a very low affinity for Mg^{2+} ions relative to Ca^{2+} ions (Urry 1978). Furthermore, the flow of calcium through the channels creates an electrical circuit which requires energy to drive the flow. The calcium currents generated constitute a major factor in the control of a great variety of biophysical and biochemical activities.

Equation (17) can also be used to explain the difference in the chemical behaviour of Na^+ and Ca^{2+} ions in biological systems. Since the K_{MA} values for the chelates of Na^+ are so much lower than for the corresponding chelates of Ca^{2+} and since the rate of dissociation of water molecules from the two cations is about the same, then the permeability to Na^+ is considerably greater than that to Ca^{2+}. Furthermore, since the extracellular concentration of Na^+ is also so much greater than that of Ca^{2+}, this has led to Na^+ becoming the major determinant of cell excitability. Ca^{2+}, on the other hand, because of its lower concentration, its lower cell permeability, its higher charge and its marked ability to form soluble complexes with many biologically important anions, has become a major regulator of the intracellular events following the excitation of the cell.

Membrane Permeability

The rate of movement of Ca^{2+} across a membrane is constrained to some extent by the permeability of the membrane to calcium and by the effect of calcium itself on that permeability. It was thought at one time that membrane impermeability was the sole determinant of the large electrochemical gradient observed across many membranes but subsequently it has been shown that most membranes, in fact, possess a finite but low permeability to Ca^{2+} ions. Thus, at most, this might help to maintain an established concentration gradient by limiting the rate of calcium leakage into the cell down the gradient. The cell would then have to expend less energy in the active extrusion of Ca^{2+} ions necessary to balance the inward rate of leakage.

A more important aspect of membrane permeability, however, is the effect of calcium on the permeability of cells to Na^+ and K^+ ions (Manery 1969). This is due to the ability of calcium to increase the rigidity of the membrane by binding to the lipid layer, a process which increases the electrical resistance of the membrane severalfold (Tobias et al. 1962). Other effects of calcium include increasing the surface pressure on the membrane (by condensing the lipid layer), neutralizing the negative charge on the lipid surface and decreasing the thickness of the lipid bilayer. The fluidity of the membrane is also reduced through the ability of calcium to stabilize the phospholipid moiety. This latter effect depends on the formation of a stable complex of Ca^{2+} with adenosine triphosphate (ATP) and on the subsequent formation of a calcium–ATP–phospholipid complex. After excitation of the membrane, Ca^{2+} is released from ATP and this renders the membrane more permeable to Na^+ ions. Thus the membrane is envisaged as passing from a calcium-associated to a calcium-dissociated state. More recently, x-ray diffraction studies have confirmed that Ca^{2+} forms tightly packed, highly ordered structures with phospholipids (Newton et al. 1978) and, because of its specific charge-to-size ratio, this results in the segregation of membrane phospholipids into discrete domains and in the conversion of the membrane to a more ordered state (Lee 1975; Papahadjopoulos et al. 1978). In fact, the plasma membrane may be viewed as a two-dimensional mosaic of integral membrane proteins embedded in a fluid lipid layer, with peripheral proteins loosely bound to either surface. Ca^{2+} can bind to this protein fraction, particularly through the –OH, >C=O and –COOH groupings of the latter (Williams 1976), and thereby increase the possibility of the occurrence of cross-linking between the proteins, with resultant effects on membrane structure and activity. The combined effect of Ca^{2+} on the phospholipid and protein moieties of the membranes is to create "gates" or "channels" which define the permeability of the membrane to various ions.

Three types of calcium channel have so far been identified. The first two operate using an ion-exchange mechanism involving either Na^+–Ca^{2+} or Ca^{2+}–Ca^{2+} exchange. The former is driven by the Na^+ gradient established by the Na^+–K^+ ATPase system (Blaustein and Nelson 1982). The significance of the calcium exchange system, which of course does not alter the intracellular concentration of Ca^{2+}, is not yet known (Baker and McNaughton 1978). These channels are found in most excitable cells and in those epithelial cells which control calcium transport (Baker 1972; Blaustein and Nelson 1982). They are referred to as the "fast channels" since the kinetics of transport are similar to that observed for the inward Na^+ current. Baker et al. (1969) have concluded that this portion of the calcium current is actually carried through the sodium channels as a result of their imperfect selectivity for Na^+ ions.

The second channel for Ca^{2+} transport is specific for that ion and is found in erythrocytes and various cultured cells (Schatzmann and Bürgin 1978; Schatzmann 1982). It involves a slow phase of calcium entry into the cell and requires energy which it derives from the hydrolysis of ATP by a specific calcium–ATPase enzyme system which also requires Mg^{2+} for its action. The biological importance of these channels will be described in a later section.

Another factor which might play a role in defining the relative permeabilities of the biological cations is the free energy required to dehydrate the hydrated ions in order to allow them to pass through the lipid membrane. The order is: $Mg^{2+} > Ca^{2+} >> Na^+ > K^+$ (Table 1.5). This may partly explain their order of permeability (P) through inert semipermeable membranes, which is: $P_K = 1$, $P_{Na} = 0.02$ and $P_{Ca} = 0.001$. Even if the cations can traverse the membrane in the hydrated form, their size will be important in determining their relative permeabilities. From Table 1.5 it can be seen that the order of permeability is inversely proportional to the radius of the hydrated ion.

Extracellular Calcium

The calcium concentration in most extracellular body fluids, particularly plasma, is kept within relatively narrow limits. In most, the total concentration of calcium (Ca_{tot}) consists of three main fractions, namely, a protein-bound fraction (CaProt) and a complexed fraction (CaL), both of which are in equilibrium with the so-called "free" or ionized fraction (Ca^{2+}). The CaL fraction consists of complexes between Ca^{2+} and phosphate, sulphate, citrate and various organic molecules which contain –COOH or –SO_4 groups. In general, the concentrations of the complexing agents are of the same order as that of calcium and their stability constants are $< 10^3$, so that a major proportion of Ca_{tot} remains in the ionized form and is available for transport via the more specific calcium-binding proteins mentioned earlier. In urine which is excessively supersaturated with calcium oxalate and/or calcium phosphate, a fourth fraction may be present in the form of crystals of these salts (Robertson et al. 1971).

Typical values for the different calcium fractions in various extracellular fluids are shown in Table 1.12. Clearly, the ionized fraction constitutes a significant proportion of the total in each fluid.

There are several methods available for measuring the concentration of ionized calcium in extracellular fluids (Robertson and Marshall 1981). After the original frog heart procedure of McLean and Hastings, most of the techniques employed colorimetry involving dyes which responded with some degree of selectivity towards Ca^{2+}. These dyes included murexide and its tetramethyl derivative (Raaflaub 1951, 1962). However, these techniques proved to yield unreliable results in protein-containing solutions and preliminary ultrafiltration was required. The development of calcium-ion-selective electrodes revolutionized the field and now most measurements of Ca^{2+} concentrations are based on this technique (Robertson and Marshall 1981).

Since the early work of McLean and Hastings (1934) on the contraction of frog heart muscle, most researchers have believed that it is the ionized calcium which is the physiologically active form of the element. Since then much evidence, from several sites in the body, has accumulated to support this hypothesis. Other workers, however, have questioned this belief and have concluded that other circulating forms of calcium are physiologically active, including the complexes of Ca^{2+} with phosphate and bicarbonate and a highly stable calcium–albumin–fatty acid complex (Toffaletti 1982).

Table 1.12. The mean percentage of total calcium in the various fractions of extracellular fluids in normal man (after Robertson and Marshall 1981)

Body fluid	State	Ca_{tot} (mmol/l)	CaProt (%)	CaL (%)	Ca^{2+} (%)
Plasma	Whole	2.2–2.6	40	10	50
	UF^a	1.3–1.6	0	16	84
Urine	Whole	1.0–7.0	0	48	52
Saliva	Whole	1.1–1.5	33	13	54
	UF	0.8–1.1	0	19	81
Bile	Common duct	1.6–4.0	48	17	35
	UF	1.2–1.6	0	33	67
	Gall bladder	2.4–12.5	30	50	20
	UF	1.0–5.2	0	71	29
CSF	Whole	4.3–5.9	10	26	64
Synovial fluid	Whole	1.2–2.4	←—— 59 ——→		41

^{a} Ultrafiltrate.

Intracellular Calcium

The total intracellular concentration of calcium has been measured in several tissues, but the ionized component has been determined in relatively few (Table 1.13). [For a review of these the reader is referred to a series of articles edited by Scarpa and Carafoli (1978) and a book by Thomas (1982).] In general, these studies show that intracellular Ca^{2+} concentrations are some 100 to 100 000 times lower than those in extracellular fluids. From the Nernst equation it is possible to calculate the electrical potential (E) produced by such a gradient. Thus

$$E = (RT/ZF)\ln ([Ca^{2+}]_e/[Ca^{2+}]_i) \tag{18}$$

where $[Ca^{2+}]_e$ and $[Ca^{2+}]_i$ are the extra- and intracellular concentrations of ionized calcium respectively. These concentrations imply that there is an electrical potential of between -60 and -150 mV just from the chemical gradients themselves. Together with the pre-existing downhill electrical potential of between -20 and -60 mV actually measured across cell membranes, the total electrochemical gradient may amount to between 80 and 210 mV. Since radioactive calcium is readily exchangeable in both directions across most cell membranes, this implies that Ca^{2+} probably enters the cell by a passive diffusion mechanism and must be pumped out by an active process which requires energy. As mentioned earlier, two such pumps have been identified, a Na^+–Ca^{2+} exchange system and a magnesium-dependent calcium–ATPase enzyme system. The first is used to eject Ca^{2+} from excitable cells and the second to pump out calcium from non-excitable cells. Recent research has shown that both systems can co-exist in the same cell.

In view of the enormous electrochemical gradients of Ca^{2+} down into the cell and the ease of diffusibility of Ca^{2+} across cell membranes, the cell has

Table 1.13. The intracellular concentrations of total and ionized calcium in various tissues (after Robertson and Marshall 1981)

Cell type	Species	Total calcium (mmol/kg wet weight)	Species	Ionized calcium (mol/l)
Cartilage	Man	8.5	—	—
Heart	Man	1.9	Frog	$<1 \times 10^{-7}$
Kidney	Man	3.5	Monkey	$<1 \times 10^{-6}$
HeLa	Man	5.0	Man	$<1 \times 10^{-6}$
Liver	Man	1.6	—	—
Smooth muscle	Man	8.3	—	—
Skeletal muscle	Man	1.3	Frog	2×10^{-7}
Placenta	Man	6.2	—	—
Lungs	Man	6.2	—	—
Nerve	Man	4.9	Squid	4×10^{-8}
Brain	Man	2.0	—	—
Spleen	Man	2.1	—	—
Skin	Man	4.3	—	—
Bone	Pig	19.5	—	—
RBC	Man	0.06	Man	$<1 \times 10^{-7}$

had to develop a set of highly sophisticated mechanisms to protect itself against untoward increases in the cytosolic concentration of calcium which might damage many of the enzyme reactions which are sensitive to that ion. Furthermore, since transport of calcium across tissue walls (for example, from gut lumen to serosal fluid) involves fluxes through the cells of which the tissue walls are comprised, it is vital that the low Ca^{2+} concentration in the cytosol be buffered to protect the cell's internal environment. Apart from the active pump mechanisms mentioned above, the cell generally has three other systems with which to protect itself (Scarpa and Carafoli 1978). First, calcium may be bound within the cytosol to various ions and molecules which have a wide range of affinities for Ca^{2+}. These include small ions, such as citrate, phosphate, creatine phosphate, adenosine diphosphate (ADP) and ATP, and larger molecules with highly selective affinities for calcium, such as the troponins, intestinal calcium-binding protein and the ubiquitous calmodulin. The second method of protection is to sequester calcium into various subcellular compartments, in particular into microsomes and mitochondria. Finally, there is the effect of calcium, mentioned earlier, on the permeability of the cell membrane.

Although the method by which the cell maintains the massive extracellular/intracellular gradient with respect to Ca^{2+} is important, it is the very low intracellular Ca^{2+} concentration itself which is the most significant feature in physiological terms. The value of $[Ca^{2+}]_i$ is so low that a small influx of Ca^{2+} ions can increase $[Ca^{2+}]_i$ quite significantly, whereas this is not the case for Na^+, K^+ or Mg^{2+} ions. This is highly advantageous for any substance which has to play a role as an intracellular "messenger" for the transmission of a signal at the cell surface. Table 1.14 contains a list of some of the reported biological phenomena in which increases in $[Ca^{2+}]_i$ appear to participate. The mechanism by which these increases operate is discussed in more detail in the next section of this chapter.

Four basic methods have evolved for measuring $[Ca^{2+}]_i$. The first involves quantitating the chemiluminescence produced by a number of photoproteins which respond selectively to Ca^{2+} ions. These include aequorin (Shimomura

Table 1.14. A list of some biological phenomena reported to depend on changes in intracellular ionized calcium

Excitation–contraction coupling in muscle
Neurotransmitter release
Cytoplasmic streaming
Control of cilia
Microtubule assembly
Membrane permeability to K^+ and to Ca^{2+}
Exocrine and endocrine gland secretion of hormones
Egg fertilization
Cell division and reproduction
Cell to cell communication
Cyclic nucleotide metabolism
Certain enzyme activities (e.g. phosphorylase kinase)
Photoluminescence activity
Excitation of rods and cones
Chromosome movement
Initiation of DNA synthesis

et al. 1963a; Ridgway and Ashley 1967), obelin (Blinks et al. 1976; Campbell et al. 1979) and halistaurin (Shimomura et al. 1963b).

The second method utilizes Ca^{2+}-sensitive microelectrodes. These are similar to the glass microelectrodes used for conventional electrical recording from cells but have been adapted to incorporate a Ca^{2+} sensor dissolved in an organic phase mounted on the electrode tip. The selective sensor ensures that the electrode responds primarily to Ca^{2+} concentrations. However, the response times of these electrodes are long (>1 s) and the electrodes tend to damage the cells (Tsien and Rink 1980).

The third method is based on the colorimetric response of certain dyes to Ca^{2+}. In the 1960s, Ohnishi and Ebashi used the colorimetric response of murexide to monitor the uptake of Ca^{2+} by isolated sarcoplasmic reticulum. Later, tetramethylmurexide was also used by the same group (Ohnishi 1978). During the 1970s, several groups demonstrated that certain azo dyes (primarily arsenazo III and antipyrylazo III) had a greater affinity for Ca^{2+} than had murexide or its derivatives and these have been widely used since for intracellular Ca^{2+} measurements (Brown et al. 1975; Thomas and Gorman 1977; Scarpa et al. 1978). However, a third generation of indicators has been developed with an even higher affinity for Ca^{2+}. These include bis(o-aminophenoxy)ethane-N,N,N',N'-tetraacetic acid (BAPTA), which is a derivative of EGTA mentioned earlier, and quin2, a methoxyquinoline derivative of BAPTA (Tsien 1980). The latter, along with its acetoxymethyl ester (quin2/AM), is now being widely used for Ca^{2+} measurements within the cell (Tsien et al. 1982).

The fourth method involves the use of various tetracyclines (in particular, chlorotetracycline) which readily penetrate cell membranes and emit a fluorescent response in the presence of Ca^{2+} ions (Chandler and Williams 1978). However, these fluorescent indicators are not specific for Ca^{2+} (the affinity for Mg^{2+} is actually higher than that for Ca^{2+}) and it is difficult to identify the origin of the signal within the cell. Because of the difficulty in interpreting the data obtained with tetracyclines, this method is not now in common usage. (See review by Caswell 1979.)

Calcium as a Second (or Third?) Messenger

Two systems have been identified which act to transmit within the appropriate cell the signal generated by a given hormone at the cell surface. These involve cyclic adenosine monophosphate (cAMP) and/or Ca^{2+} ions. The former is produced by the action of adenylate cyclase on ATP according to the reaction:

$$ATP \rightleftharpoons cAMP + PP_i \qquad\qquad\qquad (19)$$

It was proposed that the hormone (or first messenger) initially binds to a cell surface receptor and activates adenylate cyclase situated on the inner surface of the cell membrane. The cAMP produced mediates the activation of protein kinases which phosphorylate a variety of cellular substrates (enzymes, transport proteins etc.) which eventually lead to the appropriate physiological response associated with the hormone concerned. The cAMP became thought of as a "second messenger". By its very nature, however, this cAMP system is relatively slow in producing the desired response from the cell (>1 s).

Intracellular Ca^{2+} triggers such a diversity of processes in excitable and non-excitable tissues that, like cAMP, it has come to be regarded as a second messenger of hormone action. In contrast to cAMP, however, the period of action of Ca^{2+} is measured in milliseconds rather than in seconds. The reason for this, as mentioned earlier, is that Ca^{2+} can diffuse rapidly down a concentration gradient into the cell through specific calcium channels in the plasma membrane. It is estimated that several million Ca^{2+} ions can enter the cell per open channel per second. Very few enzyme-based reactions can proceed at this rate. It is this amplification factor and the short time scale of action that makes Ca^{2+} a vital second messenger for excitable cells.

It is thought that some hormones act to open the calcium channels to allow the influx of a small number of Ca^{2+} ions, sufficient to increase the intracellular concentration from 10^{-7} mol/l to about 10^{-6} mol/l. This small increase in calcium concentration is taken up by the calcium-binding protein, calmodulin, to form a calcium–calmodulin complex (CaM) as described earlier. As a result the calmodulin undergoes a conformational change which exposes a hydrophobic site, thought to be important in the subsequent interaction of CaM with its appropriate receptor protein. This produces an activated complex capable of stimulating whichever calcium-sensitive enzyme is necessary to produce the requisite biochemical or physiological response from the cell (Table 1.15). A list of the enzymes activated by CaM is shown in Table 1.16. How the CaM complex is prevented from randomly reacting with all possible receptor proteins within a given cell is not yet clear.

To terminate the activity of the protein, one possible mechanism is to activate the magnesium-dependent calcium-ATPase which pumps Ca^{2+} out of the cytosol. This will lead to dissociation of CaM and cessation of the activation of the receptor protein involved. In fact, CaM can stimulate this enzyme system and thereby bring about its own deactivation.

The two second messenger systems are not independent since CaM strongly activates both adenylate cyclase (which stimulates the production of cAMP) and phosphodiesterase (which destroys it). However, the former is situated on the inside of the cell membrane and is first to be activated; the phosphodiesterase, on the other hand, is located in the cytosol and is activated at a later stage in the

Table 1.15. Some reported calmodulin-mediated processes

Cyclic nucleotide metabolism
Glycogen metabolism
Microtubule function
Microfilament function
Phosphorylation
Secretion
Cell division
Calcium flux

Table 1.16. Some calcium–calmodulin activated enzymes

Adenylate cyclase
Cyclic nucleotide phosphodiesterase
Calcium–magnesium ATPase
Myosin light chain kinase
Phosphorylase b kinase
Glycogen synthetase kinase
NAD kinase
Ornithine decarboxylase
Phospholipase A_2
Calcium-dependent protein kinase

chain of events leading to the termination of the hormone action. Just as CaM may influence the metabolism of cAMP, so does cAMP affect the calcium second messenger system by releasing calcium from intracellular stores. This mutual interaction between cAMP and Ca^{2+} means that the latter may play a second or even third messenger role in the overall action of the hormone involved. In the case of phosphorylase kinase, for example, both messengers converge to activate the enzyme. (For reviews on calmodulin see Means and Dedman 1980; Cheung 1981; Demaille 1982.)

Finally, as mentioned earlier, the rates of binding of Ca^{2+} to the trigger proteins are very rapid. Indeed, they are close to the diffusion control limit of 10^{10}/s. Since the stability constants of the complexes are all $> 10^6$, the release rates are of the order of 10^3 to 10^4/s. This fast rate is essential for rapid relaxation of the trigger process. Mg^{2+}, on the other hand, has stability constants of about 10^4 with the trigger proteins and binding rates of $< 10^6$/s. This means that the release rate is slow ($< 10^2$/s) and that Mg^{2+} cannot act as a fast switch ion.

Calcium Ionophores

In order to study the effect of Ca^{2+} influx into the cell it is necessary to devise a method for introducing it. Direct micro-injection has been employed in large cells, such as squid axons, mast cells and salivary gland cells. However, this technique is limited in its scope. Antibiotics have now been developed which facilitate the passive movement of Ca^{2+} ions across cell membranes. These have been termed calcium ionophores. They function either as ion carriers or as pore formers, the pores remaining fixed while ions move through them (McLaughlin and Eisenberg 1975). These molecules are only selective, not specific, for Ca^{2+} ions.

To date the most widely used ionophore has been A23187 (Reed and Lardy 1972). Two molecules of this form a lipophilic complex with a single Ca^{2+} ion and the complex then migrates from the bulk phase across the cell membrane and releases its Ca^{2+} ion into the cell interior. This implies that the stability constant of the complex must be high enough to allow complexation on the outside of the membrane yet low enough for Ca^{2+} to be released again on the inside. The released Ca^{2+} is then available to stimulate or even enhance the customary action of the cell concerned.

X-537A is another such ionophore but it is less selective for Ca^{2+} than is A23187 (Pressman 1976). More recently, however, ionomycin (extracted from *Streptomyces conglobatus*) has been shown to have a greater selectivity for Ca^{2+} over Mg^{2+} than A23187 (Liu and Hermann 1978).

There are drawbacks, however, to the use of these interesting molecules. Firstly, they do not partition calcium within cells as do the natural physiological mediators; secondly, they may saturate intracellular calcium reservoirs; thirdly, they may penetrate sufficiently to release mitochondrial stores of calcium (Borle and Studer 1978); finally, they may cause soluble enzymes to be released with subsequent loss of cell viability (Chandler and Williams 1977).

Calcium Antagonists

In cardiac muscle the events leading to contraction involve (a) a rapid current into the cells carried by Na^+ ions that is responsible for the upstroke of the action potential, (b) a slow inward Ca^{2+} current related to the plateau phase and (c) an outward K^+ current related to the repolarization of the membrane (Reuter 1973). Calcium influx into the cardiac cell may be blocked by a group of agents described as "calcium antagonists". These include verapamil (Fleckenstein 1977), methoxyverapamil (D-600), nifedipine and prenylamine. Apart from the first two, the structures of these antagonists are fairly dissimilar, suggesting that their modes of action may be different (Triggle 1981) but they all appear to act by uncoupling membrane excitation from cell contraction. Clinically these drugs have proved useful as relaxants for patients with angina who have not responded to nitrate vasodilators or β-blockers, such as propranolol. Other clinically important Ca^{2+} antagonists include fendiline, diltiazem, bencyclan and cinnarizine.

Another group of potential "Ca^{2+} inhibitors" is represented by the drug trifluoperazine which binds to calmodulin with a stability constant of 10^6 (Cheung 1980). The complex is biologically inactive and is useful for checking whether or not a given Ca^{2+}-mediated process involves calmodulin.

Calcium in Other Living Organisms

Calcium has been shown to play an important role in a variety of biological processes associated with various organisms. These encompass the complete range of chemical properties of the Ca^{2+} ion outlined earlier in this chapter.

Amoeba

In amoeba, a minimum amount of calcium is required in the external medium for pinocytosis to occur. There is also an optimum level of Ca^{2+} for maximum pinocytotic activity. In *Amoeba proteus*, for example, this optimum is about 1 mmol/l. A further increase in the external Ca^{2+} concentration reduces the number of channels formed.

A large amount of calcium is present in these amoeba, amounting to 4.6 mmol/kg cells. About 18% of this is associated with the external surface, some is bound to the cytoplasmic surface, some is associated with the cytoskeleton consisting of microfilaments of actin, some is in vesicles and a small concentration (about 10^{-8} to 10^{-7} mol/l) is in the ionized form in the cytoplasm.

When an inducer of pinocytosis is bound to the surface of an amoeba it initiates a sequence of events involving Ca^{2+} which ultimately results in the uptake of the inducer into the amoeba. This is thought to involve displacement of surface Ca^{2+}, increased membrane permeability, transmission of the signal via Ca^{2+} influx into the amoeba, and transient rise in cytoplasmic Ca^{2+} concentration, interaction of Ca^{2+} with actin, channel formation, vesiculation and, finally, incorporation of the inducer.

Microbiological Systems

Growth, reproduction and sporulation in organisms all have a requirement for calcium. It is also necessary for some aspects of the locomotion of cells and for the movement of organelles within the cell (cytoplasmic streaming), and is important for the maintenance of membrane structure and cell adhesiveness. Ca^{2+} gradients participate in the stabilization of cell polarity.

Deposits of $CaCO_3$ are formed within many cells or in the cell coating of most groups of microorganisms. Calcification of algae can contribute to sedimentary deposits and reefs.

Plants

The process of plant cytokinesis has a requirement for Ca^{2+} ions. The cation is also necessary for vesicle coalescence and it plays a part in the secretory processes of the cell, in a manner similar to that observed in the stimulation–secretion coupling in gland cells and in the excitation–contraction coupling in muscle cells. In general, exocytosis appears to be triggered by a rise in intracellular Ca^{2+}. There is also some evidence for a calcium–magnesium ATPase being involved in the process of membrane fusion.

Another role for calcium in plants is in the formation of crystals of calcium oxalate or calcium carbonate. In some plants these are thought to act as storage reserves for calcium and there are reported cases of mobilization of the cation occurring after demineralization. Others believe that these crystals act as repositories for that calcium which is superfluous to the metabolic and growth requirements of the plant. A third possibility is that they are part of the cell's pH regulatory system. However, the uniformity of crystal size and habit within a given plant suggests that such crystals may be regarded as organized cell constituents, formed under strict genome control.

References

Babu YS, Sack JS, Greenhough TJ, Bugg CE, Means AR, Cook WJ (1985) Three-dimensional structure of calmodulin. Nature 315:37–40

Baker PF (1972) Transport and metabolism of calcium ions in nerve. Prog Biophys Mol Biol 24:177–223

Baker PF, McNaughton PA (1978) The influence of extracellular calcium binding on the calcium efflux from squid axons. J Physiol (Lond) 276:127–150

Baker PF, Blaustein MP, Hodgkin AL, Steinhardt RA (1969) The influence of calcium on sodium efflux in squid axons. J Physiol (Lond) 200:431–458

Best JB (1959) Some theoretical considerations concerning crystals with relevance to the physical properties of bone. Biochim Biophys Acta 32:194–202

Bianchi CP (1968) Cell calcium. Butterworth, London

Blaustein MP, Nelson MT (1982) Sodium–calcium exchange: its role in the regulation of cell calcium. In: Carafoli E (ed) Membrane transport of calcium. Academic Press, London, pp 217–236

Blinks JR, Prendergast FG, Allen DG (1976) Photoproteins as biological calcium indicators. Pharmacol Rev 28:1–93

Borle AB, Studer B (1978) Effects of calcium ionophores on the transport and distribution of calcium in isolated cells and in liver and kidney slices. J Membr Biol 38:51–72

Brown JE, Cohen LB, DeWeer P, Pinto LH, Ross WN, Salzberg BM (1975) Rapid changes of intracellular free calcium concentration: detection by metallochromic indicator dyes in squid giant axon. Biophys J 15:1155–1159

Burgess J (1978) Metal ions in solution. Ellis Horwood, Chichester

Campbell AK, Lea TJ, Ashley CC (1979) Coelenterate photoproteins. In: Ashley CC, Campbell AK (eds) Detection and measurements of free Ca^{2+} in cells. Elsevier, Amsterdam, pp 13–72

Caswell AH (1979) Methods of measuring intracellular calcium. Int Rev Cytol 56:145–181

Chandler DE, Williams JA (1977) Fluorescent probe detects redistribution of cell calcium during stimulus–secretion coupling. Nature 268:659–660

Chandler DE, Williams JA (1978) Intracellular divalent cation release in pancreatic acinar cells during stimulus–secretion coupling. 1. Use of tetracycline as a fluorescent probe. J Cell Biol 76:371–385

Cheung WY (1970) Cyclic 3′,5′-nucleotide phosphodiesterase. Demonstration of an activator. Biochem Biophys Res Commun 38:533–538

Cheung WY (1980) Calmodulin plays a pivotal role in cellular regulation. Science 207:19–27

Cheung WY (ed) (1982) Calcium and cell function, vol 1. Academic Press, New York

Conway BE (1981) Ionic hydration in chemistry and biophysics. Elsevier, Amsterdam

Cookson DJ, Levine BA, Williams RJP, Jontell M, Linde A, De Bernard B (1980) Cation binding by the rat incisor dentine phosphoprotein: a spectroscopic study. Eur J Biochem 110:273–278

Dack MRJ (1975) Solutions and solubilities, part 1. Wiley, New York

Davies CW (1962) Ion association. Butterworth, London

Debye P, Hückel E (1923) Zur Theorie der Elektrolyte. 1. Gefrierpunktserniedrigung und verwandte Erscheinungen. Physik Z 24:185–206

Demaille JG (1982) Calmodulin and calcium-binding proteins: evolutionary diversification of structure and function. In: Cheung WY (ed) Calcium and cell function, vol 2. Academic Press, New York, pp 111–144

Finlayson B (1977) Calcium stones: some physical and clinical aspects. In: David DS (ed) Calcium metabolism in renal failure and nephrolithiasis. Wiley, New York, pp 337–382

Fleckenstein A (1977) Specific pharmacology of calcium in myocardium, cardiac pacemakers and vascular smooth muscle. Annu Rev Pharmacol Toxicol 17:149–166

Furie BC, Borowski M, Keyt B, Furie B (1982) γ-Carboxyglutamic acid-containing Ca^{2+}-binding proteins. In: Cheung WY (ed) Calcium and cell function, vol 2. Academic Press, New York, pp 217–242

Hauschka PV, Lian JB, Gallop PM (1975) Direct identification of the calcium-binding amino acid, γ-carboxyglutamic acid, in mineralized tissue. Proc Natl Acad Sci USA 72:3925–3929

Klee CB, Crouch TH, Krinks MH (1979) Calcineurin: a calcium and calmodulin-binding protein of the nervous system. Proc Natl Acad Sci USA 76:6270–6273

Klee CB, Crouch TH, Richman PG (1980) Calmodulin. Annu Rev Biochem 49:489–515

Lee AG (1975) Functional properties of biological membranes: a physical-chemical approach. Prog Biophys Mol Biol 29:3–56

Lehninger AL (1975) Biochemistry, 2nd edn. Worth, New York

Levine BA, Williams RJP (1982a) The chemistry of calcium ion and its biological relevance. In: Anghileri LJ, Tuffet-Anghileri AM (eds) The role of calcium in biological systems, vol 1. CRC Press, Boca Raton, Florida, pp 3–26

Levine BA, Williams RJP (1982b) Calcium binding to proteins and anion centers. In: Cheung WY (ed) Calcium and cell function, vol 2. Academic Press, New York, pp 1–38

Liu C-M, Hermann TE (1978) Characterization of ionomycin as a calcium ionophore. J Biol Chem 253:5892–5894

Manery JF (1969) Calcium and membranes. In: Comar CL, Bronner F (eds) Mineral metabolism, vol 3. Academic Press, New York, pp 405–452

McLaughlin S, Eisenberg M (1975) Antibiotics and membrane biology. Annu Rev Biophys Bioeng 4:335–366

McLean FC, Hastings AB (1934) A biological method for the estimation of calcium ion concentration. J Biol Chem 107:337–350

Means AR, Dedman JR (1980) Calmodulin – an intracellular calcium receptor. Nature 285:73–77

Moeller T, Horwitz EP (1960) Chelation. In: Comar CL, Bronner F (eds) Mineral metabolism, vol 1A. Academic Press, New York, pp 101–118

Moreno EC, Varughese K (1981) Growth of calcium apatites from dilute solutions. J Cryst Growth 53:20–30

Neuman WF, Neuman MW (1958) The chemical dynamics of bone mineral. University of Chicago Press, Chicago

Newton C, Pangborn W, Nir S, Papahadjopoulos D (1978) Specificity of Ca^{2+} and Mg^{2+} binding to phosphatidylserine vesicles and resultant phase changes of bilayer membrane structure. Biochim Biophys Acta 506:281–287

Nordin BEC (ed) (1976) Calcium, phosphate and magnesium metabolism. Churchill Livingstone, Edinburgh

Nordin BEC, Peacock M (1969) The role of the kidney in the regulation of plasma calcium. Lancet II:1280–1283

Ohnishi ST (1978) Characterization of the murexide method: dual-wavelength spectrophotometry of cations under physiological conditions. Anal Biochem 85:165–179

Papahadjopoulos D, Portis A, Pangborn W (1978) Calcium-induced lipid phase transitions and membrane fusion. Ann NY Acad Sci 308:50–66

Potter JD, Johnson JD (1982) Troponin. In: Cheung WY (ed) Calcium and cell function, vol 2. Academic Press, New York, pp 145–173

Pressman BC (1976) Biological applications of ionophores. Annu Rev Biochem 45:501–530

Price PA, Otsuka AS, Poser JW, Kristaponis J, Ramans N (1976) Characterization of a γ-carboxyglutamic acid-containing protein from bone. Proc Natl Acad Sci USA 73:1447–1451

Raaflaub J (1951) Uber ein photometriches Verfahren zur Bestimmung des ionisierten Calciums. Z Physiol Chem 288:228–233

Raaflaub J (1962) Zur photometrichen Bestimmung des ionisierten Calciums in biologischen Flüssigkeiten, insbesondere in Harn. Z Physiol Chem 328:198–203

Rasmussen H (1972) The cellular basis of mammalian calcium homeostasis. Clin Endocrinol Metabol 1:3–20

Reed PW, Lardy HA (1972) A 23187: a divalent cation ionophore. J Biol Chem 247:6970–6977

Reuter H (1973) Divalent cations as charge carriers in excitable membranes. Prog Biophys Mol Biol 26:1–43

Ridgway EB, Ashley CC (1967) Calcium transients in single muscle fibres. Biochem Biophys Res Commun 39:229–234

Robertson WG (1969) Measurement of ionized calcium in biological fluids. Clin Chim Acta 24:149–157

Robertson WG, Marshall RW (1981) Ionized calcium in body fluids. CRC Crit Rev Clin Lab Sci 15:85–125

Robertson WG, Peacock M, Nordin BEC (1971) Calcium oxalate crystalluria and urine saturation in recurrent stone formers. Clin Sci 40:365–374

Robinson RA, Stokes RH (1959) Electrolyte solutions. Butterworth, London

Scarpa A, Carafoli E (eds) (1978) Calcium transport and cell function. Ann NY Acad Sci 307

Scarpa A, Brinley FJ, Dubyak G (1978) Antipyrylazo III, a "middle range" Ca^{2+} metallochromic indicator. Biochemistry 17:1378–1386

Schatzmann HJ (1982) The plasma membrane calcium pump of erythrocytes and other animal cells. In: Carafoli E (ed) Membrane transport of calcium. Academic Press, London, pp 41–108

Schatzmann HJ, Bürgin H (1978) Calcium in human red blood cells. Ann NY Acad Sci 307:125–146

Shimomura O, Johnson FH (1973) Further data on the specificity of aequorin luminescence to calcium. Biochem Biophys Res Commun 53:490–494

Shimomura O, Johnson FH, Saiga Y (1963a) Microdetermination of calcium by aequorin luminescence. Science 140:1339–1340

Shimomura O, Johnson FH, Saiga Y (1963b) Extraction and properties of halistaurin, a bioluminescent protein from the hydromedusan, *Halistaura*. J Cell Comp Physiol 62:9–15

Termine JD, Posner AS (1967) Amorphous/crystalline interrelationships in bone mineral. Calcif Tissue Res 1:8–23

Thomas MV (1982) Techniques in calcium research. Academic Press, London

Thomas MV, Gorman ALF (1977) Internal calcium changes in a bursting pacemaker neuron measured with arsenazo III. Science 196:531–533

Tobias JM, Agin DP, Pawlowski (1962) Phospholipid–cholesterol membrane model: control of resistance by ions on current flow. J Gen Physiol 45:989–1001

Toffaletti J (1982) Physiological importance of calcium complexes. In: Anghileri LJ, Tuffet-Anghileri AM (eds) The role of calcium in biological systems, vol 2. CRC Press, Boca Raton, Florida, pp 69–78

Triggle DJ (1981) Calcium antagonists: basic chemical and pharmacological aspects. In: Weiss GB (ed) New perspectives on calcium antagonists. Waverly Press, Baltimore, pp 1–18

Tsien RY (1980) New calcium indicators with high selectivity against magnesium and protons: design, synthesis and properties of prototype structures. Biochemistry 19:2396–2404

Tsien RY, Rink TJ (1980) Neutral carrier ion-selective microelectrodes for measurements of intracellular free calcium. Biochim Biophys Acta 599:623–638

Tsien RY, Pozzan T, Rink TJ (1982) Calcium homeostasis in intact lymphocytes: cytoplasmic free calcium monitored with a new, intracellularly trapped fluorescent indicator. J Cell Biol 94:325–334

Urry DW (1978) Basic aspects of calcium chemistry and membrane interaction: on the messenger role of calcium. Ann NY Acad Sci 307:3–26

Wasserman RH, Fullmer CS (1982) Vitamin D-induced calcium-binding protein. In: Cheung WY (ed) Calcium and cell function, vol 2. Academic Press, New York, pp 175–216

Weast RC, Astle MJ (1980) CRC Handbook of Chemistry and Physics, 61st edn. CRC Press, Boca Raton, Florida

Whedon GD, Lutwak L, Rambaut P et al. (1976) Effect of weightlessness on mineral metabolism: metabolic studies on Skylab orbital space flights. Calcif Tissue Res [Suppl] 21:423–430

Williams RJP (1976) Calcium chemistry and its relation to biological function. In: Duncan CJ (ed) Calcium in biological systems. Cambridge University Press, Cambridge, pp 1–17

Wnuk W, Cox JA, Stein EA (1982) Parvalbumins and other soluble high-affinity calcium-binding proteins from muscle. In: Cheung WY (ed) Calcium and cell function, vol 2. Academic Press, New York, pp 243–278

Chapter 2

Calcium-Regulating Hormones: Vitamin D

D. R. Fraser

Introduction

Since the discovery of vitamin D in the early years of this century, every attempt to fit it into one of the standard categories of biology or chemistry has failed. It was first described as a fat-soluble *nutrient* which prevented and cured rickets, the bone disease (Mellanby 1918; McCollum et al. 1922). Then it was discovered that vitamin D was also formed in the skin under the influence of ultraviolet light from the sun or from an ultraviolet lamp (Huldschinsky 1919; Chick et al. 1923). Although for the next 50 years vitamin D was said to be derived from these two sources, nutritional scientists continued to regard it as a nutrient and its formation in skin was considered to be merely a curiosity of nature. This concept of vitamin D as a nutrient has been accepted partly because it was found during the era of micronutrient discovery and partly also because it was so convenient to provide vitamin D by mouth.

Since the late 1970s, various studies have indicated that a person's vitamin D levels are determined mainly by exposure to sunlight (for review see Fraser 1983). Thus it appears that supplying vitamin D as a nutrient has been a practice devised only during this century. From the point of view of evolutionary biology, vitamin D is not a nutrient and therefore cannot be considered a true "vitamin".

To complicate matters, since the late 1960s it has been discovered that vitamin D is metabolically transformed before it performs its biochemical function. This metabolic change gives the molecule properties similar to those of the classical steroid hormones (Kodicek 1974). Hence vitamin D is now regarded as a precursor for a steroid hormone which acts in the regulation of whole-body calcium homeostasis. In this way its presence allows normal bone growth, and conversely, when it is absent, calcium homeostasis fails, leading to the bone defects of rickets or osteomalacia. However, even this description of vitamin D function now appears to be too restrictive. A range of new actions has been reported for the hormonal metabolite of vitamin D and these actions appear unrelated to whole-body calcium homeostasis (for review see Braidman and Anderson 1985). Thus not only is the traditional view of vitamin D as a nutrient

now seen to be erroneous, it is also apparent that the hormonal role of vitamin D
is more comprehensive than was previously believed. Any account of the biology
and function of vitamin D should now acknowledge that the true physiological
role of this substance as well as its biochemical mechanism of action are still far
from certain.

Structure and Synthesis

There are two chemical forms indicated by the term "vitamin D". These are
ergocalciferol (vitamin D_2) and cholecalciferol (vitamin D_3), which differ only in
their side-chain structure (Fig. 2.1). A third form of vitamin D was once
numbered vitamin D_1 but this turned out to be an impure mixture and the term
is no longer used.

Much of the chemistry of vitamin D was elucidated from ergocalciferol. This is
photochemically produced from the fungal or yeast sterol, ergosterol, when
activated by ultraviolet light. Ergosterol absorbs light over the wavelength range
of 250–310 nm with strong absorption maxima at 262, 271, 282 and 293.5 nm.
When a light quantum is absorbed by this molecule, the electrons of the Δ^5, Δ^7
diene are excited into an unstable state which results in the rupture of the 9,10-
bond in the sterol B ring. The product of this reaction, pre-ergocalciferol,
undergoes reversible thermal isomerization to ergocalciferol (Fig. 2.2). At body
temperature the equilibrium proportion of ergocalciferol to pre-ergocalciferol is
approximately 90 : 10 and the time taken to reach this equilibrium can be up to
12 h (Havinga 1973).

The precursor of cholecalciferol is 7-dehydrocholesterol. Because ergosterol
and 7-dehydrocholesterol are structurally identical in the region susceptible to

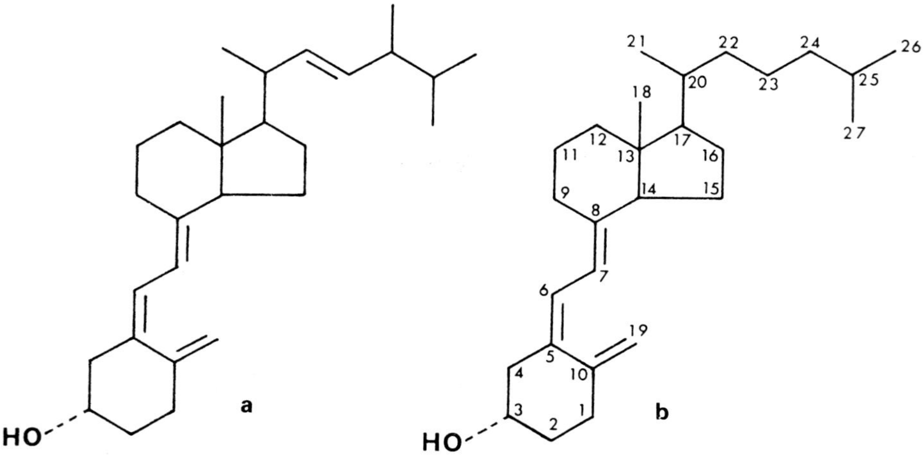

Fig. 2.1. Molecular structures of **a** ergocalciferol (vitamin D_2) and **b** cholecalciferol (vitamin D_3)

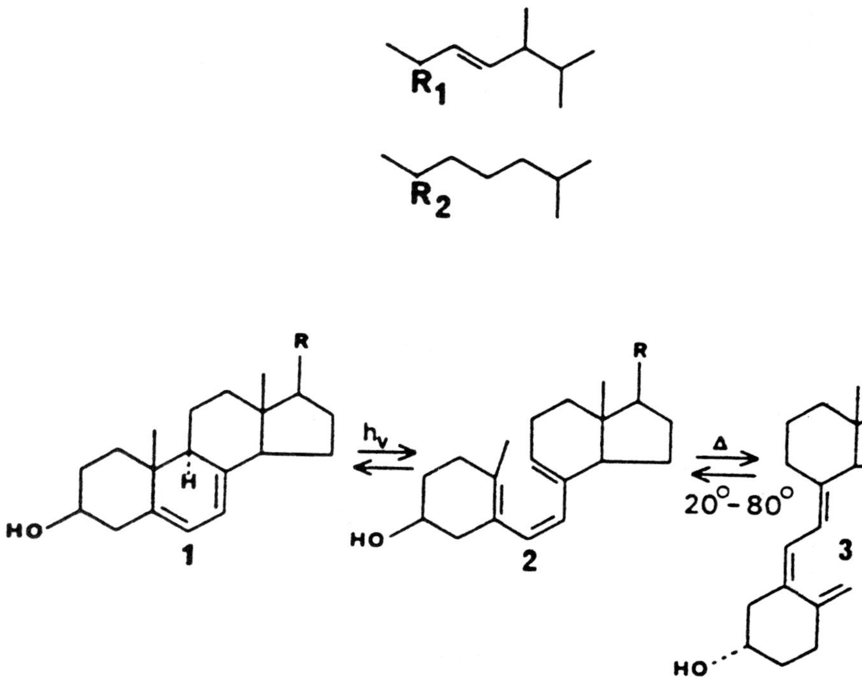

Fig. 2.2. Photochemical conversion of ergosterol or 7-dehydrocholesterol to vitamin D. R_1 = ergosterol side chain; R_2 = 7-dehydrocholesterol side chain; *1* = ergosterol or 7-dehydrocholesterol; *2* = pre-ergocalciferol or pre-cholecalciferol; *3* = ergocalciferol or cholecalciferol

photochemical change, the mechanism and conditions necessary for the formation of cholecalciferol from 7-dehydrocholesterol are essentially the same as those for the formation of ergocalciferol from ergosterol. However, 7-dehydrocholesterol is the precursor for the vitamin D found naturally in man and other animals, hence cholecalciferol is regarded as the physiological form of vitamin D in mammalian biology.

Vitamin D Formation in Skin

The production of vitamin D in skin, like the production of vitamin D by irradiation of a solution of 7-dehydrocholesterol, is a physicochemical process and requires no metabolic intervention. To understand vitamin D formation in skin it is only necessary to know (a) the location of 7-dehydrocholesterol within the skin, (b) the intensity and wavelength of ultraviolet light falling on the skin and (c) the accessibility of that light to the 7-dehydrocholesterol molecules.

7-Dehydrocholesterol is the last intermediate compound formed in the synthesis of cholesterol from acetate. In most cells the enzymic reduction of 7-dehydrocholesterol to cholesterol takes place rapidly and little of the transitory intermediate can be found. However, in the epidermis some 7-dehydrocholesterol accumulates by a mechanism not yet understood. In human skin the concentration is about 1 $\mu g/cm^2$ (Wheatley and Reinertson 1958) whereas in rat skin the concentration is as high as 4–5 $\mu g/cm^2$ (Gaylor and Sault 1964). There is a marked difference in the location of 7-dehydrocholesterol in human skin compared with its location in the skin of other animals. In the rat, 7-dehydrocholesterol appears to be synthesized in the sebaceous glands and is then secreted in sebum, becoming localized in the stratum corneum. In contrast, in human skin 7-dehydrocholesterol is concentrated, and apparently is synthesized, in the deeper layers of the epidermis. It may be that hairy animal skin, with plentiful sebaceous glands, has a greater proportion of cutaneous 7-dehydrocholesterol in the superficial layers than is the case in human skin.

Of the total extraterrestrial solar spectrum only about 5% is ultraviolet light and this proportion declines, after passing through the stratospheric ozone layer, to about 1% of the total light at sea level (Laurens 1928). The ozone layer also absorbs all radiant energy below 290 nm, hence the only sunlight capable of forming vitamin D at the Earth's surface is in the UV-B range of 290–320 nm. The intensity of light in this narrow spectral range may be diminished further by dust, smoke haze and cloud in the atmosphere (Leach et al. 1976). However, the most important factors determining the amount of activating light are latitude and season of the year. For example, the intensity of sunlight of wavelength 307.5 nm at 53°N is nearly 100 times less in winter than in the summer months (Johnson et al. 1976; Diffey 1977).

Ultraviolet light of the wavelengths which form vitamin D is also the same as that causing erythema and sunburn, and several studies have assessed the penetration of this light through human epidermis (Everett et al. 1966; Anderson and Parrish 1981; Bruls et al. 1984). The shorter the wavelength and the more pigmented the skin, the less is the transmission of activating light to the region where 7-dehydrocholesterol is concentrated. However, this penetration is enhanced if the surface of the skin is wet (Solan and Laden 1977).

It can be seen therefore, that season, latitude and skin pigmentation are three factors which are able to influence the amount of vitamin D formed. Because of this, the vitamin D status of people living in temperate regions shows a marked seasonal variation, with the highest levels occurring in late summer and the lowest at the end of winter (McLaughlin et al. 1974; Stamp and Round 1974; Stryd et al. 1979). The amount of vitamin D formed is well correlated with the extent of exposure to ultraviolet light (Beadle et al. 1980; Davie and Lawson 1980; Devgun et al. 1983). Nevertheless it is still difficult to define an exact quantitative relationship between the amount of sunlight exposure and the amount of vitamin D formed.

Clearly, the efficiency of the conversion of 7-dehydrocholesterol to vitamin D in skin is very much less than when 7-dehydrocholesterol is irradiated in solution (Bekemeier 1965). The maximum proportion of the total 7-dehydrocholesterol in human skin which can be converted to vitamin D is no more than 20% (Holick et al. 1981) and may be much less. Not only is much of the 7-dehydrocholesterol in skin inaccessible to short wavelength ultraviolet light, but also if exposure is prolonged, some of the pre-vitamin D formed is further converted to an inactive isomer, lumisterol (MacLaughlin et al. 1982).

Despite these limitations, practical experience has demonstrated that only a few hours of exposure to summer sunlight are able to produce sufficient vitamin

D to avoid deficiency for several months (Poskitt et al. 1979; Specker et al. 1985). It is also apparent that although the efficiency of the photochemical formation of vitamin D is decreased in deeply pigmented skin (Clemens et al. 1982), this is not, with reasonable sunlight exposure, a significant cause of vitamin D deficiency (Lo et al. 1986). The inefficiencies and limitations in the formation of vitamin D in skin probably had some evolutionary importance in populations living in sunny climates. It is notable that vitamin D intoxication does not occur from exposure of skin to solar ultraviolet light, no matter how extensive and prolonged that exposure may be. Conversely, vitamin D deficiency is an unlikely occurrence for both man and other animals living under natural conditions. It is only when exposure to sunlight is limited, for example following the development of urban environments, that vitamin D deficiency rickets becomes widespread.

Metabolism of Vitamin D

Since the discovery that vitamin D is metabolically transformed before it becomes biologically functional, an enormous literature has grown, cataloguing the research findings. This has been reviewed extensively by Haussler and McCain (1977), Fraser (1980a, b), DeLuca and Schnoes (1983) and Lawson (1985). Therefore, in this account, only a simple outline of the main biochemical findings will be presented.

Vitamin D undergoes two metabolic hydroxylation steps. The first of these takes place in the endoplasmic reticulum of liver cells where a side-chain hydroxylation produces the metabolite 25-hydroxyvitamin D [25(OH)D] (Blunt et al. 1968; Ponchon et al. 1969; Bhattacharyya and DeLuca 1974). This metabolite becomes the substrate for a second enzymic hydroxylation in the mitochondria of the proximal convoluted tubule cells of the kidney (Fraser and Kodicek 1970; Brunette et al. 1978). The second hydroxyl is inserted into the steroidal A ring to produce 1,25-dihydroxyvitamin D [$1,25(OH)_2D$] which is generally considered to be the functional form of vitamin D.

A number of other vitamin D metabolites have been identified in blood plasma, the most notable of these being 24,25-dihydroxyvitamin D and 25,26-dihydroxyvitamin D (see DeLuca and Schnoes 1983). These other metabolites are mainly side-chain oxidation products and their biological significance continues to be the subject of speculation. A very large number of vitamin D inactivation metabolites found in bile have been described (Avioli et al. 1967; Bell and Kodicek 1969). These metabolites have not been identified but because they have no biological activity they are regarded as evidence of a catabolic role for the liver in handling vitamin D.

Although it has been suggested that the biliary metabolites of vitamin D are part of a conservative enterohepatic circulation (Arnaud et al. 1975) it is now clear that the functional recovery of vitamin D from bile does not occur (Clements et al. 1984). Hence the liver both activates vitamin D by converting it to 25(OH)D and is the site where vitamin D is inactivated and excreted. The balance between these anabolic and catabolic processes is the main determinant of the efficiency with which vitamin D is used (Fraser 1983).

The two functional hydroxylation reactions differ not only in the anatomical sites where they occur but also in their significance for vitamin D function. It is a characteristic of steroid hormones that they all have oxygen substituents at either end of the long axis of the steroid nucleus. This enables the hormone to bind to specific receptor proteins in target cells. The specificity of this binding is further enhanced by the presence of other functional oxygen groups on the steroid molecule. The parent vitamin D molecule has only one oxygen group, at carbon 3 in the A ring. Therefore to give it the chemical features of a steroid hormone, another oxygen is required on the side chain of the molecule. The 25-hydroxylation of vitamin D creates a steroid with two polar oxygen groups. The insertion of a third oxygen by 1-hydroxylation of 25(OH)D gives the steroid its functional specificity.

The capacity of the liver to produce 25(OH)D is high, and although an increased hepatic content of unchanged vitamin D may temporarily diminish the rate of 25(OH)D production, it nevertheless is secreted into the circulation in some quantitative relationship to the supply of vitamin D. Because of this, the plasma concentration of 25(OH)D is a very good indicator of vitamin D status. In humans the normal level of 25(OH)D in plasma is 0.025–0.125×10^{-6}M (10–50 ng/ml). When the concentration falls below 0.025×10^{-6}M, this indicates little reserve of vitamin D remaining. A concentration of 0.012×10^{-6}M or less is found when there are clinical signs of vitamin D deficiency.

In contrast to 25-hydroxylation, the 1-hydroxylation reaction has an efficiently controlled and limited capacity. In general, if the concentration of 25(OH)D in plasma rises there is no increase in the output of $1,25(OH)_2$D from the kidneys. Because $1,25(OH)_2$D is the functional form of vitamin D with an endocrine role in a variety of target cells, the limited capacity for 1-hydroxylation prevents a toxic excess of $1,25(OH)_2$D being secreted by the kidney. The limited capacity of the 1-hydroxylation reaction also allows homeostatic regulation of the enzyme, so that appropriate amounts of $1,25(OH)_2$D are supplied for different physiological states. Thus, during growth, pregnancy and lactation, when there is an increased requirement for, and transport of, calcium, the 1-hydroxylase is stimulated and more $1,25(OH)_2$D is produced by the kidney. Despite this capacity for the supply of $1,25(OH)_2$D to be increased, its concentration in plasma is usually about 500 to 1000 times less than that of 25(OH)D and is in the range 0.072–0.123×10^{-9}M (30–50 pg/ml).

Vitamin D and its metabolites are transported in blood plasma in association with a specific binding protein (DBP) (for review see Haddad 1979). In human plasma, DBP is an α_1 globulin with a molecular weight of 52 000 and it represents about 6% of the total α globulin fraction (Haddad and Walgate 1976a). In 1975, it was realized that this protein was identical to the polymorphic group-specific component proteins (Gc) which had been studied extensively by geneticists as markers for gene distribution in different populations (Daiger et al. 1975). On each molecule of DBP there is one vitamin D-specific binding site. The affinity of binding is higher for 25(OH)D$_3$ ($K_d = 6.4 \times 10^{-8}$ M) than for cholecalciferol ($K_d = 4.3 \times 10^{-7}$ M) or $1,25(OH)_2$D$_3$ ($K_d = 3.4 \times 10^{-7}$ M) (Haddad and Walgate 1976b). Because of the high concentration of DBP in plasma (1–2×10^{-6} M) no more than 10% of the available binding sites contain vitamin D or its metabolites. The function of the excess binding capacity is unknown.

Apart from showing genetic polymorphism, DBP has a number of other curious features. It is synthesized in the liver and the synthesis rate increases during pregnancy and under the influence of oral contraceptives (Haddad et al. 1976). The protein is cleared rapidly from plasma ($t_{\frac{1}{2}}$ in rabbits = 1.7 d), apparently by uptake and degradation in a variety of different tissues (Haddad et al. 1981). This suggests that it is involved in the delivery of vitamin D metabolites to cells and the high affinity which DBP has for actin (Van Baelen et al. 1980; Haddad 1982) may provide a means by which interaction with cells occurs. There is some indication that DBP can associate with lymphocytes (Petrini et al. 1984) and with cells in the human placenta (Emerson et al. 1985) and this may be related to processes requiring vitamin D metabolites in those cells.

The presence of DBP in plasma, with its high affinity for 25(OH)D, enables this metabolite to remain in the circulation for long periods of time. In contrast, 1,25(OH)$_2$D, with its lower affinity for DBP, is cleared from plasma with a half-time of only a few hours compared with many days for 25(OH)D. The ability of DBP to hold 25(OH)D in extracellular fluid ensures that vitamin D levels fall only slowly when the supply of vitamin D from the environment ceases. Unlike vitamin A reserves, those of vitamin D are held outside the liver in plasma and other tissues. Despite the uncertainties about DBP function and particularly about its interaction with cells, it clearly has a role in maintaining reserves of vitamin D through the winter months in temperate regions of the world.

Regulation of Vitamin D Metabolism

A large number of endocrine and ionic factors have been shown to influence the activity of the renal 1-hydroxylase and hence to modify the production of 1,25(OH)$_2$D (for review see Fraser 1980b). These factors include parathyroid hormone (PTH), prolactin, growth hormone, insulin, glucocorticoids, sex steroids, calcium, phosphate, hydrogen ions, potassium and even 1,25(OH)$_2$D itself. Each of these is claimed to be a regulating factor and because of these multiple claims much confusion exists as to the overall manner in which 1,25(OH)$_2$D production is controlled.

The standard view has been that 1,25(OH)$_2$D is secreted in order to *stimulate* target cells in the intestinal mucosa, in the renal tubules and in bone to increase their transport of calcium. The most prominent so-called regulating factor in 1,25(OH)$_2$D formation is PTH. When the extracellular Ca^{2+} concentration falls, there is an increase in the secretion of PTH which acts in the kidney to enhance the activity of the 1-hydroxylase. It is then presumed that the effect of the extra 1,25(OH)$_2$D, by stimulating target cells, increases the extracellular Ca^{2+} concentration leading to a fall in PTH secretion. This negative control system would be a feedback loop, like those endocrine loops found linking the secretion and function of peptide hormones. However, the steroid hormones in general have longer term actions than the peptide hormones and their effects appear to *permit* the target cells to carry out particular functions rather than to *stimulate* those functions. If the action of 1,25(OH)$_2$D were viewed in this way, as a *permissive* rather than a *stimulatory* action, then the significance of the control of 1-hydroxylation would change. If the action of 1,25(OH)$_2$D were controlled at the target cell rather than by varying the central production of 1,25$(OH)_2$D in the kidney then the significance of the multitude of "regulatory factors" would decline. There is evidence that the number of "receptors" for 1,25$(OH)_2$D in target cells can be modified by the action of other hormones (see section below on vitamin D function). Thus, one way of enhancing the effect of 1,25$(OH)_2$D is to increase the number of functional binding sites in the cells in which it is acting. Then the stimulation of production of 1,25$(OH)_2$D which is found, for example, during growth, pregnancy, lactation or dietary calcium deficiency, would simply reflect the increased utilization of this renal hormone in the peripheral cells rather than a direct link between secretion and response. Such an interpretation would remove the need to find a specific biological role for each factor observed to modify the activity of the renal 1-hydroxylase.

Vitamin D Function

Most studies to determine the function of vitamin D have made use of vitamin D-deficient animals. When vitamin D is supplied to such animals, there is, after several hours delay, an increase in the absorption of calcium by the small intestine. There is also, with time, the correction of the abnormal bone structure of osteomalacia and subsequently an increased deposition of bone mineral. The delay in the intestinal response to vitamin D is now known to be partly caused by the time it takes for 1,25$(OH)_2$D to be synthesized, partly by the time it takes for this hormonal metabolite to be located in the nuclei of mucosal cells and partly by the time taken for the synthesis of specific proteins which are thought to mediate the final vitamin D actions.

This type of experimental approach has given rise to the interpretation that 1,25$(OH)_2$D is a *stimulatory* factor for calcium absorption. In vitamin D-deficiency it is inevitable that calcium absorption would be depressed and it is also inevitable that reversal of the deficiency would lead to greatly enhanced calcium absorption.

An equally valid interpretation of 1,25$(OH)_2$D function in the intestinal mucosa is that it enables the mucosal cells to have an increased *capacity* for absorption of calcium. As the vitamin D-deficient animals are also grossly deficient in calcium, then this increased absorption capacity is fully expressed. If the animals were not vitamin D-deficient and had an absorptive capacity for calcium to meet their needs, then the response to extra 1,25$(OH)_2$D would not be as marked.

Similarly, experiments to study the action of vitamin D in bone are also open to alternative interpretations. As the osteomalacic bone is removed in the early stages of repair of vitamin D deficiency bone disease (Dodds and Cameron 1938), the conclusion is widely drawn that 1,25$(OH)_2$D is a bone-mobilizing factor. This view is reinforced by observations in vitro where 1,25$(OH)_2$D is found to be a potent stimulator of osteoclastic bone resorption (Raisz et al. 1972; Reynolds et al. 1973). Yet, in vivo, there is little evidence that 1,25$(OH)_2$D is a specific stimulator of bone resorption. Because enhanced bone resorption is an essential part of the repair of osteomalacic bone, then again it is inevitable that this will occur when vitamin D or 1,25$(OH)_2$D are given to vitamin D-deficient animals. In bone culture experiments in vitro the main effect of any external factor is bone resorption and not bone formation. The limitations of the technique only allow one aspect of bone turnover to be readily demonstrated. The difference between the bone response to $1,25(OH)_2D$ in vivo and that in vitro suggests that where homeostatic control is operating, $1,25(OH)_2D$ is not primarily concerned with bone resorption. Perhaps the only interpretation that can be made at this stage is that $1,25(OH)_2D$ is one of a number of hormonal factors which coordinate the activities of cells in bone and cartilage to allow normal growth and turnover of bone to occur.

A central feature of the function attributed to $1,25(OH)_2D$ is that it acts in the nucleus of its target cells in association with a specific receptor protein (for review see Haussler 1986). The action in the nucleus is said to induce the synthesis of certain proteins which in some way modify the ability of these cells to handle calcium. The main protein which has been found to fit this interpretation is a calcium-binding protein (CaBP) with a molecular weight of about 8600 (for review see Lawson 1985). The protein is absent from the mucosal cells of the intestine of vitamin D-deficient animals and it is synthesized de novo when the animals are repleted with $1,25(OH)_2D$. Despite intensive studies on the physical properties, the cellular and subcellular location and the biosynthesis of the mRNA of this calcium-binding protein, there is still no agreement as to how it may mediate the vitamin D effect. The synthesis of other proteins, such as osteocalcin in bone (Skjodt et al. 1985) and alkaline phosphatase in intestine (Moreno et al. 1985), is also increased when cells are treated with $1,25(OH)_2D$. Again, the relationship between these proteins and the physiological function of $1,25(OH)_2D$ is not apparent. At the cellular level, the function and mechanism of $1,25(OH)_2D$ are both equally obscure.

When tissues other than the generally accepted targets for $1,25(OH)_2D$ action were examined, the specific receptor was found to be widely distributed (see Braidman and Anderson 1985). Furthermore, the addition of $1,25(OH)_2D$ to many different cell types in culture was found to depress growth, to stimulate differentiated function or to induce the secretion of hormones from endocrine cells. All of these effects have been regarded as new functions for $1,25(OH)_2D$ consequent to its binding to the intracellular receptor protein. This interpretation leads once again to confusion, for it would appear that $1,25(OH)_2D$ has two separate roles. One would be that of a key regulator in whole-body calcium homeostasis while the other would be that of a modifier of a wide range of cell functions in many different cell types.

Apart from the difficulty of understanding how one hormonal substance can have such diverse functions, it is again puzzling to understand how regulation of $1,25(OH)_2D$ formation could be the main mechanism for determining the degree of response to $1,25(OH)_2D$. If the production rate were all-important, then whenever there was an increased requirement for $1,25(OH)_2D$ to maintain calcium homeostasis, all other cells responsive to $1,25(OH)_2D$ would also show enhanced effects. Clearly this generalized modification of cell function cannot occur every time the production rate of $1,25(OH)_2D$ rises.

An alternative mechanism for regulating the effect of $1,25(OH)_2D$ in different cells would be to vary the specific binding capacity according to the response required. There is evidence that the number of receptors in bone cells (Manolagas et al. 1979) and intestinal mucosal cells (Hirst and Feldman 1982) can be diminished by glucocorticoids, that the number of receptors in intestinal cells increases at lactation (Duncan et al. 1984), that oestradiol increases the number of receptor sites in cells in the intestine (Chan et al. 1984) and uterus (Levy et al. 1984), that the number of receptors in the cells of the testes increases at puberty (Walters 1984) and that activation of T lymphocytes induces the appearance of receptors for 1,25(OH)$_2$D (Provvedini et al. 1983). These findings support the concept that the action of 1,25(OH)$_2$D is regulated at the target cell rather than by variation of the quantity produced by the kidney.

This also supports the concept that 1,25(OH)$_2$D has a permissive rather than a stimulatory role in all the cells in which it acts. The concept of such a permissive role leads to the obvious idea that there may be only one cellular function for 1,25(OH)$_2$D. It is notable that all the diverse effects attributed to 1,25(OH)$_2$D in cells in culture can be mimicked by varying the extracellular or intracellular Ca^{2+} concentration. A unifying hypothesis to draw together all these diverse observations would be that 1,25(OH)$_2$D enables cells to modify their capacity for transporting calcium. In the intestine, bone and kidney, this action would promote calcium homeostasis, whereas in other cells it would enable specialized, calcium-sensitive functions to be performed.

One specialized tissue where vitamin D may be acting is skeletal muscle (for review see Boland 1986). In vitamin D deficiency there is proximal muscle weakness of the limbs which recovers promptly within 24 h of correction of the deficiency. The mechanism of the vitamin D effect in muscle is still far from clear but its elucidation would help considerably in defining a postulated general function for vitamin D in all cells.

Vitamin D Deficiency

With the discovery of vitamin D came the opportunity for preventing rickets which had been a common disease in urban children in previous centuries. The fortification of liquid milk in USA (Weick 1967) and of National Dried Milk in Britain (Arneil 1975) apparently reduced the prevalence of rickets to very low levels in these two countries. Therefore vitamin D deficiency, as a public health problem in young children should have disappeared from affluent societies. Nevertheless, there have been reports of sporadic cases of rickets occurring in recent years in children in New Zealand (Arthur and Weston 1969), Australia (Lipson 1973), Canada (Arnaud et al. 1976) and USA (Castile et al. 1975). The affected children were usually 6–24 months of age and came from poor socioeconomic backgrounds.

However, in the UK, rickets was reported to be common once again in the 1960s and 1970s but this time the children affected were older and were well cared for (Dunnigan et al. 1962). The most characteristic feature of these children was that they were from immigrant Asian families. Although it was becoming clear that vitamin D status was determined mainly by exposure to sunlight, that skin pigmentation had minimal effect on vitamin D status and that dietary vitamin D had to be supplied at more than 10 μg/day to maintain status (see Fraser 1983), there was no indication that the UK Asian children had a markedly lower input of vitamin D (either from the diet or from formation in skin) than the Caucasian children living in the same communities (see Fraser

1981). The reason for the particular susceptibility of the Asian children was unknown.

This outbreak of vitamin D deficiency in Asians living in a northern, temperate and cloudy climate drew attention to the fact that even in the sunny countries of Asia, rickets was not a rare disease (DHSS Report on Health and Social Subjects 1980). Hence, some factor other than vitamin D supply was apparently making some populations more likely to become vitamin D deficient.

That factor is now thought to be a low intake of dietary calcium (Clements et al. 1987). Of course if there is a severe calcium deficiency during periods of growth, then bone cannot be adequately mineralized and a ricketic state will develop (Maltz et al. 1970; Kooh et al. 1977; Pettifor et al. 1978). In populations where unexplained rickets occurs there is not usually such a drastic deficiency of calcium. There is merely a low calcium content in the diet, and because the diet is also rich in phytate and dietary fibre, the availability of that calcium is low. Under these dietary conditions mild secondary hyperparathyroidism develops (Stephens et al. 1982) which would increase the production of $1,25(OH)_2D$. This extra $1,25(OH)_2D$ is now known to enhance the metabolic inactivation of $25(OH)D$ in the liver (Clements et al. 1987). If there is an adequate formation of vitamin D in skin then this increased wastage would be of no significance. However, if exposure to sunlight is limited then a low vitamin D status will develop. It should be noted that this enhanced destruction of $25(OH)D$ in the liver is not regarded as a purposeful regulation of hepatic vitamin D metabolism by $1,25(OH)_2D$. Rather, it is thought to be a secondary consequence of some primary action of $1,25(OH)_2D$ in the liver.

Vitamin D deficiency is also associated with several clinical conditions characterized by hyperparathyroidism (Stanbury 1981) or calcium malabsorption. These conditions include gastrointestinal disease (Dibble et al. 1984), intestinal resection (Compston et al. 1978), jejunoileal bypass (Teitelbaum et al. 1977), gastrectomy (Nilas et al. 1985), chronic liver disease (Dibble et al. 1984), anti-convulsant therapy (Silver et al. 1974) and gross obesity (Bell et al. 1985). In all of these abnormal states, parathyroid hormone secretion is elevated and $1,25(OH)_2D$ production is also raised. It is likely that the vitamin D deficiency which follows in these conditions is a consequence of enhanced hepatic inactivation of $25(OH)D$ under the influence of $1,25(OH)_2D$.

Whenever vitamin D deficiency rickets becomes more common, suggestions are made that the dietary intake of vitamin D should be increased. While this undoubtedly would improve vitamin D status, a recommendation to increase the oral intake of vitamin D cannot be accepted easily. Unlike most other so-called vitamins, vitamin D can cause toxic effects when taken by mouth. This route of administration apparently bypasses the mechanisms that prevent excessive formation of $25(OH)D$, the metabolite presumed to cause the toxic effects (Hughes et al. 1976). An excessive oral intake of vitamin D can give rise to hypercalcaemia with consequent calcification of soft tissues and symptomatic vomiting and diarrhoea. In this condition of hypervitaminosis D, the plasma concentration of $25(OH)D$ may rise to as high as 500 ng/ml whereas the $1,25(OH)_2D$ levels remain in the normal range.

This type of toxicity is obviously caused by pharmacological rather than nutritional sources of vitamin D. The supply of vitamin D in the usual amounts in food is very unlikely to lead to the classical hypervitaminosis state. Nevertheless, it is possible that continuous supplementation with oral vitamin D could have some long-term, unexpected toxic effect. When pigs were given dietary supplements of vitamin D in amounts somewhat higher than those needed to maintain an adequate vitamin D status, pathological changes were found in the vascular system which were very similar to the lesions of atherosclerosis in humans (Taura et al. 1979; Toda et al. 1985). The possibility of inducing such chronic pathology should be considered whenever an increase in dietary vitamin D is proposed.

References

Anderson RR, Parrish JA (1981) The optics of human skin. J Invest Dermatol 77:13–19
Arnaud SB, Goldsmith RS, Lambert PW, Go VLW (1975) 25-Hydroxyvitamin D_3: evidence of an enterohepatic circulation in man. Proc Soc Exp Biol Med 149:570–572
Arnaud SB, Stickler GG, Haworth JC (1976) Serum 25-hydroxyvitamin D in infantile rickets. Pediatrics 57:221–225
Arneil GC (1975) Nutritional rickets in children in Glasgow. Proc Nutr Soc 34:101–109
Arthur AB, Weston HJ (1969) Rickets in the welfare state. NZ Med J 70:29–31
Avioli LV, Lee SW, McDonald JE, Lund J, DeLuca HF (1967) Metabolism of vitamin D_3–^{3}H in human subjects: distribution in blood, bile, feces and urine. J Clin Invest 46:983–992
Beadle PC, Burton JL, Leach JF (1980) Correlation of seasonal variation of 25-hydroxycalciferol with UV radiation dose. Br J Dermatol 102:289–293
Bekemeier H (1965) Vitamin D der Haut Bildung und biologisches Verhalten. Int Z Vitaminforsch Beiheft 10:1–113
Bell NH, Epstein S, Greene A, Shary J, Oexmann MJ, Shaw S (1985) Evidence for alteration of the vitamin-D–endocrine system in obese subjects. J Clin Invest 76:370–373
Bell PA, Kodicek E (1969) Investigations on metabolites of vitamin D in rat bile. Separation and partial identification of a major metabolite. Biochem J 115:663–669
Bhattacharyya MH, DeLuca HF (1974) Subcellular location of rat liver calciferol-25-hydroxylase. Arch Biochem Biophys 160:58–62
Blunt JW, DeLuca HF, Schnoes HK (1968) 25-Hydroxycholecalciferol. A biologically active metabolite of vitamin D_3. Biochemistry 7:3317–3322
Boland R (1986) Role of vitamin-D in skeletal muscle function. Endocr Rev 7:434–448
Braidman IP, Anderson DC (1985) Review. Extra-endocrine functions of vitamin D. Clin Endocrinol 23:445–460
Bruls WAG, van Weelden H, van der Leun JC (1984) Transmission of UV-radiation through human epidermal layers as a factor influencing the minimal erythema dose. Photochem Photobiol 39:63–67
Brunette MG, Chan M, Ferriere C, Roberts KD (1978) Site of 1,25(OH)$_2$ vitamin D_3 synthesis in the kidney. Nature 276:287–289
Castile RG, Marks LJ, Stickler GB (1975) Vitamin D deficiency rickets. Two cases with faulty infant feeding practices. Am J Dis Child 130:964–966
Chan SDH, Chiu DKH, Atkins D (1984) Oophorectomy leads to a selective decrease in 1,25-dihydroxycholecalciferol receptors in rat jejunal villous cells. Clin Sci 66:745–748
Chick H, Dalyell EJ, Hume EM, Mackay HMM, Smith HH, Wimberger H (1923) Studies of rickets in Vienna, 1919–1922. Medical Research Council Special Report Series, no. 77, London
Clemens TL, Adams JS, Henderson SL, Holick MF (1982) Increased skin pigment reduces the capacity of skin to synthesise vitamin D_3. Lancet I:74–76
Clemens MR, Chalmers TM, Fraser DR (1984) Enterohepatic circulation of vitamin D: a reappraisal of the hypothesis. Lancet I:1376–1379
Clemens MR, Johnson L, Fraser DR (1987) A new mechanism for induced vitamin D deficiency in calcium deprivation. Nature 324:62–65
Compston JE, Ayers AB, Horton LWL, Tighe JR, Creamer B (1978) Osteomalacia after small intestinal resection. Lancet I:9–12
Daiger SP, Schanfield MS, Cavalli-Sforza LL (1975) Group-specific component (Gc) proteins bind vitamin D and 25-hydroxyvitamin D. Proc Natl Acad Sci USA 72:2076–2080

Davie M, Lawson DEM (1980) Assessment of plasma 25-hydroxyvitamin D response to ultraviolet irradiation over a controlled area in young and elderly subjects. Clin Sci 58:235–242

DeLuca HF, Schnoes HK (1983) Vitamin D: recent advances. Annu Rev Biochem 52:411–440

Department of Health and Social Security (1980) Rickets and osteomalacia. Report of the working party on fortification of food with vitamin D. Her Majesty's Stationery Office, London (Report on health and social subjects, no. 19)

Devgun MS, Johnson BE, Paterson CR (1983) Ultraviolet radiation, weather and the blood levels of 25-hydroxyvitamin D. Clin Physiol Biochem 1:300–304

Dibble JB, Sheridan P, Losowsky MS (1984) A survey of vitamin-D deficiency in gastrointestinal and liver disorders. Q J Med 53:119–134

Diffey BL (1977) The calculation of the spectral distribution of natural ultraviolet radiation under clear day conditions. Phys Med Biol 22:309–316

Dodds GS, Cameron HC (1938) Studies on experimental rickets in rats. 2. The healing process in the head of the tibia and other bones. Am J Pathol 14:273–296

Duncan WE, Walsh PG, Kowalski MA, Haddad JG (1984) Ontogenesis of the rabbit intestinal receptor for 1,25-dihydroxyvitamin D_3 – evidence for increased receptor content during late suckling and lactating periods. Comp Biochem Physiol [A] 78:333–336

Dunnigan MG, Paton JPJ, Haase S, McNicol GW, Gardner MD, Smith CM (1962) Late rickets and osteomalacia in the Pakistani community in Glasgow. Scott Med J 7:159–167

Emerson DL, Werner PA, Cheng MH, Galbraith RM (1985) Presence of Gc (vitamin D-binding protein) and interactions with actin in human placental tissue. Am J Reprod Immunol Microbiol 7:15–21

Everett MA, Yeargers E, Sayre RM, Olson RL (1966) Penetration of epidermis by ultraviolet rays. Photochem Photobiol 5:533–542

Fraser DR (1980a) Vitamin D. In: Barker BM, Bender DA (eds) Vitamins in medicine, vol 1. Heinemann, London, pp 42–146

Fraser DR (1980b) Regulation of the metabolism of vitamin D. Physiol Rev 60:551–613

Fraser DR (1981) Biochemical and clinical aspects of vitamin D function. Br Med Bull 37:37–42

Fraser DR (1983) The physiological economy of vitamin D. Lancet I:969–972

Fraser DR, Kodicek E (1970) Unique biosynthesis by kidney of a biologically active vitamin D metabolite. Nature 228:764–766

Gaylor JL, Sault FM (1964) Localization and biosynthesis of 7-dehydrocholesterol in rat skin. J Lipid Res 5:422–431

Haddad JG (1979) Transport of vitamin D metabolites. Clin Orthop 142:249–261

Haddad JG (1982) Human serum binding protein for vitamin D and its metabolites (DBP): evidence that actin is the DBP binding component in human skeletal muscle. Arch Biochem Biophys 213:538–544

Haddad JG, Walgate J (1976a) Radioimmunoassay of the binding protein for vitamin D and its metabolites in human serum. Concentrations in normal subjects and patients with disorders of mineral homeostasis. J Clin Invest 58:1217–1222

Haddad JG, Walgate J (1976b) 25-Hydroxyvitamin D transport in human plasma. Isolation and partial characterization of calcifidiol-binding protein. J Biol Chem 251:4803–4809

Haddad JG, Hillman L, Rojanasathit S (1976) Human serum binding capacity and affinity for 25-hydroxyergocalciferol and 25-hydroxycholecalciferol. J Clin Endocrinol Metab 43:86–91

Haddad JG, Fraser DR, Lawson DEM (1981) Vitamin-D plasma binding protein – turnover and fate in the rabbit. J Clin Invest 67:1550–1560

Haussler MR (1986) Vitamin D receptors: nature and function. Annu Rev Nutr 6:527–562

Haussler MR, McCain TA (1977) Basic and clinical concepts related to vitamin D metabolism and action. N Engl J Med 297:974–983, 1041–1050

Havinga E (1973) Vitamin D, example and challenge. Experientia 29:1181–1193

Hirst M, Feldman D (1982) Glucocorticoid regulation of 1,25(OH)$_2$ vitamin D_3 receptors: divergent effects on mouse and rat intestine. Endocrinology 111:1400–1402

Holick MF, MacLaughlin JA, Doppelt SH (1981) Regulation of cutaneous previtamin D_3 photosynthesis in man: skin pigment is not an essential regulator. Science 211:590–593

Hughes MR, Baylink DJ, Jones PG, Haussler MR (1976) Radioligand receptor assay for 25-hydroxyvitamin D_2/D_3 and 1α,25-dihydroxy-vitamin D_2/D_3. J Clin Invest 58:61–70

Huldschinsky K (1919) Heilung von Rachitis durch künstliche Hohensonne. Dtsch Med Wochenschr 45:712–713

Johnson FS, Mo T, Green AES (1976) Average latitudinal variation in ultraviolet radiation at the earth's surface. Photochem Photobiol 23:179–188

Kodicek E (1974) The story of vitamin D from vitamin to hormone. Lancet I:325–329

Kooh SW, Fraser D, Reilly BJ, Hamilton JR, Gall DG, Bell L (1977) Rickets due to calcium deficiency. N Engl J Med 297:1264–1266

Laurens H (1928) The physiological effects of radiation. Physiol Rev 8:1–91

Lawson DEM (1985) Vitamin D. In: Diplock AT (ed) Fat-soluble vitamins: their biochemistry and applications. Heinemann, London, pp 76–153

Leach JF, Pingstone AR, Hall KA, Ensell FJ, Burton JL (1976) Interrelation of atmospheric ozone and cholecalciferol (vitamin D_3) production in man. Aviat Space Environ Med 47:630–633

Levy J, Zuili I, Yankowitz N, Shany S (1984) Induction of cytosolic receptors for 1α-dihydroxyvitamin D_3 in the immature rat uterus by oestradiol. J Endocrinol 100:265–269

Lipson AH (1973) Epidemic rickets in Sydney. Aust Paediatr J 9:14–17

Lo CW, Paris PW, Holick MF (1986) Indian and Pakistani immigrants have the same capacity as Caucasians to produce vitamin D in response to ultraviolet irradiation. Am J Clin Nutr 44:683–685

MacLaughlin JA, Anderson RR, Holick MF (1982) Spectral character of sunlight modulates photosynthesis of previtamin D_3 and its photoisomers in human skin. Science 216:1001–1003

Maltz HE, Fish MB, Holliday MA (1970) Calcium deficiency rickets and the renal response to calcium infusion. Pediatrics 46:865–870

Manolagas SC, Anderson DC, Lumb GA (1979) Glucocorticoids regulate the concentration of 1,25-dihydroxycholecalciferol receptors in bone. Nature 277:314–315

McCollum EV, Simmonds N, Becker JE, Shipley PG (1922) Studies on experimental rickets. 21. An experimental demonstration of the existence of a vitamin which promotes calcium deposition. J Biol Chem 53:293–312

McLaughlin M, Raggatt PR, Fairney A, Brown DJ, Lester E, Wills MR (1974) Seasonal variations in serum 25-hydroxycholecalciferol in healthy people. Lancet I:536–538

Mellanby E (1918) The part played by an "accessory factor" in the production of experimental rickets. J Physiol (Lond) 52:xi–xii

Moreno J, Cortes CS, Asteggiano CA et al. (1985) Changes of intestinal alkaline phosphatase produced by cholecalciferol or 1,25-dihydroxyvitamin D_3 in vitamin D-deficient chicks. Arch Biochem Biophys 240:201–206

Nilas L, Christiansen C, Christiansen J (1985) Regulation of vitamin D and calcium metabolism after gastrectomy. Gut 26:252–257

Petrini M, Galbraith RM, Werner PAM, Emerson DL, Arnaud P (1984) Gc (vitamin D binding protein) binds to cytoplasm of all human lymphocytes and is expressed on B-cell membranes. Clin Immunol Immunopathol 31:282–295

Pettifor JM, Ross P, Wang J, Moodley G, Cowper-Smith J (1978) Rickets in children of rural origin in South Africa: is low dietary calcium a factor? J Pediatr 92:320–324

Ponchon G, Kennan AL, DeLuca HF (1969) "Activation" of vitamin D by the liver. J Clin Invest 48:2032–2037

Poskitt EME, Cole TJ, Lawson DEM (1979) Diet, sunlight, and 25-hydroxy vitamin D in healthy children and adults. Br Med J i:221–223

Provvedini DM, Tsoukas CD, Deftos LJ, Manolagas SC (1983) 1,25-Dihydroxyvitamin D_3 receptors in human leukocytes. Science 221:1181–1183

Raisz LG, Trummel CL, Holick MF, DeLuca HF (1972) 1,25-Dihydroxycholecalciferol: a potent stimulator of bone resorption in tissue culture. Science 176:768–769

Reynolds JJ, Holick MF, DeLuca HF (1973) The role of vitamin D metabolites in bone resorption. Calcif Tissue Res 12:295–301

Silver J, Davies TJ, Kupersmitt E, Orme M, Petrie A, Vajda F (1974) Prevalence and treatment of vitamin D deficiency in children on anticonvulsant drugs. Arch Dis Child 49:344–350

Skjodt H, Gallagher JA, Beresford JN, Couch M, Poser JW, Russell RGG (1985) Vitamin D metabolites regulate osteocalcin synthesis and proliferation of human bone cells in vitro. J Endocrinol 105:391–396

Solan JL, Laden K (1977) Factors affecting the penetration of light through the stratum corneum. J Soc Cosmet Chem 28:125–137

Specker BL, Valanis B, Hertzberg V, Edwards N, Tsang RC (1985) Sunshine exposure and serum 25-hydroxyvitamin D concentrations in exclusively breast-fed infants. J Pediatr 107:372–376

Stamp TCB, Round JM (1974) Seasonal changes in human plasma levels of 25-hydroxyvitamin D. Nature 247:563–565

Stanbury SW (1981) Vitamin D and hyperparathyroidism. The Lumleian lecture. J R Coll Physicians Lond 15:205–217

Stephens WP, Klimiuk PS, Warrington S, Taylor JL, Berry JL, Mawer EB (1982) Observations on the natural history of vitamin D deficiency amongst Asian immigrants. Q J Med 51:169–188

Stryd RP, Gilbertson TJ, Brunden MN (1979) A seasonal variation study of 25-hydroxyvitamin D_3 serum level in normal humans. J Clin Endocrinol Metab 48:771–775

Taura S, Taura M, Kamio A, Kummerow FA (1979) Vitamin D-induced coronary atherosclerosis in normolipemic swine: comparison with human disease. Tohoku J Exp Med 129:9–16

Teitelbaum SL, Halverson JD, Bates M, Wise L, Haddad JG (1977) Abnormalities of circulating 25-OH vitamin D after jejunoileal bypass for obesity and evidence of an adaptive response. Ann Intern Med 86:289–293

Toda T, Toda Y, Kummerow FA (1985) Coronary arterial lesions in piglets from sows fed moderate excesses of vitamin D. Tohoku J Exp Med 145:303–310

Van Baelen H, Bouillon R, De Moor P (1980) Vitamin-D binding protein (Gc-protein) binds actin. J Biol Chem 255:2270–2272

Walters MR (1984) 1,25-Dihydroxyvitamin D_3 receptors in the seminiferous tubules of the rat testis increase at puberty. Endocrinology 114:2167–2174

Weick MT (1967) A history of rickets in the United States. Am J Clin Nutr 20:1234–1241

Wheatley VR, Reinertson RP (1958) The presence of vitamin D precursors in human epidermis. J Invest Dermatol 31:51–54

Chapter 3

Calcium-Regulating Hormones: Parathyroid Hormone and Calcitonin

G. D. Aurbach

Introduction

The new technologies of biomedical science, developed since the Second World War, range from high-resolution separation techniques, radiochemical tracer methods and radioimmunoassay to cell biology and molecular biology and all these have been applied fruitfully to elucidating the nature and actions of hormones controlling calcium metabolism. The many studies utilizing these applications have spawned a substantial body of knowledge, reviewed here, describing the chemistry, biosynthesis, secretion, metabolism, physiology and mechanism of action of parathyroid hormone and calcitonin. An exhaustive and detailed outline of this field is beyond the intended scope of this volume. We have attempted, however, to provide a good general discussion and enough literature citation to give the reader an understanding of the topic and a base for further study, if so desired.

Parathyroid Hormone

Chemistry

Parathyroid hormone (PTH) is an 84-amino acid single-chain polypeptide containing no cysteine. It was first isolated in 1959, some 30 years after the first active extract of parathyroid glands was prepared by Collip (1925). His extract was prepared by boiling parathyroid glands in dilute hydrochloric acid. Development of an active extract by this means heralded the discovery years later (Keutmann et al. 1972) that the native 84-amino acid polypeptide could be hydrolysed in dilute acid to shorter active amino terminal fragments. The complete amino acid sequences of parathyroid hormone from four separate

1   35

R -A-V-S-E-I-Q-L-M-H-N-L-G-K-H-L-A-S-V-E-R-M-Q-W-L-R-K-K-L-Q-D-V-H-N-F-V-

H S-V-S-E-I-Q-L-M-H-N-L-G-K-H-L-N-S-M-E-R-V-E-W-L-R-K-K-L-Q-D-V-H-N-F-V-

B -A-V-S-E-I-Q-F-M-H-N-L-G-K-H-L-S-S-M-E-R-V-E-W-L-R-K-K-L-Q-D-V-H-N-F-V-

P -S-V-S-E-I-Q-F-M-H-N-L-G-K-H-L-S-S-L-E-R-V-E-W-L-R-K-K-L-Q-D-V-H-N-F-V-

36   70

R --S-L-G-V-Q-M-A-A-R-E-G-S-Y-Q-R-P-T-K-K-E-D-N-V-L-V-D-G-N-S-K-S-L-G-E-G--

H --A-L-G-A-P-L-A-P-R-D-A-G-S-Q-R-P-R-K-K-E-D-N-V-L-V-E-S-H-E-K-S-L-G-E-A--

B --A-L-G-A-S-I-A-Y-R-D-G-S-S-Q-R-P-R-K-K-E-D-N-V-L-V-E-S-H-Q-K-S-L-G-E-A--

P --A-L-G-A-S-I-V-H-R-D-G-G-S-Q-R-P-R-K-K-E-D-N-V-L-V-E-S-H-Q-K-S-L-G-E-A--

71   84

R --D-K-A-D-V-D-V-L-V-K-A-K-S-Q

H --D-K-A-D-V-N-V-L-T-K-A-K-S-Q

B --D-K-A-D-V-D-V-L-I-K-A-K-P-Q

P --D-K-A-A-V-D-V-L-I-K-A-K-P-Q

Fig. 3.1. Amino acid sequence for parathyroid hormone. Data shown are for rat (R), human (H), bovine (B) and porcine (P) hormones. Single letter amino acid code shown: A, alanine; D, aspartic acid; E, glutamic acid; F, phenylalanine; G, glycine; H, histidine; I, isoleucine; K, lysine; L, leucine; M, methionine; N, asparagine; P, proline; Q, glutamine; R, arginine; S, serine; T, threonine; V, valine; W, tryptophan; Y, tyrosine. (Modified from Aurbach et al. 1985)

species are now known (Fig. 3.1). The most recent to be analysed is that of the rat hormone, and its structure was deduced by sequencing a complementary DNA (cDNA) copy of the messenger ribonucleic acid (mRNA) for rat parathyroid hormone (Heinrich et al. 1984).

Structure–function studies on parathyroid hormone showed that the sequence 1–27 is the minimum required for biological activity and that removal of the first or second amino acids at the amino terminus destroys biological activity but not receptor-binding activity. A sequence representing residues 3–34 is an inhibitor of biological activity in vitro (Rosenblatt et al. 1977) and that representing residues 7–34 is a low-affinity blocker of PTH action in vivo (Horiuchi et al. 1983). Removal of even the first amino acid destroys 90% or more of the biological activity of the hormone (Potts et al. 1982).

Bioassay and Radioimmunoassay of Parathyroid Hormone

The original bioassay for parathyroid hormone was developed by Collip (1925) and depended upon a rise in serum calcium in dogs injected with gland extracts. Later in vivo assays utilized rats, chicks or quail (Aurbach et al. 1985). In vivo bioassays have now been almost entirely replaced by in vitro assays. Of these perhaps the most useful and facile are the renal adenylate cyclase assay in which parathyroid hormone stimulates cyclic AMP formation from radioactively labelled ATP (Nissenson et al. 1981) and the ROS (rat osteosarcoma) cell response which depends upon measuring cyclic AMP produced in response to parathyroid hormone added to attached cells (Lindall et al. 1983; Pines et al. 1986). Ultrasensitive cytochemical bioassays also have been developed and allow the determination of biologically active hormones present in the circulation (Chambers et al. 1978; Fenton et al. 1978). Hormones can be detected at femtogram level, corresponding to 1 : 1000 dilutions of normal human plasma. Unfortunately, this technique is cumbersome and time consuming, obviating its general applicability for clinical use.

Radioimmunoassays have been developed for parathyroid hormone from several different species. The radioimmunoassay is perhaps the most efficient means of determining parathyroid hormone levels, particularly when processing large numbers of samples. Among the currently useful assays are the "mid-region" radioimmunoassays using anti-sera directed at peptide segments in the middle of the PTH molecule (Marx et al. 1981; Mallette et al. 1982). There are potential pitfalls, however. Different antibodies are directed at different regions of the polypeptide backbone, and since as noted above, only the amino terminal region of the molecule is required for biological activity, many anti-sera recognize portions of the molecule that are biologically unimportant. Moreover, biologically inert heterogeneous fragments, as well as biologically active intact hormone, are found in the circulation. In certain clinical conditions these problems are intensified. For example, in chronic renal failure, C-terminal fragments of the molecule accumulate to a vastly greater extent than normal. However, the radioimmunoassays, with proper controls and with recognition of potential pitfalls, can be validly applied to a range of physiological and clinical uses. Serial samples under identical or controlled physiological or in vitro conditions provide a valid index for parathyroid hormone content. Moreover, newer assays, particularly amino terminal-directed radioimmunoassays, should provide even further specificity and utility for general use.

Biosynthesis

Complementary DNA (cDNA) copies of messenger RNA have been prepared from bovine, human and rat sources and provide specific probes for determining transcription and accumulation of mRNA directing biosynthesis of PTH. The gene structures of the bovine and human hormones are also known (Habener et al. 1984; Kronenberg et al. 1981). Studies on control of transcription of the gene and translation of the message are only in their infancy. To date there are indications that calcium and 1,25-dihydroxyvitamin D can influence accumulation of messenger RNA for parathyroid hormone (Russell et al. 1984).

The first stage in the biosynthesis of the polypeptide hormone itself is the formation on the ribosome of a 115-amino acid chain polypeptide called "preproPTH" (Fig. 3.2). This is a polypeptide precursor of PTH containing an additional 31 amino acid residues at the amino terminus of the hormone itself. The first 25 amino acids of preproPTH represent the leader sequence formed as the nascent polypeptide chain synthesized on the ribosome is elaborated into the cisterna of the endoplasmic reticulum. As the new peptide chain is transferred across the membranes of the endoplasmic reticulum, the prepro 25-amino acid leader sequence is hydrolysed, yielding the proform of the hormone (Cohn and MacGregor 1981; Rosenblatt 1982). As the precursor is incorporated into secretory granules, the six N-terminal amino acids of the prohormone are hydrolysed from the molecule yielding the final secretory product, the 84-amino acid parathyroid hormone molecule itself (compare Fig. 3.1 and Fig. 3.2). The prohormone is thus converted to the final product destined for secretion, the

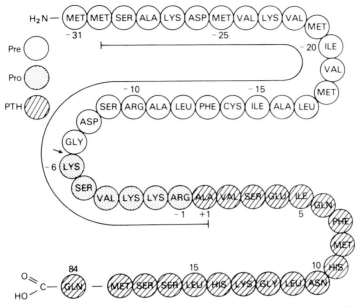

Fig. 3.2. Intermediates in the biosynthesis of parathyroid hormone. "PreproPTH" is the form of the molecule biosynthesized on the ribosomes and includes residues −31 to 84. The latter is rapidly converted to proPTH (−6 to 84) which in turn is converted to PTH (1–84), the form stored in, and released from, secretory granules. (Habener and Potts 1976)

fully active form of the hormone. The prohormone itself shows very little biological activity.

Secretion of Parathyroid Hormone

There are at least two major mechanisms that control secretion of parathyroid hormone; one uses levels of intracellular calcium, and the other is mediated by intracellular accumulation of cyclic AMP. Calcium is the classically recognized factor controlling parathyroid hormone secretion (Patt and Luckhardt 1942). Increased concentration of extracellular calcium inhibits secretion, and a fall in extracellular calcium leads to enhanced rates of secretion. The precise mechanisms whereby calcium exerts control over PTH secretion have not been identified, but recent data (Shoback et al. 1983) suggest that a rise in extracellular calcium leads to increased concentrations of intracellular calcium in the parathyroid cell. It is hypothesized that high concentrations of calcium in the parathyroid cell interact with a system(s) that inhibits secretion of the hormone. The hypothesis that extracellular calcium must be translocated to intracellular sites in order to regulate secretion is supported further by recent evidence (Fitzpatrick and Aurbach, unpublished) that calcium channel blockers prevent the calcium from controlling secretion. Moreover, calcium agonists are potent inhibitors of secretion. An apparent paradox is represented by the actions of lanthanum. Lanthanum, it is widely believed, does not enter cells. Yet lanthanum is a highly potent inhibitor of parathyroid hormone secretion, and presumably the effector transmitting the inhibitory signal of lanthanum is identical to that which recognizes calcium (see Brown 1982).

Influence of Other Ions on Secretion

Magnesium inhibits PTH secretion in a similar way to calcium (Brown 1982), although the apparent affinity of magnesium for the inhibitory site is somewhat less than is the case for calcium. However, extremely low concentrations of magnesium (and profound hypomagnesaemia in vivo) can interfere with parathyroid hormone secretion. The monovalent ions lithium and potassium (Brown et al. 1981; Dempster et al. 1982) stimulate secretion. Sodium–potassium ATPase may also be important in controlling secretion in that ouabain, an inhibitor of the ATPase, inhibits secretion (Brown et al. 1983).

Cyclic AMP-Mediated Parathyroid Hormone Secretion

Enhanced accumulation of intracellular cyclic AMP in the parathyroid cell leads to increased hormone secretion. There are several agonists that act through the adenylate cyclase–cyclic AMP system that stimulate release of parathyroid hormone. These include β-adrenergic agents, dopamine, secretin and prostaglandins of the E series (Brown 1982). Most studies on this system have been carried out in vitro with isolated parathyroid cell preparations, but dopamine, β-adrenergic catecholamines and secretin, it is reported, are effective secretagogues in vivo (see Brown 1982). Conversely, agonists that are inhibitory on the adenylate cyclase–cyclic AMP system inhibit cyclic AMP production in the parathyroid cell and diminish the rate of secretion of parathyroid hormone. Prostaglandin F ($PGF_{2\alpha}$) and α-adrenergic catecholamines belong to this class of inhibitory ligands. They act through the inhibitory guanine nucleotide regulatory protein (coupling protein) that links inhibitory ligands to inhibition of adenylate cyclase (Fitzpatrick and Aurbach, in press). A rise in the level of cAMP in the cell stimulates phosphorylation of two membrane proteins apparently linked to secretion (Lasker and Spiegel 1982).

Although these several studies clearly indicate that cyclic AMP is one intracellular regulator important in controlling parathyroid hormone secretion, calcium appears to be the more significant physiological regulator of secretion. Regardless of the concentration of cyclic AMP within the cell, calcium is capable of completely suppressing secretion. Higher concentrations of calcium are required, however, for half maximal inhibition as the cell content of cyclic AMP rises (Brown 1982).

Secretion of parathyroid hormone is dependent upon several intracellular transfer and transport phenomena. Inhibition of anion transport across the cell membrane disrupts the secretory process. Release of the hormone is inhibited by reducing external chloride or hydroxyl ion concentration or by interfering with transport of these ions. Known anion channel blockers, for example probenecid or disodium 4-acidamido-4'-isothiocyano stilbene-2,2'-disulphonate (SITS), virtually completely block hormone secretion (Brown 1982). Effects of calcium channel blockers were noted above. Cation and anion transport mechanisms may be involved in the process accounting for osmotic swelling of secretory granules prior to release of the hormone into the extracellular fluid (ECF). Several amines, including Tris, diethylamine and lysineamide, interfere with conversion of proPTH to PTH in vitro. These amines may interfere at the level of the Golgi complex (Cohn et al. 1986).

Microtubules and microfilaments are important for several functions of cells including intracellular transport and release of secretory products. The drugs colchicine and vinblastine cause disruption of microtubular and micro-filamentous function. They inhibit secretion of PTH and also interfere with conversion of proPTH to PTH in vitro. Colchicine also causes hypocalcaemia in vivo and interferes with peripheral actions of PTH (Heath et al. 1972).

The hypothesized action of vitamin D on parathyroid cell function is complex and somewhat controversial. Earlier literature contains conflicting reports showing that vitamin D or its metabolites stimulate, inhibit or are without effect on parathyroid secretion (Golden et al. 1980a,b). More recent information, however, indicates that $1,25(OH)_2D$ inhibits formation of mRNA specific for parathyroid hormone (Russell et al. 1984; Silver et al. 1985). Significant concentrations of $1,25(OH)_2D$ receptors have been identified in the parathyroid gland (see Marx et al. 1983).

Nature of Hormone Products Released

The principal biologically active product secreted from the parathyroid gland is the 84-amino acid hormone, PTH (1–84). In addition, other immunoreactive peptides are released including amino-terminal and carboxy-terminal fragments (Habener and Potts 1976). C-terminal fragments are released in a greater proportion in hypercalcaemia (Mayer et al. 1979). A fraction of secreted PTH peptides are phosphorylated. Parathyroid hormone is a substrate for protein kinases, and it is possible that phosphorylation of PTH peptides within the parathyroid cell is regulated (Goltzman 1987).

Another secretory product from the gland is PSP, a parathyroid secretory protein, molecular weight approximately 70 000 (Cohn et al. 1981, 1982; Kemper et al. 1974). It is secreted in response to calcium or magnesium in a manner similar to PTH. It is glycosylated and similar or identical to chromogranin A found in secretory granules of the adrenal medulla (Cohn et al. 1986). Just prior to secretion, carbohydrate is added and this suggests that glycosylation is involved in processing secretory granules or in the process of secretion itself. Parathyroid secretory protein also can undergo phosphorylation within the cell, and part of this protein is excreted from the cell in the phospho form (Barghava et al. 1983).

Metabolism of Parathyroid Hormone

Generally, multiple forms of radioimmunoassayable PTH are found in the circulation, but biologically active PTH in the circulation is represented virtually exclusively by PTH (1–84). Fractionation of circulating hormone shows that the major immunoreactive material in the plasma is represented predominantly by peptides in the molecular weight range 5500 to 7000 that are carboxy-terminal fragments of PTH (Habener and Potts 1976). Such peptides are biologically inert and arise from endopeptidase activity that cleaves mature PTH between amino acids 33 and 43. Parathyroid, liver and kidney cathepsin-like enzymes have been identified that are capable of cleaving PTH near or at residues 33–34 and 36–37 (Bringhurst et al. 1982; Botti and Zull 1983; Hamilton et al. 1983).

Experiments with radiolabelled PTH indicate that peptide products representing residues 34–84 and 37–84 can be produced in vivo (Segre et al. 1977). Fragments equivalent to these peptides are produced by perfused liver and generated in vitro by Kupfer cells (Bringhurst et al. 1982). Production of these metabolites is reduced following hepatectomy. Other in vitro studies show that the above two fragments are elaborated (in addition to PTH 1–84) from the parathyroid gland itself. The amino-terminal product PTH (1–33) from such a cleavage would be biologically active were it not to be metabolized further. Biologically active fragments approximating PTH (1–34) have been found upon incubating PTH (1–84) with cathepsin-containing extracts of bovine kidney and parathyroid glands (Botti and Zull 1983; Botti et al. 1981).

Although these studies allow for the possibility that biologically active fragments of PTH circulate in vivo, there is no proof that circulating metabolites of PTH are of biological significance. Essentially all circulating metabolites of PTH are biologically inert. Cytochemical bioassays (Chambers et al. 1978) show that virtually all circulating hormone in normal subjects partitions in a fashion characteristic of the native PTH (1–84). There is the possibility, however, that in uraemia smaller biologically active fragments circulate (Goltzman 1987).

Physiological Actions of Parathyroid Hormone

The most significant physiological role of parathyroid hormone is to control the concentration of calcium in the extracellular fluid (Talmage and Meyer 1976). This function is effected through activation of mechanisms that transfer calcium from bone, from glomerular filtrate and from gut to the extracellular fluid compartment. The intracellular mechanism effecting many of these functions is mediated by cyclic AMP as discussed under mechanism of action of parathyroid hormone. The actions of parathyroid hormone on the kidney and bone in effecting calcium transfer are direct whereas the transfer of calcium from the intestine to the extracellular fluid is indirect and is secondary to the influence of the hormone in enhancing formation of $1,25(OH)_2D$ from $25(OH)D$ in the kidney.

The $1,25(OH)_2D$ released into the circulation by the kidney in response to the hormone makes its way to the nuclei of target cells in the gut. Here it stimulates production of proteins (included here are calcium-binding protein and a calcium-dependent ATPase) involved in calcium transport mechanisms which facilitate movement of calcium from the gut to the extracellular fluid. In bone, the hormone acts to increase resorption of calcium from bone and to enhance transfer of calcium from bone into the extracellular fluids. These effects develop in two phases as observed in in vitro experiments (Raisz 1976). The earlier phase is manifested by release of calcium into the medium within 2 to 3 hours and does not depend upon protein synthesis. The later phase presumably involves biosynthesis of new proteins, particularly lysosomal enzymes including collagenase (Vaes 1968). Thus, the general physiological actions of parathyroid hormone which bring about an increase in calcium in the ECF proceed through two biochemically and temporally distinct phases. The earlier phase represents transport mechanisms activated rapidly by calcium- and cyclic AMP-dependent processes. The later phase, taking up to 24 hours or more to develop, depends upon protein synthesis induced in the gut secondarily to the rise in $1,25(OH)_2D$ and in bone in response to parathyroid hormone. Such processes are blocked by inhibitors of protein synthesis. The specific effects of parathyroid hormone on kidney, bone and other tissues are discussed below.

Parathyroid Hormone and the Kidney

Calcium Reabsorption

Parathyroid hormone directly affects reabsorption of calcium from the glomerular filtrate. Thus, under the influence of parathyroid hormone, at any given level of calcium load there is decreased calcium clearance; reduction in circulating parathyroid hormone leads to an increased calcium clearance. This influence of parathyroid hormone is observed clinically in states of hypo- or hyperparathyroidism (Fig. 3.3). A sharp increase in calcium clearance is noted after parathyroidectomy in humans and in experimental animals, the golden hamster being extremely sensitive to this function of PTH. Parathyroidectomy in this

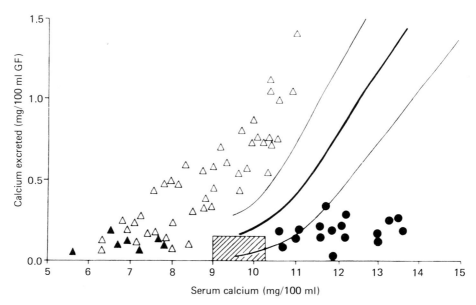

Fig. 3.3. Excretion of calcium into the urine as a function of serum calcium. *Continuous line* represents the function in normal human subjects (*dotted lines* represent SD). Hyperparathyroidism (●); hypoparathyroidism (△). Shaded area represents normal physiological range. (Nordin and Peacock 1969)

species causes such a marked loss of calcium into the urine that the attendant hypocalcaemia can be totally accounted for by calcium appearing in the urine within 60 to 120 minutes after ablation of the parathyroid glands.

The physiological mechanisms whereby calcium transport in the kidney is regulated by parathyroid hormone have been delineated only partially. It is known that calcium and sodium transport are coupled in the proximal tubule and that parathyroid hormone can influence sodium transport in that segment of the nephron. The major physiological effect of parathyroid hormone on calcium reabsorption occurs beyond the proximal tubule, however, in the thick ascending and granular portions of the distal tubule (Agus et al. 1981a, b; Dennis and Brazy 1982). Direct effects of parathyroid hormone on this segment have been identified in vitro (Bourdeau and Burg 1980), and this phenomenon is mediated through a cyclic AMP-regulated mechanism (Biddulph and Wrenn 1977; Bourdeau and Burg 1980).

Marked hypersecretion of parathyroid hormone causes an increase in calcium excretion, but this effect is secondary to hypercalcaemia (high filtered load of calcium). Even though calcium excretion is higher than normal under these circumstances, the actual clearance is reduced relative to serum calcium (Fig. 3.3).

Phosphaturic Effect

A rapid increase in the urinary excretion of phosphate is one of the first physiological effects of increased parathyroid hormone secretion to be recognized. This action represents the effects of the hormone at two distinct loci within the nephron. The phosphaturic effect may represent a direct action of PTH on a phosphate transport system or the effect might be secondary to changes in sodium or bicarbonate reabsorption. Infusion of parathyroid hormone reduces proximal tubular reabsorption of sodium and phosphate; infusion of dibutyryl cyclic AMP causes a similar effect. The overall effect is a net decrease in proximal reabsorption of phosphate. Further, the hormone has a distal tubular effect, decreasing reabsorption of phosphate there (Agus et al. 1981b). The involvement of multiple sites within the nephron in hormone-mediated phosphate transport is not unexpected in view of the large number of sites identified along the course of the nephron containing receptors for the hormone (see section on mechanism of action).

The effect of the hormone on phosphate reabsorption in the proximal tubule may be secondary to its effects on sodium. Dibutyryl cyclic AMP has a similar influence on sodium and on phosphate, and cyclic AMP is known as a regulator of sodium transport in a number of tissues (Strewler and Orloff 1977). Moreover, catecholamines, another class of hormones also acting through the cyclic AMP mechanism, produce similar effects on sodium transport in the proximal tubule. It has also been suggested that phosphaturia is secondary to changes in intraluminal pH or proximal transport of bicarbonate. Increases in pH would change the ratio of HPO_4^{2-} : $H_2PO_4^-$ and consequently decrease the rate of phosphate reabsorption; monovalently charged phosphate is more readily translocated across cell membranes than is the divalently charged ion.

Phosphate rejected through any mechanism in the proximal tubule will lead to increased excretion of phosphate into the urine because the distal tubule is less permeable to phosphate than is the proximal tubule (Agus et al. 1981b). Also contributing to the phosphaturic effect may be an overall increase in metabolism of cells along the nephron, regulated by parathyroid hormone. Activation of these cells may lead to an increased utilization of ATP, the substrate for numerous intracellular phosphorylation events. The end product of these phosphorylations and subsequent phosphatase action is inorganic phosphate, excreted to the exterior of the cells thus activated.

Other Effects on the Kidney

Enhanced excretion of bicarbonate was discovered in the earliest biological tests involving parathyroid extracts. Parathyroid hormone causes a net inhibition of bicarbonate reabsorption in the proximal tubule. The mechanism of this effect is not clear. Under conditions of marked hyperparathyroidism, a type of proximal renal tubular acidosis can develop and marked increases in bicarbonate clearance have been observed in response to parathyroid hormone infusion in patients with hereditary fructose intolerance, a condition that appears to sensitize the kidney to such actions of parathyroid hormone (Aurbach et al. 1985). The effects of parathyroid hormone on renal bicarbonate clearance may be one of the factors causing phosphaturia as noted above.

Isotonic fluid reabsorption in the proximal tubule is also impeded by parathyroid hormone. This may be secondary to decreased sodium transport in the proximal tubule since a similar effect has been observed in response to catecholamines. Beta-adrenergic catecholamines and parathyroid hormone each inhibit proximal sodium reabsorption through a cyclic AMP-mediated mechanism. The sodium thus excluded from reabsorption in the proximal tubule gives rise to a net increase in sodium delivery to the distal tubule, carrying with it an increase in associated water. The sodium is reabsorbed in the distal tubule leaving water to be excreted. This is reflected in an increase in free water clearance (Aurbach and Chase 1976; Winaver et al. 1982).

Actions of Parathyroid Hormone on Bone

Early and Late Effects

Two phases of parathyroid hormone action on bone were described above. The early phase represents mobilization of calcium from areas of bone and the enhanced transfer of this calcium into the ECF; this effect does not require new protein synthesis. The later phase is associated with an increase in synthesis of bone enzymes, particularly lysosomal enzymes that promote bone resorption and influence bone remodelling. Bone remodelling, the resorption of older regions of bone or osteons and subsequent replacement with new bone formation, is due to degradation of bone by osteoclasts and the subsequent infiltration of osteoblasts that synthesize new collagen and allow remineralization of replacement osteons. The initial effect of PTH on bone is increased resorption. This is reflected in reduced osteoblast function and enhanced osteoclast activity. Later, new bone formation is enhanced. Bone growth factors (Canellis 1985) are likely to be involved in mediating the new bone formation which occurs in response to the earlier phases of enhanced bone resorption (Tam et al. 1982).

Bone Cell Types

There are three classes of cells in bone tissue: osteoclasts, osteoblasts and osteocytes. Osteocytes are osteoblasts that have been fully encased in mineralized bone. These cells communicate through cell processes in canaliculae with other cells of their class and with osteoblasts nearer the bone surface. Osteoblasts participate in bone formation. Osteoclasts are multinucleated cells derived from monocytic precursors that are recruited to bone from other tissue sources. The osteoclast is a bone-resorbing cell and contains a complement of lysosomal enzymes that play a key role in the bone-resorptive process. Parathyroid hormone influences all three bone cell types. Administration of the hormone in vivo causes an apparent increase in the number of osteoclasts relative to osteoblasts. In general, the hormone enhances the activity of osteoclasts and diminishes the activity of osteoblasts (Cohn and Wong 1978).

Direct Actions of Parathyroid Hormone on Bone Cells

Hormonal effects on bone resorption are evident in in vitro experiments. Parathyroid hormone added to bone fragments causes enhanced bone resorption, and this enhanced osteolysis is brought about by increased activity of osteoclasts and initially by inhibition of osteoblast activity (Raisz 1976). Parathyroid hormone stimulates RNA synthesis in osteoclasts, increases the number of nuclei per osteoclast and increases the total number of osteoclasts (Peck and Klahr 1979). These changes are accompanied by increases in release of lysosomal enzymes and these increases are dependent on new RNA and protein synthesis (Eilon and Raisz 1978). Lysosomal enzymes are released rapidly from bone activated by parathyroid hormone; β-glucuronidase is released as early as, or earlier than, detectable release of calcium. Other effects of parathyroid hormone on bone include enhanced synthesis of hyaluronate, inhibition of citrate decarboxylation, inhibition of collagen synthesis and changes in alkaline phosphatase activity. The latter is detected in cytochemical assays within 3 minutes of exposure of bone in vitro to parathyroid hormone. Calcium fluxes also increase into and out of bone cells in response to parathyroid hormone (Dziak and Stern 1975).

Bone Cell Types and the Response to Parathyroid Hormone

Recent developments in the ability to separate the several types of bone cells help to clarify considerably the spectrum of biochemical responses of bone to parathyroid hormone. Differential digestion of bone in vitro with collagenase allows at least partial separation of osteoclast-like from osteoblast-like cells (Cohn and Wong 1978). The responses of these cell types can then be studied independently. Both osteoclasts and osteoblasts apparently contain receptors for parathyroid hormone whereas osteoclast-like cells respond almost exclusively to calcitonin. Characteristics of the separated cell types and their responses to parathyroid hormone and calcitonin are illustrated in Table 3.1. Note that most of the inhibitory-type effects of parathyroid hormone are exerted on osteoblast-like cells, whereas, osteoclast-like cells are stimulated by parathyroid hormone.

Table 3.1.   Effects of parathyroid hormone on bone cells (after Cohn and Wong 1978)

	"CT cells"	"PT cells"
Cyclic AMP content	↑ [a]	↑
Hyaluronate synthesis	↑ *	—
Acid phosphatase	↑ *	—
Alkaline phosphatase	—	↓
Prolyl hydroxylase	—	↓
Citrate decarboxylation	—	↓
Collagen synthesis	—	↓
Beta-glucuronidase release	↑ *	
Acetylglucosaminidase release	↑ *	

"PT cells" are analogous to osteoblasts; "CT cells" – osteoclasts. Arrows refer to effect of parathyroid hormone: ↑ = stimulatory; ↓ = inhibitory.
* Calcitonin inhibits these effects of parathyroid hormone.
[a] Calcitonin also increases cAMP content of CT cells.
[b] Other effects of parathyroid hormone include stimulating release of lysosomal enzymes, galactosidase, cathepsin and acid deoxyribonuclease.

Effects in Other Tissues

The gastrointestinal tract responds indirectly to parathyroid hormone as a consequence of the hormone's activation of the 1-α-hydroxylation system for vitamin D metabolites in the kidney. Parathyroid hormone-induced enhanced gastrointestinal absorption of calcium is secondary to the action of $1,25(OH)_2D$ formed in the kidney and acting on cells in the gut. Parathyroid hormone can also cause an acute transient hypocalcaemia that presumably reflects hormone-mediated calcium flux into cells. Other effects of parathyroid hormone include increased blood flow through the coeliac axis, increased concentrations of calcium in the mammary gland, enhanced lipolysis in isolated fat cells, increased gluconeogenesis in the liver and kidney and enhanced rates of mitosis of lymphocytes in vitro.

Mechanism of Action of Parathyroid Hormone

Specific Receptors

The actions of parathyroid hormone in specific target tissues are a consequence of its interaction with specific receptors on distinct cell types. Specific receptor sites, upon interaction with the hormone, activate one of two general intracellular messenger systems. Many of the classical effects of the hormone are mediated by activation of adenylate cyclase and the consequent increased generation of cyclic AMP within the cells. The latter intracellular messenger activates systems ultimately expressing biological actions of the hormone. Interaction with other cell types can influence distribution of intracellular calcium (Dziak and Stern 1975) and mechanisms potentially activated by calcium-sensitive protein kinases.

Mechanisms of Action in the Kidney

The nephron represents a series of distinct cell types bearing receptors for several different hormones. The work of Morel and collaborators (1982) has led to identification of specific regions of the nephron sensitive to parathyroid hormone, calcitonin, vasopressin, glucagon and catecholamines. The overall results of their studies (illustrated in Fig. 3.4) fit well with physiological studies on the action of these several hormones on the kidney. Vasopressin acts predominantly on collecting ducts, a site where vasopressin-activated adenylate cyclase was found (Fig. 3.4). Parathyroid hormone receptors are distributed in the cortical regions of proximal as well as distal tubules. In the proximal tubule, sites are found in the early convoluted as well as the "bright" portions. In the distal tubule, parathyroid hormone-sensitive adenylate cyclase is found in the granular portion and the cortical ascending limb. Calcitonin activates adenylate cyclase in the cortical ascending limb and in another site in the distal convoluted tubule near sites activated by parathyroid hormone.

The distribution found for PTH-sensitive adenylate cyclase agrees with the physiological findings that this hormone influences phosphate transport at proximal and at distal tubular sites (Klahr and Peck 1980). Within the proximal

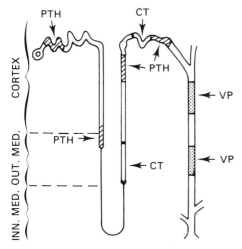

Fig. 3.4. Regions of nephron sensitive to parathyroid hormone (*PTH*), calcitonin (*CT*) and vasopressin (*VP*). (Modified from Chabardes et al. 1978)

convoluted tubule, actions of PTH also lead to enhanced activities of carbonic anhydrase and alkaline phosphatase. Glucose-6-phosphate dehydrogenase is rapidly activated in the distal convoluted tubule in response to PTH (Chambers et al. 1978). This is a cyclic AMP-mediated phenomenon and forms the basis of an ultrasensitive cytochemical bioassay (Fenton et al. 1978; Chambers et al. 1978).

Adenylate cyclase activity was utilized as the detection system for PTH-reactive receptors in the above studies on the localization of PTH-sensitive segments within the nephron (Morel et al. 1982). Direct measurement of PTH-receptor interactions has been possible in certain cell or cell membrane systems (Nissenson and Arnaud 1979; Rizzoli et al. 1983) utilizing binding of ^{125}I-labelled PTH as a ligand. Specificity of binding of the labelled ligand to receptors correlates with hormone action in that the use of competitive inhibitors concurrently blocks hormone responses and binding of ligand to the receptor. Affinity labelling of receptors has also been described (Coltrera et al. 1981). In the nephron, however, studies utilizing direct binding of radioactively labelled ligand to specific tubular sites have not yet been reported.

Mechanism of Activation of Adenylate Cyclase

Polypeptide or amine hormones control adenylate cyclase through the intermediation of guanine nucleotide regulatory (coupling) proteins of G-proteins. Several classes of G-proteins have been identified (Hurley et al. 1984), and two in particular regulate adenylate cyclase – G_s and G_i. It is the stimulatory coupling protein, G_s, that activates adenylate cyclase. The G-proteins are composed of three classes of subunit – alpha, beta and gamma. It is the alpha subunit which, after activation of receptor by hormone, dissociates from the beta/gamma subunits and interacts with the catalytic component of the adenylate cyclase enzyme itself. This is the form of the adenylate cyclase complex that actively catalyses formation of cyclic AMP. The G-protein G_i contains a beta subunit identical to the beta subunit of G_s. Activation of G_i, however, by interaction of inhibitory ligands with their receptors, inhibits adenylate cyclase activity. Recently, it has been demonstrated that in pseudohypoparathyroidism there is a deficiency of alpha subunits of G_s (Levine et al. 1980; Farfel et al. 1980). This deficiency probably reflects decreased biosynthesis of the alpha subunit due to a genetic defect in this disorder. The deficiency of the alpha subunit accounts for defective production of cyclic AMP in response to several hormones in pseudohypoparathyroidism. The complement of G_i is normal in the disorder (Downs et al. 1985).

Hormone Action in the Kidney – Control Through Cyclic AMP

Cyclic AMP accumulates in the kidney in response to parathyroid hormone and in turn activates other enzyme and ion transport systems through interaction with another class of enzymes, protein kinases. Cyclic AMP-regulated protein kinases catalyse the transfer of the gamma phosphate of ATP to a hydroxyamino acid, serine or threonine, in the acceptor protein. The kinase is composed of two classes of protein; cyclic AMP receptors and the kinase enzyme itself.

Interaction of cyclic AMP with the receptor causes dissociation of the receptor from the kinase protein with consequent activation of the latter. In the kidney, cyclic AMP-dependent protein kinase is found concentrated at the luminal surface of tubular cells (Dousa and Steiner 1978; Insel et al. 1975; Hammerman et al. 1983; Noland and Henry 1983).

Thus parathyroid hormone activation of the adenylate cyclase–cyclic AMP system in the nephron is associated with accumulation of cyclic AMP near the luminal brush border region of tubular cells and activation of the protein kinase in that region. Concentration of cyclic AMP at the luminal surface may explain the facile elaboration of cyclic AMP into luminal fluid in response to the hormone. Cyclic AMP-stimulated phosphorylation of proteins in the brush border also may account for changes in phosphate transport in the renal tubule. In addition to influencing phosphate transport within the nephron, cyclic AMP-activated protein kinases undoubtedly account for the activation in response to parathyroid hormone of alkaline phosphatase, glucose-6-phosphate dehydrogenase, gluconeogenesis and changes in the transport of sodium bicarbonate and calcium. Stimulation of calcium fluxes by cyclic AMP in renal tubule preparations has been described (Bourdeau and Burg 1980). In addition, changes in intracellular distribution of calcium may be influenced by non-cyclic-AMP-mediated mechanisms. There are suggestions that such a phenomenon may exist in bone cells (Dziak and Stern 1975).

Cellular Mechanisms in the Actions of Parathyroid Hormone on Bone

The adenylate cyclase–cyclic AMP system is clearly involved in the actions of parathyroid hormone on bone. Cyclic AMP concentrations rise in response to PTH in both osteoblast-like and osteoclast-like cells in bone (Cohn and Wong 1978). Parathyroid hormone and prostaglandins of the E series each cause a rise in cyclic AMP content of skeletal tissue incubated in vitro. These agents cause bone resorption in vitro and the action of cyclic AMP in mediating bone resorption is substantiated by the observation that dibutyryl-cAMP induces bone resorption similar to that effected by prostaglandins or parathyroid hormone. Dibutyryl-cAMP also induces lysosomal enzyme formation and elaboration by bone in vitro (Vaes 1968). Lysosomal enzymes also are involved in the resorptive process. These several observations clearly implicate cyclic AMP as an intracellular mediator in parathyroid hormone-induced bone resorption. In addition, there is evidence that calcium itself, induced in response to PTH, may be an intracellular mediator of bone resorption (Dziak and Stern 1975). Parathyroid hormone can influence calcium influx in bone cells (Marcus and Orner 1980). Moreover, certain parathyroid hormone peptide analogues show discrepant ratios of activities on cyclic AMP accumulation on the one hand and bone resorption on the other (Herrman-Erlee et al. 1983). The authors in the latter study suggested the possibility that there are two distinct classes of parathyroid hormone receptors in bone, one class governing demineralization via control of calcium influx and another class controlling adenylate cyclase activity.

Similar dichotomies in hormone receptor class have been observed in several other endocrine-regulated systems. Catecholamines and vasopressin (Fahrenholz et al. 1984) represent two hormone types showing such a dichotomy of receptor class. One class is regulated through the adenylate cyclase mechanism, the other through changes in intracellular calcium distribution and activation of calcium-sensitive protein kinases. It is not yet known whether the C-kinase system is involved in parathyroid hormone control of bone resorption. Cyclic AMP-regulated kinases are found in bone cells and undoubtedly participate in PTH-mediated effects on these cells (Livesey et al. 1982).

Another paradox in hormonal regulation of bone resorption concerns the actions of parathyroid hormone and calcitonin in regulating function of specific cell types in bone. It was noted above that parathyroid hormone-stimulated accumulation of cyclic AMP is found in both osteoblastic and osteoclastic cell types. Moreover, accumulation of cyclic AMP in response to parathyroid hormone in these cells is associated with induction of bone resorption. Calcitonin, seemingly paradoxically, counteracts bone resorption by inhibiting the action of osteoclasts through a similar or identical mechanism, that is, by activating the adenylate cyclase–cyclic AMP system. How is this paradox explained? Perhaps osteoclasts themselves can be subdivided into populations bearing parathyroid hormone receptors on the one hand and calcitonin receptors on the other. Thus each hormone might act selectively on subsets of the osteoclast population. Alternatively, other factors may control the complement of one class of receptor or the other on osteoclasts. Still other possibilities exist.

Cyclic Nucleotides in the Extracellular Fluids

Cells responding to hormones controlling cyclic nucleotide accumulation elaborate part of the cyclic nucleotide formed into the extracellular fluid. Thus, the cyclic AMP content of hepatic vein plasma reflects activation of hepatic cells by glucagon and in the adrenal vein reflects activation of adrenal cortical cells by adrenocorticotropic hormone (ACTH). Plasma cyclic AMP, then, is derived from a multitude of tissues each responding to specific hormones. Plasma cyclic AMP is cleared in part directly by glomerular filtration into the urine. In addition to plasma being a source of urinary cyclic AMP, the kidney represents a unique tissue in that some of the cells along the course of the nephron elaborate cyclic AMP directly into the luminal fluid and thence into the urine. The latter fraction of urinary cyclic AMP is the "nephrogenous" component.

Of the cyclic AMP in the urine, 50%–60% represents that cleared from plasma by glomerular filtration and the remaining 40% to 50% is the nephrogenous component (Broadus 1981). Since parathyroid hormone influences directly certain cells along the course of the nephron that contribute to the nephrogenous component of urinary cyclic AMP, this component is a parameter that reflects the rate of secretion of parathyroid hormone and the amount of parathyroid hormone in the circulation (Chase et al. 1969; Chase and Aurbach 1967). An infusion of parathyroid hormone thus causes a rapid increase in the nephrogenous component, and conversely, inhibition of parathyroid hormone secretion leads to reduced urinary cyclic AMP excretion. The parameter, urinary cyclic AMP excretion, can be expressed as the total urinary cyclic AMP in nM/100 ml of the filtrate or as the nephrogenous component. Total urinary cyclic AMP (UcAMP/dl GF) represents a simple correction of the urinary cyclic AMP to creatinine ratio for plasma creatinine ($UcAMP/U_{Cr} \times P_{Cr}$). Nephrogenous cyclic AMP represents UcAMP (concentration of cyclic AMP/100 ml of plasma). Either of these parameters can be utilized physiologically or clinically as indexes of the circulating concentration of biologically active parathyroid hormone (Broadus 1981).

Calcitonin

Chemistry

Calcitonin is a 32-amino acid polypeptide elaborated by the parafollicular or "C" cells of the thyroid gland. The "C" cells arise embryologically from the most caudal (fifth) branchial pouch and give rise to a distinct structure, the ultimobranchial body, which contains calcitonin-secreting cells. In mammals the ultimobranchial body and the medial portion of the fifth branchial pouch become incorporated into the parafollicular interstices of the thyroid. In submammalian species the ultimobranchial body remains as a separate entity.

Complete amino acid sequence information is available for nine variants of the calcitonin molecule representing the hormone from seven different species. All are 32-amino acid polypeptides with an N-terminal seven-membered disulphide ring and a C-terminus of prolineamide. These structures are remarkable in that as many as 16 of the 32 amino acids differ in the most diverse (human vs bovine) forms of the hormone (Fig. 3.5). Nevertheless, these structures share a number of common features. Six of the seven amino-terminal residues are identical among all of the congeners and the sequence variability of the middle region of the molecule is more apparent than real. An acidic residue (aspartic acid or glutamic acid) is found uniformly at position 15 and the only other acidic residue is found at position 30. Basic residues also are limited to a few positions. Aromatic residues may exist at positions 12, 13, 16, 19, 22 or 27 but have never been found within the amino-terminal 11 residues. At least one aromatic amino acid is found in all variants, but some contain neither tryptophan nor tyrosine. A unique feature of the ovine molecule is that it contains three tyrosines. The most potent of the congeners are the fish and chicken calcitonins and these molecules are imnmunologically similar as well. A high degree of immunological cross-reactivity is found between the rat and human hormones which show a high degree of homology in amino acid sequence.

Structure–Function Relationships

Despite the wide evolutionary changes in amino acid sequence for the calcitonin molecules, it is apparent that certain characteristics are required for biological activity. The 32-amino acid chain is required to be virtually complete for biological activity. Deletion of serine from the seven-membered amino-terminal di-cysteine ring can be effected with no loss of biological activity (Potts and Aurbach 1976; Schwartz et al. 1981). Removal of even one amino acid at position 16 (Findlay et al. 1983, 1985), however, destroys 80% or more of the activity, and shortening the chain in any other way causes almost total loss of activity. Methionine, when located at position 8 immediately adjacent to the

The figure is an alignment table of calcitonin amino acid sequences. The complete human sequence is given; for the other calcitonins only residues differing from human are printed and dashes (–) mark residues identical to human. Residue positions are numbered in even steps (2, 4, … 32). The alignment is reproduced below, split into residues 1–16 and 17–32.

Residues 1–16

Species	1	2	3	4	5	6	7	8	9	10	11	12	13	14	15	16
Eel	–	Ser	–	–	–	–	–	Val	–	–	Lys	Lys	Ser	–	Glu	Leu
Salmon I	–	Ser	–	–	–	–	–	Val	–	–	Lys	Lys	Ser	–	Glu	Leu
Salmon II	–	Ser	–	–	–	–	–	Val	–	–	Lys	Lys	Ser	–	–	Leu
Salmon III	–	Ser	–	–	–	–	–	Met	–	–	Lys	Lys	Ser	–	–	Leu
Human	Cys	Gly	Asn	Leu	Ser	Thr	Cys	Met	Leu	Gly	Tyr	Thr	Thr	Gln	Asp	Phe
Rat	–	–	–	–	–	–	–	–	–	–	Ala	–	Trp	Arg	Asn	Leu
Porcine	–	Ser	–	–	–	–	–	Val	–	Ser	Ala	–	Trp	Lys	–	Leu
Bovine	–	Ser	–	–	–	–	–	Val	–	Ser	Ala	–	Trp	Lys	–	Leu
Ovine	–	Ser	–	–	–	–	–	Val	–	Ser	Ala	–	Trp	Lys	–	Leu

Residues 17–32

Species	17	18	19	20	21	22	23	24	25	26	27	28	29	30	31	32
Eel	His	–	Leu	Gln	–	Tyr	–	Arg	–	Asp	Val	–	Ala	–	Thr	–
Salmon I	His	–	Leu	Gln	–	Tyr	–	Arg	–	Asn	Thr	–	Ser	–	Thr	–
Salmon II	His	–	Leu	Gln	–	–	–	Arg	–	Asn	Thr	–	Ala	–	Val	–
Salmon III	His	–	Leu	Gln	–	–	–	Arg	–	Asn	Thr	–	Ala	–	Val	–
Human	Asn	Lys	Phe	His	Thr	Phe	Pro	Gln	Thr	Ala	Ile	Gly	Val	Gly	Ala	Pro-NH$_2$
Rat	–	–	–	–	–	–	–	–	–	Ser	–	–	Gly	–	Thr	–
Porcine	–	Asn	–	–	Arg	–	Ser	Gly	Met	Gly	Phe	–	Pro	Glu	Thr	–
Bovine	–	Asn	Tyr	–	Arg	–	Ser	Gly	Met	Gly	Phe	–	Pro	Glu	Thr	–
Ovine	–	Asn	Tyr	–	Arg	Tyr	Ser	Gly	Met	Gly	Phe	–	Pro	Glu	Thr	–

Fig. 3.5. Amino acid sequences for the calcitonins. The entire sequence is shown for human calcitonin. For other calcitonins, the only residues shown are those that differ from human calcitonin (*dashes* indicate residues identical to those in the human molecule). (From Aurbach et al. 1985)

heptapeptide ring, represents a site of potential inactivation through oxidation. Conversion of the methionine to methionine sulphone at this locus destroys the biological activity. When located at position 25 however, oxidation of methionine is without effect on biological activity. Substitution of asparagine for the aspartic acid at position 15 in bovine calcitonin enhances biological potency.

The high biological potency of the fish hormones as compared to other calcitonins (CTs) is of interest. Their amino acid sequences show particular residues (at positions 11, 13, 17, 19, 20 and 24) characteristic of the most active molecules. Charge is one possible feature important for high potency. Salmon calcitonin shows the highest net positive charge. Deamidation of carboxyl-terminal proline (with consequent increase in negative charge in the molecule) causes a decrease in biological activity. Areas of increased positive charge may be important in binding to receptors. The loss of activity with deletion of leucine-16 (salmon) or phenylalanine-16 (human) may also imply that hydro-phobicity at that position is important for biological activity (Potts and Aurbach 1976).

Biological Assay

In vivo bioassays generally are based on the hypocalcaemic effect. The simplest bioassays depend on subcutaneous injection of test material in rats, although the intravenous route has been adapted for some assays. Generally the minimum amount of hormone detected by in vivo bioassays is 0.1–1.0 mU. Useful in vitro assays can use adenylate cyclase measurements (Aurbach and Chase 1976) or the ability of calcitonin in vitro to stimulate cyclic AMP production in cells (Moseley et al. 1982). Several of these systems are sensitive to the subnanogram range for salmon calcitonin.

Biosynthesis of Calcitonin

The availability of cDNA copies of mRNA coding for calcitonin has allowed studies on the calcitonin gene leading to interesting new knowledge. The calcitonin gene includes five introns separating six exons. Within the mRNA coding for calcitonin itself are represented four of the six exons. Calcitonin, however, is not the sole product coded for by the calcitonin gene. Another protein, calcitonin gene-related peptide (CGRP), is expressed predominantly in the central nervous system. The precursor mRNA for CGRP has represented within it five of the six exons of the calcitonin gene. Specific control processes allow expression of calcitonin itself in the C cells whereas expression of the gene product in the central nervous system is represented only by CGRP. The control processes involved are illustrated in Fig. 3.6. In the C-cell, mRNA is translated to a large precursor polypeptide (136 amino acids for the rat hormone). This protein is then cleaved to yield three peptides, an N-terminal peptide, calcitonin and a carboxy-terminal peptide (CTP-1). In the central nervous system translation of mRNA yields an identical amino-terminal peptide, CGRP, and a different carboxy-terminal peptide, CTP-2 (Birnbaum et al. 1984). The

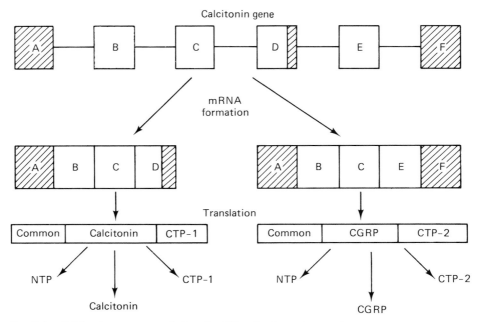

Fig. 3.6. Calcitonin gene transcription. In "C" cells, flow of transcription is towards splicing of pro-mRNA segments to produce the peptide precursor represented by *N*-terminal peptide (*NTP*), calcitonin and *C*-terminal peptide-1 (*CTP-1*). In the central nervous system the precursor peptide is composed of NTP, CGRP (calcitonin gene-related peptide) and CTP-2

physiological significance of CGRP is unknown. In the human being there are apparently CGRP-related genes (Hoppener et al. 1984). It is of interest but uncertain significance that both the parathyroid hormone gene and the calcitonin/CGRP gene are located on chromosome 11 (Hoppener et al. 1984).

Secretion of Calcitonin

The secretion of calcitonin, like that of parathyroid hormone, is controlled by the two intracellular messengers calcium and cyclic AMP. Unlike parathyroid hormone secretion, however, calcitonin secretion is stimulated by calcium. C cells bear receptors for several agonists capable of activating the adenylate cyclase–cyclic AMP system. These include glucagon (in dogs), cholecystokinin, gastrin and cerulein. (Pentagastrin and calcium are the principle secretagogues useful in testing for medullary carcinoma.) Other than calcium, the actions of the secretagogues, which also include β-adrenergic catecholamines, enhance secretion by increasing intracellular cyclic AMP content (Peck and Klahr 1979).

The presence of calcium in the extracellular fluid also influences the C-cell content of calcitonin as well as the rate of release of calcitonin into the circulation. Calcitonin in some species is important for controlling blood calcium after a meal (see Cooper et al. 1978). Calcium instilled in the stomach of rats normally causes little or no change in plasma calcium but in thyroidectomized animals causes hypercalcaemia. It is likely that a gastrointestinal hormone is released in response to calcium or a calcium-containing meal and that this gastrointestinal hormone is a secretagogue for calcitonin. In pigs, gastrin appears to be the secretagogue of gastrointestinal origin responsible for stimulating calcitonin release (Cooper et al. 1978). In the rat the corresponding gastrointestinal secretagogue has yet to be identified. In adult man, there is no proven significance of calcitonin for calcium metabolism and no corresponding gastrointestinal–calcitonin axis has been established.

In the circulation, immunoreactive calcitonin exists in multiple forms including calcitonin monomer, oxidized monomer, a dimer and possibly a precursor of calcitonin with a molecular weight of approximately 12 000 (Tobler et al. 1983).

Physiology of Calcitonin

Actions on Bone

Calcitonin was first identified as a hypocalcaemic agent. It brings about hypocalcaemia through inhibition of bone resorption, an effect observed in vitro as well as in vivo (Munson 1976; Talmage et al. 1983). Calcitonin also inhibits the direct action of parathyroid hormone on release of calcium from bone in vitro. The actions of calcitonin on bone also include reduction in alkaline phosphatase and pyrophosphatase activity and inhibition of hydroxyproline production. Calcitonin also produces hypophosphataemia as a consequence of inhibition of bone resorption plus promotion of phosphate entry into bone. There is also a further modest effect of calcitonin on the kidney to effect phosphaturia (Munson 1976).

The activity of calcitonin in vivo to produce hypocalcaemia reflects predominantly inhibition of calcium resorption from bone and is independent of any effect on kidney, gastrointestinal tract or parathyroid gland. Studies with radiocalcium kinetics as well as in vitro data substantiate this function of calcitonin. The predominant cell affected by calcitonin is the osteoclast which shows morphological changes including a decrease in the ruffled border characteristic of this cell type (Holtrop 1978; Jones and Boyd 1978; Chambers and Magnus 1982). Enzymatic functions of osteoclasts, particularly those stimulated by parathyroid hormone, are also inhibited by calcitonin. The effects of virtually all agents that stimulate bone resorption in vitro, including parathyroid hormone, vitamin A, vitamin D metabolites, dibutyryl cyclic AMP and prostaglandins, are inhibited by calcitonin. The resorptive effects inhibited by calcitonin include lysosomal enzyme changes, mineral release and degradation of collagen. Physiologically, the greater the endogenous rate of bone resorption, the greater the effect calcitonin produces.

Mechanism of Action

Calcitonin is another hormone that controls bone and mineral metabolism through the intermediation of intracellular messenger cyclic AMP. Cyclic AMP content of bone as well as kidney in vitro increases in response to calcitonin, and dibutyryl cyclic AMP mimics the effect of calcitonin in vitro on bone. Moreover, the order of potency of the several calcitonin congeners from different species shows a close parallelism between effects in vivo on hypocalcaemia and in vitro on cyclic AMP production (Potts and Aurbach 1976). This parallelism in dose–response function for both physiology and biochemistry supports the concept that cyclic AMP is a mediator of the action of calcitonin.

Receptors for Calcitonin

The discussion on cyclic AMP as the mediator of calcitonin action implied that both bone and kidney must harbour specific receptors for calcitonin. Within the kidney, receptors for calcitonin are located in the medullary ascending thick limb of the nephron as well as in the "bright" segment of the distal cortical tubule (Fig. 3.4). In bone cells, it is the osteoclast that bears receptors for calcitonin and responds upon interaction of calcitonin with these specific receptors with an increase in cyclic AMP production.

With the advent of radiolabelled ligand, it became possible to identify directly calcitonin receptors on cells. Radioiodinated salmon calcitonin binds with high affinity to membranes in these tissues and dissociates very slowly from the receptor (Marx et al. 1973). Calcitonins from several different species, however, show the same order of potency in inhibiting labelled calcitonin binding to receptors as in stimulating adenylate cyclase in vitro and effecting hypocalcaemia in vivo. Receptors for calcitonin have also been identified in a number of cell lines in tissue culture. Of interest is that even though calcitonin has undergone a remarkable degree of modification throughout evolution (human calcitonin differs from salmon calcitonin by as many as 16 amino acids in a molecule that totals only 32), the receptors for calcitonin show essentially no change in characteristics among species as diverse as rat and human.

References

Agus ZS, Wasserstein A, Goldfarb S (1981a) PTH, calcitonin, cyclic nucleotides, and the kidney. Ann Rev Physiol 41:583–595
Agus ZS, Goldfarb S, Wasserstein A (1981b) Calcium transport in the kidney. Rev Physiol Biochem Pharmacol 90:155–169
Aurbach GD, Chase LR (1976) Cyclic nucleotides and the biochemical actions of parathyroid hormone and calcitonin. In: Greep RO, Astwood EB (eds) American Physiological Society handbook of physiology, section 7. Endocrinology, vol 3. Parathyroid gland. Williams and Wilkins, Baltimore, pp 353–381
Aurbach GD, Marx SJ, Spiegel AM (1985) Parathyroid hormone, calcitonin and the calciferols. In: Wilson JD, Foster DW (eds) Williams' textbook of endocrinology, 7th edn. Saunders, Philadelphia, pp 1137–1217
Bhargava G, Russell J, Sherwood LM (1983) Phosphorylation of parathyroid secretory protein. Proc Natl Acad Sci USA 80:878–881
Biddulph DM, Wrenn RW (1977) Effects of parathyroid hormone on cyclic AMP, cyclic GMP, and efflux of calcium in isolated renal tubules. J Cyclic Nucleotide Res 3:129–138

Birnbaum RS, Mahoney WC, Burns DM, O'Neil JA, Miller RE, Roos BA (1984) Identification of procalcitonin in a rat medullary thyroid carcinoma cell line. J Biol Chem 10:2870–2874

Botti RE Jr, Zull JE (1983) Identification of an ATP-activated endopeptidase from rat kidney which catalyzes cleavage of parathyroid hormone to fragments identical to those produced in the rat kidney in vivo. Endocrinology 112:393–395

Botti RE, Jr, Heath E, Frelinger AL, Chuang J, Roos BA, Zull JE (1981) Specific cleavage of bovine parathyroid hormone catalyzed by an endopeptidase from bovine kidney. J Biol Chem 256: 11483–11488

Bourdeau JR, Burg MB (1980) Effect of PTH on calcium transport across the cortical thick ascending limb of Henle's loop. Am J Physiol 239:F121–126

Bringhurst FR, Segre GV, Lampman GW, Potts JT Jr (1982) Metabolism of parathyroid hormone of Kupffer cells: analysis by reverse-phase high-performance liquid chromatography. Biochemistry 21:4252–4258

Broadus AE (1981) Nephrogenous cyclic AMP. Recent Prog Horm Res 37:667–701

Brown EM (1982) Parathyroid secretion in vivo and in vitro. Regulation by calcium and other secretagogues. Miner Electrolyte Metab 8:130–150

Brown EM, Adragna N, Gardner DG (1981) Effect of potassium on PTH secretion from dispersed bovine parathyroid cells. J Clin Endocrinol 53:1304–1306

Brown EM, Jones P, Adragna N (1983) Effects of [^{3}H]ouabain binding, ^{86}Rb uptake, cellular sodium and potassium, and parathyroid hormone secretion in dispersed bovine parathyroid cells. Endocrinology 113:371–378

Canalis E (1985) Effect of growth factors on bone cell replication and differentiation. Clin Orthop 193:246–263

Chambers DJ, Schafer DH, Laugharn JA Jr et al. (1978) Dose-related activation by PTH of specific enzymes in various regions of the kidney. In: Copp DH, Talmage RV (eds) Endocrinology of calcium metabolism. Excerpta Medica, Amsterdam, p 216

Chambers TJ, Magnus CJ (1982) Calcitonin alters behaviour of isolated osteoclasts. J Pathol 136:27–39

Chase LR, Aurbach GD (1967) Parathyroid function and the renal excretion of 3′,5′-adenylic acid. Proc Natl Acad Sci USA 8:518–525

Chase LR, Melson GL, Aurbach GD (1969) Pseudohypoparathyroidism: defective excretion of 3′,5′-AMP in response to parathyroid hormone. J Clin Invest 48:1832–1844

Cohn DV, MacGregor RR (1981) The biosynthesis, intracellular processing and secretion of parathormone. Endocr Rev 2:1–26

Cohn DV, Wong GL (1978) The actions of parathormone, calcitonin and 1,25-dihydroxycholecalciferol on isolated osteoclast- and osteoblast-like cells in culture. In: Copp DH, Talmage RV (eds) Endocrinology of calcium metabolism. Excerpta Medica, Amsterdam p 241

Cohn DV, Morrissey MM, Hamilton JW, Shofstall RE, Smardo FL, Chu LL (1981) Isolation and partial characterization of secretory protein I from bovine parathyroid glands. Biochemistry 7: 4135–4140

Cohn DV, Zangerle R, Fischer-Colbric R, Chu LLH, Elting JJ, Hamilton JW, Winkler H (1982) Similarity of secretory protein I from parathyroid gland to chromogranin A from adrenal medulla. Proc Natl Acad Sci US 79: 6056–6059

Cohn DV, Kumarasamy R, Ramp WK (1986) Intracellular processing and secretion of parathyroid gland proteins. Vitam Horm 43:283–316

Collip JB (1925) Extraction of a parathyroid hormone which will prevent or control parathyroid tetany and which regulates the level of blood calcium. J Biol Chem 63:395–438

Coltrera MD, Potts JT Jr, Rosenblatt M (1981) Identification of a renal receptor for parathyroid hormone of photoaffinity radiolabelling using a synthetic analogue. J Biol Chem 256:10555–10559

Cooper CW, Bolman RM III, Linehan WM, Wells SA Jr (1978) Interrelationships between calcium, calcemic hormones and gastrointestinal hormones. Recent Prog Horm Res 34:259–283

Dempster DW, Tobler PH, Olles P, Born W, Fischer JA (1982) Potassium stimulates parathyroid hormone release from perfused parathyroid cells. Endocrinology 111:191–195

Dennis VW, Brazy PC (1982) Divalent anion transport in isolated renal tubules. Kidney Int 22:498–506

Dousa TP, Steiner AL (1978) Immunofluorescent localization of cyclic nucleotides in the nephron. In: Copp DH, Talmage RV (eds) Endocrinology of calcium metabolism. Excerpta Medica, p 221

Downs RW, Sekura RD, Levine MA, Spiegel AM (1985) The inhibitory adenylate cyclase coupling protein in pseudohypoparathyroidism. J Clin Endocrinol Metab 61:351–354

Dziak R, Stern PH (1975) Calcium transport in isolated bone cells. 3. Effects of parathyroid hormone and cyclic 3′,5′-AMP. Endocrinology 97:1281–1287

Eilon G, Raisz LG (1978) Comparison of the effects of stimulators and inhibitors of resorption on the release of lysozomal enzymes and radioactive calcium from fetal bone in organ culture. Endocrinology 103:1969–1975

Fahrenholz F, Boer R, Crause P, Fritzsch G, Grzonka Z (1984) Interactions of vasopressin agonists and antagonists with membrane receptors. Eur J Pharmacol 13:47–58

Farfel Z, Brickman AS, Kaslow HR, Brothers VM, Bourne HR (1980) Deficiency of receptor–cyclase coupling protein in pseudohypoparathyroidism. N Engl J Med 303:237–242

Fenton S, Somers S, Heath DA (1978) Preliminary studies with the sensitive cytochemical assay for parathyroid hormone. Clin Endocrinol 9:381–384

Findlay DM, Michelangeli VP, Orlowski RC, Martin TJ (1983) Biological activities and receptor interactions of des-Leu[16] salmon and des-Phe[16] human calcitonin. Endocrinology 112: 1288–1291

Findlay DM, Michelangeli VP, Martin TJ, Orlowski RC, Seyler JK (1985) Conformational requirements for activity of salmon calcitonin. Endocrinology 117: 801–805

Golden P, Greenwalt A, Martin K, Bellorin-Font E, Mazey R, Klahr S, Slatopolsky E (1980a) Lack of a direct effect of 1,25-dihydroxycholecalciferol on parathyroid hormone secretion by normal bovine parathyroid glands. Endocrinology 107: 602–607

Golden P, Greenwalt A, Mazey R, Martin K, Slatopolsky E (1980b) Does vitamin D or its metabolites directly affect the release of PTH? Contrib Nephrol 18:135–138

Goltzman D (1987) Recent progress in hormone research [in press]

Habener JF, Potts JT Jr (1976) Chemistry, biosynthesis, secretion and metabolism of parathyroid hormone. In: Greep RO, Astwood EB (eds) American Physiological Society handbook of physiology, section 7. Endocrinology, vol 3, Parathyroid gland. Williams and Wilkins, Baltimore, p 353

Habener JF, Rosenblatt M, Potts JR Jr (1984) Parathyroid hormone: biochemical aspects of biosynthesis, secretion, action and metabolism. Physiol Rev 64: 985–1053

Hamilton JW, Jilka RL, MacGregor RR (1983) Cleavage of parathyroid hormone to the 1–34 and 35–84 fragments of cathepsin D-like activity in bovine parathyroid gland extracts. Endocrinology 113:285–292

Hammerman MR, Hansen VA, Morrissey JJ (1983) Cyclic AMP-dependent protein phosphorylation and dephosphorylation alter phosphate transport in canine renal brush border vesicles. Biochim Biophys Acta 755:10–16

Heath DA, Palmer JS, Aurbach GD (1972) The hypocalcemic action of colchicine. Endocrinology 90:1589–1593

Heinrich G, Kronenberg HM, Potts JT Jr (1984) Gene encoding parathyroid hormone. Nucleotide sequence of the rat gene and deduced amino acid sequence of rat preproparathyroid hormone. J Biol Chem 259:3320–3323

Holtrop ME, King GJ, Raisz LG (1978) Factors influencing osteoclast activity as measured by ultrastructural morphometry. In: Copp DH, Talmage RV (eds) Endocrinology of calcium metabolism. Excerpta Medica, Amsterdam, p 91

Hoppener JW, Steenbergh PH, Zandberg J et al. (1984) Localization of the polymorphic human calcitonin gene on chromosome 11. Hum Genet 66:309–312

Horiuchi N, Holick MF, Potts JT Jr, Rosenblatt M (1983) A parathyroid hormone inhibitor in vivo: design and biological evaluation of a hormone analog. Science 220: 1053–1055

Hurley JB, Simon MI, Teplow DB, Robishaw JD, Gilman AG (1984) Homologies between signal transducing G proteins and rat gene products. Science 226:860–862

Insel P, Balakir R, Sacktor B (1975) Binding of cyclic AMP to renal brush-border membranes. J Cyclic Nucleotide Res 1:107–122

Jones SJ, Boyde A (1978) Scanning electron microscopy of bone cells in culture. In: Copp DH, Talmage RV (eds) Endocrinology of calcium metabolism. Excerpta Medica, Amsterdam, p 97

Kemper B, Habener JF, Rich A, Potts JT Jr (1974) Parathyroid secretory protein: discovery of a major calcium-dependent protein. Science 187: 167–169

Keutmann HT, Dawson BF, Aurbach GD, Potts JT Jr (1972) A biologically active amino-terminal fragment of bovine parathyroid hormone prepared by dilute acid hydrolysis. Biochemistry 13:1646–1652

Klahr S, Peck WA (1980) Cyclic nucleotides in bone and mineral metabolism. 2. Cyclic nucleotides and the renal regulation of mineral metabolism. Adv Cyclic Nucleotide Res 13:133–180

Kronenberg HM, McDevitt BE, Hendy GM, Mulligan RC, Szoka PR, Habener JF, Rich A, Potts JT Jr (1981) Studies of parathyroid hormone biosynthesis using recombinant DNA technology. In: Cohn DV, Talmage RV, Matthews JL (eds) Hormonal control of calcium metabolism. Excerpta Medica, Princeton, pp 5–18

Lasker RD, Spiegel AM (1982) Endogenous substrates for cAMP-dependent phosphorylation in dispersed bovine parathyroid cells. Endocrinology 111:1412–1414

Levine MA, Downs RW Jr, Singer MJ Jr, Marx SJ, Aurbach GD, Spiegel AM (1980) Deficient activity of guanine nucleotide regulatory protein in erythrocytes from patients with pseudohypoparathyroidism. Biochem Biophys Res Commun 94:1319–1324

Lindall AW, Elting J, Ells J, Roos BA (1983) Estimation of biologically active intact parathyroid hormone in normal and hyperparathyroid sera by sequential N-terminal immunoextraction and midregion radioimmunoassay. J Clin Endocrinol 57:1007–1014

Livesey SA, Kemp BE, Re CA, Partridge NC, Martin TJ (1982) Selective hormonal activation of cyclic AMP-dependent protein kinase isoenzymes in normal and malignant osteoblasts. J Biol Chem 257:14983–14987

Mallette LE, Tuma SN, Berger RE, Kirkland JL (1982) Radioimmunoassay for the middle region of human parathyroid hormone using an homologous antiserum with a carboxy-terminal fragment of bovine parathyroid hormone as radioligand. J Clin Endocrinol Metab 54:1017–1024

Marcus R, Orner FB (1980) Parathyroid hormone as calcium ionophore: tests of specificity. Calcif Tissue Int 32:207–211

Marx SJ, Woodard C, Aurbach GD, Glossman H, Keutmann HT (1973) Renal receptors for calcitonin: binding and degradation of hormone. J Biol Chem 248:4797–4802

Marx SJ, Sharp ME, Krudy A, Rosenblatt M, Mallette LE (1981) Radioimmunoassay for the middle region of human parathyroid hormone: studies with a radioiodinated synthetic peptide. J Clin Endocrinol Metab 53:76–84

Marx SJ, Liberman UA, Eil C (1983) Calciferols: actions and deficiencies in action. Vitam Horm 40:235–308

Mayer GP, Keaton JA, Hurst JG, Habener JF (1979) Effects of plasma calcium concentration on the relative proportion of hormone and carboxyl fragments in parathyroid venous blood. Endocrinology 104:1778–1784

Morel F, Chabardes D, Imbert-Teboul M, Le Bouffant F, Hus-Citharel A, Mont'egut M (1982) Multiple hormonal control of adenylate cyclase in distal segments of the rat kidney. Kidney Int [Suppl] 11:555–567

Moseley JM, Findlay DM, Martin TJ, Gormon JJ (1982) Covalent cross-linking of a photoactive derivative of calcitonin to human breast cancer cell receptors. J Biol Chem 257:5846–5851

Munson PL (1976) Physiology and pharmacology of thyrocalcitonin. In: Greep RO, Astwood EB (eds) American Physiological Society handbook of physiology, section 7. Endocrinology, vol 3, Parathyroid gland. Williams and Wilkins, Baltimore, p 443

Nissenson RA, Arnaud CP (1979) Properties of the parathyroid hormone receptor–adenylate cyclase system in chicken renal plasma membranes. J Biol Chem 254:1469–1475

Noland TA Jr, Henry HL (1983) Protein phosphorylation in chick kidney: response to parathyroid hormone, cyclic AMP, calcium and phosphatidylserine. J Biol Chem 258:538–546

Nordin BEC, Peacock M (1969) Role of the kidney in regulation of plasma calcium. Lancet II: 1280–1283

Patt HM, Luckhardt AB (1942) Relationship of low blood calcium to parathyroid secretion. Endocrinology 31:384–392

Peck WA, Klahr S (1979) Cyclic nucleotides in bone and mineral metabolism. Adv Cyclic Nucleotide Res 11:89–130

Pines M, Santora A, Gierschik P, Menczel J, Spiegel A (1986) The inhibitory guanine nucleotide regulatory protein modulates agonist-stimulated cAMP production in rat osteosarcoma cells. Bone Miner 1:15–26

Potts JT Jr, Aurbach GD (1976) Chemistry of the calcitonins. In: Greep RO, Astwood EB (eds) American Physiological Society handbook of physiology, section 7. Endocrinology, vol 3. Parathyroid gland. Williams and Wilkins, Baltimore, p 423

Potts JT Jr, Kronenberg HM, Rosenblatt M (1982) Parathyroid hormone: chemistry, biosynthesis and mode of action. Adv Protein Chem 35:323–396

Raisz LG (1976) Mechanisms of bone resorption. In: Greep RO, Astwood EB (eds) American Physiological Society handbook of physiology, section 7. Endocrinology, vol 3. Parathyroid gland. Williams and Wilkins, Baltimore, p 117

Rizzoli RE, Somerman M, Murray TM, Aurbach GD (1983) Binding of radioiodinated parathyroid hormone of cloned bone cells. Endocrinology 113:1832–1838

Rosenblatt M (1982) Pre-proparathyroid hormone, proparathyroid hormone, and parathyroid hormone: the biologic role of hormone structure. Clin Orthop 170:260–276

Rosenblatt M, Callahan EN, Mahaffey JE, Pont A, Potts JT Jr (1977) Parathyroid hormone inhibitors: design, synthesis, and biologic evaluation of hormone and analogues. J Biol Chem 252: 5847–5851

Russell J, Silver J, Sherwood LM (1984) The effects of calcium and vitamin D metabolites on cytoplasmic mRNA coding for pre-proparathyroid hormone in isolated parathyroid cells. Trans Assoc Am Physicians 97:296–303

Schwartz KE, Orlowski RC, Marcus R (1981) des-Ser^2 salmon calcitonin, a biologically potent synthetic analog. Endocrinology 108: 831–835

Shoback D, Thatcher J, Leombruno R, Brown E (1983) Effects of extracellular Ca^{++} and Mg^{++} on cytosolic Ca^{++} and PTH release in dispersed bovine parathyroid cells. Endocrinology 113:424–426

Segre GV, Niall HD, Sauer RT, Potts JT Jr (1977) Edman degradation of radioiodinated parathyroid hormone: application to sequence analysis and hormone metabolism in vivo. Biochemistry 16:2417–2427

Silver J, Russell J, Sherwood LM (1985) Regulation by vitamin D metabolites of messenger ribonucleic acid for preproparathyroid hormone in isolated bovine parathyroid cells. Proc Natl Acad Sci USA 82:4270–4273

Strewler GJ, Orloff J (1977) Role of cyclic nucleotides in the transport of water and electrolytes. Adv Cyclic Nucleotide Res 8:311–361

Talmage RV, Meyer RA Jr (1976) Physiological role of parathyroid hormone. In: Greep RO, Astwood EB (eds) American Physiological Society handbook of physiology, section 7. Endocrinology, vol 3. Parathyroid gland. Williams and Wilkins, Baltimore, p 343

Talmage RV, Cooper CW, Toverud SU (1983) The physiological significance of calcitonin. Bone Mineral Res 1: 74–143

Tam CS, Heersche JNM, Murray TM, Parsons JA (1982) Parathyroid hormone stimulates the bone apposition rate independently of its resorptive action: differential effects of intermittent and continuous administration. Endocrinology 110: 506–512

Tobler PH, Tschopp FA, Dambacher MA, Born W, Fisher JA (1983) Identification and characterization of calcitonin forms in plasma and urine of normal subjects and medullary carcinoma patients. J Clin Endocrinol Metab 57:749–754

Vaes G (1968) Parathyroid hormone-like action of N^6-2'-O-dibutyryladenosine-3',5'-(cyclic)-monophosphate on bone explants in tissue culture. Nature 219:939–940

Winaver J, Chen TC, Fragola J, Robertson G, Slatopolsky E, Puschett JB (1982) Alterations in renal tubular water transport induced by parathyroid hormone: evidence for both antidiuretic hormone-mediated and independent effects. J Lab Clin Med 99:457–473

Chapter 4

Calcium-Regulating Hormones: General

P. H. Adams

Introduction

Several hormones, other than those directly involved in calcium homeostasis (parathyroid hormone, vitamin D, calcitonin), exert effects on bone and calcium metabolism. During the normal growth and development of animals and man, the bones increase in size, change in shape (modelling) and undergo changes in their internal structure and chemical composition. In immature animals, the growth of the skeleton is accompanied by a progressive bodily retention of calcium which extends to the onset of maturity; thereafter bone tissue is lost, especially in women after the menopause. All these changes are brought about by the activity of bone cells (osteoblasts, osteoclasts) and their behaviour is governed by a variety of internal and external factors.

Among the internal factors, growth hormone, adrenal glucocorticoids, thyroid hormone and the sex hormones are necessary for the normal growth of the skeleton and the maintenance of bone balance (skeletal homeostasis). The precise cellular events which underly their effects are not fully understood: thyroid hormone, glucocorticoids and insulin appear to act directly on bone cells; growth hormone exerts its effects through growth hormone-dependent intermediary compounds (somatomedins); and sex steroids appear to act indirectly but the mechanisms involved have not been identified. Irrespective of their mode of action, deficiencies or excesses of these humoral factors disturb bone balance and alter calcium metabolism directly or indirectly through changes in the circulating concentrations of calcium-regulating hormones.

Adrenal Corticosteroids

The cells of the adrenal cortex synthesize a variety of steroid compounds from a common precursor (cholesterol), and the final hormonal products can be classified according to their principal metabolic effects as glucocorticoids (cortisol), mineralocorticoids (corticosterone, aldosterone) or sex steroids (androstenedione, androgen and oestrogen). The amounts of androgen and oestrogen derived from the adrenals are small in comparison with the secretions of these hormones from the testis and ovary, but in women assume greater importance after the menopause. In terms of bone and calcium metabolism, cortisol is the most important and most studied of the adrenal corticosteroids.

Glucocorticoids

There is abundant evidence from clinical and experimental studies that states of glucocorticoid excess cause losses of compact and trabecular bone, and in immature animals (including man), delayed growth and maturation of the skeleton. The adverse effects on bone mass predispose to osteoporosis and easy fracture of the vertebral bodies and long bones. This is the case for patients with endogenous overproduction of cortisol (Cushing's disease), and for patients receiving glucocorticoids for therapeutic purposes. In man the development of glucocorticoid-induced osteoporosis is related to the total dose received and duration of treatment (Hahn 1978; Cannigia et al. 1981; Dykman et al. 1985), to age and sex and to the presence of underlying disease. Children and postmenopausal women are the most vulnerable groups; patients with rheumatoid disease are particularly at risk because they lose more bone than non-rheumatoid patients receiving the same amount of corticosteroid (Hahn 1978).

In glucocorticoid-induced osteoporosis there is a greater deficiency of bone in the axial than the appendicular skeleton (Seeman et al. 1982), which could be evidence for a preferential effect of glucocorticoids on trabecular bone. The finding that more bone is lost from the distal radius (which contains a moderate amount of trabecular bone) than the mid-shaft (which contains very little) during glucocorticoid therapy is also regarded as evidence for a differential effect of glucocorticoid on trabecular bone (Hahn et al. 1974; Adinoff and Hollister 1983).

The mechanisms underlying the accelerated loss of bone induced by glucocorticoids are complex, but there is general agreement that reduced bone formation as a result of impaired osteoblastic activity is particularly important (Sissons 1960; Riggs et al. 1966; Jowsey and Riggs 1970; Bressot et al. 1979). Opinions differ on the degree of osteoclastic activity. Sissons (1960) noted a marked paucity of osteoclasts in Cushing's disease, whereas Jowsey and Riggs (1970) found increased bone resorption as assessed by quantitative microradiography. More recent quantitative histomorphometric studies have shown modest increases in measures of osteoclastic bone resorption in Cushing's disease and with glucocorticoid therapy (Bressot et al. 1979). These variable findings in man are clearly related to the techniques used, but the finding of reduced bone formation is supported by experimental studies.

The effect of glucocorticoids on bone formation can be attributed to the direct effect they have on the osteoblast series. Chen et al. (1977) found corticosteroid receptors in cytoplasmic extracts of bone cells derived from foetal rat calvaria, and noted that glucocorticoids (particularly the synthetic analogue dexamethasone) inhibited bone cell growth in tissue culture. Jee et al. (1970) found that glucocorticoids depressed the formation of osteoblasts from precursor cells in rats and rabbits at all the dose levels used. However, the effects on osteoclastic activity were different; low doses stimulated osteoclastic activity and bone resorption, whereas much higher doses depressed progenitor cell proliferation of the osteoblast and osteoclast series with no evidence of increased bone resorption. Yasumura (1976) studied intact and thyroparathyroidectomized rats and concluded that glucocorticoids depressed bone resorption. Studies made in vitro are in broad agreement with these observations; glucocorticoids inhibit osteoblastic activity and fail to stimulate bone resorption (Raisz 1965; Raisz et al. 1972; Raisz et al. 1978).

The primary effect of glucocorticoids on bone therefore appears to be an inhibition of osteoblastic and osteoclastic activity. The increased bone resorption observed in rats given low doses of steroids was found to be dependent on parathyroid hormone (Jee et al. 1970), and was prevented by prior parathyroidectomy; this is not a universal experience (Kukreja et al. 1974). Secondary hyperparathyroidism might be relevant to the modest increase in bone resorption observed in man, but an alternative or contributory factor could be deficiencies of sex steroids. Bone resorption is increased in oestrogen and androgen deficiencies; glucocorticoid excess appears to inhibit testicular and ovarian function (Iannaccone 1959; Gabrilove et al. 1974; Doerr and Pirke 1976; Reid et al. 1985).

Although excessive amounts of circulating glucocorticoids induce a negative bone balance, the total and ionized serum calcium concentrations are generally normal, albeit in the lower part of the normal range (Soffer et al. 1961; Ross et al. 1966; Bressot et al. 1979; Hahn et al. 1981). Occasional patients with Cushing's disease have moderate degrees of hypercalcaemia, which can be attributed to co-existing primary hyperparathyroidism or the disease itself (Soffer et al. 1961; Ross et al. 1966). Experimental studies suggest that glucocorticoids have a serum calcium-lowering effect. Stoerk et al. (1963) found that parathyroidectomized rats treated with hydrocortisone had a lower serum calcium and required more parathyroid extract to maintain normocalcaemia than parathyroidectomized rats given none; intact rats treated with hydrocortisone had a normal serum calcium which, by implication, must have been achieved through increased parathyroid activity. Although the precise mechanisms for this apparent hypocalcaemic effect have not been identified, there is good evidence that glucocorticoids reduce the intestinal absorption of calcium. In man, glucocorticoids reduce the net intestinal absorption of calcium with faecal calcium sometimes exceeding the dietary intake (Slater et al. 1959). The intestinal absorption of radiocalcium (which largely reflects the active transport component) is also reduced (Gallagher et al. 1973; Hahn et al. 1981), an effect which varies with the dose and particular glucocorticoid used (Klein et al. 1977; Cannigia et al. 1981).

Experimental studies are in general agreement with these observations: glucocorticoids reduce the active transport of calcium across the duodenum and the levels of calcium removed from the intestine by diffusion (Harrison and Harrison 1960; Finkelstein and Schachter 1962; Kimberg et al. 1971). These intestinal responses are the reverse of those induced by vitamin D, but there is no convincing evidence that glucocorticoids act by interfering with the further metabolism of vitamin D or the subcellular localization of $1,25(OH)_2D$ in the intestinal mucosa, or impede the formation of the vitamin D-dependent calcium-binding protein within intestinal cells (Kimberg et al. 1971; Favus et al. 1973; Lukert et al. 1973; Seeman et al. 1980). Moreover glucocorticoids have not been shown to reduce the circulating concentration of $1,25(OH)_2D$ in adult man; on the contrary there is some evidence that they increase the serum concentration of $1,25(OH)_2D$, albeit that it remains within the normal range (Hahn et al. 1981). The available evidence indicates that glucocorticoids reduce calcium absorption through direct effects on the transport mechanisms within intestinal cells; glucocorticoids also reduce the duodenal transport of cations (iron) not dependent on vitamin D for their absorption (Kimberg et al. 1971). In contrast to all these findings in the proximal small intestine, experimental studies indicate that glucocorticoids stimulate the active transport of calcium across the terminal ileum and distal colon in rats (Kimberg et al. 1971; Lee 1983) but not in chicks (Fox and Heath 1981).

The urinary excretion of calcium is normal or increased in states of glucocorticoid excess. Patients with Cushing's disease have a propensity to form renal stones (calcium phosphate) which is likely to be related to hypercalciuria and alkalinity of the urine (Sprague et al. 1956; Soffer et al. 1961; Ross et al. 1966). In Cushing's disease the absolute amount of calcium excreted is a direct function of the urinary free cortisol (and hence the severity of the disease), and the fasting renal excretion of calcium (an indirect measure of bone resorption) is greater than normal (Findling et al. 1982). Although increased losses of calcium from bone are mainly responsible for the development of hypercalciuria, glucocorticoids are also thought to diminish the renal tubular reabsorption of calcium (Laake 1960), but this is not a consistent finding (Crilly et al. 1979). There is no evidence for a direct effect of cortisol on the renal handling of calcium, at least in acute experiments in man (Lemann et al. 1970). Considerations of the possible renal effects of glucocorticoids need to take into account concomitant changes in the renal handling of sodium (Pechet et al. 1959).

Glucocorticoids have been shown to stimulate the secretion of parathyroid hormone in organ culture (Au 1976); in man rapid infusions of hydrocortisone induce brisk but transient increases in the serum concentration of parathyroid hormone, and then in the absence of measurable changes in the serum calcium (Fucik et al. 1975). A substantial and sustained increase in parathyroid activity has been found in conjunction with long-term glucocorticoid therapy, and this is thought to be a secondary phenomenon in response to intestinal malabsorption of calcium and excessive losses of calcium in the urine (Suzuki et al. 1983). This secondary hyperparathyroidism is related to the dose of steroid used and is reversible with intravenous calcium (Fucik et al. 1975; Lukert and Adams 1976; Bressot et al. 1979). Despite these findings, biochemical and other features of hyperparathyroidism are not consistently found in patients with Cushing's disease or in those receiving long-term glucocorticoid therapy (Seeman et al. 1980; Findling et al. 1982).

Relatively few studies have been made of the effects of cortisol deficiency on calcium metabolism. Hypercalcaemia is a well-recognized, and not uncommon, clinical feature of primary adrenal failure (Walser et al. 1963). Experimental studies have confirmed this finding in rabbits, dogs and cats (Walser et al. 1963; Myers et al. 1964; Jowsey and Simons 1968), and variable responses to bilateral adrenalectomy have been found in rats, some studies showing hypercalcaemia (Raman 1970) and others not (Walser et al. 1963). This hypercalcaemic response is likely to be related to cortisol deficiency itself, but it should be recognized that there is a global deficiency of adrenal corticosteroids in primary adrenal failure in man; some of the observed effects on calcium metabolism (especially the renal excretion of calcium) are directly related to the deficiencies of sodium and water which accompany lack of mineralocorticoids. The same stricture can be applied to some of the experimental observations, few of which were designed to reveal the specific effects of cortisol deficiency alone. The studies of Jowsey and Simons (1968) in dogs were designed with this specific purpose. They showed that cortisol deficiency in dogs caused significant hypercalcaemia, which persisted with a low-calcium diet and in the absence of the parathyroid glands, but which was dependent on the presence of thyroxine.

The cause of the hypercalcaemia in adrenalectomized animals and man is complex, and can be largely attributed to the effects of haemoconcentration (from concomitant hypovolaemia), increased affinity of the serum proteins for calcium and an increase in the concentration of the complexed fraction of the serum calcium (Walser et al. 1963). Whether the level of the biologically important ionized fraction of the serum calcium is also increased is uncertain. Few direct measurements of serum ionic calcium have been made clinically or experimentally; bilateral adrenalectomy has been shown to increase the serum ionized calcium concentration in rats (Raman 1970) but not in dogs (Walser et al. 1963; Myers et al. 1964). Muls et al. (1982) described a patient with primary adrenal failure (Addison's disease) who presented with hypercalcaemia, and in whom they found a raised serum ionized calcium concentration. Moreover, the fact that the onset of primary adrenal insufficiency in patients with hypoparathyroidism is accompanied by a rise in the serum calcium (sometimes to supranormal levels) and the relief of hypocalcaemic tetany (Papadatos and Klein 1954; Farrell et al. 1976; Walker and Davies 1981), provides indirect evidence that lack of adrenal corticosteroids is accompanied by an increase in the ionized fraction of the serum calcium in man. Further, this is consistent with experimental findings which show that the hypercalcaemia of adrenal insufficiency is not dependent on parathyroid hormone (Myers et al. 1964; Jowsey and Simons 1968).

Although experimental studies seem to indicate that bilateral adrenalectomy promotes the active transport of calcium across the duodenum (Kimberg et al. 1971), there is no evidence for enhanced intestinal absorption of calcium in patients with adrenal insufficiency (Pechet et al. 1959). Moreover clinical and experimental studies show that the hypercalcaemia cannot be prevented with a low-calcium diet (Walser et al. 1963; Jowsey and Simons 1968; Farrell et al. 1976). Hypercalciuria has been reported in occasional patients with hypercalcaemia (Pedersen 1967), but in general the renal excretion of calcium is reduced, particularly in relation to the prevailing serum calcium (Walker and Davies 1981; Muls et al. 1982). This reduction in calcium excretion arises through the combined effects of a reduced filtered load of calcium (because of a reduction in glomerular filtration rate) and enhanced renal tubular reabsorption of calcium; both effects are related to hypovolaemia.

Few studies have been made of the effects of cortisol deficiency on bone. Jowsey and Simons (1968) reported increased bone resorption in the cortisol-deficient dog, and enhanced bone resorption has been inferred from clinical studies of patients with adrenal insufficiency (Pedersen 1967; Farrell et al. 1976; Downie et al. 1977; Walker and Davies 1981; Muls et al. 1982). The combination of a reduced capacity to excrete calcium and increased bone resorption is sufficient to explain the development of hypercalcaemia in adrenal insufficiency.

Mineralocorticoids

Mineralocorticoids (aldosterone, desoxycorticosterone) act on distal parts of the nephron where they promote the renal tubular reabsorption of sodium in exchange for potassium and hydrogen ions. These effects occur rapidly in response to acute infusions of mineralocorticoids in animals and man, but they are not accompanied by changes in the renal handling of calcium (Massry et al. 1967; Lemann et al. 1970). The effects of more prolonged administration are different; sodium retention is followed within a few days by an increase in sodium excretion (the "escape" phenomenon) which is accompanied by a rise in the renal excretion of calcium (Massry et al. 1968; Suki et al. 1968). These responses are related to expansion of the extracellular space which leads to reductions in the reabsorption of sodium and calcium at more proximal sites of the renal tubule.

Suki et al. (1968) have shown experimentally that this increase in calcium excretion can be prevented with dietary restriction of sodium, and persists in the presence of a low-calcium diet. The findings in man are in broad agreement with these observations. Rastegar et al. (1972) showed that in patients with primary aldosteronism (Conn's syndrome) the renal excretion of calcium varied directly with the intake of sodium, and increased disproportionately in response to a high-sodium diet compared with control subjects – responses which they related to expansion of the extracellular space rather than to direct effects of aldosterone. Although the urinary output of calcium varies directly with the dietary sodium intake (and hence sodium status), patients with primary aldosteronism show no disturbance of external calcium balance, and the intestinal absorption of calcium is normal (Milne et al. 1956; Gale et al. 1960). Surgical correction of primary aldosteronism has been accompanied by the virtual disappearance of calcium from the urine, and in some patients with a concomitant reduction in the intestinal absorption of calcium (Milne et al. 1956; Gale et al. 1960); the renal response might be related to sodium deficiency and excessive proximal tubular reabsorption of calcium, but the intestinal malabsorption of calcium is unexplained.

Thyroid Hormone

The thyroid hormones (thyroxine, tri-iodothyronine) are necessary for the normal growth and development of the skeleton. Lack of thyroid hormone retards bone growth and development, and interferes with the normal ossification of the epiphyses (epiphyseal dysgenesis). Thyroid hormones also exert important effects on the mature skeleton; the metabolic activity of bone cells varies directly with the state of thyroid function (Jowsey and Detenbeck 1969). Although thyroid hormones are not directly implicated in the regulation of the serum calcium, variations in thyroid status cause perturbations of the serum calcium which arise mainly through induced changes in the behaviour of bone cells. Serum calcium homeostasis is also disturbed, at least as judged from the response to acutely induced hypercalcaemia and hypocalcaemia. Patients with hyperthyroidism dispose of an intravenous calcium load, and achieve normocalcaemia after infusions of calcium and edetic acid more rapidly than control subjects (Adams et al. 1967; Adams et al. 1969; Lim et al. 1969). The reverse obtains in hypothyroidism; restoration of normocalcaemia after acutely induced hypocalcaemia and hypercalcaemia is delayed (Adams et al. 1968; Riggs et al. 1968; Adams et al. 1969; Jowsey and Detenbeck 1969). The calcaemic (bony) response to administered parathyroid hormone also varies directly with thyroid status; the response is exaggerated in hyperthyroidism and blunted in hypothyroidism (Harrison et al. 1964; Jowsey and Detenbeck 1969; Castro et al. 1975). These and other findings suggest that thyroid hormone is necessary for the full expression of the effects of parathyroid hormone on bone.

It is well recognized that patients with hyperthyroidism lose compact and trabecular bone, and are vulnerable to easy fracture (Fraser et al. 1971). This bone lesion has been variously described as osteomalacia, osteitis fibrosa and osteoporosis, but there is now general agreement that this is an osteoporosis characterized by increased bone resorption and bone formation (Adams et al. 1967). These changes in bone metabolism are directly related to the severity of the hyperthyroidism and are reversible with correction of the disease (Mosekilde et al. 1977; Mosekilde and Melsen 1978), but restoration of the deficits in compact and trabecular bone mass is often incomplete, especially in postmeno-pausal women (Fraser et al. 1971; Toh et al. 1985).

Thyroid hormones almost certainly stimulate bone resorption through a direct effect on bone cells, but the precise cellular events which underly the response are unknown. The exaggerated calcaemic response to parathyroid extract found in patients with hyperthyroidism might suggest that their bone cells are unduly sensitive to the effects of endogenous parathyroid hormone (Castro et al. 1975). Nonetheless experimental studies indicate that the increased bone resorption is not dependent on the presence of the parathyroid glands (Adams and Jowsey 1967). Thyroid hormones have also been shown to stimulate bone resorption in tissue culture, a response which differs in its timing and magnitude from that induced with parathyroid hormone, and which is inhibited by cortisol, phosphate and propranolol but not indomethacin (Mundy et al. 1976).

These skeletal changes are accompanied by a negative calcium balance and a tendency to hypercalcaemia which is directly related to the severity of the disease (Baxter and Bondy 1966; Adams et al. 1967; Mosekilde and Christensen 1977). This change in the serum calcium is attributable to a rise in the ionized calcium fraction (Frizel et al. 1967; Burman et al. 1976), but because the serum albumin concentration tends to be reduced in hyperthyroidism, total serum calcium values uncorrected for this change are commonly normal (Adams et al. 1967). The hypercalcaemia of hyperthyroidism is generally of modest degree, and seldom sufficient to cause symptoms. More severe symptomatic hypercal-caemia should suggest the presence of alternative or additional causes, especially primary hyperparathyroidism; there is a well-recognized association between hyperthyroidism and primary hyperparathyroidism (Parfitt and Dent 1970).

As with the effects of hyperthyroidism on bone, the development of hypercalcaemia is not dependent on the presence of the parathyroid glands (Adams and Jowsey 1967). Moreover patients with hyperthyroidism tend to have reduced parathyroid activity (Harden et al. 1964; Adams et al. 1967; Bouillon and De Moor 1974; Mosekilde and Christensen 1977; MacFarlane et al. 1982), a predictable response, given the inverse relation which normally obtains between the serum ionized calcium concentration and parathyroid hormone secretion. This conditioned state of hypoparathyroidism largely explains the tendency to hyperphosphataemia and enhanced renal tubular reabsorption of phosphate found in hyperthyroidism (Harden et al. 1964; Adams et al. 1967). The formation of $1,25(OH)_2D$ is reduced, through the combined effects of the induced hypoparathyroidism, hypercalcaemia and hyperphosphataemia, and as a result some patients have subnormal serum concentrations of $1,25(OH)_2D$ (Bouillon et al. 1980; Jastrup et al. 1982; MacFarlane et al. 1982); the formation of precursor $25(OH)D$ is not disturbed, and might even be accelerated (Velentzas 1983).

The development of hypercalcaemia must arise through an increase in the net entry of calcium into the extracellular space, and in hyperthyroidism this is best explained through an increase in the net release of calcium from bone. Thyroid hormones tend to reduce the intestinal absorption of calcium, although this has not been a consistent finding in animals or man. Thyroid hormones have been shown to reduce the active transport of calcium across the duodenum in young (Finkelstein and Schachter 1962; Friedland et al. 1965) but not mature rats (Krawitt 1967). Although this difference could suggest that age modifies the response, the active transport of calcium normally falls with age (Adams et al. 1974) and, depending on the technique used, any reduction induced by thyroid hormone would be more difficult to detect in older animals. Estimates of the intestinal absorption of radiocalcium in man have also varied, some showing malabsorption of calcium and others not (Adams et al. 1967; Shafer and Gregory 1972; Haldimann et al. 1980). The results from external calcium balance measurements have been more consistent; in hyperthyroidism, excretion of calcium in the faeces tends to be increased and sometimes exceeds the dietary intake (Cook et al. 1959). The endogenous component of the faecal calcium is probably not increased, and calcium absorption cannot be improved with vitamin D (Cook et al. 1959). Thus some patients with hyperthyroidism have true intestinal malabsorption of calcium, which could be a direct reflection of reduced circulating concentrations of $1,25(OH)_2D$.

The renal excretion of calcium in hyperthyroidism tends to be increased, sometimes to supranormal levels. This increased output of calcium is directly related to the severity of the disease (Mosekilde and Christensen 1977), and is the result of an increased filtered load of calcium coupled with reduced renal tubular reabsorption (Laake 1959). Despite the common finding of hypercalciuria, renal stones and nephrocalcinosis are not features of hyperthyroidism. The excretion of calcium in the sweat is also increased (Harden 1964).

Hypothyroidism is accompanied by a marked reduction in bone turnover, but the mineralization of bone collagen is normal (Bordier et al. 1967; Jowsey and Detenbeck 1969). The serum concentration of calcium is generally normal, but as a group, patients with hypothyroidism have a lower serum calcium than controls (Adams et al. 1968); the concentration of the ionized fraction of the serum calcium tends to be reduced (Frizel et al. 1967), and the serum concentration of parathyroid hormone tends to be increased (Bouillon and De Moor 1974; Bouillon et al. 1980). Exceptional patients develop hypercalcaemia which can be corrected with a low-calcium diet or with thyroxine (Lowe et al.

1962; Adams et al. 1968). The response to acute and chronic calcium deprivation is abnormal, indicating an impaired ability to mobilize calcium efficiently from bone. In hypothyroidism the fall in the serum calcium induced acutely with edetic acid is greater than normal, and restoration of normocalcaemia is delayed; both responses are reversible with correction of the thyroid deficiency (Adams et al. 1968; Goldsmith et al. 1968; Jowsey and Detenbeck 1969). More prolonged calcium deprivation, induced by a low-calcium diet or oral sodium phytate, causes a significant reduction in the serum calcium, sometimes to subnormal levels (Adams et al. 1968). The calcaemic response to exogenous parathyroid hormone is also impaired, which suggests that lack of thyroid hormone reduces the sensitivity of bone cells to the effects of parathyroid hormone (Jowsey and Detenbeck 1969; Castro et al. 1975).

These collected findings indicate that the primary effect of hypothyroidism on serum calcium homeostasis is to reduce the capacity of parathyroid hormone to mobilize calcium from bone; the observed increase in parathyroid activity and the ensuing fall in the serum ionized calcium concentration can be regarded as secondary consequences (Adams et al. 1968; Adams et al. 1969). In hypothyroidism, normocalcaemia is achieved largely through enhanced renal tubular reabsorption of calcium, and possibly increased absorption of calcium across the intestine. Secondary hyperparathyroidism probably explains the observed reduction in the renal tubular reabsorption of phosphate, especially since this is reversible with small infusions of calcium (Adams et al. 1968). The increased intestinal absorption of calcium (Lekkerkerker et al. 1971) could be related to changes in the metabolism of vitamin D. In contrast to the findings in hyperthyroidism, the circulating concentration of $1,25(OH)_2D$ tends to be increased, sometimes to supranormal levels (Bouillon et al. 1980). This increase in serum $1,25(OH)_2D$ is reversible with correction of the thyroid deficiency, and almost certainly arises indirectly through the combined effects of hyperparathyroidism, hypophosphataemia and the tendency to hypocalcaemia which stimulate $25(OH)D$-1-α-hydroxylase activity. Increased production of $1,25(OH)_2D$ could explain the increased sensitivity to vitamin D found experimentally and in patients with hypothyroidism (Goormatigh and Handovsky 1938; Fanconi and Chastonay 1950; Wilkins 1957). The renal excretion of calcium tends to be reduced (Aub et al. 1929; Robertson 1942) because of a reduced filtered load (the glomerular filtration rate falls in hypothyroidism) and possibly enhanced renal tubular reabsorption of calcium mediated by parathyroid hormone. Despite this reduction in calcium excretion, some patients with long-standing hypothyroidism develop nephrocalcinosis and renal stones, and then sometimes in association with hypercalcaemia and osteosclerosis (Bateson and Chander 1965).

Hormones of the Anterior Pituitary Gland

Growth Hormone

Growth hormone (somatotrophin) is necessary for the normal growth and development of the skeleton. In childhood, deficiency of growth hormone slows growth and development and leads to dwarfism; excessive secretion of growth hormone accelerates growth and leads to gigantism. In adults, excessive growth hormone secretion (acromegaly) is associated with an increase in the turnover of compact and trabecular bone (Riggs et al. 1972a). These skeletal responses to growth hormone are the indirect response to the action of a series of growth-promoting polypeptides (somatomedins), which are growth hormone dependent, and which act directly on cartilage and bone. Large doses of growth hormone stimulate bone turnover and increase bone mass in dogs (Harris et al. 1969), and stimulate the secretion of parathyroid hormone in intact rats (Lancer et al. 1976), but not in man (Aloia et al. 1976). Hypophysectomy in rats leads to a fall in the serum concentration of $1,25(OH)_2D$ (Spanos et al. 1978b; Yeh and Aloia 1984; Yeh et al. 1986), and abolishes the increase in the serum concentration of $1,25(OH)_2D$ which accompanies phosphate deprivation (Gray and Garthwaite 1985); both responses can be reversed with growth hormone.

Growth hormone has a direct stimulatory effect on the activity of 25(OH)D 1-α-hydroxylase (Spanos et al. 1981). The serum concentration of $1,25(OH)_2D$ tends to be increased in patients with acromegaly and falls to normal with cure of the disease (Eskildsen et al. 1979; Brown et al. 1980). Whether growth hormone directly influences the intestinal absorption of calcium, or has an indirect effect through its effects on the serum concentration of $1,25(OH)_2D$ is uncertain. In man the administration of growth hormone is accompanied by an increase in the net intestinal absorption of calcium, an effect which persists for days after the hormone is withdrawn (Beck et al. 1960; Henneman et al. 1960). Experimental studies of the effects of hypophysectomy on the intestinal absorption of calcium have yielded conflicting results. Finkelstein and Schachter (1962) found in rats that hypophysectomy initially stimulated and later depressed calcium transport across the duodenum, but this has not been confirmed. When expressed in terms of mucosal weight, the active transport of calcium has been found to be normal (Yeh and Aloia 1984) or increased (Krawitt et al. 1977), even though the serum concentration of $1,25(OH)_2D$ is reduced (Yeh and Aloia 1984). In rats, hypophysectomy does not reduce the concentration or binding characteristics of the $1,25(OH)_2D$ receptor protein in the intestinal mucosa, which suggests that hypophysectomy influences factors controlling calcium absorption other than $1,25(OH)_2D$ (Yeh et al. 1986).

That $1,25(OH)_2D$ might not be directly implicated in the improved intestinal absorption of calcium which follows the administration of growth hormone is supported by observations in man. In children with growth hormone deficiency, the administration of growth hormone was accompanied by an increase in fractional calcium absorption across the intestine and a fall in the serum concentration of $1,25(OH)_2D$; this reciprocal response is difficult to explain, but could suggest that in growth hormone deficiency there is a resistance to the effects of $1,25(OH)_2D$ at the intestinal level (Chipman et al. 1980). Gertner et al. (1979) also studied children with growth hormone deficiency, and found that the improved growth caused by growth hormone replacement therapy was not accompanied by changes in parathyroid activity or vitamin D metabolism. The administration of growth hormone to humans is consistently accompanied by an increase in the renal excretion of calcium (Henneman et al. 1960; Beck et al. 1960; Hanna et al. 1961) but whether this reflects a direct effect of growth hormone on the kidney is unknown. Acute studies in dogs have shown no effect of growth hormone on the renal clearances of calcium and phosphate, or the glomerular filtration rate (Westby et al. 1977).

Bone turnover is increased in acromegaly, and the shafts of the long bones increase in girth through stimulation of periosteal new bone formation; compact bone mass is preserved although the bone becomes more porous (Jowsey and Gordan 1971; Riggs et al. 1972a). Similar changes occur in trabecular bone, but here bone resorption exceeds bone formation and trabecular bone mass therefore falls. The net effect is a redistribution of bony tissue from trabecular to compact bone but in whole body terms skeletal mass is probably preserved (Riggs et al. 1972a). There is certainly little evidence to support the commonly held view that patients with acromegaly are at increased risk of developing osteoporosis (Nadarajah et al. 1968; Riggs et al. 1972a). Calcium balance is usually in equilibrium in acromegaly, but this depends critically on the dietary intake of calcium. The net intestinal absorption of calcium is normal at all levels of calcium intake, but when the dietary calcium falls below a critical level, the absolute amount of calcium absorbed is insufficient to offset the excessive and obligatory losses of calcium in the urine (Pearson et al. 1960; Nadarajah et al. 1968). The serum calcium tends to be increased and some patients develop hypercalcaemia which can be attributed to co-existing primary hyperparathyroidism or the disease itself (Nadarajah et al. 1968).

In normocalcaemic patients, cure of acromegaly is accompanied by a fall in the serum calcium, but not to subnormal levels. Whether there is an increase in the concentration of the ionized fraction of the serum calcium in acromegaly is not known; the complexed fraction might be increased because large doses of growth hormone cause increases in the serum concentration (and urinary excretion) of citrate (Henneman and Henneman 1960). As noted above, the urinary excretion of calcium is increased, and presumably reflects increased losses of calcium from bone, since it does not vary with variation in the dietary calcium (Nadarajah et al. 1968).

Prolactin

Although prolactin has anabolic and growth-promoting properties, there is no evidence that it has an important role in the growth and development of the skeleton. In mammals the principal target tissue is the mammary gland, and prolactin is necessary for the initiation and maintenance of lactation. The serum concentration of prolactin rises during the first 2 months of human pregnancy, and thereafter increases progressively to term; this physiological hyperprolactinaemia persists with lactation. Similar changes in prolactin secretion occur during the reproductive cycle in birds. These increases in the secretion of prolactin are accompanied by an increase in the serum concentration of $1,25(OH)_2D$, and in laying hens also by hypercalcaemia (Spanos et al. 1976; Kumar et al. 1979; Brown et al. 1980; Steichen et al. 1980).

The hormonal events which occur during reproduction are complex, and the precise role of prolactin in the accompanying changes in maternal calcium metabolism is uncertain, at least in mammals. During much of human pregnancy, the increased serum concentration of $1,25(OH)_2D$ can be explained by an oestrogen-induced increase in the serum concentration of vitamin D-binding protein (Bouillon et al. 1981), but this does not seem to be true for rats

(Bouillon et al. 1979). Only in the last month of human pregnancy, when there is unequivocal evidence for an increase in the absorption of calcium across the maternal intestine, is there an increase in the concentration of the biologically available, free fraction of the serum 1,25(OH)$_2$D (Bouillon et al. 1981). Although there is clear evidence from in vitro studies of chick renal tubules for a direct stimulatory effect of prolactin on the biosynthesis of 1,25(OH)$_2$D (Spanos et al. 1981), the role of prolactin in the changes in vitamin D metabolism found during mammalian reproduction is uncertain. Prolactin could have a more specific role in the increased intestinal absorption of calcium which accompanies lactation, through effects on the metabolism of vitamin D; the serum concentration of 1,25(OH)$_2$D is increased during lactation in rats and man (Boass et al. 1977; Kumar et al. 1979). The observation that bromocriptine (which inhibits the secretion of prolactin) reduces the serum 1,25(OH)$_2$D in lactating rats (Spanos et al. 1978b) could be evidence for a direct role of prolactin, but a similar response occurs after parathyroidectomy (Pike et al. 1979). In rats the observed changes in the serum 1,25(OH)$_2$D levels might well be related to variations in the serum concentrations of calcium and parathyroid hormone; in this species there is an inverse relationship between the serum concentrations of calcium and 1,25(OH)$_2$D during pregnancy and lactation (Pike et al. 1979).

The responses to administered prolactin vary among the species and with the source of the prolactin. In chicks, prolactin increases the serum concentrations of calcium and 1,25(OH)$_2$D (Spanos et al. 1976). In rats, prolactin causes increases in the serum concentration and renal excretion of calcium (Mahajan et al. 1974) but no stimulation of 1-α-hydroxylase activity (Matsumoto et al. 1979). This calcaemic response occurs in vitamin D-deficient and vitamin D-replete rats, in the absence of the parathyroid glands and in the presence of a low-calcium diet (Robinson et al. 1975; Pahuja and deLuca 1981), findings which indicate that the effect arises through increased mobilization of calcium from bone. Prolactin may stimulate the active transport of calcium across the duodenum in vitamin D-replete rats (Mainoya 1975a,b; Pahuja and deLuca 1981); this response is thought to be related to the accompanying stimulation of the duodenal transport of sodium, but clearly is not mediated by vitamin D.

Studies in patients with hyperprolactinaemia due to causes other than pregnancy and lactation (for example, due to pituitary tumours) have shown deficiencies of compact and trabecular bone which are related to the severity of the accompanying oestrogen deficiency but not to the degree of hyperprolacti-naemia (Klibanski et al. 1980; Schlechte et al. 1983; Cann et al. 1984; Koppelman et al. 1984). Despite this clear evidence of bone loss, patients with hyperprolactinaemia show no changes in the serum concentrations of calcium, 1,25(OH)$_2$D or parathyroid hormone, and have normal intestinal absorption of calcium (Adams et al. 1979; Klibanski, et al. 1980; Kumar et al. 1980; Schlechte et al. 1983; Koppelman et al. 1984).

Prolactin appears to mobilize calcium from bone in several species, and has a direct stimulatory effect on the biosynthesis of 1,25(OH)$_2$D in birds. Apart from a possible role played by prolactin in the changes in calcium and vitamin D metabolism which accompany pregnancy and lactation in concert with other hormones, there is no evidence that prolactin exerts significant independent effects on calcium metabolism in man.

Insulin

Apart from the well-recognized effects of insulin on carbohydrate and fat metabolism, insulin has anabolic and growth-promoting properties. These anabolic effects apply to bone; insulin stimulates the synthesis of bone collagen in vitro through a direct effect on osteoblasts (Raisz et al. 1978). Insulin deficiency retards the normal growth and development of the skeleton, and leads to deficiencies of compact and trabecular bone which predispose to easy fracture (Levin et al. 1976; Rosenbloom et al. 1977; McNair et al. 1978). This loss of bone arises in part through reduced bone formation, and in its early and more rapid phase is accompanied by increased urinary losses of calcium (McNair et al. 1979). The renal excretion of calcium increases after a carbohydrate meal, and this response might be related in part to glucose-induced increases in the serum concentration of insulin. Experimental studies in dogs have shown that insulin acts directly on the kidney, and reduces the reabsorption of sodium and calcium in the proximal renal tubule; however, the amount of calcium ultimately excreted does not change (Defronzo et al. 1976). In acute experiments in man, infusions of insulin diminished the renal excretion of sodium through enhanced reabsorption at the distal tubule, but unlike the case in dogs, the urinary output of calcium was increased (Defronzo et al. 1975).

The urinary output of calcium tends to rise in uncontrolled diabetes mellitus, and is then directly related to the accompanying glycosuria (Raskin et al. 1978). Apart from this effect on urinary calcium, no obvious disturbance of calcium or vitamin D metabolism has been found in patients with diabetes mellitus, well controlled or not; the intestinal absorption of calcium is normal (Heath et al. 1979). In these respects rats differ from man. In rats, experimentally induced diabetes mellitus leads to marked disturbances of calcium and vitamin D metabolism which vary with the experimental conditions. In the short term (5–12 d), insulin deficiency is accompanied by hypocalcaemia, secondary hyperparathyroidism, hypercalciuria and reductions in the serum concentration of $1,25(OH)_2D$ and the active transport of calcium across the duodenum (Schneider et al. 1977a,b; Schedl et al. 1978; Charles et al. 1981). In the long term (7 weeks), rats with experimentally induced diabetes mellitus develop hypercalcaemia, suppressed parathyroid activity, increased intestinal absorption and urinary excretion of calcium and a marked reduction in bone turnover (Hough et al. 1981). Apart from the hypercalciuria, which is only partly reversible, all these abnormalities are corrected with insulin irrespective of the duration of the experiment. The reason for these different results is not clear; neither fully accord with the experience in man.

Insulin appears to play a permissive role in the full expression of the activity of $25(OH)D$ 1-α-hydroxylase, and appears to be necessary for the stimulatory effect of parathyroid hormone on the biosynthesis of $1,25(OH)_2D$ by cultured chick kidney cells (Henry 1981). Insulin also plays a permissive role in the stimulation of $1,25(OH)_2D$ production and the demineralization of the skeleton induced by phosphate deprivation in rats (Matsumoto et al. 1986). There is no evidence that insulin itself stimulates the activity of 1-α-hydroxylase, and whether it has a role in vitamin D metabolism in man is not known.

Sex Steroids

Oestrogens

The principal circulating oestrogens in humans are oestradiol and oestrone. Most of the circulating oestradiol is synthesized and secreted by the ovarian follicles under the influence of pituitary gonadotrophins; oestrone, which is a biologically weaker oestrogen than oestradiol, is formed mainly from the peripheral conversion of androstenedione (of ovarian and adrenal origin) in peripheral tissues (fat and muscle). No oestrogen receptors have been detected in mammalian bone cells, yet there is unequivocal evidence that variations in oestrogen status have an important bearing on bone turnover and bone balance, and in the young on the maturation of the skeleton (Silberberg and Silberberg 1971). Although the failure to find oestrogen receptors in bone cells might be related to lack of suitable techniques for detecting them, the changes in bone and calcium metabolism observed during states of oestrogen deficiency and excess must at present be regarded as secondary phenomena. The same is true for the effects of oestrogen on calcium-regulating hormones.

Numerous clinical and experimental studies have shown that oestrogen deficiency delays the maturation of the skeleton, and interferes with the normal growth of the bones (Silberberg and Silberberg 1971). In mature rats, oophorectomy leads to losses of trabecular and compact bone, although this response varies with the calcium content of the diet; the lower the dietary calcium the greater the effect (Hodgkinson et al. 1978). Most studies of the effects of oestrogen deficiency on bone and mineral metabolism in man have been made in relation to the menopause. There is general agreement that the onset of the menopause (natural and artificial) is accompanied by an imbalance in bone turnover which leads to accelerated losses of compact and trabecular bone. There is a considerable variation in the rate of loss of bone among postmenopausal women of the same age (Adams et al. 1970; Smith et al. 1976) which is related in part to variations in oestrogen status (Lindsay et al. 1979; Johnston et al. 1985). These changes in bone behaviour are accompanied by a decrease in the intestinal absorption of calcium and an increase in the urinary calcium (Heaney et al. 1978a,b). The serum concentration of calcium and the fasting urinary excretions of calcium and hydroxyproline are increased compared with premenopausal values (Young and Nordin 1967; Gallagher and Nordin 1973; Johnston et al. 1985; Stock et al. 1985); these findings are consistent with a net increase in bone resorption.

The precise mechanisms responsible for this increased bone resorption have not been established with certainty, but the effect is likely to be mediated by parathyroid hormone. Patients with hypoparathyroidism lose little bone after the menopause (Hossain et al. 1970). In rats, treatment with oestrogens protects the skeleton from the bone-resorbing effects of administered parathyroid hormone (Orimo et al. 1972). Oestrogens also reduce bone resorption and improve calcium balance in patients with primary hyperparathyroidism (Gallagher and Nordin 1973). These collected findings suggest that in vivo oestrogens modify the sensitivity of bone-resorbing cells to the effects of parathyroid hormone. All the changes in bone turnover, bone mass and calcium metabolism found in oestrogen deficiency can be prevented or reversed with oestrogen therapy, although the response varies with the nature and dose of the oestrogen used (Gallagher and Nordin 1973; Heaney et al. 1978a,b; Lindsay et al. 1980; Gallagher 1981; Christiansen et al. 1982). There is however no evidence that this favourable response to oestrogens on bone resorption is a direct effect; physiological amounts of oestrogen fail to inhibit bone resorption in vitro (Caputo et al. 1976; Liskova 1976).

The observed increases in the intestinal absorption and renal tubular reabsorption of calcium are thought to be mediated by alterations in the circulating concentrations of the calcium-regulating hormones (parathyroid hormone, $1,25(OH)_2D$) induced indirectly by the change in bone balance (Heaney 1981; Riggs et al. 1981). This might be the explanation in long-term studies, but treatment with oestrogen has been shown to reduce bone resorption in postmenopausal women within a few weeks of starting treatment, and then without measurable changes in the serum concentrations of parathyroid hormone, calcitonin or free $1,25(OH)_2D$ (Selby et al. 1985; Stock et al. 1985). In birds, treatment with oestrogen is accompanied by increases in the serum concentrations of calcium and $1,25(OH)_2D$, and increased activity of 1-α-hydroxylase (assessed in vitro) (Castillo et al. 1977); in rats, pharmacological doses of oestrogens also increase the concentrations of $1,25(OH)_2D$ in plasma, intestine and kidney (Baksi and Kenny 1978). These are likely to be indirect effects; oestrogens have not been shown to stimulate 1-α-hydroxylase activity in isolated preparations of chick kidney (Spanos et al. 1978a; Trechsel et al. 1979).

Short- and long-term courses of oestrogen therapy have also been found to increase the serum concentration of $1,25(OH)_2D$ in women before and after the menopause (Gallagher et al. 1980; Stock et al. 1985). Whether this simply reflects a concomitant increase in the serum concentration of vitamin D-binding protein has not been established (Bouillon et al. 1981); in the short term, the raised serum concentration of $1,25(OH)_2D$ can be accounted for entirely on this basis (Selby et al. 1985).

Studies of the effects of oestrogens on the intestinal absorption of calcium in postmenopausal women have yielded conflicting results, some showing increased intestinal absorption (Cannigia et al. 1970; Heaney et al. 1978a; Gallagher et al. 1980) and others not (Nordin et al. 1981; Selby et al. 1985). The reason for these different findings is not apparent, but might be related to differences in the duration of oestrogen therapy and in the techniques used to make the measurement. Gallagher et al. (1980) gave postmenopausal women conjugated oestrogens and found a significant direct relationship between the increments in intestinal calcium absorption and the serum concentration of $1,25(OH)_2D$, and suggested that both findings were causally related; however, the possible effect of the administered oestrogen on the serum concentration of vitamin D-binding protein was not considered. Although the precise relationship between oestrogen status and vitamin D metabolism has not been established in man, oestrogens could modify intestinal calcium absorption through effects on the number of intestinal receptors for $1,25(OH)_2D$; in rats the number of intestinal receptors for $1,25(OH)_2D$ falls within a few weeks of oophorectomy (Chan et al. 1984).

Although oestrogens have not been shown to exert direct effects on the renal handling of calcium, calcium excretion rises after the menopause. This increase in calcium excretion can be attributed in part to the increased mobilization of calcium from bone, and arises through an increase in filtered load (because of the increase in the serum calcium) and a reduction in the renal tubular reabsorption of calcium (Selby et al. 1985). The reduced renal tubular reabsorption of calcium found in postmenopausal women, and the improvement found with oestrogen-replacement therapy have been attributed to variations in parathyroid activity (Riggs et al. 1981). In postmenopausal women, long-term (6 months) oestrogen therapy is accompanied by an increase in the serum concentration of parathyroid hormone (Riggs et al. 1972b), but the renal tubular reabsorption of calcium has been found to increase within a few weeks of starting treatment, and at a time when there is no evidence for increased parathyroid activity (Selby et al. 1985; Stock et al. 1985).

Progesterone and Progestogens

Unlike oestrogens, receptors for progesterone have been found in bone cells (Chen et al. 1977). Progesterone competes with dexamethasone for glucocorticoid receptors, and in this way could modulate the effects of corticosteroids on bone. This effect of progesterone could be relevant to the accelerated loss of bone which accompanies the menopause (Gallagher 1981). Johnston et al. (1979) found that postmenopausal women losing bone at the fastest rates had significantly lower serum concentrations of progesterone than those losing bone at a slower rate. Relatively few studies have been made of the effects of progesterone (or progestogens) on bone and calcium metabolism. Albright and Reifenstein (1948) found that progesterone had no effect on external calcium balance, the intestinal absorption of calcium or the serum calcium in patients with postmenopausal osteoporosis. The synthetic progestogens in current usage are derived from progesterone (dihydroxyprogesterone, medroxyprogesterone, gestronol hexanoate) or testosterone (ethisterone, norethisterone). Gestronol prevents loss of bone in postmenopausal women but has no effect on the fasting urinary excretion of hydroxyproline (and hence bone resorption), which could suggest that gestronol stimulates bone formation (Lindsay et al. 1978). Norethisterone acts differently, and reduces bone resorption in postmenopausal women and in patients with primary hyperparathyroidism; bone formation also falls with long-term treatment (Gallagher 1981; Nordin et al. 1981; Selby et al. 1985). In postmenopausal women the effects on the serum concentration and urinary excretion of calcium are similar to those induced with oestrogen (Gallagher 1981; Nordin et al. 1981).

Experimental studies in chicks have failed to show a sizeable independent effect of progesterone on vitamin D metabolism (Castillo et al. 1977); in man, treatment with norethisterone has not been shown to influence the serum concentrations of $1,25(OH)_2D$, vitamin D-binding protein or parathyroid hormone, at least in the short term (Selby et al. 1985). This lack of effect on vitamin D metabolism is consistent with the finding that norethisterone does not stimulate the intestinal absorption of calcium in postmenopausal women (Nordin et al. 1981; Selby et al. 1985). Progesterone could play a role in the changes in vitamin D metabolism (and thereby calcium metabolism) found during reproduction. In birds, the combination of progesterone and oestradiol stimulates 1-α-hyroxylase activity to a greater extent than oestradiol alone (Tanaka et al. 1978).

Androgens

The male sex steroids (testosterone, androstenedione) are necessary for the normal growth and development of the skeleton. In immature animals and humans, lack of androgens delays skeletal maturation and prolongs the period available for bone growth (Silberberg and Silberberg 1971). Experimental studies in mature male rats have shown that castration causes osteoporosis with loss of compact and trabecular bone (Wink and Felts 1980). Osteoporosis is also a well-recognized feature of hypogonadism in adult man. Treatment with testosterone is accompanied by significant retentions of calcium, phosphorus and nitrogen (Albright and Reifenstein 1948), and in hypogonadal males with increased bone formation (Baran et al. 1978). The effects of testosterone (and anabolic non-virilizing steroids) have also been examined in postmenopausal women; treatment with testosterone improves calcium balance through enhanced intestinal absorption and reduced urinary excretion of calcium (Nordin et al. 1981). The effects on bone are similar to those induced with oestrogen: there is a sustained reduction in bone resorption; bone formation is unchanged during the first few months of treatment, but falls in the long term (Riggs et al. 1972b; Nordin et al. 1981). These bony responses are probably indirect effects; physiological amounts of androgen fail to inhibit bone resorption in tissue culture (Caputo et al. 1976).

Unlike treatment with oestrogens, treatment with androgens has no effect on the serum concentrations of calcium or parathyroid hormone (Riggs et al. 1972b). The reduction in the urinary excretion of calcium arises in part through the reduced mobilization of calcium from bone, and possibly through enhanced renal tubular reabsorption of calcium (Riggs et al. 1972b). Experimental studies in chicks have not shown that testosterone exerts independent effects on the further metabolism of vitamin D (Castillo et al. 1977), but like progesterone, testosterone acts synergistically with oestradiol in stimulating 1-α-hyroxylase activity (Tanaka et al. 1976).

References

Adams ND, Garthwaite TL, Gray RW, Hagen TC, Lemann J (1979) The interrelationships among prolactin, 1,25-dihydroxyvitamin D, parathyroid hormone in humans. J Clin Endocrinol Metab 49:628–630

Adams PH, Jowsey J (1967) Bone and mineral metabolism in hyperthyroidism: an experimental study. Endocrinology 81:735–740

Adams PH, Jowsey J, Kelly PJ, Riggs BL, Kinney VR, Jones JD (1967) Effects of hyperthyroidism on bone and mineral metabolism in man. Q J Med 36:1–16

Adams PH, Chalmers TM, Riggs BL, Jones JD (1968) Parathyroid function in spontaneous primary hypothyroidism. J Endocrinol 40:467–475

Adams PH, Chalmers TM, Calder IM (1969) Calcium regulation in thyroid disease. In: Tissus calcifiés. Société d'édition d'enseignement supérieur, Paris, pp 197–200

Adams PH, Davies GT, Sweetnam P (1970) Osteoporosis and the effects of ageing on bone mass in elderly men and women. Q J Med 39:601–615

Adams PH, Hill LF, Wain D, Taylor CM (1974) The effects of undernutrition and its relief on intestinal calcium transport in the rat. Calcif Tissue Res 16:293–304

Adinoff AD, Hollister JR (1983) Steroid-induced fractures and bone loss in patients with asthma. N Engl J Med 309:265–268

Albright F, Reifenstein EC (1948) The parathyroid glands and metabolic bone disease. Williams and Wilkins, Baltimore

Aloia JF, Zanzi I, Ellis K et al. (1976) Effects of growth hormone in osteoporosis. J Clin Endocrinol 43:992–999

Au WYW (1976) Cortisol stimulation of parathyroid hormone secretion by rat parathyroid glands in organ culture. Science 193:1015–1017

Aub JC, Bauer W, Heath C, Ropes M (1929) Studies of calcium and phosphorus metabolism: the effects of thyroid hormone and thyroid disease. J Clin Invest 7:97–137

Baksi SN, Kenny AD (1978) Does oestradiol stimulate in vivo production of 1,25-dihydroxyvitamin D$_3$ in the rat? Life Sci 22:787–792

Baran DT, Bergfeld MA, Teitelbaum SL, Avioli LV (1978) Effect of testosterone therapy on bone formation in an osteoporotic hypogonadal male. Calcif Tissue Res 26:103–106

Bateson EM, Chander S (1965) Nephrocalcinosis in cretinism. Br J Radiol 38:581–584

Baxter JD, Bondy PK (1966) Hypercalcaemia of thyrotoxicosis. Ann Intern Med 65:429–442

Beck JC, McGarry EE, Dyrenfurth I, Morgan RO, Bird ED, Venning EH (1960) Primate growth hormone studies in man. Metabolism 9:699–737

Boass A, Toverud SU, McCain TA, Pike JW, Haussler MR (1977) Elevated serum levels of 1,25-dihydroxycholecalcifereol in lactating rats. Nature 267:630–632

Bordier P, Miravet L, Matrajt H, Hioco D, Ryckewaert A (1967) Bone changes in adult patients with abnormal thyroid function. Proc R Soc Med 60:1132–1134

Bouillon R, De Moor P (1974) Parathyroid function in patients with hyper- and hypothyroidism. J Clin Endocrinol Metab 38:999–1004

Bouillon R, Vandoren G, Van Baelen H, De Moor P (1979) Immunochemical measurement of the vitamin D-binding protein in rat serum. Endocrinology 102:1710–1715

Bouillon R, Muls E, De Moor P (1980) Influence of thyroid function on the serum concentration of 1,25-dihydroxyvitamin D$_3$. J Clin Endocrinol Metab 51:793–797

Bouillon R, Van Aasche FA, Van Baelen H, Heyns W, De Moor P (1981) Influence of the vitamin D-binding protein on the serum concentration of 1,25-dihydroxyvitamin D$_3$. J Clin Invest 67:589–596

Bressot C, Meunier PJ, Chapuy MC, Lejeune E, Edouard C, Darby AJ (1979) Histomorphometric profile, pathophysiology and reversibility of corticosteroid-induced osteoporosis. Metab Bone Dis Rel Res 1:303–311

Brown DJ, Spanos E, MacIntyre I (1980) Role of pituitary hormones in regulating renal vitamin D metabolism in man. Br Med J 280:277–278

Burman KD, Monchik JM, Earll JM, Wartofsky L (1976) Ionised and total serum calcium and parathyroid hormone in hyperthyroidism. Ann Intern Med 84:668–671

Cann CE, Martin MC, Genant HK, Jaffe RB (1984) Decreased spinal mineral content in amenorrheic women. J Am Med Assoc 251:626–629

Cannigia A, Genari C, Borrella G et al. (1970) Intestinal absorption of calcium-47 after treatment with oral oestrogen–gestogen in senile osteoporosis. Br Med J iv:30–32

Cannigia A, Nuti R, Lore F, Vattimo A (1981) Pathophysiology of the adverse effects of glucoactive corticosteroids on calcium metabolism in man. J Steroid Biochem 15:153–161

Caputo CB, Meadows D, Raisz LG (1976) Failure of oestrogens and androgens to inhibit bone resorption in tissue culture. Endocrinology 98:1065–1068

Castillo L, Tanaka Y, Deluca HF, Sunde ML (1977) The stimulation of 25-hydroxyvitamin D$_3$-1α-hydroxylase by oestrogen. Arch Biochem Biophys 179:211–217

Castro JH, Genuth SM, Klein L (1975) Comparative response to parathyroid hormone in hyperthyroidism and hypothyroidism. Metabolism 24:839–848

Chan SDH, Chiu DKH, Atkins D (1984) Oophorectomy leads to a selective decrease in 1,25-dihydroxycholecalciferol receptors in rat jejunal villous cells. Clin Sci 66:745–748

Charles MA, Tirunaguru P, Zolock DT, Morrissey RL (1981) Duodenal calcium transport and calcium binding protein levels in experimental diabetes mellitus. Miner Electrolyte Metab 5:15–22

Chen TL, Aronow L, Feldman D (1977) Glucocorticoid receptors and inhibition of bone cell growth in primary culture. Endocrinology 100:619–628

Chipman JJ, Zerwekh J, Nicar M, Marks J, Pak CYC (1980) Effect of growth hormone administration: reciprocal changes in serum 1α,25-dihydroxyvitamin D and intestinal calcium absorption. J Clin Endocrinol Metab 53:321–324

Christiansen C, Christensen MS, Larsen NE, Transbol I (1982) Pathophysiological mechanisms of estrogen effect on bone metabolism: dose–response relationships in early postmenopausal women. J Clin Endocrinol Metab 55:1124–1130

Cook PB, Nassim JR, Collins J (1959) The effects of thyrotoxicosis upon the metabolism of calcium, phosphorus and nitrogen. Q J Med 28:505–529

Crilly RG, Marshall DH, Nordin BEC (1979) Metabolic effects of corticosteroid therapy in postmenopausal women. J Steroid Biochem 11:429–433

Defronzo RA, Cooke CR, Andres R, Faloona GR, Davis PJ (1975) The effect of insulin on renal handling of sodium, potassium, calcium and phosphate in man. J Clin Invest 55:845–855

Defronzo RA, Goldberg M, Agus ZS (1976) The effects of glucose and insulin on renal electrolyte transport. J Clin Invest 58:83–90

Doerr P, Pirke KM (1976) Cortisol-induced suppression of plasma testosterone in normal adult males. J Clin Endocrinol Metab 43:622–629

Downie WW, Gunn A, Paterson CR, Howie GF (1977) Hypercalcaemic crisis as presentation of Addison's disease. Br Med J i:145–146

Dykman TR, Gluck OS, Murphy WA, Hahn TJ, Hahn BH (1985) Evaluation of factors associated with glucocorticoid-induced osteopenia in patients with rheumatic diseases. Arthritis Rheum 228:361–368

Eskildsen PC, Lund B, Sorensen OH, Lund B, Bishop JE, Norman AW (1979) Acromegaly and vitamin D metabolism: effect of bromocriptine treatment. J Clin Endocrinol Metab 49:484–486

Fanconi G, Chastonay E (1950) Cited by Bateson EM, Chander S (1965) Nephrocalcinosis in cretinism. Br J Radiol 38:581–584

Farrell PM, Rikkers H, Moel D (1976) Cortisol–dihydrotachysterol antagonism in a patient with hypoparathyroidism and adrenal insufficiency: apparent inhibition of bone resorption. J Clin Endocrinol Metab 42:953–957

Favus MJ, Kimberg DV, Millar GN, Gershon E (1973) Effects of cortisone administration on the metabolism and localisation of 25-hydroxycholecalciferol in the rat. J Clin Invest 52:1328–1335

Findling JW, Adams ND, Lemann J, Gray RW, Thomas CJ, Tyrrell BJ (1982) Vitamin D metabolites and parathyroid hormone in Cushing's syndrome: relationship to calcium and phosphorus homeostasis. J Clin Endocrinol Metab 54:1039–1044

Finkelstein JD, Schachter D (1962) Active transport of calcium by intestine: effects of hypophysectomy and growth hormone. Am J Physiol 203:873–880

Fox J, Heath H (1981) Retarded growth rate caused by glucocorticoid treatment or dietary restriction: associated changes in duodenal, jejunal and ileal calcium absorption in the chick. Endocrinology 108:1138–1141

Fraser SA, Anderson JB, Smith DA, Wilson GM (1971) Osteoporosis and fractures following thyrotoxicosis. Lancet I:981–983

Friedland JA, Williams GA, Bowser N, Henderson WJ, Hoffeins E (1965) Effect of hyperthyroidism on intestinal absorption of calcium in the rat. Proc Soc Exp Biol Med 120:20–23

Frizel D, Malleson A, Marks V (1967) Plasma levels of ionised calcium and magnesium in thyroid disease. Lancet I:1360–1361

Fucik RF, Kukreja SC, Hargis GK, Bowser EN, Henderson WJ, Williams GA (1975) Effect of glucocorticoids on function of the parathyroid glands in man. J Clin Endocrinol Metab 40:152–155

Gabrilove JL, Nicolis GL, Sohval AR (1974) The testis in Cushing's syndrome. J Urol 112:95–99

Gale I, Jory HI, Mulligan L, Woollen JW (1960) Primary aldosteronism. Am J Med 28:311–322

Gallagher JC (1981) Biochemical effects of estrogen and progesterone on calcium metabolism. In: Deluca HF, Frost HM, Jee WSS, Johnston CC, Parfitt AM (eds) Osteoporosis: recent advances in pathogenesis and treatment. University Park Press, Baltimore, pp 231–238

Gallagher JC, Nordin BEC (1973) Oestrogens and calcium metabolism. Front Horm Res 2:98–117

Gallagher JC, Aaron J, Horsman A, Wilkinson R, Nordin BEC (1973) Corticosteroid osteoporosis. Clin Endocrinol Metab 2:355–368

Gallagher JC, Riggs BL, Deluca HF (1980) Effect of estrogen on calcium absorption and serum vitamin D metabolites in postmenopausal osteoporosis. J Clin Endocrinol Metab 51:1359–1364

Gertner JM, Horst RL, Broadus AE, Rasmussen H, Genel M (1979) Parathyroid function and vitamin D metabolism during human growth hormone replacement. J Clin Endocrinol Metab 49:185–188

Goldsmith RE, King LR, Zalme E, Bahr GK (1968) Serum calcium homeostasis in radioiodine treated thyrotoxic subjects as measured by ethylenediamine-tetra-acetate infusion. Acta Endocrinol 58:565–577

Goormatigh N, Handovsky H (1938) Cited by Bateson EM, Chander S (1965) Nephrocalcinosis in cretinism. Br J Radiol 38:581–584

Gray RW, Garthwaite TL (1985) Activation of renal 1,25-dihydroxyvitamin D_3 synthesis by phosphate deprivation: evidence for a role for growth hormone. Endocrinology 116:189–193

Hahn TJ (1978) Corticosteroid-induced osteopenia. Arch Intern Med 138:882–885

Hahn TJ, Boisseau VC, Avioli LV (1974) Effect of chronic corticosteroid administration on diaphyseal and metaphyseal bone mass. J Clin Endocrinol Metab 39:274–282

Hahn TJ, Halstead LR, Baran DT (1981) Effects of short term glucocorticoid administration on intestinal calcium absorption and circulating vitamin D metabolite concentrations in man. J Clin Endocrinol Metab 52:111–115

Haldimann B, Kaptein EM, Singer FR, Nicoloff ST, Massry SG (1980) Intestinal calcium absorption in patients with hyperthyroidism. J Clin Endocrinol Metab 51:995–997

Hanna S, Harrison MT, MacIntyre I, Fraser R (1961) Effects of growth hormone on calcium and magnesium metabolism. Br Med J ii:12–15

Harden RM (1964) Calcium excretion in thermal sweat in thyrotoxicosis. J Endocrinol 28:153–157

Harden RM, Harrison MT, Alexander WD, Nordin BEC (1964) Phosphate excretion and parathyroid function in thyrotoxicosis. J Endocrinol 28:281–288

Harris W, Heaney RP, Jowsey J, Haywood E, Cockin J (1969) The effect of growth hormone on skeletal metabolism in adult dogs. J Bone Joint Surg [Am] 51:807 (abstract)

Harrison HE, Harrison HC (1960) Transfer of Ca^{45} across intestinal wall in vitro in relation to action of vitamin D and cortisol. Am J Physiol 199:265–271

Harrison MT, Harden RM, Alexander WD (1964) Some effects of parathyroid hormone in thyrotoxicosis. J Clin Endocrinol Metab 24: 214–217

Heaney RP (1981) Unified concept of the pathogenesis of osteoporosis: updated. In Deluca HF, Frost HM, Jee WSS, Johnston CC, Parfitt AM (eds) Osteoporosis: recent advances in pathogenesis and treatment. Unversity Park Press, Baltimore, pp 369–372

Heaney RP, Recker RR, Saville PD (1978a) Menopausal changes in calcium balance performance. J Lab Clin Med 92:953–963

Heaney RP, Recker RR, Saville PD (1978b) Menopausal changes in bone remodelling. J Lab Clin Med 92:964–970

Heath H, Lambert PW, Service FJ, Arnaud CD (1979) Calcium homeostasis in diabetes mellitus. J Clin Endocrinol Metab 49:462–466

Henneman DH, Henneman PH (1960) Effect of human growth hormone on levels of blood and urinary carbohydrate and fat metabolites in man. J Clin Invest 39:1239–1245

Henneman PH, Forbes AP, Moldawer M, Dempsey EF, Carroll EL (1960) Effects of human growth hormone in man. J Clin Invest 39:1223–1238

Henry HL (1981) Insulin permits parathyroid hormone stimulation of 1,25-dihydroxyvitamin D_3 production in cultured kidney cells. Endocrinology 108:733–735

Hodgkinson A, Aaron JE, Horsman A, McLachlan MSF, Nordin BEC (1978) Effect of oophorectomy and calcium deprivation on bone mass in the rat. Clin Sci Mol Med 54:439–446

Hossain M, Smith DA, Nordin BEC (1970) Parathyroid activity and postmenopausal osteoporosis. Lancet I:809–811

Hough S, Avioli LV, Bergfeld MA, Fallon MD, Slatopolsky E, Teitelbaum SL (1981) Correction of abnormal bone and mineral metabolism in chronic streptozotocin-induced diabetes mellitus in the rat by insulin therapy. Endocrinology 108:2228–2234

Iannaccone A, Gabrilove JL, Sohval AR, Soffer LJ (1959) The ovaries in Cushing's syndrome. N Engl J Med 261:775–780

Jastrup B, Mosekilde L, Melsen F, Lund B, Lund B, Sorensen OH (1982) Serum levels of vitamin D metabolites and bone remodelling in hyperthyroidism. Metabolism 31:126–132

Jee WSS, Park HZ, Roberts WE, Kenner GH (1970) Corticosteroid and bone. Am J Anat 129:477–479

Johnston CC, Norton JA, Khairi RA, Longcope C (1979) Age related bone loss. In: Barzel US (ed) Osteoporosis II. Grune and Stratton, New York, pp 91–100

Johnston CC, Hui SL, Witt RM, Appledorn R, Baker RS, Longcope C (1985) Early menopausal changes in bone mass and sex steroids. J Clin Endocrinol Metab 61:905–911

Jowsey J, Detenbeck LC (1969) Importance of thyroid hormones in bone metabolism and calcium homeostasis. Endocrinology 85:87–95

Jowsey J, Gordan G (1971) Bone turnover and osteoporosis. In: Bourne GH (ed) The biochemistry and physiology of bone, vol 3. Academic Press, New York, pp 201–238

Jowsey J, Riggs BL (1970) Bone formation in hypercortisolism. Acta Endocrinol 63:21–28

Jowsey J, Simons GW (1968) Normocalcaemia in relation to cortisone secretion. Nature 217:1277–1278

Kimberg DV, Baerg RD, Gershon E, Graudusius RT (1971) Effect of cortisone on the active transport of calcium by the small intestine. J Clin Invest 50:1309–1321

Klein RG, Arnaud SB, Gallagher JC, De Luca HF, Riggs BL (1977) Intestinal calcium absorption in exogenous hypercortisonism. J Clin Invest 60:253–259

Klibanski A, Neer RM, Beitins IZ, Ridgway EC, Zervas NT, McArthur JW (1980) Decreased bone density in hyperprolactinemic women. N Engl J Med 303:1511–1514

Koppelman MCS, Kurtz DW, Morrish KA et al. (1984) Vertebral body mineral content in hyperprolactinemic women. J Clin Endocrinol Metab 59:1050–1053

Krawitt EL (1967) Duodenal calcium transport in hyperthyroidism. Proc Soc Exp Biol Med 125: 417–419

Krawitt EL, Kunin AS, Sampson HW, Bacon BF (1977) Effect of hypophysectomy on calcium transport by rat duodenum. Am J Physiol 232:E229–E233

Kukreja SC, Bowser EN, Hargis GK, Henderson WJ, Williams GA (1974) Mechanisms of steroid induced osteopenia: role of parathyroid glands. Clin Res 22:618A

Kumar R, Cohen WR, Silva P, Epstein FH (1979) Elevated 1,25-dihydroxyvitamin D plasma levels in normal human pregnancy and lactation. J Clin Invest 63:342–344

Kumar R, Abboud CF, Riggs BL (1980) The effect of elevated prolactin levels on plasma 1,25-dihydroxyvitamin D and intestinal absorption of calcium. Mayo Clin Proc 55:51–53

Laake H (1959) Clinical investigations of renal calcium clearance. Acta Med Scand 165:71–80

Laake H (1960) The action of corticosteroids on the renal reabsorption of calcium. Acta Endocrinol 34:60–64

Lancer SR, Bowser EN, Hargis GK, Williams GA (1976) The effect of growth hormone on parathyroid function in rats. Endocrinology 98:1289–1293

Lee DBN (1983) Unanticipated stimulatory action of glucocorticoids on epithelial calcium absorption. Effect of dexamethasone on rat distal colon. J Clin Invest 71:322–328

Lekkerkerker JFF, Van Woudenberg F, Beekhuis H, Doorenbos H (1971) Enhancement of calcium absorption in hypothyroidism. Isr J Med Sci 7:399–400

Lemann JF, Piering WF, Lemon EJ (1970) Studies of the acute effects of aldosterone and cortisol on the inter-relationship between renal sodium, calcium and magnesium excretion in normal man. Nephron 7:117–130

Levin ME, Boisseau VC, Avioli LV (1976) Effects of diabetes mellitus on bone mass in juvenile and adult-onset diabetes. N Engl J Med 294:241–245

Lim P, Jacob E, Khoo OT (1969) Handling of induced hypercalcaemia in hyperthyroidism. Br Med J ii:715–717

Lindsay R, Hart DM, Purdie D, Ferguson MM, Clark AS, Kraszewski A (1978) Comparative effects of oestrogen and a progestogen on bone loss in postmenopausal women. Clin Sci Mol Med 54:193–195

Lindsay R, Hart DM, Manolagas S, Anderson DC, Coutts JR, MaClean A (1979) Sex steroids in pathogenesis and prevention of postmenopausal osteoporosis. In: Barzel US (ed) Osteoporosis II. Grune and Stratton, New York, pp 161–177

Lindsay R, Hart DM, Forrest C, Baird C (1980) Prevention of spinal osteoporosis in oophorecto-mised women. Lancet II:1151–1154

Liskova M (1976) Influence of estrogen on bone resorption in organ culture. Calcif Tissue Res 22:207–218

Lowe CE, Bird ED, Thomas WC (1962) Hypercalcaemia in myxoedema. J Clin Endocrinol Metab 22:261–267

Lukert BP, Adams JS (1976) Calcium and phosphorus homeostasis in man: effect of corticosteroids. Arch Intern Med 136:1249–1253

Lukert BP, Stanbury SW, Mawer EB (1973) Vitamin D and intestinal transport of calcium: effect of prednisolone. Endocrinology 93:718–722

MacFarlane IA, Mawer EB, Berry J, Hahn J (1982) Vitamin D metabolism in hyperthyroidism. Clin Endocrinol 17:51–59

Mahajan KK, Robinson CJ, Horrobin DF (1974) Prolactin and hypercalcaemia. Lancet I:1237–1238

Mainoya JR (1975a) Further studies on the action of prolactin on fluid and ion absorption by the rat jejunum. Endocrinology 96:1158–1164

Mainoya JR (1975b) Effects of bovine growth hormone, human placental lactogen and ovine prolactin and intestinal fluid and ion transport in the rat. Endocrinology 96:1165–1170

Massry SG, Coburn JW, Chapman LW, Kleeman CR (1967) The acute effect of adrenal steroids on the interrelationship between the renal excretion of sodium, calcium and magnesium. J Lab Clin Med 70:563–570

Massry SG, Coburn JW, Chapman LW, Kleeman CR (1968) The effect of long term deoxycortico-sterone acetate administration on the renal excretion of calcium and magnesium. J Lab Clin Med 71:212–219

Matsumoto T, Horiuchi N, Suda T, Takahashi H, Shimazawa E, Ogata E (1979) Failure to demonstrate stimulatory effect of prolactin on vitamin D metabolism in vitamin D-deficient rats. Metabolism 28:925–927

Matsumoto T, Kawanobe Y, Ezawa I, Shibuya N, Hata K, Ogata E (1986) Role of insulin in the increase in serum 1,25-dihydroxyvitamin D concentrations in response to phosphorus deprivation in streptozotocin-induced diabetic rats. Endocrinology 118:1440–1444

McNair P, Madsbad S, Christiansen C, Faber OK, Transbol I, Binder C (1978) Osteopenia in insulin treated diabetes mellitus: its relation to age at onset, sex and duration of disease. Diabetologia 15:87–90

McNair P, Madsbad S, Christensen MS et al. (1979) Bone mineral loss in insulin-treated diabetes mellitus: studies in pathogenesis. Acta Endocrinol 90:463–472

Milne MD, Muehrcke RC, Aird I (1956) Primary aldosteronism. Q J Med 26:317–333

Mosekilde L, Christensen MS (1977) Decreased parathyroid function in hyperthyroidism: interrelationships between serum parathyroid hormone, calcium–phosphorus metabolism and thyroid function. Acta Endocrinol 84:566–575

Mosekilde L, Melsen F (1978) Effect of antithyroid treatment on calcium–phosphorus metabolism in hyperthyroidism. 2. Bone histomorphometry. Acta Endocrinol 87:751–758

Mosekilde L, Melsen F, Bagger JP, Myhre-Jensen O, Sorensen NS (1977) Bone changes in hyperthyroidism: interrelationships between bone histomorphometry, thyroid function and calcium–phosphorus metabolism. Acta Endocrinol 85:515–525

Muls E, Bouillon R, Boelaert J et al. (1982) Etiology of hypercalcaemia in a patient with Addison's disease. Calcif Tissue Int 34:523–526

Mundy GR, Shapiro JL, Bandelin JG, Canalis EM, Raisz LG (1976) Direct stimulation of bone resorption by thyroid hormones. J Clin Invest 58:529–534

Myers WPL, Rothschild EO, Lawrence W (1964) Adrenal factors in plasma calcium regulation. In: Blackwood HJJ (ed) Bone and tooth. Pergamon Press, Oxford, pp 193–206

Nadarajah A, Hartog M, Redfern B et al. (1968) Calcium metabolism in acromegaly. Br Med J iv:797–801

Nordin BEC, Marshall DH, Francis RM, Crilly RG (1981) The effects of sex steroids and corticosteroid hormones on bone. J Steroid Biochem 15:171–174

Orimo H, Fujita T, Yoshikawa M (1972) Increased sensitivity of bone to parathyroid hormone in ovariectomised rats. Endocrinology 90:760–763

Pahuja DN, Deluca HF (1981) Stimulation of intestinal calcium transport and bone calcium mobilisation by prolactin in vitamin D deficient rats. Science 214:1038–1039

Papadatos C, Klein R (1954) Addison's disease in a boy with hypoparathyroidism. J Clin Endocrinol Metab 14:653–660

Parfitt AM, Dent CE (1970) Hyperthyroidism and hypercalcaemia. Q J Med 39:171–187

Pearson E, Soroff HS, Prudden JF, Schwartz M (1960) Studies on growth hormone. 5. Effect on the mineral and nitrogen balances of burned patients. Am J Med Sci 239:17–26

Pechet MM, Bowers B, Bartter FC (1959) Metabolic studies with a new series of 1,4-diene steroids. 1. Effects in Addisonian subjects of prednisone, prednisolone, and the 1,2-dehydro analogues of corticosterone, desoxycorticosterone, 17-hydroxy-11-desoxycorticosterone, and 9-fluorocortisol. J Clin Invest 38:681–690

Pedersen KO (1967) Hypercalcaemia in Addison's disease. Acta Med Scand 191:691–698

Pike JW, Parker JB, Haussler MR, Boass A, Towerud SH (1979) Dynamic changes in circulating 1,25-dihydroxyvitamin D during reproduction in rats. Science 204:1427–1429

Raisz LG (1965) Bone resorption in tissue culture. Factors influencing the response to parathyroid hormone. J Clin Invest 44:103–116

Raisz LG, Trummel CL, Wener JA, Simmons H (1972) Effect of glucocorticoids on bone resorption in tissue culture. Endocrinology 90:961–967

Raisz LG, Canalis EM, Dietrich JW, Kream BE, Gworek SC (1978) Hormonal regulation of bone formation. Recent Prog Horm Res 34:335–348

Raman A (1970) Effect of adrenalectomy on ionic and total plasma calcium in rats. Horm Metab Res 2:181–183

Raskin P, Stevenson MRM, Barilla DE, Pak CYC (1978) The hypercalciuria of diabetes mellitus: its amelioration with insulin. Clin Endocrinol 9:329–335

Rastegar A, Agus Z, Connor TB, Goldberg M (1972) Renal handling of calcium and phosphate during mineralocorticoid "escape" in man. Kidney Int 2:279–286

Reid IR, Ibbertson HK, France JT, Pybus J (1985) Plasma testosterone concentrations in asthmatic men treated with glucocorticoids. Br Med J 291:574

Riggs BL, Jowsey J, Kelly PJ (1966) Quantitative microradiographic study of bone remodelling in Cushing's syndrome. Metabolism 15:773–780

Riggs BL, Jones JD, Arnaud CD (1968) Effect of calcium infusion on serum calcium levels in hypothyroidism. Metabolism 17:747–750

Riggs BL, Randall RV, Wahner HW, Jowsey J, Kelly PJ, Singh M (1972a) The nature of the metabolic bone disorder in acromegaly. J Clin Endocrinol Metab 34:911–918

Riggs BL, Jowsey J, Goldsmith RS, Kelly PJ, Hoffman DL, Arnaud CD (1972b) Short- and long- term effects of estrogen and synthetic anabolic hormone in postmenopausal osteoporosis. J Clin Invest 51:1659–1663

Riggs BL, Gallagher JC, Deluca HF (1981) Disordered systemic regulation of mineral homeostasis as a cause of osteoporosis. In: Deluca HF, Frost HM, Jee WSS, Johnston CC, Parfitt AM (eds) Osteoporosis: recent advances in pathogenesis and treatment. University Park Press, Baltimore, pp 353–358

Robertson JD (1942) Calcium and phosphorus excretion in thyrotoxicosis and myxoedema. Lancet I:672–676

Robinson CJ, Mahajan KK, Horrobin DF (1975) Some effects of prolactin on calcium metabolism. J Endocrinol 65:27P

Rosenbloom AL, Lezotte DC, Weber FT et al. (1977) Diminution of bone mass in childhood diabetes. Diabetes 26:1052–1055

Ross EJ, Marshall-Jones P, Friedman M (1966) Cushing's syndrome: diagnostic criteria. Q J Med 35:149–192

Schedl HP, Heath H, Wenger J (1978) Serum calcitonin and parathyroid hormone in experimental diabetes: effects of insulin treatment. Endocrinology 103:1368–1373

Schlechte JA, Sherman B, Martin R (1983) Bone density in amenorrheic women with and without hyperprolactinemia. J Clin Endocrinol Metab 56:1120–1123

Schneider LE, Nowosielski LM, Schedl HP (1977a) Insulin-treatment of diabetic rats: effects on duodenal calcium absorption. Endocrinology 100:67–73

Schneider LE, Schedl HP, McCain T, Haussler MR (1977b) Experimental diabetes reduced circulating 1,25-dihydroxyvitamin D in the rat. Science 196:1452–1454

Seeman E, Kumar R, Hunder GG, Scott M, Heath H, Riggs BL (1980) Production, degradation and circulating levels of 1,25-dihydroxyvitamin D in health and in chronic glucocorticoid excess. J Clin Invest 66:664–669

Seeman E, Wahner HW, Offord KP, Kumar R, Johnson WJ, Riggs BL (1982) Differential effects of endocrine dysfunction on the axial and the appendicular skeleton. J Clin Invest 69:1302–1309

Selby PL, Peacock M, Barkworth SA, Brown WB, Taylor GA (1985) Early effects of ethinyloestradiol and norethisterone treatment in post-menopausal women on bone resorption and calcium regulating hormones. Clin Sci 69:265–271

Shafer R, Gregory DH (1972) Calcium malabsorption in hyperthyroidism. Gastroenterology 63:235–239

Silberberg M, Silberberg R (1971) Steroid hormones and bone. In: Bourne GH (ed) The biochemistry and physiology of bone, vol 3. Academic Press, New York, pp 401–484

Sissons HA (1960) Osteoporosis of Cushing's syndrome. In: Rodahl K, Nicolson JT, Brown EM (eds) Bone as a tissue. McGraw-Hill, New York, pp 3–17

Slater JDH, Heffron PF, Vernet A, Nabarro JDN (1959) Clinical and metabolic effects of dexamethasone. Lancet I:173–177

Smith DM, Khairi MRA, Norton J, Johnston CC (1976) Age and activity effects on rate of bone mineral loss. J Clin Invest 58:716–721

Soffer LJ, Iannoccone A, Gabrilove JL (1961) Cushing's syndrome: a study of fifty patients. Am J Med 30:129–146

Spanos E, Pike JW, Haussler MR et al. (1976) Circulating 1α-dihydroxyvitamin D in the chicken: enhancement by injection of prolactin during egg laying. Life Sci 19:1751–1756

Spanos E, Barrett DI, Chong KT, MacIntyre I (1978a) Effect of oestrogen and 1,25-dihydroxy-cholecalciferol on 25-hydroxycholecalciferol metabolism in primary chick kidney-cell cultures. Biochem J 174:231–236

Spanos E, Barrett DI, MacIntyre I, Pike JW, Safilian JW, Haussler MR (1978b) Effect of growth hormone on vitamin D metabolism. Nature 273:246–247

Spanos E, Brown DJ, Stevenson JC, MacIntyre I (1981) Stimulation of 1,25-dihydroxycholecalciferol production by prolactin and related peptides in intact renal cell preparations in vitro. Biochim Biophys Acta 672:7–15

Sprague RG, Randall RV, Salassa RM et al. (1956) Cushing's syndrome. Arch Intern Med 98:389–397

Steichen JJ, Tsang RC, Gratton TL, Hamstra A, Deluca HF (1980) Vitamin D homeostasis in the perinatal period: 1,25-dihydroxyvitamin D in maternal, cord and neonatal blood. N Engl J Med 302:315–319

Stock JL, Coderre JA, Mallette LE (1985) Effects of a short course of estrogen on mineral metabolism in postmenopausal women. J Clin Encodrinol Metab 61:595–600

Stoerk KC, Peterson AC, Jelinek VC (1963) The blood calcium lowering effect of hydrocortisone in parathyroidectomised rats. Proc Soc Exp Biol Med 114:690–695

Suki WN, Schwettmann RS, Rector FC, Seldin DW (1968) Effect of chronic mineralocorticoid administration on calcium excretion in the rat. Am J Physiol 215(1):71–74

Suzuki Y, Ichikawa Y, Saito E, Homma M (1983) Importance of increased urinary calcium excretion in the development of secondary hyperparathyroidism of patients under glucocorticoid therapy. Metabolism 32:151–156

Tanaka Y, Castillo L, Deluca HF (1976) Control of renal vitamin D hydroxylases in birds by sex hormones. Proc Natl Acad Sci USA 73:2701–2706

Tanaka Y, Castillo L, Wineland MJ, Deluca HF (1978) Synergistic effect of progesterone, testosterone, and estradiol in the stimulation of chick renal 25-hydroxyvitamin D_3 1α-hydroxylase. Endocrinology 103:2035–2039

Toh SH, Claunch BC, Brown PH (1985) Effect of hyperthyroidism and its treatment on bone mineral content. Arch Intern Med 145:883–886

Trechsel U, Bonjour JP, Fleisch H, Monod A, Portenier J (1979) Regulation of the metabolism of 25-hydroxyvitamin D_3 in primary cultures of chick kidney cells. J Clin Invest 64:206–217

Velentzas GC (1983) Some observations on vitamin D_3 metabolism in thyrotoxicosis. Acta Vitaminol Enzymol (Milano) 5:159–163

Walker DA, Davies M (1981) Addison's disease presenting as a hypercalcaemia crisis in a patient with idiopathic hypoparathyroidism. Clin Endocrinol 14:419–423

Walser M, Robinson BHB, Duckett JW (1963) The hypercalcaemia of adrenal insufficiency. J Clin Invest 42:456–464

Westby GR, Goldfarb S, Goldberg M, Agus ZS (1977) Acute effects of bovine growth hormone on renal calcium and phosphorus excretion. Metabolism 26:525–530

Wilkins L (1957) The diagnosis and treatment of endocrine disorders of childhood and adolescence. Blackwell, Oxford

Winks CS, Felts WJL (1980) Effects of castration on the bone structure of male rats: a model of osteoporosis. Calcif Tissue Int 32:77–82

Yasumura S (1976) Effect of adrenal steroids on bone resorption in rats. Am J Physiol 230:90–93

Yeh JK, Aloia JF (1984) Effect of hypophysectomy and 1,25-dihydroxyvitamin D on duodenal calcium absorption. Endocrinology 114:1711–1717

Yeh JK, Aloia JF, Vaswani AN, Semla H (1986) Effect of hypophysectomy on the occupied and unoccupied binding sites for 1,25-dihydroxyvitamin D_3. Bone 7:49–53

Young M, Nordin BEC (1967) Effects of natural and artificial menopause on plasma and urinary calcium and phosphorus. Lancet II:118–120

Chapter 5

Gastrointestinal Absorption of Calcium

F. Bronner

Introduction

Calcium phosphate in the vertebrate skeleton is a solid that constitutes the repository of over 99% of the body's calcium. The path of calcium salts through the body involves a series of transformations of state, from solid to liquid (as in digestion), from liquid to solid (as in bone salt deposition) and back from solid to liquid (as in bone salt resorption). Like other cations, calcium also moves across and between cell membranes and in the process enters and leaves compartments whose calcium concentrations may differ by as much as four orders of magnitude. To be able to maintain such enormous thermodynamic gradients, the body has developed exquisite controls. Consequently, a plasma calcium concentration of 2.5 mM in a given individual is likely to vary by less than 5% over a 24-hour period (Bronner 1982). However, the net amount of calcium absorbed from the intestine in a rat over a 24-hour period may exceed 70 mg, or nearly 100 times the amount that circulates in its bloodstream. The intracellular free calcium ion concentration is 0.1–0.2 μM in most cells. This also applies to the mucosal cells of the intestine which transport calcium from the lumen to the blood and lymph.

Thus, calcium absorption represents an impressive balancing act on the part of the intestinal cells and tissue: calcium, often in bound or solid form, must be solubilized from the ingesta. In a rat on a high calcium intake, the luminal calcium concentration can reach 100 mM, even if a fivefold dilution of the ingesta by body fluids is taken into account (Zornitzer and Bronner 1971). Some of that calcium must then move through the mucosal cells at a rate that does not raise the free intracellular calcium ion concentration to dangerous levels. Calcium that enters the body fluids, having either been pumped out by the intestinal cells or moved into the fluids through the intercellular spaces, must then be cleared from blood at a rate that permits the total plasma calcium concentration to remain virtually invariant at 2.5 mM and the plasma ionized calcium concentration to remain at about 1.25 mM. In what follows, the process by which calcium is transported across and between intestinal cells will be discussed in detail and the nutritional and physiological implications evaluated.

Calcium Absorption

General Statement

As chyme enters the small intestine, it is subject to mechanical action due to peristalsis and to chemical action by intestinal enzymes, largely peptidases. As a result, calcium is released and, along with many other products of digestion, is carried across the intestinal epithelium into the lymph and blood plasma. The central fact of transmural calcium movement, to be detailed now, is that it involves two major processes; the first is a saturable one, taking place via a transcellular route, that is subject to physiological and nutritional regulation via the vitamin D endocrine system and takes place largely in the proximal intestine, i.e. the duodenum and upper jejunum. The second process is a non-saturable one, where calcium moves down its concentration gradient from lumen to the body fluids, probably by a paracellular route. That route is not subject to the kind of direct endocrine regulation to which the saturable process is subjected. Calcium movement via the non-saturable route occurs all along the intestine, and is independent of whether calcium is also moved by the saturable process. The saturable process, on the other hand, is subject to down-regulation.

Calcium absorption, kinetically speaking, is a rate (mass/time). The selection of units for mass and time is a matter of convenience and tradition. For rodents, the usual unit is mg Ca/day; for humans, it can be either mg Ca/day or g Ca/day. Transcellular movement can also be conveniently expressed as μmol/h/g tissue (wet) or as mol/s/cell. The non-saturable movement is given as a fraction per unit time; in rats it averages 0.16/h (Bronner et al. 1986). This means that 16% of the calcium present in the lumen is absorbed by this route in a 1-hour period.

Methods of Analysis

Net Intestinal Calcium Absorption

This represents the difference between the amount that enters the intestine via the food and the quantity lost in the stool. The amount ingested can be determined by weighing the food and analysing its calcium content. Estimates of the calcium content of standard foods, based on published tables, tend to be quite accurate. To measure faecal calcium output requires chemical analysis of the stool, collected between markers, e.g. carmine red ingested at the beginning and end of the collection period. The length of the period depends on the species. In humans, a 4-day period would be a minimum, with a 6-day period adequate. In rats, a satisfactory period is 60–72 hours. To obtain reliable estimates, the subject being tested should have had an adaptive period prior to the test. Alternatively the experimental diet should simulate the pretest intake.

Faecal calcium is made up of unabsorbed food calcium and unabsorbed endogenous calcium. The latter is calcium that has entered the lumen of the intestine via succus entericus, bile juices and so on. A portion of the calcium entering the intestine via this route is recycled and re-enters the body. The amount of endogenous calcium lost in the stool can be estimated by radioactive labelling of the plasma calcium (Bronner 1973, 1979, 1982). By combining a mass balance approach with radiocalcium studies one can estimate the true calcium absorption, v_a, and endogenous calcium excretion, v_{ndo}, according to the following relationship:

$$v_i - v_F = S_i = v_a - v_{ndo} \qquad (1)$$

where v_i = calcium intake; v_F = faecal calcium excretion; S_i = net intestinal calcium absorption; v_a = true calcium absorption, i.e. the amount of food calcium that has left the intestine; and v_{ndo} = the faecal endogenous calcium, i.e. the quantity of endogenous calcium that has not been reabsorbed. All of these values are rates, e.g. mg Ca/day.

The amount of endogenous calcium lost by humans via the stool, v_{ndo}, is approximately the same as that lost in the urine, v_u (Hall et al. 1969). In the rat, v_{ndo} is five to ten times greater than v_u at normal or high levels of calcium intake (Sammon et al. 1970). In the cow, the ratio of $v_{ndo} : v_u$ is about 10 : 1 (Ramberg et al. 1970). Since the amount of calcium excreted in the stool is dependent on the amount ingested, the fraction of stool calcium that is endogenous is variable. In man and the rat, with typical calcium intakes of 700 mg/day and 70 mg/day, respectively, v_{ndo} would constitute about 15% of the total faecal calcium output.

True Calcium Absorption, v_a

This is the result of two processes: a saturable process subject to feedback regulation, and a non-saturable process. Their relationship may be expressed as follows:

$$v_a = \frac{V_m \, [Ca]}{K_m + [Ca]} + b[Ca] \qquad (2)$$

where V_m = maximum rate of calcium absorption via the saturable process, expressed in mmol/day; [Ca] = luminal calcium concentration, mM; K_m = the luminal calcium concentration at which $V_m/2$ is attained, mM; b = fraction absorbed per day by the non-saturable route.

Equation (2) assumes that the saturable process can be described by a simple Michaelis–Menten relationship. Evidence in support of this comes from in situ loop (Bronner et al. 1986) and perfusion studies (Miller et al. 1984) and will now be reviewed.

In the in situ loop procedure, a buffer solution containing calcium is instilled in a ligated segment of the intestine (duodenum, jejunum, ileum) and the per cent absorbed is determined as a function of time and concentration of the calcium remaining in the intestine. As can be seen from Fig. 5.1, absorption of calcium in a rat duodenal loop is nearly complete in 10–20 min when the calcium concentration is below 10 mM, but is slower at higher calcium concentrations. At 150 mM, only about half the available calcium is absorbed in 2.5 h. If calcium removal from the loop were to involve only a non-saturating process, e.g. simple diffusion, the fractional rate of extraction would be independent of calcium concentration and all curves should be superimposable. This is not the case and a saturable process is thus involved.

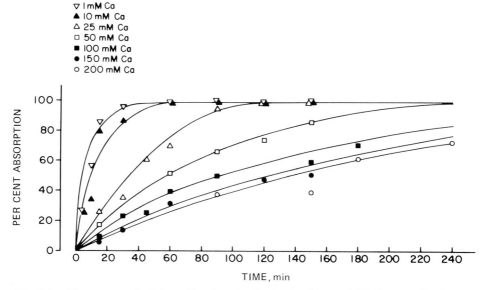

Fig. 5.1. Time course of calcium efflux from in situ duodenal loops. Male Sprague–Dawley rats, 40–50 days of age, 120–150 g BW, had been placed on a semisynthetic diet (1.5% calcium and 1.5% phosphorus; Pansu et al. 1983b) for about 10 days. Two days before the experiment, the animals were placed on a low-calcium diet (0.05% calcium, 0.2% phosphorus). Calcium, at the various concentrations shown in the graph, was instilled in the intestinal lumen. Efflux, estimated from the amount of ^{45}Ca lost in the indicated time period, is shown as per cent absorbed. Each point represents the mean estimate of two to six loops; average standard error, 7%.

Analysis of efflux at 25 mM, on the assumption that it can be represented by a Michaelis–Menten function, yields V_m = 22 μmol/h/loop and K_m = 3.9 mM. At higher concentrations efflux is faster than predicted, i.e. is clearly a combination of saturable and non-saturable processes. (Bronner et al. 1986)

When this situation was analysed (Bronner et al. 1986) assuming that it could be represented by a Michaelis–Menten function, integrated between the limits of $[Ca]_0$ at $t = 0$ and $[Ca]_t$ at $t = t$, where [Ca] represents the luminal calcium concentration, it turned out that at luminal calcium concentrations below 50 mM, such a function was a reasonable description of the events depicted in Fig. 5.1, with a V_m of 22 μmol/h/loop and K_m = 3.9 mM. However, at concentrations above 50 mM [Ca], the experimental points diverged widely from the theoretical curve, indicating a much faster experimental than theoretical efflux. In other words, at high luminal calcium concentrations a substantial contribution to calcium efflux comes from a component other than that which saturates at low calcium concentrations.

Wasserman and Taylor (1969), on theoretical grounds, had previously suggested that calcium transport in the duodenum was the result of a saturable and a non-saturable process. The existence of both processes was demonstrated experimentally in the rat (Zornitzer and Bronner 1971; Pansu et al. 1981, 1983a,b). Pansu and colleagues also showed that the saturable process is vitamin D-dependent, occurs primarily in the duodenum and upper jejunum, can be induced by exogenous vitamin D in vitamin D-deficient animals and decreases with age. It should be noted that these conclusions, now firmly established, had already been suggested by the studies of Schachter and colleagues (Schachter and Rosen 1959; Schachter et al. 1960). However, the earlier authors did not deal with the non-saturable process, which does not vary with age, nutritional status or pregnancy, and which is quantitatively similar throughout the length of the intestine.

Experiments in which the intestine was perfused with solutions whose calcium concentration was kept constant during the perfusion have yielded values for K_m more directly, but failed to reveal the existence of a non-saturable component, because they were conducted over a more restricted concentration range (Miller et al. 1984). The values that were calculated (Bronner et al. 1986) for Sprague–Dawley rats as reported by Miller et al. (1984) were $K_m = 1.6$ mM and $V_m = 16$ μmol/h/g; they agree quite well with those calculated from the in situ experiments.

Active Transport

As shown in Fig. 5.1, the in situ loop preparation is ideal for determining rates of absorption. The direct experimental demonstration of active transport, i.e. term 1 of Eq.(2), requires a preparation like the everted sac (Wilson and Wiseman 1954). Adapted for calcium transport studies by Martin and DeLuca (1969), it has been widely utilized for the evaluation of the effect of vitamin D on active calcium transport. In this procedure an intestinal segment is everted, filled with buffer, ligated and placed in the same buffer solution, with the outside volume sufficiently large for it to constitute an infinite reservoir (Roche et al. 1986b). When the outside solution is oxygenated, a sac made up of duodenal, but not of ileal, tissue will concentrate calcium in the serosal (inside) fluid against a moderate gradient (Pansu et al. 1983b). The process is inhibited by substituting nitrogen for oxygen. The ratio of the calcium concentration of the inside to outside solution at the end of the experiment has been used widely to identify the existence of an active calcium transport process. However, a detailed kinetic analysis of calcium transport in the everted sac has only recently been carried out (Bronner et al. 1986). Figure 5.2 shows the time course of calcium transport in an everted duodenal sac preparation and the amount of calcium accumulated in the serosal fluid of everted intestinal sacs prepared from duodenum and ileum.

From the time course it is evident that after a delay, calcium accumulates in the serosal fluid in near-linear fashion, so that a rate can be calculated from the amount that accumulates in the serosal fluid during the experimental period. This rate can be seen (Fig. 5.2b) to have risen sharply with increasing buffer calcium concentration, to have reached a maximum at 1 mM [Ca] and thereafter to have dropped until the serosal fluid [Ca] equalled the mucosal concentration at 6 mM. This behaviour can be explained (Bronner et al. 1986) by an analysis of what happens to the calcium concentration of the serosal fluid in ileal sacs (lower curve, Fig. 5.2b). As can be seen, it falls at all buffer concentrations during the incubation, the calcium entering the tissue (Roche et al. 1986b). If it is assumed that the process of calcium accumulation in the tissue is similar in duodenum and ileum, then the movement of calcium into the serosal fluid can be calculated as the sum of the measured net accumulation of calcium in the duodenal sac plus the measured reverse movement of calcium seen in the ileum. The upper curve in Fig. 5.2b represents the sum of these two processes and can be seen to

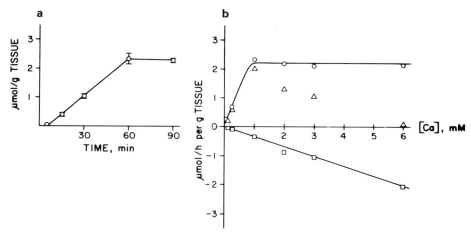

Fig. 5.2. Calcium transport in everted intestinal sacs. **a** Time course of calcium accumulation in the serosal fluid of everted duodenal sacs for male Sprague–Dawley rats (BW: 120–160 g) that had been on a low-calcium, semisynthetic diet (0.06% calcium, 0.2% phosphorus) for about 1 week. **b** Calcium accumulation in the serosal fluid of everted intestinal sacs from duodenum or ileum. The lower line represents calcium lost from the serosal fluid in the ileal preparation; the middle data points ($\triangle$) represent calcium accumulated in the duodenum. The upper line is the algebraic difference between duodenal and ileal transfer rates. A V_m of 2.2 μmol/h/g (wet weight) and a K_m of 0.35 mM are appropriate. (Adapted from Roche et al. 1986b)

resemble a simple saturation function with V_m = 2.2 μmol/h/g and K_m = 0.35 mM. Both values differ from the corresponding loop values by an order of magnitude.

Since both types of events, calcium absorption from the intact loop and calcium accumulation in the serosal fluid, describe the same phenomenon, i.e. luminal calcium loss, one would have expected identical kinetic parameters for both. Indeed, calcium transport at low luminal calcium concentrations given by the ratio V_m/K_m is essentially identical in both situations (loop: V_m/K_m = 22.3/3.85 = 5.8/h; sac: V_m/K_m 2.2/0.35 = 6.3/h).

In the sac, a rise in the ambient calcium concentration appears to be associated with the emergence of a rate-limiting step, perhaps inhibition of the basolateral calcium–magnesium-ATPase (see below) by calcium accumulated at the serosal surface. In the loop no such rate-limiting step is encountered, as the free calcium concentration at the serosal surface will not much exceed that of the body ($\leq$ 1.5 mM [Ca]). Thus, events in the intestinal loop and the everted sacs seem consistent.

Passive Transport

The non-saturable process has so far been evaluated only by the in situ loop procedure. A listing of the various rates that have been reported is shown in Table 5.1. The average value of 0.16/h is characteristic of male rats between the ages of 40 days and 19 weeks. Figure 5.3 shows that the non-saturable process is the same in the vitamin D-deficient duodenum and ileum and that treatment with 1,25-dihydroxyvitamin D$_3$ [1,25(OH)$_2$D$_3$], the active vitamin D metabolite, induces active transport in the duodenum, but has no effect on the non-saturable component, which is similar in intensity throughout the small intestine.

Table 5.1. Non-saturable calcium transport (fractional rate per hour)

Location in intestine	Physiological or nutritional status	Rate (per hour)	Reference
Duodenum	7-week-old +D rats, high-calcium diet	0.14	Zornitzer and Bronner (1971)
	19-week-old +D rats, high-calcium diet	0.13	Zornitzer and Bronner (1971)
	8-week-old +D rats, low-calcium diet	0.16	Zornitzer and Bronner (1971)
	8-week-old +D rats, intermediate-calcium diet	0.20	Pansu et al. (1981)
	7-week-old −D rats, high-calcium diet	0.12	Pansu et al. (1983b)
Jejunum	4-week-old +D rats, high-calcium diet	0.13	Pansu et al. (1983b)
Ileum	4-week-old +D rats, high-calcium diet	0.14	Pansu et al. (1983b)

Nellans and Kimberg (1979a, b) have shown that in the rat ileum most of calcium flux is proportional to the voltage applied to the tissue mounted in an Ussing chamber (Ussing and Zerahn 1951), and they suggested that a large proportion of this voltage-dependent flux involves a paracellular pathway. It is difficult to carry out comparable studies in the duodenum. However, the nature of the non-saturable pathway may be inferred from the following considerations.

Transcellular calcium movement involves entry into the cell via the brush border, diffusion through the cytoplasm and extrusion at the basolateral pole of the enterocyte. These stages will be discussed in greater detail below. The rate of calcium diffusion within the cell can be estimated with the aid of the following equation:

$$F/A = \frac{1}{L} \times D_{Ca} \times \Delta[Ca] \tag{3}$$

where F/A = flux per unit area; L = length of the mucosal cell, 10 μm; D_{Ca} = the diffusion coefficient of calcium in water at 37°C, 2.1×10^{-5} cm^2/s; $\Delta[Ca]$ = the difference in the free calcium concentration between the brush border and basolateral poles of the cell. Since the mean free calcium concentration of a mucosal cell is about 100 μM (Bikle et al. 1985), the difference in free calcium concentrations between the two poles of the cell probably does not exceed 200 μM.

Equation (3) when solved for F per mucosal cell, yields a flux rate of 1.6×10^{-18} moles/s.

When an intestinal loop is filled with 1 ml of a 200 mM calcium solution, its rate of loss by the non-saturable process alone would be 0.16×200 or 32 μmol/h

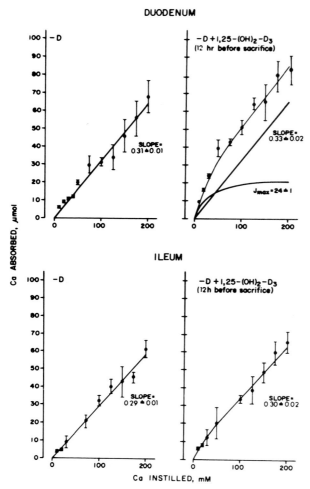

Fig. 5.3. Calcium absorption in duodenum and ileum of vitamin D-deficient rats before and after treatment with 1,25-dihydroxyvitamin D_3. Male weanling Sprague–Dawley rats were placed on a vitamin D-deficient diet (1.5% calcium, 1.5% phosphorus) and about 4 weeks later, when they weighed 116±4 g and their mean plasma calcium level was 5.6±0.1 mg/dl, calcium absorption studies by the in situ loop procedure were initiated. The *left* panels show calcium absorption of the duodenum and ileum of the vitamin D-depleted animals ($n=72$), the *right* panels show calcium absorption in the duodenum and ileum of vitamin D-deficient animals that had received 3 ng 1,25$(OH)_2D_3$ by i.p. injection 12 hours before sacrifice. (Pansu et al. 1983b)

(see also Fig. 5.1). Since an intestinal loop contains 10 to 50 × 10^6 mucosal cells, the non-saturable transfer rate per cell would be 1.78 × 10^{-16} moles/s, or 100 times faster than the theoretical transcellular diffusion rate. In other words, the rate at which calcium would diffuse through the cell is far slower than has been determined experimentally. No such discrepancy would exist if calcium moves down its chemical gradient by a paracellular route. Thus, both electrical measurements and kinetic considerations lead to the conclusion that passive calcium movement is paracellular.

Further support for this notion comes from experiments in which hyperosmolar solutions (900 mOsm) were instilled in jejunal loops (Pansu et al. 1976). Tripling the osmolarity, regardless of the chemical nature of the solution, led to a near-threefold increase in the rate of calcium loss from the loop. The most likely explanation is that the patency of the tight and intermediate junctions was increased by the significant swelling brought about when fluid entered the lumen in response to the osmotic gradient. Alterations in intercellular structure, brought about by instilling 1 mM ethylenediaminetetraacetate (EDTA) solutions in the inestinal lumen of rats (Cassidy and Tidball 1967), have been shown markedly to increase the rate of transmural movement of phenol red, a substance that moves between but not across cells. Moreover, if hyperosmolar solutions are instilled intraluminally, the cells would be compressed and transcellular movement would be slower rather than faster.

Recent experiments (Bronner and Spence 1987), carried out in intestinal sacs that were right-side out, i.e. *not* everted, have confirmed that the non-saturable rate of calcium transfer is indeed 16% per hour throughout the length of the intestine, regardless of the calcium concentration employed. These experiments have also shown that the rate of loss of phenol red from the sacs equals that of calcium, i.e. 16%/h. This supports the notion that the non-saturable flux of calcium in the intestine is indeed paracellular, inasmuch as phenol red does not move transcellularly (Cassidy and Tidball 1967).

Active Calcium Transport

Calcium Entry

The intestinal cell has a highly polar organization, with the brush-border membrane located at the luminal side, from which calcium enters. Entry is down an electrochemical gradient, inasmuch as the calcium concentration in the lumen during feeding can easily exceed 10 mM, whereas the intracellular free calcium concentration is less that 1 μM. Moreover, the cell interior is negative with respect to the intestinal lumen (-35 mV).

Studies of brush-border membrane vesicles have been made in order to understand the entry process (Rasmussen et al. 1979; Miller and Bronner 1981). These studies have shown calcium entry to be a time-dependent, concentration-dependent process that occurs in the absence of exogenous energy. Once inside the vesicle, calcium becomes bound to its inner aspect (Miller et al. 1982). The K_m of calcium uptake as measured with the brush-border preparation is ~1.1 mM, a value comparable to that derived for transmural movement.

However, the maximum velocity of the entry step is some 4 nmol/min/mg vesicle protein or 6.7×10^{-18} mol/s/cell, a value nearly two orders of magnitude below the measured transmural V_m, 1.2×10^{-16} mol/s/cell (assuming 50×10^6 cells per loop). This could be the result of channel or carrier inactivation in the process of vesicle preparation. Alternatively, it is also possible the vesicle preparation has provided evidence for a slower, parallel system of uncertain function. A third possibility is that evaluation of the true initial rate of calcium uptake requires kinetic methods of far greater speed than have been used heretofore.

As yet, it is uncertain whether calcium entry is via a calcium channel or facilitated transport. If calcium were to enter the brush border via channels like those identified in the guinea pig ventricular cell (Hess and Tsien 1984), one can calculate (Bronner et al. 1986) that an intestinal cell would need 873 channels to handle calcium entry at conditions of maximum transport (1.2×10^{-16} nmol/s). This is not an unreasonable number and not one that would make it likely that entry is the limiting step in transmural calcium transport.

If calcium were to enter the mucosal cell via a carrier, with a typical turnover number of 1000/s (Stein 1986), the number of carriers would need to be 3.7×10^5/cell. This again is not an unreasonable number. Therefore, unless the turnover number is unusually low, neither the number of carriers nor the number of channels would be limiting for the experimental V_m.

Calcium entry into brush-border membrane vesicles has been shown to be modified by the vitamin D status of the animal donor (Rasmussen et al. 1979; Miller and Bronner 1981), but the overall effect is modest (20%–30%).

A brush-border membrane-related particulate complex, of which a part is vitamin D-dependent, has been reported by Kowarski and Schachter (1980) who have assigned it a transport role (Schachter and Kowarski 1982). Miller and colleagues (1979) have identified a membrane-bound, vitamin D-dependent calcium-binding protein from the brush border which differs in several characteristics from the cytosolic calcium-binding protein (CaBP). That protein could constitute the monomer of the larger protein of Kowarski and Schachter (1980), but definitive information on either protein is not available.

In conclusion, calcium entry into the duodenal cell, enhanced only 20%–30% by the action of vitamin D, does not appear to be a rate-limiting step for transcellular calcium transport.

Intracellular Calcium Movement

Once calcium has entered the cell, it must move through the cytoplasm to the basolateral membrane where it is extruded. It can only move across the cell along a concentration gradient. According to Eq. (3), a probable maximum rate of diffusion in a mucosal cell would be 1.6×10^{-18} mol/s. The experimental rate of transcellular calcium movement, V_m, however, is 1.22×10^{-16} moles/s/cell, or over 70 times greater than the theoretical value. The discrepancy arises because the intracellular calcium gradient between the two poles of the cell is too low to sustain transcellular calcium transport at the measured rate.

If there existed an intracellular soluble calcium-binding protein, calcium that entered the cell would bind to that ligand. If that ligand had a concentration to match the calcium concentration that, on the basis of Eq. (3), would be needed to equal the observed transcellular flux, the soluble ligand would act like a calcium ferry, as it would amplify the small cellular calcium gradient.

Kretsinger et al. (1982) have proposed just such a mechanism and have suggested that the soluble, vitamin D-dependent calcium-binding protein (CaBP, mol. wt. $\simeq 9000$ in mammals; mol. wt. $\simeq 28\,000$ in birds) fulfils such a role. Feher (1983) has experimented with the chick CaBP and demonstrated that it can act to facilitate calcium diffusion in an experimental set-up like an Ussing chamber. Pansu and collaborators (1981, 1983a) have published findings that show that a close linear relationship exists between the V_m of transmural transport in rats, measured with in situ intestinal loops, and the CaBP content of the intestine, evaluated in parallel experiments (Fig. 5.4a). These data show that active transport is totally suppressed when CaBP reaches non-detectable limits and that this relationship holds true throughout the length of the rat intestine. Moreover, a similar linear relationship was found to exist when exogenous 1,25-dihydroxyvitamin D_3 [$1,25(OH)_2D_3$] was administered to rats (Roche et al. 1986a). In these experiments, CaBP and active calcium transport, assessed with the aid of everted duodenal sacs, were found to vary in parallel, regardless of the vitamin D status of the animal (Fig. 5.4**b**). Finally, it has been calculated (Bronner et al. 1986) that the presence of CaBP in the rat intestine can account for a 60-fold facilitation of transport, close to the 70-fold needed to account for the discrepancy between experiment and theory. The calculation took into account the prevailing free calcium concentration, the CaBP content of the duodenal cell (0.2–0.4 mM, Ueng et al. 1979; Thomasset et al. 1982), the K_d for CaBP (0.3 μM) and the fact that CaBP diffuses through the cell more slowly than free calcium.

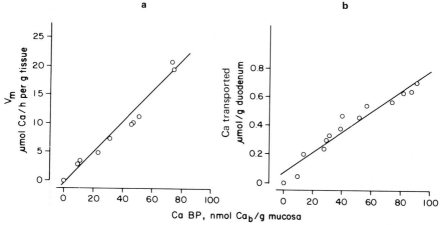

Fig. 5.4. The relationship between intestinal calcium transport and CaBP content. **a** V_m, calculated from in situ duodenal, jejunal and ileal loop experiments (Pansu et al. 1981, 1983b) shown as a function of CaBP content. The equation describing the relationship is:

$$V_m = -0.59 + 0.26 \text{ CaBP}$$

units as shown, $r = 0.98$ ($n = 10$). **b** Calcium transport, as evaluated from everted duodenal sac experiments (90 min incubation, 0.25 mM calcium), shown as a function of CaBP content (Roche et al. 1986). The equation describing the relationship is:

$$Ca_{transport} = 0.070 + 0.00714 \text{ CaBP}$$

units as shown, $r = 0.97$ ($n = 14$). (Bronner et al. 1986)

Other soluble calcium-binding proteins will also facilitate calcium transport. For calmodulin with a cellular concentration of about 0.06 mM (Thomasset et al. 1981) such facilitation can be calculated to be 10% of that due to CaBP (Bronner et al. 1986). No other calcium-binding protein has been reported to equal CaBP or calmodulin in concentration. Hence facilitation of intracellular calcium transport can logically and quantitatively be assigned to CaBP. Moreover, no soluble calcium-binding protein other than CaBP has been found to be vitamin D-dependent. It seems reasonable to conclude, therefore, that intracellular calcium movement is the rate-limiting step in transcellular calcium movement in the intestinal cell and is facilitated by the vitamin D-dependent CaBP.

In 1969 Wasserman and Taylor discussed in detail how vitamin D might influence transcellular calcium movement and concluded from an analysis of their experiments and those of Harrison and Harrison (1963) and of Holdsworth (1965) that "at least one effect of vitamin D is to increase the diffusional permeability of the intestine to Ca. The high efficiency of the Ca pump in the normal or vitamin D-replete animal in contrast to the rachitic animal may be only an indirect consequence of this permeability change, that is, allowing more Ca to penetrate to the site of the active transport system (Harrison and Harrison 1963)". It is interesting that CaBP, discovered by Wasserman, appears to be the molecule responsible for this increase in calcium diffusion in the duodenal cell.

Recent experiments have shown (Pansu et al. 1987) that the concentration-dependent inhibition by theophylline of active calcium transport is due to theophylline-mediated inhibition of calcium binding by CaBP. This inhibition is thought to be due to theophylline binding to CaBP. In turn this would distort the CaBP molecule and diminish its capacity to bind calcium. These studies therefore support the concept that CaBP functions as a calcium ferry.

Calcium Extrusion

For a cell to be able to maintain an intracellular free calcium concentration below 1 μM, when the extracellular free calcium concentration approximates 1 mM, means there must exist in all cells an efficient mechanism to extrude calcium. Two such structures have been identified: the Ca/Mg-ATPase and the Na^+/Ca^{2+} exchanger.

Intracellular organelles, such as mitochondria, the Golgi apparatus or the rough endoplasmic reticulum, help buffer calcium that enters the cell and may contribute to the efficiency of intracellular calcium homeostasis by binding calcium at a rate that exceeds the rate of calcium release from these organelles. This could be true particularly for mitochondria that possess the capacity to accumulate calcium, probably by a charge-uncompensated movement (Carafoli and Crompton 1978), and to release it by a sodium-dependent efflux (Crompton et al. 1976). However, a low intracellular free calcium concentration can in the steady state only be assured by a well-functioning, efficient calcium extrusion system. In the intestinal transporting cell, that system is the Ca/Mg-ATPase; the quantitative importance of the Na^+/Ca^{2+} exchanger in intestinal cells has been questioned (Nellans and Popovitch 1984). Moreover, the exchanger does not seem to be subject to regulation by vitamin D, the major regulator of the saturable transcellular transport of calcium in the intestine.

The history and nature of the Ca/Mg-ATPase have been widely described (Schatzmann 1975, 1982) and the structure and function of the sarcoplasmic Ca/Mg-ATPase in particular are known in great detail (Carafoli 1984; Inesi 1985; MacLennan et al. 1985; Brandl et al. 1986). The intestinal Ca/Mg-ATPase has been partially purified, localized to the basolateral membrane exclusively, is stimulated by calmodulin, inhibited by vanadate, has a K_m of 0.25 μmol, a V_m of 40 nmol Ca/min/mg protein and is a polymeric peptide with a molecular weight of 120–140 $\times$ 10^3 (Ghijsen et al. 1982). The Ca/Mg-ATPase of red blood cell membranes and heart sarcolemma appears to have a hydrophobic constituent with a molecular weight of 35 $\times$ 10^3 that is not involved in calcium transport, whereas the transient fragment, with a molecular weight of 90 $\times$ 10^3, when reconstituted into liposomes, transports calcium in a calmodulin-stimulated process (Carafoli 1984). A detailed model of how the sarcoplasmic Ca/Mg-ATPase may bind, gate and translocate calcium has been published (Brandl et al. 1986).

Are the kinetic properties of the intestinal Ca/Mg-ATPase consistent with a role in transmural calcium transport? Schiffl and Binswanger (1980) found that the activity of the enzyme in mucosal homogenates correlated closely and linearly with active calcium transport of everted intestinal sacs from rats fed varying levels of calcium and of differing vitamin D status. Subsequently Ghijsen and colleagues (1982) refined the enzyme measurement, correlated it with calcium uptake determinations on isolated basolateral membrane vesicles and calculated that active calcium transport by the intestine might well be accounted for by the enzyme-driven calcium extrusion process.

From the mean Ca/Mg-ATPase activity at saturating levels of calcium in normal rats fed on a 0.9% calcium, 0.8% phosphorus diet (Schiffl and Binswanger 1980) one can calculate a transport capacity of 100 μmol Ca/h/g duodenum wet weight (Bronner et al. 1986). This is five times the maximum value that has been calculated for the loop (Fig. 5.1) and demonstrates that enzyme activity is not limiting under these dietary conditions.

However, the value for K_m is 10^4 times lower than the corresponding value for transmural transport, calculated at 3.9 mM (Fig. 5.1). This can be explained by recalling that for the enzyme, which acts at the inner aspect of the basolateral membrane, the relevant calcium level is the intracellular calcium concentration, 1 μM, whereas in transmural transport it is the calcium level at the extracellular, serosal surface, ~1 mM, that is relevant.

In vitamin D deficiency, the activity of the Ca/Mg-ATPase is reduced to about one-third of what it is at maximum vitamin D levels (Schiffl and Binswanger 1980; Ghijsen et al. 1982; Ghijsen et al. 1986). From the data of Schiffl and Binswanger (1980) one can calculate (Bronner et al. 1986) that in nephrectomized animals on a high-calcium diet, when the production of the active vitamin D metabolite, $1,25(OH)_2D_3$, is virtually totally suppressed, the basal level of enzyme activity still suffices to transport calcium out of the cell. The threefold increase in enzyme activity effected by vitamin D may therefore simply prevent enzyme action from becoming rate-limiting under conditions of maximum transmural calcium flux. Whatever the mechanism by which vitamin D causes the activity of the intestinal Ca/Mg-ATPase to increase, quantitative analysis of transmural calcium transport indicates that the enzyme-driven process of cellular calcium extrusion by itself does not constitute the rate-limiting step of transcellular calcium movement.

Regulation of Active Calcium Transport by Vitamin D

As already briefly discussed in the preceding section, vitamin D affects four steps involved in transcellular calcium transport; entry across the brush border, intracellular diffusion, interaction with fixed intracellular calcium-binding sites in a variety of organelles and extrusion at the basolateral pole by the Ca/Mg-ATPase. However, the quantitative importance of vitamin D for each of these steps is very different – it is minor for the entry step and major for the diffusion process (Table 5.2).

Table 5.2.   Effects of vitamin D on transcellular calcium transport in the intestine

Step	Mechanism or structure	Effect of vitamin D	Mechanism
Entry across brush border	Down electrochemical gradient via channel or carrier	Enhances by 20%–30%	Possibly via integral calcium-binding protein
Binding to fixed cellular sites (buffering)	Golgi apparatus, RER mitochondria	Enhances by 100%	Unknown
Intracellular movement	Diffusion	Facilitates diffusion up to 100-fold	Biosynthesis of soluble CaBP ($M_r \simeq$ 9000 in mammals, $M_r \simeq$ 28 000 in birds) which acts as calcium ferry
Extrusion	Pumping against gradient; Ca/Mg-ATPase; Na^+/Ca^{2+} exchanger	Increases Ca/Mg-ATPase 200%–300%	Unknown

Two general mechanisms for the action of vitamin D in the intestine have been proposed. The first is a biosynthetic action, similar to that of steroid hormones, where the vitamin D metabolite, $1,25(OH)_2D_3$, interacts with an intracellular receptor protein. The receptor–vitamin D complex then induces a series of transcriptional and translational events that lead to the synthesis of a protein. For the intestinal CaBP, many of these steps have been studied in detail. Its synthesis is considered to be totally dependent on vitamin D, although specifics of the regulatory role played by vitamin D, whether purely transcriptional or also post-transcriptional, are still under investigation (for review, see Henry and Norman 1984; Kumar 1984).

The second general mechanism, termed the liponomic action of vitamin D, has been proposed to account for findings (Spencer et al. 1978; Thomasset et al. 1979) that indicated that intestinal calcium transport was stimulated in vitamin D-deficient tissue before initiation of CaBP synthesis or under conditions when protein synthesis was inhibited (Bikle et al. 1978; Rasmussen et al. 1982). Other investigators, however were unable to separate transport stimulation from protein synthesis (Bronner et al. 1982; Bishop et al. 1983).

The liponomic hypothesis is based on the observation (Rasmussen et al. 1982) that reorganization of choline phosphoglyceride fatty acids occurred with an increase in linoleic and arachidonic acids when vitamin D-deficient chicks were treated with $1,25(OH)_2D_3$. These lipid changes were found to precede increases in calcium transport (Matsumoto et al. 1981). Conceivably, a vitamin D-induced increase in the activity of the deacylation–reacylation cycle could alter the phospholipid composition of the membrane without implicating RNA and protein synthesis. However, even if the 20%–30% increase in calcium entry via the brush border were due to phospholipid-related changes in membrane fluidity, it can hardly account for the up to 100-fold increase in transcellular V_m that occurs in parallel with a comparable increase in CaBP (Fig. 5.4). All available data indicate that the membrane effect of vitamin D is indeed quite small. Moreover, the existence of a vitamin D-dependent calcium-binding protein that may be integral to the brush-border membrane, as reported by Kowarski and Schachter (1980) and Miller et al. (1979), may represent an alternative mechanism by which vitamin D might act on calcium entry. Such a protein could either be a part of a calcium channel or constitute a carrier.

Putkey and Norman (1983) have reported that vitamin D has an effect on protein composition and core material structure of the chick intestinal brush-border membrane. The importance of these findings for intestinal calcium transport is not known, but these effects also are unlikely to be rate-limiting for calcium entry.

It is interesting to note that vitamin D has an enhancing effect on calcium binding by intestinal cells. Thus, it was found (Bronner et al. 1983) that isolated duodenal cells from vitamin D-replete rats took up twice as much calcium in vitro than such cells from vitamin D-deficient animals. Calcium uptake by ileal cells, whether from vitamin D-deficient or -replete animals, was about the same as that by duodenal cells from vitamin D-deficient animals.

It should be noted that if vitamin D acted only to enhance cellular calcium entry, to increase intracellular calcium diffusion and to magnify energy-dependent calcium extrusion, the net effect of vitamin D repletion would be to lower free and hence bound calcium. Yet the opposite occurs, i.e. total cellular calcium is raised. Indeed, in duodenal cells from vitamin D-replete rats the level of bound calcium is much higher than the CaBP concentration and, of course, much higher than the free calcium concentration.

This increased buffering capacity for calcium of mucosal cells from vitamin D-replete animals may be explained as follows. If an animal that is adapted to a low calcium intake and, therefore, has a high absorptive capacity, unexpectedly feeds on a high-calcium food, down-regulation of the high transport capacity will not occur for several hours (Freund and Bronner 1975). The higher calcium-binding capacity associated with full vitamin D expression characteristic of the animal on a low-calcium diet may help protect the mucosal cells involved in transcellular calcium transport from becoming flooded with excess free calcium. Weiser and colleagues (Freedman et al. 1977; Weiser et al. 1981) have shown that treating vitamin D-deficient rats with vitamin D stimulates calcium uptake by Golgi bodies and other organelles. This intracellular calcium binding may well represent the mechanism of the increase in calcium uptake shown by isolated mucosal cells (Bronner et al. 1983).

Although vitamin D repletion is associated with a two- to threefold increase in Ca/Mg-ATPase activity (Schiffl and Binswanger 1980; Ghijsen et al. 1982), the mechanism by which this increase is effected is not known. It seems unlikely that biosynthesis of the enzyme is directly dependent on or influenced by vitamin D. Such a possibility would require that synthesis of the Ca/Mg-ATPase of the intestinal cell depend at least in part on a gene not implicated in the synthesis of this ATPase in other cells. Alternatively, enzyme function may depend on a vitamin D-induced regulator protein. Reports have been published to suggest that renal CaBP (mol. wt. $\sim$ 28 000) may stimulate the basolateral Ca/Mg-ATPase in kidney (Freund and Christakos 1985). Moreover, addition of exogenous calmodulin does not appear to stimulate the intestinal enzyme dramatically (Nellans and Popovitch 1981), nor have Ghijsen et al. (1982) been able to inhibit their membrane enzyme preparation with trifluoperazine. On the other hand, trifluoperazine addition leads to dose-dependent inhibition of active calcium transport in everted duodenal sacs (Roche et al. 1986b) and also of ATP-dependent calcium uptake by basolateral membrane vesicles.

Ghijsen et al. (1986) have recently shown that calmodulin can indeed stimulate the intestinal Ca/Mg-ATPase. They have also shown that intestinal CaBP does not stimulate calcium transport by basolateral membrane vesicles, even though they confirmed that the amount of the enzyme is increased under conditions of vitamin D repletion. Thus the molecular nature of the vitamin D-dependent stimulation of Ca/Mg-ATPase activity remains to be elucidated.

From the above discussion and the summary in Table 5.2 of the action of vitamin D on transcellular calcium transport it thus appears that although vitamin D has a pleiotropic effect, its major action is via the transcriptionally mediated synthesis of the CaBP that acts like a calcium ferry.

Developmental Aspects

Rats are born without an active calcium transport system (Pansu et al. 1983a; Dostal and Toverud 1984), and without functional CaBP in their intestines (Ueng et al. 1979). Since the number of receptors for $1,25(OH)_2D_3$ in the proximal intestine is also low (Halloran and DeLuca 1980, 1981), it is not surprising that treatment of newborn rats with $1,25(OH)_2D_3$, although it led to a rise in plasma calcium, failed to increase the CaBP content in their intestines (Ueng et al. 1979). Starting at about 5 days of age, CaBP becomes detectable in the duodenum (Ueng et al. 1979) and the tissue content of the protein rises till the animals are about 30 days old (Fig. 5.5). Thereafter, CaBP levels decline gradually. Active calcium transport, V_m, rises and falls in these animals completely in parallel with these changes in CaBP levels (Fig. 5.5).

The developmental time course of the non-saturable route is quite different. At first, virtually all of the calcium is absorbed by this route, but when the animals are 30 days old, i.e. when active transport is at its peak, on the average only about 17% of the luminal calcium content is transported by the passive, non-saturable route (Fig. 5.5) and this fraction then remains unchanged (see also Table 5.1).

Henning (1981) has described how the intestine of the newborn rat acts virtually like a sieve, with all substances being absorbed fully and without discrimination. This seems to be true also for calcium in the rat before it is

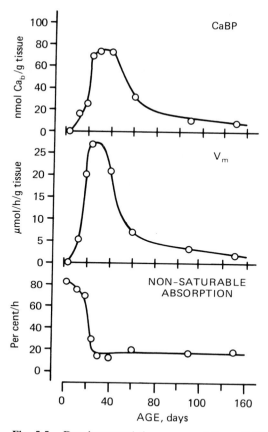

Fig. 5.5. Developmental time course of tissue CaBP content, active calcium transport (V_m) and non-saturable calcium transport in rat duodenum. The mothers of the suckling rats (aged 3–19 days) were fed a vitamin D-replete, high-calcium, semisynthetic diet (1.5% calcium, 1.5% phosphorus), as were the rats older than 21 days. (Adapted from Pansu et al. 1983a)

weaned. At about 2 weeks after weaning, the non-saturable process becomes stabilized at around 16%/h. It is possible that the high rate of passive absorption in the newborn rat represents a high degree of patency of the tight junction and that paracellular flow of calcium becomes restricted as the tight junction matures and becomes tighter. This may indeed be true, at least in part, inasmuch as pregnancy and thyroid hormone status have been shown to affect the structure of the tight junction (Pitelka et al. 1973; Tice et al. 1975). However, literature has also accumulated that indicates that in the suckling rat there exists an additional route for nutrient absorption, namely an endocytic one. In the very young rat, antibody absorption occurs by endocytosis, with antibodies first reacting with specific receptors located on the outer cell membrane (Morris 1978). Although antibody absorption seems to be confined largely to the jejunum and ileum, evidence for the existence of endocytotic vesicles in the upper intestine has been reported by Pansu et al. (1983a), Kirshtein and Bauman (1980) and Drisdale (1983). Moreover, the time course of disappearance of these vacuoles coincides with the time course of the decrease in non-saturable calcium transport in the rat under 30 days of age. In the 10–20-day-old suckling rat there may exist, therefore, three calcium transport processes: an emerging calcium-specific active transport process that is transcellular; a non-specific endocytic process, the disappearance of which parallels the emergence of the calcium-specific active one; and a paracellular process that is largely age-independent. Studies are needed to differentiate between the paracellular and endocytic pathways.

The decline in the active calcium transport process that occurs in the rat after the age of 30 days (Fig. 5.5) is likely to be due to two causes: a calcium-dependent down-regulation and an age-dependent decrease. Figure 5.6, like Fig. 5.5, shows that intestinal calcium absorption peaked and then declined as a function of age, whether the animals were fed a high- or low-calcium diet. However, the age at which absorption peaked and then declined differed in the two diet groups. The age-dependent decline in active calcium absorption and in CaBP level has also been described by Armbrecht and colleagues (1980).

Dostal and Toverud (1984) have done a detailed study of the effect of vitamin D on calcium absorption in vitamin D-replete and vitamin D-deficient rat sucklings. Their findings indicate that the effect of vitamin D repletion on calcium absorption is negligible in 11-day-old rats and barely detectable in 18-day-old animals, in conformity with the studies by Pansu et al. (1983a). In weaned rats, on the other hand, vitamin D repletion markedly increased active and, therefore, total calcium transport. Dostal and Toverud (1984) also showed,

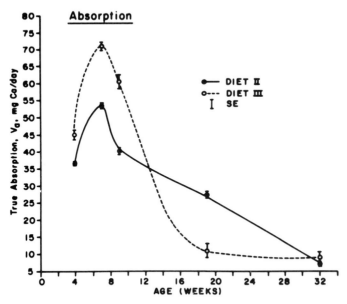

Fig. 5.6. Calcium absorption in rats as a function of age. Weanling male Sprague–Dawley rats were placed on one of two vitamin D-replete semisynthetic regimens (II, 0.5% calcium, 0.5% phosphorus; III, 1.5% calcium, 1.5% phosphorus) and their calcium absorption was measured by a combination of radioisotope and balance procedures (Sammon et al. 1970) at the indicated ages. (Chang and Bronner, unpublished)

in conformity with the data shown in Fig. 5.5, that the non-saturable component dropped to about one-third of its value between 15 and 35 days of age.

Analysis of Whole Animal Absorption Studies

True calcium absorption is a curvilinear function of calcium intake in the rat, whereas endogenous faecal calcium output appears to be a simple linear function of intake (Fig. 5.7). If true calcium absorption, as measured in vivo, is indeed the sum of a saturable transcellular and of a non-saturable paracellular movement, it should be possible to represent the relationship between v_a and v_i as such a sum.

Hurwitz et al. (1969), who had measured calcium absorption by a combination balance and kinetic method in vitamin D-replete (+D) and vitamin D-deficient (−D) rats that had been fed a semisynthetic diet that contained 0.5% calcium and 0.5% phosphorus, obtained the following relationships:

$$v_a = 3.4 + 0.51 \, v_i \qquad +D \qquad (4)$$
$$v_a = -8.6 + 0.55 \, v_i \qquad -D \qquad (5)$$

where v_a = true calcium absorption (mg/day) and v_i = calcium intake (mg/day), studied in the range of 45–115 mg/day for the +D and in the range of 18–85 mg/day for the −D animals.

The difference in calcium absorption between Eqs. (4) and (5) is largely due to the difference in intercepts. Inspection of v_a as a function of v_i in Fig. 5.7

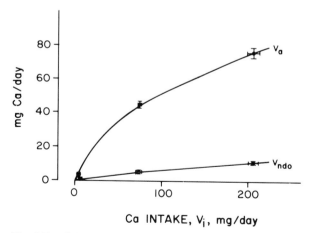

Fig. 5.7. Calcium absorption, v_a, and endogenous faecal calcium excretion, v_{ndo}, as a function of calcium intake, v_i, in normal Sprague–Dawley rats ($\simeq 200$ g BW). The values are means obtained from three groups of animals, placed on a low-calcium (0.05% calcium, 0.2% phosphorus; $n = 25$), intermediate-calcium (0.5% calcium, 0.5% phosphorus; $n = 30$) and a high-calcium (1.5% calcium, 1.5% phosphorus; $n = 22$) semisynthetic diet for 10 days. The values for v_i, v_a and v_{ndo} were obtained in 3-day balance and radioisotope experiments. (Adapted from Sammon et al. 1970)

suggests that the intercepts of Figs. 5.4 and 5.5 represent the saturable, vitamin D-dependent step, as already assumed in the discussion of Eq. (2), above, whereas the slope, virtually the same in the two equations, represents the non-saturable component. Let us now examine whether the experimental findings justify the formulation in Eq. (2).

According to that reasoning, the mean slope, 0.53, would represent the fraction of ingested calcium that is absorbed via the non-saturable route. In the rat, this has been calculated to average 0.16/h (see Table 5.1), so that 0.53/0.16 h^{-1} would represent the sojourn time of calcium in the small intestine, i.e. 3.3 h, a value that accords well with the experimental value reported by Marcus and Lengemann (1962).

The difference in intercepts of Eqs. (4) and (5), i.e. 12 mg/day, may be considered equal to V_m in the experiment of Hurwitz et al. (1969). If, as a first approximation for K_m, we use the value of 3.8 mM from the loop studies, this translates to a value for V_m of 9.5 mg/day, assuming a daily water intake of 20 ml and a fivefold dilution of the ingesta by body fluids. Equation (2) then becomes

$$v_a = \frac{12 \times v_i}{9.5 + v_i} + 0.53 \, v_i \tag{6}$$

In an earlier study, vitamin D-replete rats fed on a similar diet with a calcium intake of 71.2 mg/day were found to absorb 41.3 mg/day (Bronner and Aubert 1965), whereas in a later study (Sammon et al. 1970) rats with a calcium intake of 72.8 mg/day absorbed 44.5 mg/day. The respective values predicted from Eq. (6) are 48.3 (10.6 by the saturable route and 37.7 by the non-saturable route) and 49.2 (10.6 + 38.6) mg/day, a very satisfactory agreement considering the uncertainties in deriving Eq. (6).

The first term of Eqs. (2) and (6) represents the vitamin D-dependent component which, at low values for v_i, is large in relation to the second term. At high calcium intakes, on the other hand, v_a would approach 0.53 v_i $(b \cdot v_i)$ asymptotically, with the first term negligibly small. Thus, without vitamin D, the rat would absorb only half of the calcium offered at all calcium intake levels, whereas with vitamin D, when the dietary calcium is low, it can absorb it all.

The same reasoning would apply to calcium absorption in other species, including man, though the numerical values and relative importance of active and non-saturable calcium absorption may vary in different species and at different ages. In the pig, for example, significant amounts of CaBP, presumably indicative of active calcium absorption, are found much more distally than in the rat (Arnold et al. 1975). In the hamster, active calcium transport is appreciable in the ileum, with the ileal V_m in fact twice as great as the duodenal V_m (Schedl et al. 1981). The reason for this curious reversal is unclear. If the amount of calcium in the food supply is uncertain, it may be more appropriate to extract calcium actively at the proximal end of the intestine. If, however, the food calcium supply is assured, active calcium extraction at the end of the journey through the intestine may be more efficient.

The same reasoning might be applied to the observation that vitamin D-dependent active calcium transport occurs in the colon (Favus et al. 1980, 1981; Lee et al. 1980). On the basis of the good agreement that exists between experimental and theoretical values for v_a one might infer that the contribution to overall calcium absorption made by active transport in the colon is small and probably has limited functional significance.

In summary it appears that quantitative values of the saturable and non-saturable components of transmural calcium absorption, as derived from in vitro studies, are consistent with in vivo measurements. At the same time, these values emphasize the quantitative importance of the non-saturable component. For this reason, in vivo studies concerned with regulation should be carried out under conditions of varying dietary calcium levels. Studying absorption at different calcium intakes will not only show how a given factor alters the regulatable component, but at the same time will clarify the practical importance that can be attributed to modifying the regulatable component.

Altering Calcium Absorption

Calcium absorption may be modulated or altered in three general ways: by increasing or decreasing the amount of calcium available for absorption, by altering active calcium transport and by changing paracellular calcium flow. In what follows, each of these three general approaches will be discussed by way of examples, but no attempt will be made to list all conditions or situations that have been reported to alter calcium absorption.

Calcium Availability

Increasing or decreasing calcium intake is the most general way to influence the amount absorbed, since the luminal calcium concentration is a direct function of calcium intake and since, according to Eq. (2), both active and passive calcium transport are functions of the luminal calcium concentration.

Another way to influence calcium absorption is to alter calcium availability. In other words, if calcium was ingested in the form of a salt or compound that would not be digestible or solubilized, the effect would be equivalent to diminishing the amount ingested. However, determination of the degree of unavailability of a given calcium salt has proved difficult and controversial. Partly this is true because the chemically complicated situation that prevails in the gut lumen does not correlate with predictions based on what happens in simple, dilute solutions. A good example is the current controversy regarding the absorption of $CaCO_3$ by achlorhydric patients. Bo-Linn and associates (1984), using a novel lavage technique, found that net calcium absorption in these patients from a 1063-mg calcium meal was 45%, whether the gastric pH was 7.4 or 3.0. These investigators speculate that the pH in the intestine is sufficiently low to solubilize the $CaCO_3$.

Recker (1985), on the other hand, measured calcium absorption by an isotope technique and found that control subjects absorbed citrate and carbonate equally well, at a level of about 24% of the total, an estimate more closely in accord with what would be predicted from balance studies. While the achlorhydric subjects absorbed only one-fifth as much calcium as the controls when offered $CaCO_3$ as such, their calcium absorption approached that of the controls when the carbonate was given with food. Interestingly, the absorption of calcium from citrate by the achlorhydric subjects was nearly double that of the controls, an observation that remains as yet unexplained.

Ingestion of oxalate or phytate has traditionally been associated with situations where calcium intake is low. However, the practical significance of this is uncertain, perhaps because the amounts of phytate or oxalate ingested may not in practice bind enough of the calcium available to bring about an effective lowering of calcium intake. Alternatively, it is conceivable the complex itself is absorbed via the paracellular pathway, albeit more slowly than the calcium ion.

In the 1950s Bronner and colleagues (1954, 1956) in one of the earliest human studies utilizing ^{45}Ca, fed 13-year-old boys two experimental breakfasts, one containing about 90 mg calcium, the other approximately 240 mg calcium. Some of the subjects received enough phytate in oatmeal to bind 90 mg calcium; the control subjects received farina. An additional group received farina plus sodium phytate equivalent to the amount of oatmeal phytate. 180 g of oatmeal, a typical breakfast portion, contains about 100 mg phytate phosphorus. This, in turn, reduced calcium absorption as if the amount of calcium fed had been reduced by 23 mg, i.e. by 25% when intake was 90 mg, but by a negligible 10% at the higher calcium intake level. Sodium phytate, as opposed to food phytate, had double the effect; 100 mg phosphorus in the form of sodium phytate caused an equivalent of about 50 mg dietary calcium to become unavailable. Thus, it takes about 4 mg of food phytate phosphorus to make 1 mg food calcium unavailable. Unless, therefore, total phytate phosphorus intake equals at least 600 mg daily, with calcium intake in the low normal range of 600 mg/day, the practical effect of phytate in the diet is undoubtedly small.

In summary, calcium availability involves two possible situations: one where the actual amount of calcium available for transport is less than predicted from elemental analysis; the other, as yet more speculative, where substances or molecules in the lumen interact with calcium so as to alter its paracellular diffusion, thereby increasing or diminishing the rate of non-saturable calcium transport (see p. 116).

Factors or Situations that Alter Active Calcium Transport

Since vitamin D is the major regulator of active calcium transport, conditions that alter the vitamin D supply or modify vitamin D metabolism can be expected to affect active calcium transport. Thus, vitamin D deficiency ultimately leads to the suppression of active transport (Figs. 5.3 and 5.4). Vitamin D deficiency may be the result of nutritional deficiency, of metabolic conditions or defects that impair or prevent the synthesis of $1,25(OH)_2D_3$, the active vitamin D metabolite, or of intestinal conditions that impair the absorption of vitamin D or its congeners. These situations are listed in Table 5.3.

Large doses of vitamin D will increase active, and therefore, total calcium absorption. Hall et al. (1969) have reported that after 4 weeks of daily treatment with 150 000 units vitamin D, true calcium absorption in two postmenopausal women increased from 49% to 70% and from 30% to 47%, respectively, net calcium absorption having increased by about 200 mg daily in each subject. On the basis of Eqs. (3) and (4), this increase in absorption must have resulted from

Table 5.3. Examples of vitamin D deficiency with diminished calcium absorption (adapted from Levine et al. 1982)

Low vitamin D intake
  Nutritional rickets
  Osteomalacia

Low vitamin D absorption
  Steatorrhoea
  Gluten enteropathy
  Blind loop syndrome
  Malabsorption syndrome
  Patients with jejuno-ileal bypass surgery
  Anti-convulsant therapy

Impaired synthesis of 1,25(OH)$_2$D$_3$
  Renal failure
  Treatment with diphosphonate
  Vitamin D-resistant rickets
  Diabetes mellitus
  Hypoparathyroidism

an increase in active calcium transport; calcium intakes of the patients were the same before and after vitamin D treatment.

Varying calcium intake has two effects that tend to work in opposite directions: active transport is high under conditions of low calcium intake and is down-regulated when calcium intake goes up. Passive transport varies directly with intake; it goes up when intake increases and down when intake decreases.

Regulation of active transport is mediated by 1,25(OH)$_2$D$_3$, with increased hydroxylation of the sterol at carbon 1 occurring in renal cells in response to a drop in plasma and intracellular calcium. Thus, a fractional drop in plasma calcium is associated with an increase in the renal production of 1,25(OH)$_2$D$_3$ (Kumar 1984); this, in turn, leads to a rise in the circulating level of 1,25(OH)$_2$D$_3$ and, therefore, in the quantity of 1,25(OH)$_2$D$_3$ available to interact with the receptors of the intestinal cells. If plasma calcium rises, the production of 1,25(OH)$_2$D$_3$ will be down-regulated and less 1,25(OH)$_2$D$_3$ will be available to interact with intestinal receptors. High levels of 1,25(OH)$_2$D$_3$ lead to high levels of CaBP (Roche et al. 1986b) and of active calcium transport (Fig. 5.4). Parathyroid hormone, whose release is inversely proportional to the plasma calcium level, also causes the renal cells to produce more 1,25(OH)$_2$D$_3$, but the mechanism is unknown.

In addition to the systemic regulation of active calcium transport by 1,25(OH)$_2$D$_3$, a second, local mechanism has been hypothesized to play a role in CaBP induction (Bronner and Buckley 1982). According to this hypothesis, based on experiments with whole animals and isolated intestinal cells (Singh and Bronner 1982), a rise in intracellular calcium represses full CaBP expression, whereas low intracellular calcium levels derepress it.

From the viewpoint of regulation, the effects of vitamin D and calcium transport represent the two paths of a control loop: vitamin D is the leading function and active calcium transport varies directly with the expression of 1,25(OH)$_2$D$_3$, the active vitamin D metabolite. Vitamin D action is most directly expressed via the amount of CaBP synthesized, as well as via effects on calcium entry and extrusion (Table 5.2).

Calcium intake, on the other hand, represents the feedback path of the control loop: when calcium intake is down, active transport is increased via the increased expression of $1,25(OH)_2D_3$, described above. When calcium intake is raised, active transport is down-regulated via the diminished expression of vitamin D action.

Active calcium transport involves calcium entry at the brush border, facilitated calcium diffusion through the cytosol and extrusion via the Ca/Mg-ATPase. In addition to modulating active calcium transport via the vitamin D system or via agents that modify vitamin D metabolism or expression (see also Table 5.3), it is possible to alter active calcium transport by means of drugs that alter one of the three steps directly.

The calcium channel blocker verapamil, at concentrations above 10^{-5} M, reduces active calcium transport in everted duodenal sacs (Roche 1982). Verapamil, at concentrations of 0.3 to 10 mM, also inhibits the active component of calcium absorption as evaluated by in situ loop preparations (Fox and Green 1986). Interestingly, neither nifedipine nor diltiazem had an effect under these conditions (Fox and Green 1986). Verapamil also inhibits calcium uptake by brush-border membrane vesicles (Miller and Bronner 1981).

Theophylline, apparently by inhibiting calcium binding by CaBP, inhibits active calcium transport in everted duodenal sacs (Roche et al. 1986a; Pansu et al. 1987). The drug neither inhibits calcium entry, as evaluated by uptake by brush-border membrane vesicles, nor calcium extrusion as determined by ATP-stimulated calcium uptake by basolateral membrane vesicles; rather it inhibits calcium binding by CaBP and therefore prevents CaBP from exerting its calcium ferrying action (Pansu et al. 1987).

Trifluoperazine inhibits active calcium transport (Favus et al. 1985; Roche et al. 1986b) by a concentration-dependent effect on the ATP-dependent process of calcium extrusion, as evaluated by its inhibition of calcium uptake by inside-out basolateral membrane vesicles. The drug has no effect on uptake of calcium by brush-border membrane vesicles, i.e. on the entry step of transcellular calcium transport (Roche et al. 1986b).

In summary, active calcium absorption may be inhibited or stimulated by altering the amount or expression of vitamin D, either directly or indirectly, or by agents or drugs that act directly on one of the three steps of transcellular calcium transport – entry, intracellular diffusion or extrusion.

Factors or Situations that Alter Passive Calcium Transport

Of the many factors that have been reported to alter calcium absorption, some are unlikely to act directly on active, transcellular transport, although in most instances, experimental evidence to that effect is missing. Examples of factors which are unlikely to affect the active component include lactose, amino acids and medium-chain triglycerides, all of which have been reported to enhance calcium absorption.

Lactose in the diet has long been known to enhance calcium retention. The lactose effect on intestinal calcium absorption is most pronounced in the distal segment of the small intestine (Lengemann 1959; Lengemann et al. 1959; Dupuis and Fournier 1964). Pansu and colleagues (1979, 1981) showed that a

30%-lactose diet had an effect equivalent to raising the calcium intake of rats from 0.44% to 0.7%. Efficiency of total calcium absorption was decreased and the CaBP content of the duodenum was markedly diminished, with the effect greater in young than in old rats, i.e. greater under conditions when active calcium transport is proportionately more important and when down-regulation of active transport has a proportionately greater effect.

How can lactose, acting particularly in the distal small intestine, i.e. the region where active calcium transport is unimportant, lead to an increase in the total amount of calcium absorbed and retained? A reasonable explanation is that the presence of lactose in the intestinal lumen leads to an increase in the amount of calcium that is absorbed by the non-saturable route, i.e. the paracellular pathway. That, in turn, will lead to a down-regulation of active calcium transport, while the total calcium absorbed increases due to the lactose-induced increase in the non-saturable, paracellular flow.

Lactose under these conditions might act either chemically or osmotically to alter the tight and intermediate junctions between the epithelial cells so as to permit increased calcium flow. A non-specific, osmotic effect seems the most likely explanation, inasmuch as Wasserman and Taylor (1969) have reported that instillation of 0.25 M solutions of fructose, xylose and saffinose led to a marked rise in calcium efflux from ligated ileal loops. Pansu et al. (1976) reported similar results when hyperosmolar solutions of sodium chloride or glucose were instilled in their intestinal loop preparations.

L-lysine and L-arginine, but also other amino acids, have been shown to enhance calcium absorption (Wasserman et al. 1956; Wasserman and Taylor 1969). Wasserman and Taylor (1969) argue that increased solubilization of luminal calcium is not the entire explanation. That the action is partly mediated via an effect on paracellular flow can be argued in the light of the report (Martinez-Palomo and Erlij 1975) that 50–100 mM lysine added to the apical solution bathing urinary hemibladders from *Bufo marinus* increased the permeability of the gap junctions of that tissue reversibly.

A modification of the functional space between mucosal cells is a possible explanation for the observation (Kehayoglou et al. 1968) that medium-chain triglycerides were found to enhance calcium absorption in the rat. That report, like most others in the literature, did not differentiate between active and passive calcium transport. An effect of triglycerides on active transport seems unlikely, though an effect on calcium entry at the brush border cannot be ruled out (Rasmussen et al. 1982; O'Doherty 1979).

The substances mentioned above are but a sampling of the many that may act on paracellular calcium movement by modifying passage of the cation through the tight and intermediate junctions between cells. Whether these effects are purely physical, i.e. the space between cells is distended, or whether the cell surfaces are also changed chemically is largely unknown. Whatever the mechanism, further study of paracellular transfer of calcium in the intestine seems indicated, particularly in the light of possible therapeutic applications (see over).

Therapeutic Considerations

There are two ways to increase calcium absorption: firstly, by increasing active transport, e.g. by the administration of vitamin D, and secondly, by increasing the amount carried across by the non-saturable process, that is, by increasing the amount of calcium ingested.

While there can be little doubt that vitamin D administration will increase calcium absorption, it does not normally result in an increase in calcium retention. This is illustrated by the work of Hall et al. (1969) where calcium absorption in two normal women was increased by 59% on the average, following 4 weeks of daily treatment with 150 000 units of vitamin D, but where the balance was unchanged, because urinary calcium output increased to the extent by which calcium absorption had increased. This had been previously observed in patients (Saville et al. 1955) and was subsequently confirmed in experimental animals given $1,25(OH)_2D_3$ (Bonjour et al. 1975).

An increase in calcium intake will lead to a rise in the amount absorbed, even though active absorption will be down-regulated. Thus, from Recker's (1985) study of a group of postmenopausal women one can calculate that raising the calcium intake from 500 to 1000 mg/day will cause fractional absorption to drop from about 0.24 to 0.21, with absorption rising from 120 to 210 mg/day. From the work of Hall et al. (1969), who studied similar subjects, one can estimate that for every 100 mg increase in net calcium absorption, the urinary calcium output will increase by nearly one-quarter of the net absorbed. Hence by doubling calcium intake, one should, in these subjects, achieve a net gain in the calcium balance of 70–80 mg/day. How long such a gain would be sustained is, however, uncertain.

Agents that enhance the magnitude of calcium entry or extrusion are unlikely to be important therapeutically, since, as discussed in this chapter, neither of these steps appears limiting. On the other hand, agents that inhibit these steps may have treatment possibilities. For example, the gastrointestinal vasoactive peptide, VIP, has recently been reported to inhibit active calcium transport by inhibiting extrusion from the basolateral membrane (Roche et al. 1985). It may, therefore, prove to be a useful drug in conditions of idiopathic hyperabsorption of calcium.

The simplest and generally the safest method for increasing calcium absorption is to increase intake. Even though this manoeuvre will lead to down-regulation of active calcium transport, an increase in absorption and retention generally occurs, at least in the short term. If, as a result of future research, it becomes possible to modify the paracellular path as well, the latter approach will become important, particularly in situations of intraluminal feeding, where a simple increase in calcium concentration may not be feasible or practical. From the nutritionist's point of view, an increase in calcium intake still seems the safest and most direct approach to increasing calcium absorption.

References

Armbrecht HJ, Zenser TV, Gross CJ, Davis BB (1980) Adaptation to dietary calcium and phosphorus restriction changes with age in rat. Am J Physiol 239:E322–327

Arnold BM, Kuttner M, Willis DM, Hitchman AJW, Harrison JE, Murray TM (1975) Radioimmunoassay studies of intestinal calcium-binding protein in the pig. 2. The distribution of intestinal CaBP in pig tissues. Can J Physiol Pharmacol 53:1135–1140

Bikle DD, Zolock DT, Morissey RL, Herman RH (1978) Independence of 1,25-dihydroxyvitamin D_3-mediated calcium transport from *de novo* RNA and protein synthesis. J Biol Chem 253:484–488

Bikle DD, Shoback DM, Munson SJ (1985) 1,25-Dihydroxyvitamin D increases the intracellular free calcium concentration of duodenal epithelial cells. In: Norman AW, Schaefer K, Grigoleit H-T, Herrath D V (eds) Vitamin D chemical, biochemical and clinical update. Sixth workshop on vitamin D, Merano, Italy. Walter de Gruyter, Berlin

Bishop CW, Kendrick NC, DeLuca HF (1983) Induction of calcium-binding protein before 1,25-dihydroxyvitamin D_3 stimulation of duodenal calcium uptake. J Biol Chem 258:1305–1310

Bo-Linn GW, Davis GR, Buddrus DJ, Morawski SG, Ana CS, Fortran JS (1984) An evaluation of the importance of gastric acid secretion in the absorption of dietary calcium. J Clin Invest 70:640–647

Bonjour JP, Trechsel U, Fleisch H, Schenk R, DeLuca HF, Baxter LA (1975) Action of 1,25-dihydroxyvitamin D_3 and a diphosphonate on calcium metabolism in rats. Am J Physiol 229:402–408

Brandl CJ, Green NM, Kozczak B, MacLennan DH (1986) Two Ca^{2+} AtPase genes: homologies and mechanistic implication of deduced amino acid sequences. Cell 44:597–607

Bronner F (1973) Kinetic and cybernetic analysis of calcium metabolism. In: Irvin JT (ed) Calcium and phosphorus. Academic Press, New York, pp 149–186

Bronner F (1979) Clinical investigations of calcium metabolism. In: Kunin A, Simmons D (eds) Skeletal research – an experimental approach. Academic Press, New York, pp 243–261

Bronner F (1982) Calcium homeostasis. In: Bronner F, Coburn JW (eds) Disorders of mineral metabolism, vol 2. Academic Press, New York, pp 43–97

Bronner F, Aubert JP (1965) Bone metabolism and regulation of the blood calcium level in rats. Am J Physiol 209:887–890

Bronner F, Buckley M (1982) The molecular nature of 1,25(OH)$_2$-D$_3$-induced calcium-binding protein biosynthesis in the rat. In: Massry JM, Ritz E (eds) Regulation of phosphate and mineral metabolism. Plenum Press, New York, pp 355–360

Bronner F, Freund T (1975) Intestinal CaBP: A new quantitative index of vitamin D deficiency in the rat. Am J Physiol 229:689–694

Bronner F, Spence K (1987) Non-saturable Ca transport in the rat intestine is via the paracellular pathway. In: Bronner F, Peterlik M (eds) Cellular calcium and phosphate transport in health and disease. Alan R Liss, New York (in press)

Bronner F, Harris RS, Maletskos CJ, Benda CE (1954) Studies in calcium metabolism. Effect of food phytates on calcium^{45} uptake in children on low-calcium breakfasts. J Nutr 54:523–542

Bronner F, Harris RS, Maletskos CJ, Benda CE (1956) Studies in calcium metabolism. Effect of food phytates on calcium^{45} uptake in boys on a moderate calcium breakfast. J Nutr 59:393–406

Bronner F, Lipton J, Pansu D, Buckley M, Singh R, Miller A III (1982) Molecular and transport effects of 1,25-dihydroxyvitamin D_3 in rat duodenum. Fed Proc 41:61–65

Bronner F, Pansu D, Bosshard A, Lipton JH (1983) Calcium uptake by isolated rat intestinal cells. J Cell Physiol 116:322–328

Bronner F, Pansu D, Stein WD (1986) An analysis of intestinal calcium transport across the intestine. Am J Physiol 250:G561–569

Carafoli E (1984) Molecular, mechanistic, and functional aspects of the plasma membrane calcium pump. In: Bronner F, Peterlik M (eds) Epithelial calcium and phosphate transport: molecular and cellular aspects. Alan R Liss, New York, pp 13–18

Carafoli E, Crompton M (1978) The regulation of intracellular calcium. In: Bronner F, Kleinzeller A (eds) Current topics in membranes and transport, vol 10. Academic Press, New York, pp 151–216

Cassidy MM, Tidball CS (1967) Cellular mechanisms of intestinal permeability alterations produced by chelation depletion. J Cell Biol 32:685–698

Crompton M, Capano M, Carafoli E (1976) The sodium induced efflux of calcium from heart mitochondria. A possible mechanism for the regulation of mitochondrial calcium. Europ J Biochem 69:453–462

Dostal LA, Toverud SU (1984) Effect of vitamin D_3 on duodenal calcium absorption in vivo during early development. Am J Physiol 246:G528–534

Drisdale D (1983) Ultrastructural localization of calcium by electro-probe x-ray microanalysis in the small intestine of suckling rats. Tissue Cell 15:417–428

Dupuis Y, Fournier PL (1964) Etude comparée de l'action de la vitamine D et du lactose sur les échanges calciques durant la vie du rat. CR Acad Sci[D] (Paris) 258:2906–2909

Favus MJ, Kathpalia SC, Coe FL, Mond E (1980) Effects of diet calcium and 1,25-dihydroxyvitamin D_3 on colon calcium active transport. Am J Physiol 238:G75–78

Favus MJ, Kathpalia SC, Coe FL (1981) Kinetic characteristics of calcium absorption and secretion by rat colon. Am J Physiol 240:G350–354

Favus MJ, Angerd-Backman E, Breyer MD, Coe FL (1985) Effects of trifluoperazine, ouabain, and ethacrynic acid on intestinal calcium transport. Am J Physiol 244:G111–115

Feher JJ (1983) Facilitated calcium diffusion by intestinal calcium-binding protein. Am J Physiol 244:C303–307

Fox J, Green D (1986) Direct effects of calcium channel blockers on duodenal calcium transport in vivo. Eur J Pharmacol 129:159–164

Freedman RA, Weiser M, Isselbacher KJ (1977) Calcium translocation by Golgi and lateral-based membrane vesicles from rat intestine: decrease in vitamin D-deficient rats. Proc Natl Acad Sci USA 74:3612–3616

Freund TS, Bronner F (1975) Regulation of intestinal calcium-binding protein by calcium intake in the rat. Am J Physiol 228:861–869

Freund TS, Christakos S (1985) Enzyme modulation by renal calcium binding protein. In: Norman AW, Schaefer K, Grigoleit H-T, Herrath D v (eds) Vitamin D chemical, biochemical and clinical update. Sixth workshop on vitamin D, Merano, Italy. Walter de Gruyter, Berlin

Ghijsen WEJM, Dejon MD, van Os CH (1982) ATP-dependent calcium transport and its correlation with Ca^{2+}-ATPase activity in basolateral plasma membranes of rat duodenum. Biochim Biophys Acta 689:327–336

Ghijsen WEJM, Van Os CH, Heizman CW, Murer H (1986) Regulation of duodenal Ca^{2+} pump by calmodulin and vitamin D-dependent Ca^{2+}-binding protein. Am J Physiol 251:G223–229

Hall BD, MacMillan DR, Bronner F (1969) Vitamin D-resistant rickets associated with high fecal endogenous calcium output: a report of two cases. Am J Clin Nutr 22:448–457

Halloran BP, DeLuca HF (1980) Calcium transport in small intestine during early development: role of vitamin D. Am J Physiol 239:G473–479

Halloran BP, DeLuca HF (1981) Appearance of the intestinal cytosolic receptor for 1,25-dihydroxy-vitamin D_3 during neonatal development in the rat. J Biol Chem 256:7338–7342

Harrison HE, Harrison HC (1963) Theories of vitamin D action. In: Wasserman RH (ed) The transfer of calcium and strontium across biological membranes. Academic Press, New York, pp 229–252

Henning SJ (1981) Postnatal development: coordination of feeding, digestion and metabolism. Am J Physiol 241:G199–214

Henry HL, Norman AW (1984) Vitamin D: metabolism and biological actions. Annu Rev Nutr 4:493–520

Hess P, Tsien RW (1984) Mechanism of ion permeation through calcium channels. Nature 309:453–458

Holdsworth ES (1965) Vitamin D_3 and calcium absorption in the chick. Biochem J 96:475–483

Hurwitz S, Stacey RE, Bronner F (1969) Role of vitamin D in plasma calcium regulation Am J Physiol 216:254–262

Inesi G (1985) Mechanism of calcium transport. Annu Rev Physiol 47:573–601

Kehayoglou CK, Williams HS, Whimster WF, Holdsworth C (1968) Calcium absorption in the normal, bile duct-ligated and cirrhotic rat, with observations on the effect of long- and medium-chain triglycerides. Gut 9:597–603

Kirshtein BE, Bauman VK (1980) Pinocytic transport of calcium through the intestinal epithelium of new-born rats. Tsitologiia 22:57–60

Kowarski S, Schachter D (1980) Intestinal membrane calcium-binding protein: vitamin D-dependent membrane component of the intestinal calcium transport mechanism. J Biol Chem 255:10834–10840

Kretsinger RH, Mann JE, Simmonds JG (1982) Model of the facilitated diffusion of calcium by the intestinal calcium binding protein. In: Norman AW, Schaefer K, Herrath D v, Gringoleit H-G (eds) Vitamin D. Chemical, biochemical and clinical endocrinology of calcium metabolism.

Walter de Gruyter, Berlin, pp 233–246

Kumar R (1984) Metabolism of 1,25-dihydroxyvitamin D₃. Physiol Rev 64:478–504

Lee DBM, Walling MW, Gafter U, Silis V, Coburn JW (1980) Calcium and inorganic phosporus transport in rat colon. Dissociated response to 1,25-dihydroxyvitamin D₃. J Clin Invest 65:1326–1331

Lengemann FW (1959) The site of action of lactose in the enhancement of calcium utilization. J Nutr 69:23–27

Lengemann FW, Wasserman RH, Comar CL (1959) Studies on the enhancement of radio Ca and radio Sr absorption by lactose in the rat. J Nutr 68:443–445

Levine BS, Walling MW, Coburn JW (1982) Intestinal absorption of calcium: its assessment, normal physiology, and alterations in various disease states. In: Bronner F, Coburn JW (eds) Disorders of mineral metabolism. Academic Press, New York, pp 103–180

MacLennan DH, Brandl CJ, Korczak B, Green NM (1985) Amino acid sequence of Ca^{2+} + Mg^{2+} dependent ATPase from rabbit muscle sarcoplasmic reticulum, deduced from its complementary DNA sequence. Nature 316:696–700

Marcus CS, Lengemann FW (1962) Absorption of Ca^{45} and Sr^{85} from solid and liquid food at various levels of the alimentary tract of the rat. J Nutr 77:155–160

Martin DL, DeLuca HF (1969) Calcium transport and the role of vitamin D. Arch Biochem Biophys 134:139–148

Martinez-Paloma A, Erlij D (1975) Structure of tight junctions in epithelia with different permeability. Proc Natl Acad Sci USA 72:4487–4491

Matsumoto T, Fontaine D, Rasmussen H (1981) Effect of 1,25-dihydroxyvitamin D₃ on phospholipid metabolism in chick duodenal mucosal cell: relationship to its mechanism of action. J Biol Chem 256:3354–3360

Miller A III, Bronner F (1981) Calcium uptake in isolated brush-border vesicles from rat small intestine. Biochem J 196:391–401

Miller A III, Ueng T-H, Bronner F (1979) Isolation of a vitamin D-dependent, calcium-binding protein from brush borders of rat duodenal mucosa. FEBS Lett 103:319–322

Miller A III, Li ST, Bronner F (1982) Characterization of calcium binding to brush border membranes from rat duodenum. Biochem J 208:773–782

Miller D, Conway T, Grawe T, Schedl H (1984) Intestinal Ca transport in the spontaneously hypertensive (SH) rat: response to Ca depletion. Fed Proc 43:727

Morris IG (1978) The receptor hypothesis of protein ingestion. In: Hemmings WF (ed) Antigen absorption by the gut. MTP Press, Lancaster, pp 3–22

Nellans HN, Kimberg DV (1979a) Anomalous secretion of calcium in rat ileum: role of the paracellular pathway. Am J Physiol 236:E473–E481

Nellans HN, Kimberg DV (1979b) Intestinal calcium transport absorption, secretion and vitamin D. In: Crane RK (ed) International review of physiology. Gastrointestinal physiology III. University Park Press, Baltimore, pp 227–261

Nellans HN, Popovitch JR (1981) Calmodulin-regulated, ATP-driven calcium transport by basolateral membranes of rat small intestine. J Biol Chem 256:9932–9936

Nellans HN, Popovitch JR (1984) Role of sodium in intestinal calcium transport. In: Bronner F, Peterlik M (eds) Epithelial calcium and phosphate transport: molecular and cellular aspects. Alan R Liss, New York, pp 301–306

O'Doherty PJA (1979) 1,25-Dihydroxyvitamin D₃ increases the activity of the intestinal phosphatidylcholine deacylation–reacylation cycle. Lipids 14:75–77

Pansu D, Chapuy MC, Milani M, Bellaton C (1976) Transepithelial calcium transport enhanced by xylose and glucose in the rat jejunal ligated loop. Calcif Tissue Res [Suppl] 21:45–52

Pansu D, Bellaton C, Bronner F (1979) Effect of lactose on duodenal calcium-binding protein and calcium absorption. J Nutr 109:509–512

Pansu D, Bellaton C, Bronner F (1981) The effect of calcium intake on the saturable and non-saturable components of duodenal calcium transport. Am J Physiol 240:G32–37

Pansu D, Bellaton C, Bronner F (1983a) Developmental changes in the mechanisms of duodenal calcium transport in the rat. Am J Physiol 244:G20–G26

Pansu D, Bellaton C, Roche C, Bronner F (1983b) Duodenal and ileal calcium absorption in the rat and effects of vitamin D. Am J Physiol 244:G696–700

Pansu D, Bellaton C, Roche C, Bronner F (1987) Theophylline inhibits active Ca transport in rat intestine by inhibiting Ca binding by CaBP. In: Bronner F, Peterlik M (eds) Cellular calcium and phosphate transport in health and disease. Alan R Liss, New York [in press]

Pitelka DR, Hamamoto ST, Duafala JG, Nemanic MK (1973) Cell contacts in the mouse mammary gland. 1. Normal gland in postnatal development and the secretory cycle. J Cell Biol 56:797–818

Putkey JA, Norman WA (1983) Vitamin D. Its effect on the protein composition and core material structure of the chick intestinal brush-border membrane. J Biol Chem 258:8971–8978

Ramberg CF, Phang JM, Kronfeld DS (1970) A compartmental model of calcium metabolism in cows. In: Anderson JJB (ed) Parturient hypocalcemia. Academic Press, New York, pp 119–134

Rasmussen H, Fontaine O, Max EE, Goodman DP (1979) The effect of 1(alpha)-hydroxyvitamin D_3 administration on calcium transport in chick intestine brush-border membrane vesicles. J Biol Chem 254:2993–2999

Rasmussen H, Matsumoto J, Fontaine D, Goodman DBP (1982) The role of changes in membrane lipid structure in the action of $1,25(OH)_2D_3$. Fed Proc 41:72–77

Recker RR (1985) Calcium absorption and achlorhydria. N Engl J Med 313:70–73

Roche C (1982) Effect of the calcium antagonists, verapamil, diltiazem and nifedipine, on calcium absorption in the rat. DHE Thesis Physiology, University Claude Bernard, Lyon, France

Roche C, Bellaton C, Pansu D, Bronner F (1985) Vasoactive intestinal peptide (VIP) decreases active duodenal Ca transport directly. In: Abstracts 7th annual scientific meeting American Society for Bone and Mineral Research, Washington DC, 1985

Roche C, Bellaton C, Pansu D, Bronner F (1986a) Inhibition by theophylline of CaBP-mediated active calcium transport in rat intestine. In: Abstracts 7th international workshop on calcified tissues, Ein Gedi, Israel, 1986, p 69

Roche C, Bellaton C, Pansu D, Miller A III, Bronner F (1986b) Localization of vitamin D-dependent active Ca^{2+} transport and relationship to CaBP. Am J Physiol 28:G314–G320

Sammon PF, Stacey R, Bronner F (1970) Role of parathyroid hormone in calcium homeostasis and metabolism. Am J Physiol 218:479–485

Saville PD, Nassim R, Stevenson FH, Mulligan L, Carey M (1955) The Fanconi Syndrome – metabolic studies on treatment. J Bone Joint Surg [Br] 37:529–539

Schachter D, Kowarski S (1982) Isolation of the protein 1M Cal, a vitamin D-dependent membrane component of the intestinal transport mechanism for calcium. Fed Proc 41:84–87

Schachter D, Rosen SM (1959) Active transport of ^{45}Ca by the small intestine and its dependence on vitamin D. Am J Physiol 196:357–362

Schachter D, Dowdle EB, Schenker H (1960) Active transport of calcium by the small intestine of the rat. Am J Physiol 198:263–268

Schatzmann HJ (1975) Active calcium transport and Ca^{2+} activated ATPase in human red cells. In: Bronner F, Kleinzeller A (eds) Current topics in membranes and transport, vol 6. Academic Press, New York, pp 126–168

Schatzmann HJ (1982) The plasma membrane calcium pump of erythrocytes and other animal cells. In: Carafoli (ed) Membrane transport of calcium. Academic Press, New York, pp 41–108

Schedl HP, Miller DL, Iskandarani M (1981) Small intestinal site, calcium transport kinetics, and calcium homeostasis in the hamster. In: Bronner F, Peterlik M (eds) Calcium and phosphate transport across biomembranes. Academic Press, New York, pp 151–154

Schiffl H, Binswanger U (1980) Calcium ATPase and intestinal calcium transport in uremic rats. Am J Physiol 238:G424–G428

Singh RP, Bronner F (1982) Duodenal calcium binding protein: induction by 1,25-dihydroxyvitamin D_3 in vivo and in vitro. Indian J Exp Biol 20:107–111

Spencer R, Chapman M, Wilson PW, Lawson DEM (1978) The relationship between vitamin D-stimulated calcium transport and intestinal calcium-binding protein in the chicken. Biochem J 170:93–101

Stein WD (1986) Transport and diffusion across cell membranes. Academic Press, New York

Thomasset M, Cuisinier-Gleizes P, Mathieu H (1979) 1,25-Dihydroxycholecalciferol: dynamics of the stimulation of duodenal calcium-binding protein, calcium transport and bone calcium mobilization in vitamin D- and calcium-deficient rats. FEBS Lett 107:91–94

Thomasset M, Molla A, Parkes CO, Demaille J (1981) Intestinal calmodulin and calcium-binding protein differ in their distribution and in the effect of vitamin D steroids on their concentration. FEBS Lett 127:13–16

Thomasset M, Parkes C-O, Cuisinier-Gleizes P (1982) Rat calcium-binding proteins: distribution, development and vitamin D-dependence. Am J Physiol 243:E483–E488

Tice LW, Wollman SH, Carter RC (1975) Changes in tight junctions of thyroid epithelium with changes in thyroid activity. J Cell Biol 66:657–663

Ueng T-H, Golub EE, Bronner F (1979) The effect of age and 1,25-dihydroxyvitamin D_3 treatment on the intestinal calcium-binding protein of suckling rats. Arch Biochem Biophys 196:624–630

Ussing HH, Zerahn K (1951) Active transport of sodium as the source of electric current in the short-circuited isolated frog skin. Acta Physiol Scand 23:110–127

Wasserman RH, Taylor AN (1969) Some aspects of the intestinal absorption of calcium with special reference to vitamin D. In: Comar CL, Bronner F (eds) Mineral metabolism – an advanced treatise, vol 3. Academic Press, New York, pp 321–403

Wasserman RH, Comar CL, Nold MM (1956) The influence of amino acids and other organic compounds on the gastrointestinal absorption of calcium[45] and strontium[89] in the rat. J Nutr 59:371–383

Weiser MM, Bloor JH, Dasmahapatra A, Freedman RA, MacLaughlin JA (1981) Vitamin D-dependent rat intestinal Ca^{2+} transport. Ca^{2+} uptake by Golgi membranes and early nuclear events. In: Bronner F, Peterlik M (eds) Calcium and phosphate transport across biomembranes. Academic Press, New York, pp 264–273

Wilson TH, Wiseman G (1954) The use of sacs of everted small intestine for the study of the transference of substances from the mucosal to the serosal surface. J Physiol (Lond) 123:116–125

Zornitzer AE, Bronner F (1971) In situ studies of calcium absorption in rats. Am J Physiol 220:1261–1265

Chapter 6

Renal Excretion of Calcium

M. Peacock

Introduction

Calcium, like sodium, potassium and magnesium, the other biologically important cations in the group I and IIa elements of the periodic table, is not catabolized by the human body and its total content in tissues is set by the balance between that absorbed and that excreted. In the form of complexes calcium is, in general, well tolerated by tissues but tolerance to the ionized form is limited. Because ionized calcium is utilized in a wide range of cellular mechanisms, efficient excretory systems have evolved to avoid, on the one hand, the development of toxic levels and on the other, low levels injurious to normal function.

Calcium Excretion

Removal of ionized calcium from ECF occurs by two disparate mechanisms (Fig. 6.1). The first involves sequestration of calcium into a non-ionic, metabolically inactive form. In the skeleton, calcium is complexed with phosphate. This complex both forms the major calcium reservoir and acts as the mechanical support for osteoid. In plasma, about half of the calcium is protein bound and has no apparent biological function apart from buffering changes in the concentration of ionized calcium in the ECF. In cells, complexed calcium, located in several organelles, allows the ion to play its vital role as a transmembrane and intracellular messenger (Rasmussen and Barrett 1984; Berridge 1985).

The second mechanism removes ionized calcium completely from the body by two main pathways (Fig. 6.1). The minor route is via the gastrointestinal tract and skin. Calcium is carried passively in gut and skin secretions at a

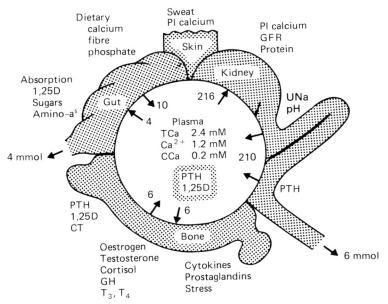

Fig. 6.1.   Schema of the organs (skin, gut, bone, kidney) responsible for calcium excretion from the ECF in the non-pregnant, non-lactating healthy adult. The amounts of calcium transported are in mmol/24 h; the main modulators for each organ and the calcium fractions in plasma and their endocrine regulators are shown. (*Pl*, plasma; *PTH*, parathyroid hormone; 1,25D, calcitriol; *CT*, calcitonin; *TCa*, total calcium; *Ca²⁺*, ionized calcium; *CCa*, complexed calcium; *UNa*, urinary sodium; *GFR*, glomerular filtration rate.)

concentration set by the level of calcium in the ECF. The major excretory route is via the kidney although during pregnancy and lactation substantial amounts of calcium are lost to the foetus and to milk.

Modulation of Calcium Excretion

The mechanisms for removal of calcium from the ECF are modulated by a wide variety of factors (Fig. 6.1). In addition, ECF ionized calcium concentration is regulated by a sophisticated homeostatic system. The concentration of ionized calcium in the ECF controls the secretion of parathyroid hormone and calcitriol through negative "feedback" and a rise in their concentration stimulates calcium transport to restore the ionized calcium level (Peacock 1980; Habener et al. 1984). Calcitriol increases resorption of bone and absorption by gut whereas parathyroid hormone increases bone turnover and reabsorption by the kidney. Equally importantly, parathyroid hormone also stimulates the secretion of calcitriol from the kidney, the major organ for calcitriol production (Haussler and McCain 1977). The kidney, therefore, is not only the key organ in calcium excretion and conservation but also occupies a central position in the endocrine regulation of calcium homeostasis. The kidney's double role in calcium metabolism thus ensures that the excretion and conservation of calcium is firmly linked via endocrine regulation to calcium requirements, and consequently kidney disease frequently presents as a disturbance in calcium metabolism.

Form and Measurement of Calcium in Urine

In health there is marked variation in both calcium and water excretion throughout the 24 hours and free water clearance has no effect on calcium excretion (Thorn 1960). Consequently, the range of calcium concentrations considered normal is wide. About 50% of the urinary calcium is normally present in the ionized form (Fig. 6.2) and the remainder is complexed mainly with citrate, sulphate, phosphate and oxalate (Robertson 1969).

Urine Calcium Measurement

Methods based on atomic absorption spectrophotometry and colorimetry are available which give precise and accurate measurements of calcium in biological fluids including urine (Robertson and Marshall 1979). It is important before analysis that all calcium which has precipitated in vivo or after voiding is redissolved, and to ensure this the urine should be collected in a small quantity of acid. Ionized calcium is difficult to measure accurately, either with ion-selective electrodes or indirectly by tetramethyl murexide dye, since the

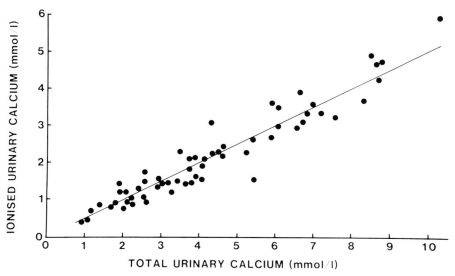

Fig. 6.2. The relationship between measured ionized calcium and total calcium concentration in 24-hour urine from 59 normal subjects ($r = 0.95$). (With permission of Robertson 1969)

concentration range of interfering urinary ions such as sodium and hydrogen is so wide that the measurement lacks precision (Robertson 1969).

Urine Saturation with Calcium Salts

Some calcium complexes, such as citrate, are highly soluble in urine whereas others, such as calcium oxalate and calcium phosphate, are relatively insoluble and may form crystals in the urine of healthy subjects (Robertson et al. 1969). In certain circumstances the crystals grow and aggregate and may eventually form urinary stones. The formation and growth of calcium oxalate crystals and calcium phosphate crystals are determined in part by the saturation of urine with the corresponding salts and this is defined by their activity product (Robertson et al. 1968). The activity product of calcium oxalate (K_{CaOx}), for example, is:

$$K_{CaOx} = [Ca^{2+}] \times f_{Ca} \times [Ox^{2-}] \times f_{Ox}$$

where f is the activity coefficient and [] represents the concentration of the enclosed ion. The concentration is the free ionized concentration as distinct from total concentration. To avoid the difficulty of measuring the free concentration of the ions in urine, the activity product of the calcium salt may be calculated from a knowledge of the total concentrations of calcium and oxalate and the concentrations of all their corresponding complexes (Robertson et al. 1968; Robertson 1969).

The saturation of urine by a calcium salt occurs in any one of three zones (Fig. 6.3): an undersaturated region in which crystals of the salt dissolve, a metastable region in which the salt crystals grow and a labile region in which crystals spontaneously precipitate. The solubility product of the salt separates the undersaturated from the metastable region and the formation product separates the metastable from the labile region. In healthy subjects the urine saturation level with calcium phosphate tends to be in the undersaturated zone (Robertson et al. 1968) but may lie in any of these three regions mainly because of the wide range in urinary pH and the affect this has on the activity product (Fig. 6.3). On the other hand, the level of calcium oxalate saturation, which is much less dependent on pH, has a narrower range and lies in the metastable or, less often, in the labile region (Robertson et al. 1968) (Fig. 6.3).

However, the propensity of urine to form crystals of a calcium salt and for these to grow, aggregate and form stones is dependent not only on the saturation levels of the salt but also on the activity of urinary inhibitors (Robertson et al. 1976) and promotors (Hallson and Rose 1979) of crystallization. A measure of the capacity of urine to form calcium stones which takes into account the saturation levels and the activity of the inhibitors has been devised (Robertson et al. 1978). Using such a measure on 24-hour urine collections it has been shown that children are less likely to form stones than adults and women are less likely to form stones than men (Robertson et al. 1980). In patients with recurrent primary stone formation the value relates to the recurrence rate and predicts the outcome of treatments aimed at preventing stone recurrence (Peacock and Robertson 1988). The amount of calcium in urine also relates to variations in stone incidence but its predictive value is poor (Peacock and Robertson 1988) and although the presence of calcium in urine is essential for calcium stone

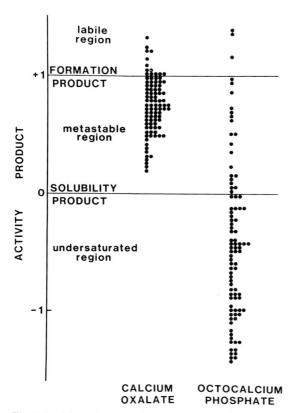

Fig. 6.3. The activity products of calcium oxalate and octocalcium phosphate in all 2-hourly urines collected during 24 hours in nine normal men. The formation product and solubility product have been set at one and zero respectively and the three zones of urine saturation are indicated in relation to these two products.

formation it is never the only, and rarely the main, risk factor (Robertson et al. 1978; Robertson and Peacock 1980; Peacock 1982).

Expression of Urinary Calcium

Urinary calcium can be expressed in a number of ways. The concentration of calcium in the urine is the relevant expression when saturation and calcium stone formation are being considered, the loss of calcium in urine in absolute quantities in relation to dietary intake is the most relevant for nutritional purposes; and to quantify its role in calcium homeostasis it is necessary to express the urinary calcium excretion in relation to plasma calcium concentration and renal function.

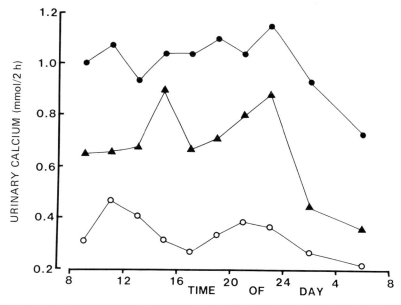

Fig. 6.4. Mean urinary calcium excretion in all 2-hourly urines collected during 24 hours in ten normal subjects (○), ten patients with primary calcium stone disease (▲) and 27 patients with stone disease due to primary hyperparathyroidism (●). Meals were taken at 9, 13 and 18 h.

Table 6.1. Variability of urinary calcium

	n	Mean	SD of differences between pairs	% Error
Fasting urinary calcium/ creatinine (molar)[a]				
Normal men	19	0.169	0.099	60.9
Postmenopausal women	40	0.341	0.183	56.0
Calcium stone formers	48	0.260	0.142	73.1
Primary hyperparathyroidism	37	0.506	0.454	50.8
Renal TmCa (mmol/l glomerular filtrate[a]*)*				
Normal men	19	2.111	0.165	7.8
Postmenopausal women	40	1.926	0.154	8.0
Calcium stone formers	48	1.961	0.136	6.9
Primary hyperparathyroidism	37	2.254	0.277	12.3
24-hour urinary calcium (mmol)[b]				
Women	16	4.812	0.207	4.3
24-hour urinary creatinine (mmol)[b]				
Women	16	12.828	1.008	7.86
24-hour urinary volume (l)[b]				
Women	16	1.910	0.342	17.91

[a] Outpatients.
[b] Inpatients on balance.

Twenty-Four-Hour Excretion

The rate of calcium excretion normally varies throughout the day and night. It is highest during the hours of activity and falls during sleep reaching its nadir before breakfast (Fig. 6.4). These variations are largely due to the ingestion of food, particularly to its calcium content, since there is no substantial intrinsic cycle to calcium excretion (Heaton and Hodgkinson 1963; Hodgkinson and Heaton 1965). A 24-hour urinary collection from breakfast one day to before breakfast the following day integrates these variations in excretion and is a useful and widely used expression (Nordin et al. 1967). In subjects on a fixed dietary intake with the collections made on a metabolic ward, the urinary calcium excretion in 24 hours is remarkably constant, having a coefficient of variation of 4.3% (Table 6.1) (Peacock and Marshall 1986). However, the less strict the dietary control and the supervision of collections, the greater is the day-to-day variation.

It might be thought that a 24-hour urine collection is inappropriate for studies in stone formation since it obscures the peaks in the saturation levels triggering the crystalluria which may be the first step in stone formation (Robertson et al. 1969). Although normally there are wide swings in urine saturation over a 24-hour period (Fig. 6.3) there is no single collection or series of collections which appears to discriminate between stone formers and controls better than the 24-hour collection itself (Hodgkinson et al. 1971; Peacock et al. 1976).

Postprandial Excretion

Following an oral load of a calcium salt (Peacock et al. 1968) or a meal containing calcium (Heaton and Hodgkinson 1963), in healthy subjects the urinary calcium excretion increases over the following 3–4 hours, thereafter decreasing exponentially over a period of 6–12 hours depending on the amount of calcium absorbed (Fig. 6.5). This response is due mainly to a rise in plasma calcium concentration (Fig. 6.5). The greater the calcium load or its absorption, the greater the rise in plasma calcium (Fig. 6.5), and the calcium excreted with time in response to a standard calcium load can be used as a test of absorption providing that skeletal uptake is normal (Peacock et al. 1968; Broadus et al. 1978). Throughout the absorptive and postabsorptive phase, calcium homeostasis is perturbed and plasma calcium concentration, its homeostatic regulators and the urinary calcium excretion are continuously changing.

Fasting Excretion

In contrast to postprandial excretion, the urinary calcium after an overnight fast of at least 12 hours represents calcium removed from the body stores to maintain plasma calcium concentration (Nordin et al. 1972a) and is assumed to be the calcium released by net bone resorption (Gallagher et al. 1972). During this phase, plasma calcium concentration, its homeostatic regulators and urinary calcium excretion are relatively stable and are generally assumed to be basal. However, the length of fast required to return calcium homeostasis to this basal state is set by the efficiency of absorption and the clearance by the kidney. In

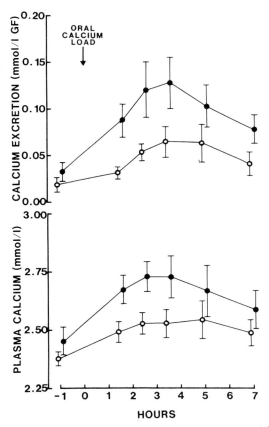

Fig. 6.5. The mean (± 2 SE) change in plasma calcium concentration and calcium excretion in relation to time after an oral load of calcium in nine normal subjects (○) and nine patients with absorptive hypercalciuria and stone disease (●)

hyperabsorptive states with normal renal function or normal absorptive states with impaired glomerular filtration, a 12-hour fast is insufficient to clear the absorbed calcium and a longer period of fasting is necessary (Peacock et al. 1968) (Figs. 6.4 and 6.5). On the other hand, prolonged fasting, as might be undertaken in the treatment of obesity, is associated with metabolic changes which eventually result in hypercalciuria. The coefficient of variation on this estimation is high unless there is strict control of the timing of the last meal and of the urinary collection (Table 6.1).

Calcium Clearance

It is axiomatic that a clearance is measured over a period in which the filtered load is either constant or changing regularly with time and the urine collection truly represents the urine formed during the clearance period. In clinical practice calcium clearance measured from periods of less than an hour are imprecise except in subjects who are water loaded or catheterized.

An accurate clearance can be measured using the urine collected from 8 a.m. to 10 a.m. and the blood taken at 9 a.m., providing that the absorbed calcium from the previous meal has been completely cleared from the ECF. A clearance calculated from a 24-hour urine collection is meaningless and from postprandial urine inaccurate because of changes in plasma calcium during these times (Figs. 6.4 and 6.5). Glomerular filtration rate may be measured accurately but inconveniently by inulin infusion. If endogenous creatinine clearance is calculated from the same samples used for calcium clearance, the procedure is simplified and does not lose much in accuracy. Calcium clearance (Cca) may then be expressed as a fraction of the clearance of endogenous creatinine (Ccr), which simplifies to

$$\frac{\text{Urine calcium} \times \text{plasma creatinine}}{\text{Urine creatinine} \times \text{plasma calcium}}$$

and obviates the need to time the urine collections.

Body Weight Correction

However urinary calcium is expressed, the range of values obtained is wide. Some of this variation is due to body size, which determines not only dietary intake and absorption but also the rate of glomerular filtration. Correcting calcium excretion for body weight, or body weight × height, allows small and large individuals to be compared (Nordin et al. 1967; Bulusu et al. 1970) (Fig. 6.6). For nutritional purposes it is conventional to express calcium excretion corrected for body weight (Knapp 1947). The justification for doing so is not clear and in terms of the calcium stores, the absolute excretion is of more relevance.

Creatinine Correction

Probably a more useful correction than that for body weight is to express urinary calcium as a ratio to the creatinine excreted in the same urine (Nordin et al. 1967; Bulusu et al. 1970) (Fig. 6.6). This has two main advantages. It not only corrects for lean body mass, but since creatinine is excreted at a relatively constant rate it corrects for errors in the timing of urine collections, which may be substantial in non-catheterized subjects. It should be noted however that it is subject to variation in the dietary intake of flesh protein and to variation from rapid rates of loss of muscle mass.

Variation in Daily Calcium Excretion

Urinary calcium excretion varies in an individual on a free diet from day to day and also among individuals and populations. The variability reflects not only the

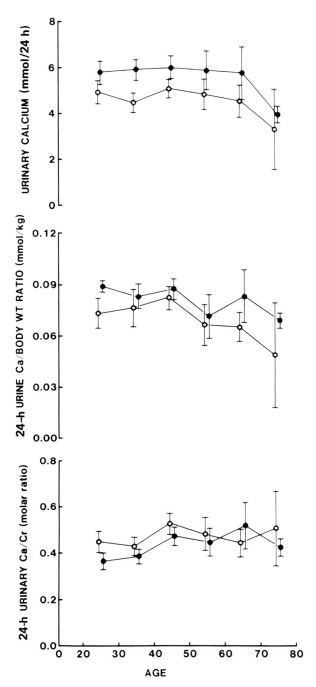

Fig. 6.6.   Mean (± 1 SE) 24-hour urinary calcium in 104 males (●) and 142 females (○) in relation to age: *top panel* in mmol; *middle panel*, as a ratio to body weight; *bottom panel*, as a ratio to creatinine excretion in the same urine.

wide range in the calcium content and composition of food but also the wide range in calcium requirement.

Age

Urine calcium excretion is lower in children than adults (Fig. 6.7). Although the calcium intake of a child may be as high as that of an adult and the absorption more efficient, the major fraction of the absorbed calcium is retained in bone and not excreted in urine. Calcium retention varies with the rate of skeletal growth and increases from about 1.5 mmol/day at age 3 to about 10 mmol/day at puberty; subsequently it decreases to less than 1 mmol/day (Leitch and Aitken 1959). Throughout childhood there is a slight increase in calcium excretion with age which becomes more marked after puberty, and by the age of 20 the mean urinary calcium level has reached the adult range (Fig. 6.7). The calcium to creatinine ratio, on the other hand, is not age-related and in childhood is similar to that of the adult (Figs. 6.6 and 6.7). From ages 20 to 65 urinary calcium

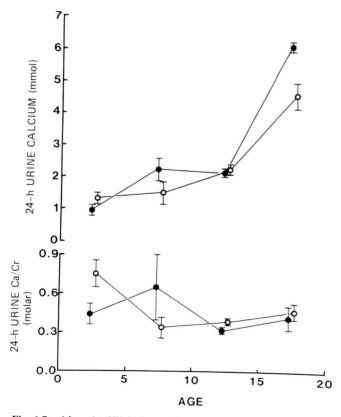

Fig. 6.7. Mean (± SE) 24-hour urinary calcium in 155 boys (●) and 122 girls (○) in relation to age: *top panel*, in mmol; *bottom panel*, as a ratio to creatinine excretion in the same urine.

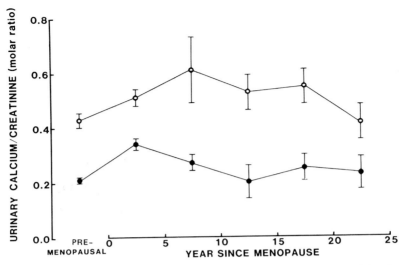

Fig. 6.8. The mean (± 1 SE) 24-hour (○) and fasting (●) urinary calcium/creatinine ratio in relation to age since menopause in 91 women. The values in 212 premenopausal women are shown.

excretion is relatively constant in men. In women there is a rise in both 24-hour and fasting urinary calcium after the menopause (Fig. 6.8). After the age of 65 calcium excretion falls progressively with age. Although a decrease in dietary intake may play a role in this fall, the major causes are a fall in glomerular filtration rate and a decrease in calcium absorption which in turn is caused by both vitamin D deficiency and insufficiency from decreased renal function (Francis et al. 1983a; Peacock et al. 1983; Francis et al. 1984).

Sex

Before puberty there is little difference in the calcium excretion of girls as compared to boys (Fig. 6.7). Thereafter women have, on average, lower urinary calcium than men throughout life (Fig. 6.6). The lower urinary calcium reflects a lower dietary calcium intake, and although the glomerular filtration rate is lower in women than men there is no evidence to suggest that either their absorption of calcium is less efficient or their setting of plasma calcium to the basal concentration of parathyroid hormone and calcitriol is different. This, however, may not be the case after the menopause since it appears that oestrogen regulates both the rate of bone resorption and the secretion threshold for parathyroid hormone and calcitriol (Selby et al. 1985; Selby and Peacock 1986). The postmenopausal rise in urinary calcium due to a rise in bone resorption reverses with oestrogen treatment without altering plasma parathyroid hormone and calcitriol concentrations. (During pregnancy and lactation urinary calcium does not decrease (Kerr et al. 1962), which in view of the increased calcium requirement and the increased plasma oestrogen levels over these periods appears paradoxical, but this may be because there is both an increase in glomerular filtration rate (Lindheimer and Katz 1981) and an increase in calcium absorption due to a rise in plasma bound and free calcitriol levels (Bouillon et al. 1977).)

Although oestrogen deficiency in postmenopausal women causes a rise in urinary calcium, the fivefold changes in plasma oestrogen levels during the menstrual cycle are not associated with changes in calcium excretion (Tjellsen et al. 1983). The plasma oestrogen level affecting urinary calcium appears therefore to be lower than the levels occurring at menstruation. On the other hand, treatment with oral contraceptives in premenopausal women (Goulding and McChesney 1977) and with replacement oestrogen in postmenopausal women (Nordin et al. 1982) reduces calcium excretion to low levels, probably indicating a pharmacological inhibition of bone resorption.

Diet

Urinary calcium excretion is positively related to both the dietary and absorbed calcium (Knapp 1947; Peacock et al. 1967; Marshall et al. 1976; Lemann et al. 1979) (Fig. 6.9). The relationship is steeper in subjects with good absorption than in those with normal absorption (Peacock et al. 1967) and since some healthy subjects have calcium "hyperabsorption", the distribution of urinary calcium at any calcium intake is skewed towards the higher values (Knapp 1947) (Fig. 6.9).

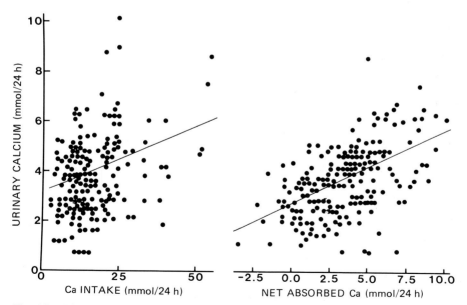

Fig. 6.9. The relationship between 24-hour urinary calcium and dietary calcium intake (*left panel*) and net absorbed calcium (*right panel*) in 212 calcium balances on 84 normal subjects taken from the published literature. (With permission from Marshall et al. 1976).

In addition to the calcium content of food, dietary factors altering calcium availability, such as phosphate (Farquharson et al. 1931; Malm 1953; Lotz et al. 1964; O'Brien et al. 1967; Heyburn et al. 1982) and fibre (Shah et al. 1980; Rao et al. 1982) and affecting absorption, such as sugars (Wasserman and Taylor 1969; Condon et al. 1970) and amino acids (Wasserman and Taylor 1969), influence the relationship between urinary calcium and dietary calcium intake. Dietary sodium (King et al. 1964; Kleeman et al. 1964; Edwards and Hodgkinson 1965a; Modlin 1966; Phillips and Cooke 1967) and protein (Sherman 1920; McCance et al. 1942; Anand and Linkswiler 1974; Linkswiler et al. 1974; Margen et al. 1974; Robertson et al. 1979b) increase urinary calcium but the effect is probably mediated in the kidney rather than the gut.

Season/Climate

In northern latitudes urinary calcium increases during summer and decreases in winter (Robertson et al. 1974; Robertson et al. 1975) (Fig. 6.10). These changes may depend on seasonal changes in diet. However, they may also reflect changes in absorption since they coincide with changes in plasma calcidiol which are due to seasonal fluctuations in sunlight exposure (Stamp and Round 1974; Peacock 1984). Although calcidiol does not directly effect absorption over the range induced by sunlight exposure such changes in calcidiol concentration cause small but significant rises in calcitriol concentration which in turn increase calcium absorption and urinary calcium (Peacock et al. 1982). The same mechanism is probably involved in the increase in urinary calcium excretion occurring in healthy subjects moving from places of low sunlight exposure to ones of high sunlight exposure (Parry and Lister 1975).

Geography

Urinary calcium excretion varies widely with geographic location and ranges from about 2 mmol/24 h in the Eskimo and Bantu to over 6 mmol/24 h in many industrialized societies (Nordin et al. 1967; Robertson 1985). Some of this variation may be due to racial differences in calcium metabolism but most is likely to be due to variations in the composition and calcium content of the diet. Although a low calcium intake may be responsible for the low urinary calcium of non-industrialized societies, a high calcium intake is not the sole cause of the high urine calcium in the industrialized societies, and dietary protein and fibre appear to be equally important (Robertson 1985). The changes in calcium excretion in populations with time are also probably due more to changes in protein and fibre intake than to changes in calcium intake (Robertson et al. 1979a).

Renal Excretion of Calcium

The calcium excreted in urine is that fraction of the filtered calcium not reabsorbed by the kidney since there is no evidence that calcium is secreted by

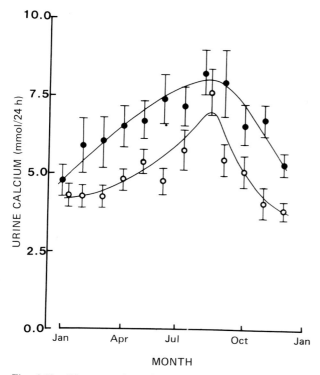

Fig. 6.10. The mean (± SE) 24-hour urinary calcium in 160 healthy men (●) and 250 healthy women (○) in relation to the month of the year of the urine collection.

the tubule. The relationship between the filtered load of calcium and its reabsorption is such that under basal conditions about 97% of the filtered load is reabsorbed. However, at higher filtered loads only about 50% of the increase in filtered calcium is reabsorbed (Kleeman et al. 1961; Peacock and Nordin 1968; Peacock et al. 1969; Mioni et al. 1971) (Fig. 6.11).

Filtered Load of Calcium

The amount of calcium reabsorbed is dependent on tubular function, whereas the filtered load is determined by both the plasma concentration of ultrafiltrable calcium and the rate of glomerular filtration.

Ultrafiltrable Calcium

About 44% of the total calcium in plasma is bound to protein, normal range 1.2–1.5 mM, and for all practical purposes does not pass the glomerular

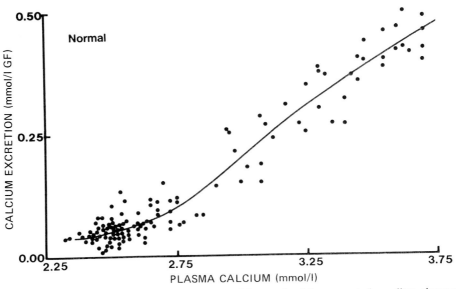

Fig. 6.11. The relationship between plasma calcium and calcium excreted per litre glomerular filtrate during intravenous infusions of calcium gluconate in nine normal subjects. (Filtered load of calcium per litre glomerular filtrate = plasma Ca mM/l × 0.56.)

membrane (Robertson and Peacock 1968; Pederson 1969; Marshall 1976). The remainder is completely filtered and comprises a complexed fraction, normal range 0.2–0.4 mM, composed of calcium phosphate, citrate and carbonate, and an ionized fraction with normal range 1.12–1.28 mM (Robertson and Peacock 1968; Pedersen 1970). All three fractions can be measured but measurements of total and ionized plasma calcium are the easiest to obtain in practice and on duplicate samples have a coefficient of variation of 3% and 4% respectively (Peacock and Marshall 1986). A number of formulae (Parfitt 1969b; Payne et al. 1973) have been proposed to correct for variations in plasma protein level, for example:

corrected plasma calcium (mM) = measured plasma calcium (mM) − 0.02 × [measured − mean normal albumin (g/l)] − 0.007 × [measured − mean normal globulin (g/l)]

For most clinical purposes the corrected total plasma calcium × 0.56 can be used as the measure of ultrafiltrable calcium per litre of glomerular filtrate.

Small changes in plasma phosphate, citrate and pH alter the complexed fraction only marginally (Marshall and Hodgkinson 1983) but pH has an effect on the binding affinity of the plasma proteins for calcium such that acidosis increases and alkalosis decreases the ionized fraction (Robertson and Peacock 1968; Pederson 1970; Marshall 1976). A circadian rhythm in ultrafiltrable plasma calcium levels follows the pattern of eating. After the menopause there is a rise in plasma ultrafiltrable calcium; conversely, during pregnancy, as the oestrogen levels increase, the plasma ionized calcium falls.

Glomerular Filtration Rate

Differences in glomerular filtration rate among individuals and between the sexes and changes in the rate with age are paralleled by differences in calcium excretion indicating that calcium intake and absorption and renal function are normally linked. It is generally held that glomerular filtration rate is maintained at a relatively constant level by autoregulation and that changes in glomerular filtration rate are balanced by compensatory changes in tubular reabsorption (Brenner et al. 1981). However, in some species, such as the dog, large changes in glomerular filtration rate occur in response to protein ingestion (Pitts 1944; O'Connor and Summerville 1976). Although the effect is less marked in humans it does occur and is accompanied by increased calcium excretion (Pullman et al. 1954; Margen et al. 1974; Robertson et al. 1979b; Arora et al. 1985).

During pregnancy there is an increase in glomerular filtration rate (Lindheimer and Katz 1981) and despite a fall in ionized plasma calcium level (Tan et al. 1972) and an increase in calcium requirement, the urinary calcium excretion tends to increase. Diseases such as insulin-dependent diabetes cause increased glomerular filtration (Dirzel and Schwartz 1967; Morgensen 1976) and decreased bone mass (Levin et al. 1976; Wiske et al. 1982) suggesting excessive urinary calcium loss. On the other hand, pathological impairment of filtration decreases proportionately the ability of the kidney to clear calcium from the ECF (Fig. 6.12) which combined with calcium malabsorption (Liu and Chu 1943; Stanbury and Lumb 1962; Cochran and Nordin 1971; Peacock et al. 1977) due to calcitriol deficiency (Haussler and McCain 1977; Peacock et al. 1980) results in low urinary calcium excretion (Hodgkinson and Pyrah 1958; Lichtwitz et al. 1960; Kaye and Silverman 1965; Cochran and Nordin 1971). It would appear therefore that glomerular filtration rate is normally more variable than generally believed and that changes in filtration may result in significant changes in urinary calcium excretion.

Tubular Reabsorption

Calcium reabsorption takes place along the length of the tubule with the proximal tubule reabsorbing the bulk of the calcium and the distal tubule regulating the amount finally excreted (Table 6.2). Numerous regulatory and non-regulatory factors influence the reabsorptive mechanism (Robertson 1976; Sutton and Dirks 1981; Massry 1982) (Table 6.3). At reduced levels of filtered load, calcium never completely disappears from the urine, at normal filtered loads over 95% is reabsorbed and at increased filtered loads a constant proportion of the increase continues to be reabsorbed (Fig. 6.11). Since the glomerular filtration rate varies within and among individuals, the range of the relationship between calcium reabsorbed and filtered is reduced by expressing it per litre of glomerular filtrate. If endogenous creatinine clearance (CCr) is used as a measure of glomerular filtration rate the relationship, filtered calcium (CaF) = calcium reabsorbed (CaR) + calcium excreted (CaE), can be established relatively simply from any simultaneously collected blood and urine samples: CaF = CaR + CaE in mmol/min becomes, CaF/CCr = CaR/CCr + CaE/CCr, where CaF/CCr is the ultrafiltrable calcium concentration in mM; CaE/CCr simplifies to UCa × PlCr/UCr where UCa is the urinary calcium, PlCr is the

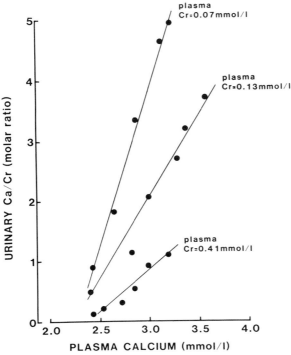

Fig. 6.12. The relationship between urinary calcium excretion and plasma calcium concentration (i.e. filtered load) in three subjects with different rates of glomerular filtration (PlCr 0.07, 0.13 and 0.41 mM).

plasma creatinine and UCr is the urinary creatinine, all in mM; and CaR/CCr is CaF/CCr − CaE/CCr. Tubular reabsorption can therefore be assessed clinically from four relatively simple measurements; the plasma ultrafiltrable calcium concentration (or 56% of the total plasma calcium concentration), the plasma creatinine concentration, and the urinary calcium and creatinine concentration.

Table 6.2. Calcium reabsorption in the renal tubule (Lassiter et al. 1963)

Tubule site	Tubular fluid calcium	Tubular fluid inulin	Calcium reabsorption (%)
	Glomerular fluid calcium	Glomerular fluid inulin	
Proximal early	1.0	1.0	0.0
Proximal late	1.1	3.0	63
Distal early	0.7	5.0	23
Distal late	0.6	20.0	11

Tubular Reabsorptive Mechanism in the Nephron

About 60% of calcium in the glomerular filtrate is reabsorbed in the proximal tubule (Lassiter et al. 1963; Frick et al. 1965; Duarte and Watson 1967;

Table 6.3. Factors affecting tubular reabsorption of calcium

Decreased reabsorption	Increased reabsorption
Increased calcium load	Decreased calcium load
Increased sodium load	Decreased sodium load
Increased ECF volume	Decreased ECF volume
Decreased parathyroid hormone	Increased parathyroid hormone
Diuretics except thiazide	Thiazide diuretic
Acidosis	Alkalosis
Cardiac glycosides	
Phosphate depletion	High phosphate intake
Increased magnesium load	Calcidiol
Vasoactive molecules: acetylcholine, catecholamines, bradykinin, angiotensin, ADH	
Calcitonin infusion	
Increased anion load: citrate, EDTA	
Insulin	
Glucose	
Glucagon	

Murayama et al. 1972; Agus et al. 1973; Jamison et al. 1974). Evidence suggests that 20% of this is transported actively and the remainder passively with reabsorbed sodium (Ullrich et al. 1976). Throughout the proximal tubule, reabsorption of calcium and sodium is similar, strongly suggesting that calcium and sodium share a common transport mechanism. Between the end of the proximal tubule and the ascending limb of the loop of Henle a further 10% of the filtered calcium is reabsorbed (DeRouffignac et al. 1973; Jamison et al. 1974). In the thick ascending limb 20% of the filtered calcium is reabsorbed and in the distal tubule through to the collecting duct there is active reabsorption of calcium which is independent of sodium and electropotential difference (Lassiter et al. 1963; Agus et al. 1973; Edwards et al. 1973).

The energy-dependent mechanisms responsible for calcium reabsorption are unknown. It has been suggested that sodium transport through the tubule cell is linked to the active expulsion of calcium at the plasma basolateral membrane (Taylor and Winghager 1979). In addition, the basolateral membrane contains a calcium-sensitive ATPase (Kinne et al. 1978) suggesting that at least two mechanisms are involved in the regulation of calcium transport from the tubule cell to the ECF. The role of the vitamin D-induced calcium-binding protein in calcium reabsorption is not clear (Kretsinger 1980; Wasserman and Fullmer 1982) but even if it is not directly involved in active transport it clearly plays an important role in maintaining the cytosolic calcium concentration in the face of large calcium fluxes through the cell.

Sodium

Tubular transport of calcium has many features in common with tubular transport of sodium: secretion does not occur, a constant proportion of the increase in filtered load above basal levels is reabsorbed, the proximal tubule is responsible for the bulk of the transport and reabsorption in the distal tubule is hormonally regulated.

The hormone regulating calcium reabsorption is parathyroid hormone whereas that regulating sodium is aldosterone; hypoparathyroidism and Addison's disease, primary hyperparathyroidism and Conn's syndrome and secondary hyperparathyroidism and secondary aldosteronism are analogues. Changes in sodium reabsorption are frequently followed by changes in calcium reabsorption. In health there is a positive relationship between urinary sodium and calcium excretion (Fig. 6.13) and increased sodium excretion following high salt intake (King et al. 1964; Kleeman et al. 1966; Phillips and Cooke 1967) or saline infusion (Chen and Neuman 1955; Walser 1961) is almost always accompanied by decreased calcium reabsorption, even to the extent of producing a negative calcium balance. Furthermore, the calciuric effect of sodium is sustained over prolonged periods (Brickman et al. 1971). Conversely, increased calcium excretion following intravenously infused calcium causes urinary loss of sodium (Massry et al. 1968; DiBona 1971) (Fig. 6.14).

In chronic hypercalcaemia the loss may be so substantial that it leads to sodium depletion and a fall in ECF volume and glomerular filtration rate. The subsequent decrease in sodium excretion is associated with a paradoxical increase in tubular reabsorption of calcium which results in the hypercalcaemia progressively increasing (Fig. 6.15a, b). This hypercalcaemic effect of sodium depletion is mediated in the proximal tubule, and the hypercalcaemia which may occur in hypocorticosteroidism probably has a similar pathogenesis. It is particularly common when distal tubular reabsorption of calcium is reduced as in vitamin D poisoning or malignant hypercalcaemia (Fig. 6.15a, b), whereas it is rare when distal tubular reabsorption of calcium is increased by parathyroid hormone as in primary hyperparathyroidism, except in those patients with

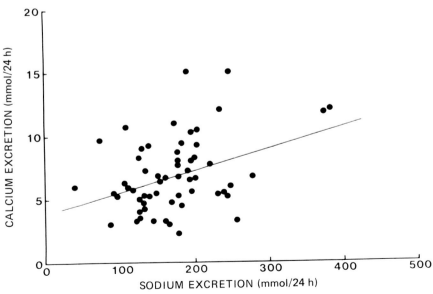

Fig. 6.13. The relationship between 24-hour urine calcium and sodium excretion in normal subjects on a free diet.

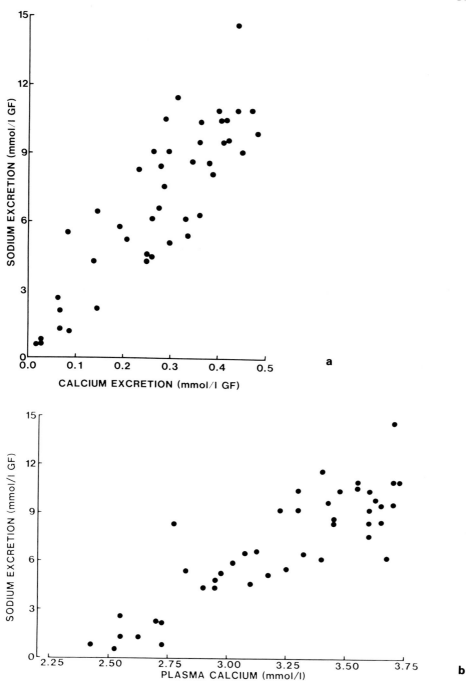

Fig. 6.14. The relationship between urinary sodium and calcium excretion (**a**) and urinary sodium excretion and plasma calcium (**b**) in eight normal subjects during an intravenous infusion of calcium gluconate.

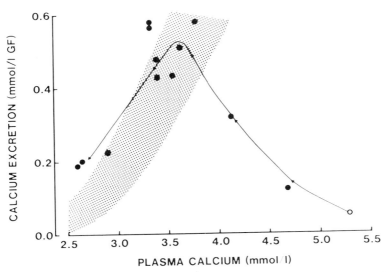

Fig. 6.15.a. The relationship between calcium excretion and plasma calcium (i.e. filtered calcium load) in a hypoparathyroid patient admitted with hypercalcaemia (○) due to vitamin D intoxication during treatment (●) with 4–6 day saline only. The *arrowed line* indicates the change in relationship with time during treatment and the *shaded area* the normal relationship (see Fig. 6.11).

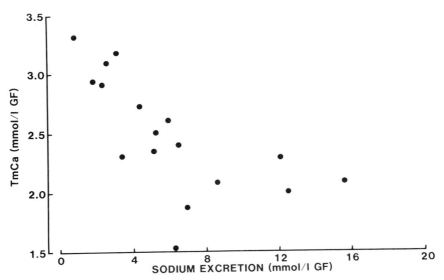

Fig. 6.15.b. The relationship between tubular reabsorption of calcium (TmCa – see p. 150) and urinary sodium excretion in the same patient as in Fig. 6.15.a treated with saline during which time the TmCa decreased from 3.4 to 1.5 mmol/1 GF.

plasma calcium concentrations over 4 mM. A plasma calcium of 3.5 mM causing a urinary calcium excretion of 0.4 mmol/l GF (Fig. 6.11) requires a sodium excretion of over 7 mmol/l GF (Fig. 6.14) to prevent an increased tubular reabsorption of calcium induced by sodium depletion. The effects of changes in sodium load and ECF volume on calcium reabsorption are difficult to separate but they result in similar changes (DiBona 1971).

The intimate link between sodium and calcium reabsorption is also illustrated by the action of diuretics. The osmotic (Wesson 1962) and mercurial (Parfitt 1969a) diuretics and frusemide and ethacrynic acid (Demartini et al. 1967; Duarte 1968; Antoniou et al. 1969; Eknoyan et al. 1970), acting on the proximal tubule, decrease both sodium and calcium reabsorption. On the other hand, thiazides acting on the distal tubule reduce sodium but increase calcium reabsorption (Lamberg and Kuhlback 1959; Higgins et al. 1964; Duarte and Bland 1965; Sotornik et al. 1969; Brickman et al. 1972), the latter effect occurring over prolonged periods (Yendt et al. 1966) and possibly even in the absence of parathyroid hormone (Quamme et al. 1975; Porter et al. 1978) although this is disputed (Brickman et al. 1972; Parfitt 1972). Aldosterone increases sodium reabsorption in the distal tubule but has no effect on calcium (Lemann et al. 1970). The evidence therefore strongly suggests that sodium and calcium reabsorption share a common mechanism in the proximal tubule but in the distal tubule, where reabsorption is under hormonal regulation, the two reabsorptive mechanisms are disparate.

Parathyroid Hormone

Parathyroid hormone has a marked effect on calcium reabsorption and is its main endocrine regulator (Kleeman et al. 1961; Widrow and Levinsky 1962; Peacock et al. 1969; Biddulph et al. 1970; Agus et al. 1971; Biddulph 1972) (Fig. 6.16). In its absence, tubular reabsorption basally is reduced and plasma calcium falls until the flow of calcium to the ECF from bone and gut equals the urinary loss. As plasma parathyroid hormone levels increase, tubular reabsorption in the basal state rises and calcium excretion relative to the filtered load is low since most is returned to the ECF to maintain or increase plasma calcium levels. There are receptors for parathyroid hormone in both proximal and distal tubules and parathyroid hormone-activated adenyl cyclase is present in all segments of the nephron in which calcium reabsorption increases with parathyroid hormone levels. Whereas cAMP increases both calcium and sodium reabsorption in the distal tubule, parathyroid hormone increases only calcium reabsorption (Costanzo and Windhager 1978; Burnatowska et al. 1977).

The effect of parathyroid hormone on calcium reabsorption, as assessed by clearance studies, appears to take several hours (Froeling and Bijvoet 1974). However, a decrease in tubular reabsorption is seen immediately following parathyroidectomy (Fig. 6.17) and the steeper slope of excretion with increasing filtered load in hyperparathyroid subjects as compared to hypoparathyroid patients (Fig. 6.16) is probably due to suppression of endogenous parathyroid hormone during the 5 hours of calcium infusion.

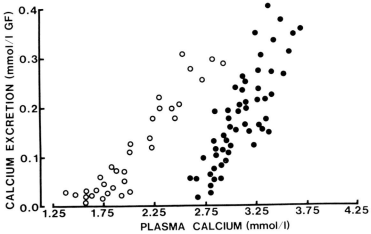

Fig. 6.16. The relationship between calcium excretion and plasma calcium (i.e. filtered calcium) in hypoparathyroid patients (○) and patients with mild primary hyperparathyroidism (●) during an infusion of calcium gluconate.

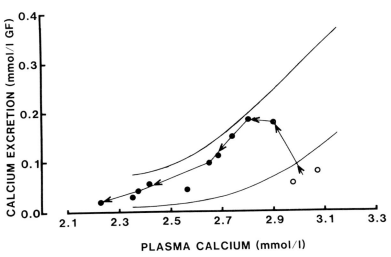

Fig. 6.17. The relationship between calcium excretion and plasma calcium (i.e. filtered calcium load) in a patient with primary hyperparathyroidism before (○) and immediately following parathyroidectomy (●). The *arrowed line* indicates the change in relationship with time and the solid lines the normal relationship (see Fig. 6.11).

Miscellaneous Factors

pH. Calcium excretion is increased by acidosis (Bogert and Kirkpatrick 1922; Greenberg et al. 1966; Lemann et al. 1967; Gray et al. 1973) and reduced by alkalosis (Parfitt et al. 1964; Edwards and Hodgkinson 1965b). A low plasma pH increases urinary calcium in at least three ways: the affinity of plasma proteins for calcium is lowered (Robertson and Peacock 1968; Marshall 1976) and net bone resorption is increased (Lemann et al. 1979), both of which increase the filtered calcium load, and tubular reabsorption of calcium is reduced (Sutton et al. 1979). The mechanisms by which acid acts on bone and kidney are unknown. It does not appear to work through either parathyroid hormone or calcitriol (Lemann et al. 1979), but its effects are reversed by alkalosis, as are its effects on the calcium affinity for plasma protein.

Magnesium. Hypermagnesaemia reduces calcium reabsorption (Walser 1969; Massry et al. 1970) probably by a direct action on the tubule and by inhibition of parathyroid hormone secretion (Massry et al. 1970). Severe hypomagnesaemia may result in hypocalcaemia with reduced calcium excretion (Selby et al. 1984a). An increase in calcium excretion increases magnesium excretion (Fig. 6.18).

Phosphate. Taken orally (Bernstein and Newton 1966) or infused intravenously (Herbert et al. 1966; Coburn et al. 1971), phosphate lowers urinary calcium. The effect is probably due to a decrease in both absorbed and filtered calcium resulting in stimulation of parathyroid hormone secretion (Sherwood et al. 1966). However, there is some evidence that phosphate also increases calcium transport in the distal tubule (Coburn and Massry 1970). Hypercalciuria occurs in phosphate depletion (Lotz et al. 1964). This is due to increased absorption of calcium because of its increased availability from digestion, to decreased parathyroid hormone secretion because of mild hypercalcaemia (Goldfarb et al. 1977) and possibly to increased calcitriol due to phosphate depletion (Haussler and McCain 1977). There may be also a tubular component to the hypercalciuria (Quamme et al. 1976).

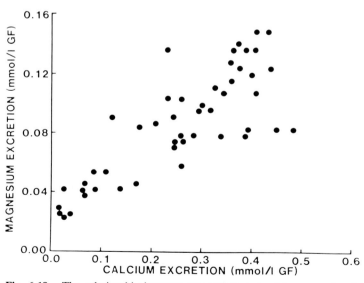

Fig. 6.18. The relationship between magnesium and calcium excretion in eight normal subjects during an intravenous calcium gluconate infusion.

Calcitonin. This causes a decrease in calcium reabsorption (Bijvoet et al. 1968; Ardaillou et al. 1969; Cochran et al. 1970; Paillard et al. 1972). The clinical relevance is however not clear, particularly since patients with medullary carcinoma of the thyroid have a normal excretion of calcium (Melvin and Tashjian 1968).

Calcitriol. Although calcidiol has been reported to increase calcium reabsorption (Sutton et al. 1975), an effect of calcitriol or other D metabolites on calcium reabsorption is not clinically obvious. The major effect of calcitriol on urinary calcium appears to be mediated through changes in filtered load.

Corticosteroids and Mineralocorticoids. Both the corticosteroids and the mineralocorticoids do little to urine calcium when given acutely (Lemann et al. 1970). Chronic administration may increase calcium excretion (Collins et al. 1962) but the reason for the calcium increase is not certain.

Growth Hormone. This hormone appears to increase urinary calcium (Haymovitz and Horwith 1964). The reason for the increase is not certain but its effect is likely to be on glomerular filtration and plasma calcium concentration.

Maximum Tubular Reabsorption (TmCa)

The intravenous infusion of calcium salts into healthy subjects shows that even at the highest plasma calcium concentration it is possible to achieve with safety a total plasma calcium concentration of 3.75 mM (2.1 mM ultrafiltrable calcium). About 50% of the increase in filtered load continues to be reabsorbed with no evidence that reabsorption is close to a maximum (Kleeman et al. 1961; Peacock and Nordin 1968; Mioni et al. 1971) (Fig. 6.11).

Calcium infusions in patients with hypoparathyroidism and primary hyperparathyroidism lead to a similar response but in addition show that the threshold above which calcium excretion rises linearly with filtered load is set by the prevailing concentration of plasma parathyroid hormone concentration (Peacock et al. 1969) (Fig. 6.16). These data are compatible with a reabsorptive mechanism which has two components, one of which is unsaturable and the other saturable. In anatomical terms the first component occurs in the proximal tubule, probably sharing a common mechanism with sodium, and the second occurs in the distal tubule where the level of calcium saturation is set by parathyroid hormone and the reabsorptive mechanism is independent of sodium. Such an arrangement is similar to the transport system for calcium in the gut which has a diffusion component and an active saturable transport component regulated by calcitriol (D. M. Marshall 1976; Wilkinson 1976).

On the other hand, the data have been used to support the concept that there is a maximum tubular reabsorptive capacity for calcium (TmCa) with a calculated normal value of 1.875 mmol/l GF and a K_m of 0.15 mmol/l GF, equivalent to a stability constant of 6×10^3/M (Marshall et al. 1972; D. M. Marshall 1976). Despite the limitations on this calculation caused by the narrow range of values of filtered and excreted calcium and the fact that infused calcium suppresses parathyroid hormone and causes sodium depletion during the period of study, the calculation of TmCa from simultaneously collected blood and urine is biochemically meaningful and clinically useful (Figs. 6.15b and 6.21). TmCa can be calculated from the expression:

$$TmCa \ (mmol/l \ GF) = \frac{UFCa - CaE}{1 - 0.08 \log_e (UFCa/CaE)}$$

where UFCa = ultrafiltrable plasma calcium (mol/l) and CaE = calcium excreted in mmol/l GF.

Renal Components of Hypercalcaemia

Hypercalcaemia may arise from increased absorption, resorption or reabsorption. Clinically, more than one mechanism is usually involved and in the kidney a change in the glomerular filtration rate in addition to an abnormality in reabsorption is common. The extent to which the kidney contributes to the hypercalcaemia in any individual case can be resolved as shown in Fig. 6.19. Given the plasma calcium and creatinine concentrations and the urinary calcium and creatinine excretion, the calcium excretion per litre glomerular filtrate is plotted against the plasma calcium, and the tubular component is measured as the distance on the plasma calcium axis of this relationship from the normal relationship. The glomerular filtration and resorption components can then, in turn, be calculated as described in Fig. 6.19.

The same information can be obtained using the three variables TmCa, CCa/CCr and total plasma calcium from a nomogram (Selby et al. 1984) or using an algorithm best solved by computer.

Regulation of Plasma Calcium by the Kidney

The concentration of ionized calcium in plasma is controlled by the calcium-regulating hormones acting on calcium transport in the gut, bone and kidney. In health a perturbation of plasma ionized calcium concentration is adjusted by an integrated response in these three target organs induced by changes in the concentration of the calcium-regulating hormones. The rate of recovery is determined both by the amount and the rate of calcium transported into or out of ECF. However, the degree and rate of response in the three target organs are not quantitatively the same nor is the cause of the perturbation and the health of the target organs irrelevant to the response evoked in each of them.

Gut

At normal dietary intakes, about 6 mmol of calcium is absorbed, net, each day into the ECF (Fig. 6.1). This can be increased within 24 hours by increased

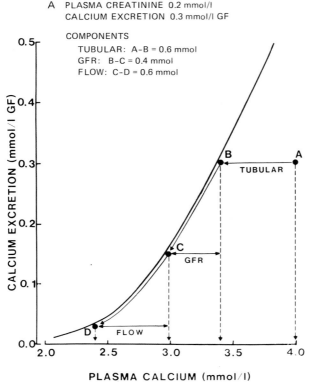

PLASMA CALCIUM   4.0 mmol/l
A   PLASMA CREATININE   0.2 mmol/l
CALCIUM EXCRETION   0.3 mmol/l GF

COMPONENTS
TUBULAR: A–B = 0.6 mmol
GFR: B–C = 0.4 mmol
FLOW: C–D = 0.6 mmol

PLASMA CALCIUM (mmol/l)

Fig. 6.19.   Schema of the steps for resolving the renal components in a case of hypercalcaemia using the relationship between calcium excretion and filtered calcium load in normal subjects (see Fig. 6.11). A = essential biochemistry needed. Step 1. Plot point A; Step 2. Move point A to point B on the normal slope A–B = tubular component (mmol); Step 3. Move point B up or down the normal line depending on the ratio of the plasma creatinine to 0.1 mM (mean value in normals), B–C = GFR component (mmol); Step 4. Move point C to D (i.e. 2.4 mM, the mean basal value in normals), D–C = the flow of calcium from absorption and resorption (mmol).

calcitriol secretion (Peacock 1984) which has a plasma half-life of about 12 hours (Mawer et al. 1976; Peacock et al. 1980) and a biological half-life, in terms of absorption, of 24 hours. In the absence of dietary calcium there is an obligatory loss of about 4 mmol/day from the ECF by endogenous secretion. The amount of calcium transported by the gut can therefore range from about +10 mmol to −4 mmol per 24-h period which is sufficient to replace 40% or remove 20% of the normal ECF calcium. Endogenous secretion is continuous whereas calcium absorption is intermittent, normally taking place over a period of about 6 hours (Nordin 1976) within a 24-hour period. However, the bone calcium reservoir, normally about four times the ECF calcium reservoir (Marshall et al. 1972), prolongs the effect of calcium absorption and it takes about 12 hours to clear the calcium from the ECF taken in from normal absorption (Nordin 1976).

Bone

In parallel with its action on absorption, calcitriol stimulates resorption of bone without giving rise to increased mineralization. In the absence of parathyroid hormone, as in hypoparathyroidism, calcitriol at normal plasma concentrations maintains a normal plasma calcium level albeit with a greatly increased loss of calcium in the urine. In hypoparathyroid patients maintained at a normal plasma calcium level with calcitriol, the fasting urinary calcium is about 5 mmol/day which represents the calcium flow from the calcium bone reservoir and to some extent prolonged hyperabsorption (Heyburn and Peacock 1977). On stopping calcitriol, plasma calcium has a half-life of about 2–3 days depending on calcium intake and renal function (Kanis and Russell 1977). The plasma concentration of calcitriol which leads to bone resorption is within or just above its normal plasma concentration limits, but the amount of bone resorbed depends on the amount of calcium absorbed, with calcium from bone making up the deficit which is not supplied from the gut (Meierehofer et al. 1983).

In hypoparathyroid patients taking 1–2 μg calcitriol orally per day and a normal dietary intake of calcium, bone resorption, as judged by hydroxyproline excretion, does not increase unless the plasma calcium is forced above the normal range (Heyburn and Peacock 1977). This greater sensitivity of the gut as compared to bone to calcitriol may only be apparent because of a local action of oral calcitriol on gut (Francis and Peacock 1987).

In the absence of adequate amounts of calcitriol and in the presence of increased levels of parathyroid hormone, as in vitamin D-deficient osteomalacia, rates of bone resorption and mineralization are very high since the effect of parathyroid hormone on resorption is to stimulate bone turnover. Despite this high resorption rate, plasma calcium is low and urinary calcium very low, indicating that in the absence of calcitriol the calcium content of the bone reservoir is negligible and that the calcium involved in bone turnover cannot contribute to the maintenance of the plasma calcium level. Indeed if calcium is infused intravenously in such patients the infused calcium is taken up by bone and is not available for plasma calcium homeostasis (Marshall et al. 1972). Similarly, when calcitriol is used to treat such patients the absorbed calcium is taken up by bone and only when the bone turnover starts to fall does the absorbed calcium begin to contribute to maintaining the plasma calcium level (Peacock 1984).

In the presence of increased parathyroid hormone levels the net amount of calcium moving from or to the ECF due to bone turnover is zero because of the firm coupling between mineralization and resorption. The coupling process is "resorption-driven" (Frost 1973) and the cycle of activity from stimulation of resorption to mineralization of new bone takes about 100 days with complete mineralization following resorption by several weeks' delay (Frost 1973). The overall effect of a rise in parathyroid hormone is the resorption of a package or quantum of bone, the calcium from which then becomes available for the maintenance of plasma calcium. Eventually, however, the bone returns to balance but at an increased rate of turnover. Parathyroid hormone's action on bone, therefore, is to regulate the rate of bone turnover and not plasma calcium concentration. Its effect, however, on calcitriol secretion and subsequently on calcium absorption is quantitatively greater and more rapid than its effect on bone. It has been suggested that parathyroid hormone regulates the level of plasma calcium by setting a concentration threshold at which calcium can move to and from bone (Massry 1982). However, no convincing evidence for such a mechanism has been produced.

The Kidney

By virtue of range in the setting of tubular reabsorption by parathyroid hormone (Fig. 6.16), the kidney can excrete from 3% to 30% of its filtered calcium load which in 24 hours is 23% to 250% of the ECF calcium (Fig. 6.1). Furthermore, the response in tubular reabsorption induced by parathyroid hormone occurs within hours (Fig. 6.17). Therefore in terms of both the quantity of calcium excreted and the rate of response the kidney is normally the major transport organ regulating plasma calcium concentration (Nordin and Peacock 1969; Nordin et al. 1972b; Nordin 1976).

Hypercalciuria

Apart from increasing the risk of crystalluria and stone formation, a high urinary calcium does no damage. However, in the absence of an obvious cause, such as hypercalcaemia, hypercalciuria often indicates the presence of certain states or diseases. The ability, therefore, to classify and elucidate the pathogenesis of a hypercalciuria is always of importance (Nordin et al. 1972a; Pak et al. 1974). The term hypercalciuria is used to describe a urinary calcium greater than normal, irrespective of how it is expressed. Since the distribution of calcium excretion in the healthy population is always skewed by high values, the upper limit of normal is never easily defined (Hodgkinson and Pyrah 1958; Bulusu et al. 1970; Robertson and Morgan 1972) (Fig. 6.20). This difficulty is compounded by the normal variation in urinary calcium excretion with factors such as body size, sex, age, geography, season and diet. Therefore, calcium excretion can only be defined as being hypercalciuric relative to an upper value for urinary calcium calculated from a uniform population of normal subjects or relative to a measured variable such as dietary calcium intake or plasma calcium concentration (Nordin et al. 1967).

It should also be noted that a disease may increase calcium excretion within the normal range and the absence of hypercalciuria does not exclude a significant effect of disease on calcium metabolism. Although the term is most frequently used for the calcium present in a 24-hour urine collection, it is equally valid to use it for other expressions such as the calcium/creatinine ratio or the calcium excretion per litre of glomerular filtrate.

In healthy adults in calcium balance the 24-hour urinary calcium is the sum of the net calcium absorbed from diet and resorbed from bone. The source of a hypercalciuria is therefore either diet or bone. The kidney also acts as a source

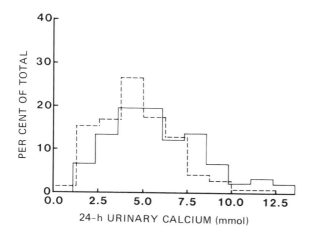

24-h URINARY CALCIUM (mmol)

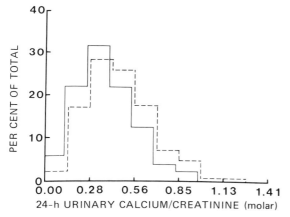

24-h URINARY CALCIUM/CREATININE (molar)

Fig. 6.20.   The distribution of urine calcium in 24-hour collections from 104 men (–) and 142 women (–––) on a free diet. *Top panel*, expressed in mmols; *bottom panel*, expressed as a ratio to creatinine excretion in the same urine.

of the hypercalciuria either by decreasing reabsorption or increasing the filtration rate, but these effects are only transitory unless the calcium lost in the urine eventually stimulates absorption or resorption or both. Despite this proviso, hypercalciuria is most usefully classified according to its source, namely gut, bone or kidney (Table 6.4). Various attempts have been made to subdivide this classification further according to the physiological mechanism responsible for the increased calcium excretion and the aetiological factors acting on these mechanisms (Nordin et al. 1972a, b; Coe et al. 1973; Pak et al. 1974; Broadus et al. 1978; Coe et al. 1982) (Table 6.4). None of these attempts have been entirely satisfactory however, because of the overlap in the biochemistry with the normal population and the difficulty in establishing accurate and precise criteria for its diagnosis (Peacock 1982; Robertson and Peacock 1985; Peacock and Robertson 1988).

Table 6.4.   Classification of hypercalciuria

Organ	Physiological process	Aetiological factors	
Gut	Diet	↑ Calcium ↓ Fibre ↓ Phosphate	
	Absorption	↑ 1,25(OH)$_2$D ↑ Sugars ↑ Protein (?)	↓ PO$_4$ ↑ PTH
Bone	Resorption	↑ Immobilization ↑ PTH ↑ 1,25(OH)$_2$D	↑ Thyroxine ↑ Acidosis ↓ Oestrogen
Kidney	Glomerular filtration	↑ Protein	
	Tubular reabsorption	↑ Sodium ↑ Acidosis ↓ PTH	

↑, increase; ↓, decrease, leads to hypercalciuria.

Gut

Diet

Urinary calcium is positively related to dietary calcium and high calcium intakes result in hypercalciuria (Fig. 6.9). Even on normal intakes however hypercalciuria may occur due to an increase in calcium availability such as occurs with low dietary phosphate (Lotz et al. 1964) and fibre (Shah et al. 1980; Rao et al. 1982) or with high protein (Sherman 1920; McCance et al. 1942; Anand and Linkswiler 1974; Linkswiler et al. 1974; Margen et al. 1974; Robertson et al. 1979b) or salt intake (King et al. 1964; Kleeman et al. 1964). Studies aimed at classifying a hypercalciuria must therefore be performed on subjects under a fixed dietary regimen of average composition and under these conditions dietary hypercalciuria disappears.

Absorption

In most circumstances there is a positive relationship between plasma calcitriol concentration and calcium absorption (Peacock et al. 1980). The absorptive hypercalciuria of primary calcium stone disease (Haussler and McCain 1977; Peacock and Robertson 1988), primary hyperparathyroidism (Broadus et al. 1980; Peacock and Robertson 1988) and sarcoidosis (Adams et al. 1983) is due to raised plasma levels of calcitriol. Hypophosphataemia has been suggested as the stimulus for the raised plasma calcitriol levels in primary calcium stone disease (Gray et al. 1977; Haussler and McCain 1977). However, although plasma phosphate may be involved in some patients, it cannot be the only factor (Peacock 1982; Robertson and Peacock 1985; Peacock and Robertson 1988) and the pathogenesis of the raised calcitriol in primary stone formers remains controversial. In absorptive hypercalciuria the 24-hour urine calcium is higher than the calcium excretion after an overnight fast (Table 6.5) and is higher than normal at all dietary intakes (Peacock et al. 1968; Coe et al. 1982) after an oral load of calcium (Peacock et al. 1968; Broadus et al. 1978) (Fig. 6.5). The hyperabsorption appears to be specific for calcium and is not part of a general hyperabsorptive state. Certain amino acids and sugars also promote calcium absorption. As a cause of absorptive hypercalciuria however, their role has not been clearly defined.

Table 6.5. Diseases affecting calcium excretion

	Hypercalciuria				Hypocalciuria		
	24-hour Ca/Cr	Fasting Ca/Cr	TmCa		24-hour Ca/Cr	Fasting Ca/Cr	TmCa
Malignancy	↑	↑	↓	Hypoparathyroidism	↓	↓	↓
Thyrotoxicosis	↑	↑	↓	Renal failure	↓	↓	↓
Calcium stone	↑	N↑	N↓	Vitamin D deficiency osteomalacia	↓	↓	↑
1° Hyperpara-thyroidism	↑	↑	↑	Small-bowel disease	↓	↓	↑
Oestrogen deficiency	↑	↑	N	Sodium depletion	↓	↓	↑
Vitamin D intoxication	↑	↑	↓	Addison's disease	↓	↓	↑
Sarcoidosis and granuloma	↑	↑N	↑N	3° Hyperpara-thyroidism	↓	↓	↓
Immobilization	↑	↑	↓	Familial hyperpara-thyroidism	N	↓	↑
Acidosis	↑	↑	↓	Alkalosis	↓	↓	↑
Phosphate depletion	↑	↑	↓	Thiazide	↓	↓	↑
Diuretics (except thiazide)	↑	↑	↓	High phosphate intake	↓	↓	N↑
Acromegaly	↑	↑	↓	Cellulose phosphate	↓	↓	N↑
Parenteral nutrition	↑	↑	↓	Magnesium deficiency	↓	↓	↓

↑, increase; ↓, decrease; N, normal.

Bone

Resorption

Hypercalciuria due to an increase in net bone resorption is best illustrated by the amount of calcium excreted in urine after an overnight fast. If net bone resorption is sufficiently high to suppress calcitriol, the calcium excretion after fasting becomes equal to that of the 24-hour urine. The high rate of net bone resorption should always be confirmed by a corresponding rise in hydroxypro-line/creatinine ratio in the urine after an overnight fast. It can also be confirmed, but with more difficulty, by showing negative calcium balance, increased resorption on histomorphometry of biopsied bone and skeletal demineralization as measured by sequential densitometry at a suitable skeletal site. Frank osteoporosis may eventually develop.

Kidney

Renal hypercalciuria results either from an increase in glomerular filtration rate or from a decrease in tubular reabsorption.

Glomerular Filtration

Hypercalciuria due to an increase in glomerular filtration rate is difficult to diagnose since it relies on accurate measurements of both plasma calcium and glomerular function. An increased glomerular filtration rate is clearly a factor in the higher calcium excretion by large individuals since plasma calcium is not related to lean body mass whereas 24-hour urinary calcium is. On the other hand dietary factors such as protein intake (Robertson et al. 1979b), and physiological states such as physical activity (Heaton and Hodgkinson 1963; Loutit and Papworth 1965) change glomerular filtration and may affect calcium excretion.

Tubular Reabsorption

In tubular hypercalciuria the calcium excretion per unit of glomerular filtration in relation to the filtered plasma calcium is abnormally high. The abnormality may be shown by measuring calcium clearance, providing that the plasma calcium concentration and glomerular filtration rate are stable. Ideally the slope of excretion with changes in filtered load should be established during an intravenous calcium infusion (Peacock and Nordin 1968). In hypercalcaemia this is contraindicated because of the toxicity of calcium.

Since tubular hypercalciuria can only be sustained by causing secondary hyperparathyroidism, evidence for the latter state is necessary to establish the diagnosis. This in itself presents a diagnostic problem (Habener et al. 1984) because of the difficulty of measuring the bioactive parathyroid hormone in plasma and because the biological response in plasma phosphate and calcitriol, calcium absorption and bone resorption are present in other forms of hypercalciuria. The major source of error in making this diagnosis is that the subject is not fasted long enough to make sure all the absorbed calcium is cleared from the ECF. This is particularly important in subjects with hyperabsorption of calcium. If due precautions are not taken, absorptive hypercalciuria is readily misdiagnosed as tubular or resorptive hypercalciuria. Cellulose phosphate given to the patient the night before investigation to reduce the absorption of calcium is useful in these subjects (Knebel et al. 1985).

Diseases Affecting Urinary Calcium

A variety of diseases may increase or decrease calcium excretion but because of the wide range of urinary calcium found in health, only the mean value may be different from normal (Table 6.5, Fig. 6.21). In each case the disease acts on calcium transport in the gut, bone or kidney.

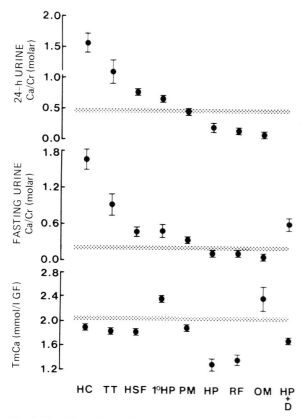

Fig. 6.21. Mean (± 1 SE) 24-hour urinary calcium/creatinine (*top panel*), fasting urinary calcium/creatinine (*middle panel*) and calculated TmCa (*bottom panel*) in patients with selected diseases affecting calcium excretion. Each group comprised 12 patients and the normal mean ± 1 SE, shown as a *shaded band*, was established in 216 healthy subjects. *HC*, hypercalcaemia of cancer; *TT*, thyrotoxicosis; *HSF*, hypercalciuric stone formers; *1°HP*, primary hyperparathyroidism; *PM*, postmenopausal; *HP*, hypoparathyroidism; *RF*, renal failure; *OM*, vitamin D-deficient osteomalacia; *HP+D*, hypoparathyroidism treated with calcitriol.

Hypercalciuria

Secondary Cancer in Bone

Tumours arising from a wide spectrum of tissues (Fiskin et al. 1980) may cause severe degrees of hypercalciuria which is usually diagnosed because of the presence of hypercalcaemia. The major cause of the hypercalciuria is release of calcium to the ECF following bone destruction due to a variety of agents produced by the tumour. The 24-hour and the fasting calcium/creatinine ratios are high and similar (Fig. 6.21). Tubular reabsorption of calcium is reduced in the majority of cases (Fig. 6.21) and calcitriol and parathyroid hormone activities are usually suppressed (Godsell et al. 1986). Hypercalcaemia usually develops as a result of the very high rate of bone destruction in combination with a decreased glomerular filtration rate and frequently an increased tubular calcium reabsorption due to sodium depletion. In some cases there may be an increase in tubular reabsorption of calcium arising from the action of hormonal products of the tumour itself (Ralston et al. 1984).

It cannot be overemphasized, however, that a tumour-induced increase in reabsorption can only be diagnosed in the presence of a urinary sodium excretion appropriate for the plasma calcium concentration and calcium excretion (Figs. 6.14 and 6.15). In rare cases the hypercalciuria is due to increased calcium absorption following increased production of calcitriol.

Thyrotoxicosis

When hypercalciuria occurs along with thyrotoxicosis (Gordan et al. 1974) it is due to an increased rate of net bone resorption. The 24-hour and fasting calcium/creatinine ratios are increased and similar, the tubular reabsorption of calcium is reduced (Fig. 6.21) and there is suppression of calcium absorption, plasma calcitriol and parathyroid hormone. The degree of hypercalciuria, however, does not correlate closely with the plasma levels of thyroxine, suggesting that the effect on bone represents an excessive sensitivity to thyroxine or the presence of bone-resorbing factors other than thyroxine (Francis et al. 1983b). The latter seems more likely since the response to anti-thyroid treatment takes longer in bone than in other tissues.

Calcium Stone Formers

About 30%–50% of primary calcium stone formers have hypercalciuria due to hyperabsorption of calcium which is the result of increased plasma calcitriol concentrations (Peacock and Robertson 1988). In most cases the fasting calcium excretion and the tubular reabsorption of calcium are in the normal range. However, if the most hypercalciuric subjects are considered (Fig. 6.21), the fasting calcium excretion is increased and the tubular reabsorption decreased. The hydroxyproline excretion, however, is normal and the parathyroid hormone level suppressed. The high fasting urinary calcium can be reduced by prolonged fasting (Peacock et al. 1968) or with cellulose phosphate (Knebel et al. 1985). The fasting hypercalciuria therefore is similar to that observed in normal subjects and hypoparathyroid patients taking oral calcitriol (Meierhofer et al. 1983) and does not represent tubular or resorptive hypercalciuria. Some investigators have reported that tubular hypercalciuria is present in a substantial number of calcium stone formers. They have based their opinion on the presence of a low tubular reabsorption of calcium and a raised plasma parathyroid hormone concentration (Coe et al. 1973) but this is not the general experience.

Primary Hyperparathyroidism

The 24-hour urinary calcium, fasting calcium excretion and tubular reabsorption of calcium are increased in primary hyperparathyroidism (Fig. 6.21). The 24-hour urine calcium is in general higher than the fasting calcium excretion because of increased calcium absorption (Peacock 1975). The variation in the activity of the disease and the high prevalence of renal impairment are however reflected in the wide range of urinary calcium excretions.

Oestrogen Deficiency

Following the menopause, the fall in plasma oestrogen levels is associated with resorptive hypercalciuria (Gallagher et al. 1972). The increase in calcium excretion is not marked and the mean increase lies within the normal range (Figs. 6.8 and 6.21). The increase in urinary calcium excretion appears to continue for 15–20 years after the menopause (Fig. 6.8).

Hypocalciuria

Hypoparathyroidism

The 24-hour urinary calcium, fasting calcium excretion and tubular reabsorption are low in hypoparathyroidism (Fig. 6.21). With vitamin D treatment the 24-hour urinary calcium and fasting calcium excretion rise into the hypercalciuric range even before plasma calcium is normalized (Heyburn and Peacock 1977) (Fig. 6.21). Tubular reabsorption of calcium (Fig. 6.21) also rises but remains low even when plasma calcium is normal.

Renal Failure

The 24-hour and fasting urine calcium are low and the tubular reabsorption of calcium is reduced despite secondary hyperparathyroidism. The cause of the low urinary calcium is the decrease in the glomerular filtration rate and calcium absorption often aggravated by hypocalcaemia and osteomalacia. The decreased tubular reabsorption per litre of glomerular filtrate is probably due to a combination of a high sodium and calcium load per nephron and acidosis.

Vitamin D-Deficient Osteomalacia

The 24-hour and fasting calcium excretions are low and the tubular reabsorption is increased. The former are due to calcium malabsorption and osteomalacia and the latter to hypocalcaemia and secondary hyperparathyroidism (Peacock 1984).

References

Adams JS, Sharma DP, Gacad MA, Singer FR (1983) Metabolism of 25-hydroxy vitamin D_3 by cultured pulmonary alveolar macrophages in sarcoidosis. J Clin Invest 72:1856–1860

Agus ZS, Puschett JB, Senesky D, Goldberg M (1971) Mode of action of parathyroid hormone and cyclic adenosine 3',5'-monophosphate on renal tubular phosphate reabsorption in the dog. J Clin Invest 50:617–626

Agus ZS, Gardner LB, Beck LH, Goldberg M (1973) Effects of parathyroid hormone on renal tubular reabsorption of calcium, sodium and phosphate. Am J Physiol 224:1143–1148

Anand CR, Linkswiler HM (1974) Effect of protein intake on calcium balance of young men given 500 mg calcium daily. J Nutr 104:695–700

Antoniou LD, Eisner GM, Slotkoff LM, Lilienfield LS (1969) Relationship between sodium and calcium transport in the kidney. J Lab Clin Med 74:410–420

Ardaillou R, Fillastre JP, Milhaud G, Rousselet F, DeLaunay F, Richet G (1969) Renal excretion of phosphate, calcium and sodium during and after a prolonged thyrocalcitonin infusion in man. Proc Soc Exp Biol Med 131:56–60

Arora B, Selby Pl, Norman RW, Peacock M, Robertson WG (1985) The effects of an increased intake of various constituents of a high animal protein diet on the risk of calcium oxalate stone formation. In: Schwille PO, Smith LH, Robertson WG, Vahlensieck W (eds) Urolithiasis and related clinical research. Plenum, New York, pp 85–88

Bernstein DS, Newton R (1966) The effect of oral sodium phosphate on the formation of renal calculi and on idiopathic hypercalciuria. Lancet II:1105–1107

Berridge MJ (1985) Calcium: a universal second messenger. Triangle 24(3/4):79–90

Biddulph DM (1972) Influence of parathyroid hormone and the kidney on maintenance of blood calcium concentration in the golden hamster. Endocrinology 90:1113–1118

Biddulph DM, Hirsch PF, Cooper CW, Munson PL (1970) Effect of thyroparathyroidectomy and parathyroid hormone on urinary excretion of calcium and phosphate in the golden hamster. Endocrinology 87:1346–1350

Bijvoet OLM, Van der Sluys Veer, Jansen AP (1968) Effects of calcitonin on patients with Paget's disease, thyrotoxicosis or hypercalcaemia. Lancet I:876–881

Bogert LJ, Kirkpatrick EE (1922) Studies in inorganic metabolism. 3. The effects of acid-forming and base-forming diets upon calcium metabolism. J Biol Chem 54:375–386

Bouillon R, van Baelen H, DeMoor P (1977) 25-Hydroxyvitamin D and its binding protein in maternal and cord serum. J Clin Endocrinol Metab 45:679–684

Brenner BM, Ichikawa I, Deen WM (1981) Glomerular filtration. In: Brenner BM, Rector FC (eds) The kidney. Saunders, Philadelphia, pp 289–327

Brickman AS, Massry SG, Coburn JW (1971) Calcium deprivation and renal handling of calcium during repeated saline infusions. Am J Physiol 220:44–48

Brickman AS, Massry SG, Coburn JW (1972) Changes in serum and urinary calcium during treatment with hydrochlorothiazide. Studies on mechanisms. J Clin Invest 51:945–954

Broadus AE, Dominquez M, Bartter FC (1978) Pathophysiological studies in idiopathic hypercalciuria: use of an oral calcium tolerance test to characterize distinctive hypercalciuria subgroups. J Clin Endocrinol Metab 47:751–760

Broadus AE, Horst RL, Lang R, Littledike ET, Rasmussen H (1980) The importance of circulating 1,25 dihydroxy vitamin D in the pathogenesis of hypercalciuria and renal stone formation in primary hyperparathyroidism. N Engl J Med 302:421–425

Bulusu L, Hodgkinson A, Nordin BEC, Peacock M (1970) Urinary excretion of calcium and creatinine in relation to age and body weight in normal subjects and patients with renal calculi. Clin Sci 38:601–612

Burnatowska MA, Harris CA, Sutton RAC, Dirks JH (1977) Effects of PTH and cAMP on renal handling of calcium, magnesium and phosphate in the hamster. Am J Physiol 233:F514

Chen PS, Neuman WF (1955) Renal excretion of calcium by the dog. Am J Physiol 180:623–631

Coburn JW, Massry SG (1970) Changes in serum and urinary calcium during phosphate depletion: studies on mechanisms. J Clin Invest 49:1619–1629

Coburn JW, Hartenbower DL, Massry SG (1971) Modification of calciuretic effect of extracellular volume expansion by phosphate infusion. Am J Physiol 220:377–383

Cochran M, Nordin BEC (1971) The causes of hypocalcaemia in chronic renal failure. Clin Sci 40:305–315

Cochran M, Peacock M, Sachs G, Nordin BEC (1970) Renal effects of calcitonin. Br Med J i:135–137

Coe FL, Canterbury JM, Firpo JJ, Reiss E (1973) Evidence for secondary hyperparathyroidism in idiopathic hypercalciuria. J Clin Invest 52:134–142

Coe FL, Favus MJ, Crockett T et al. (1982) Effects of low-calcium diet on urine calcium excretion, parathyroid function and serum 1,25(OH)$_2$D$_3$ levels in patients with idiopathic hypercalciuria. Am J Med 72:25–32

Collins EJ, Garrett ER, Johnston RL (1962) Effect of adrenal steroids on radiocalcium metabolism in dogs. Metabolism 11:716–726

Condon JR, Nassim JR, Millard FJC, Hilbe A, Stainthorpe EM (1970) Calcium and phosphorus metabolism in relation to lactose tolerance. Lancet I: 1027–1029

Costanzo LS, Windhager EE (1978) Calcium and sodium transport by the distal convoluted tubule of the rat. Am J Physiol 235:F492

Demartini FE, Briscoe AM, Ragan C (1967) Effect of ethacrynic acid on calcium and magnesium excretion. Proc Soc Exp Biol Med 124:320–324

DeRouffignac C, Morel F, Moss N, Roinel N (1973) Micropuncture study of water and electrolyte movements along the loop of Henle in Psammomys with special reference to magnesium, calcium and phosphorus. Pfluegers Arch 334:309–326

DiBona G (1971) Effect of hypercalcaemia on renal tubular sodium handling in the rat. Am J Physiol 220:49–53

Dirzel J, Schwartz M (1967) Abnormally increased glomerular filtration rate in short-term insulin-treated diabetic subjects. Diabetes 16:264–267

Duarte CG (1968) Effects of ethacrynic acid and furosemide on urinary calcium, phosphate and magnesium. Metabolism 17:867–876

Duarte CG, Bland JH (1965) Calcium, phosphorus and uric acid clearances after intravenous administration of chlorothiazide. Metabolism 14: 211–219

Duarte CG, Watson JF (1967) Calcium reabsorption in proximal tubule of the dog. Am J Physiol 212:1355–1360

Edwards BR, Baer PG, Sutton RAL, Dirks JH (1973) Micropuncture study of diuretic effects on sodium and calcium reabsorption in the dog nephron. J Clin Invest 52:2418–2427

Edwards NA, Hodgkinson A (1965a) Metabolic studies in patients with idiopathic hypercalciuria. Clin Sci 29:143–154

Edwards NA, Hodgkinson A (1965b) Studies of renal function in patients with idiopathic hypercalciuria. Clin Sci 29:327–338

Eknoyan G, Suki WN, Martinez-Maldonado M (1970) Effect of diuretics on urinary excretion of phosphate, calcium and magnesium in thyroparathyroidectomised dogs. J Lab Clin Med 76: 257–266

Farquharson RF, Salter WT, Aub JC (1931) Studies of calcium and phophorus metabolism. 13. The effect of ingestion of phosphates on the excretion of calcium. J Clin Invest 10:251–269

Fisken RA Heath DA, Bold AM (1980) Hypercalcaemia–A hospital survey. Q J Med 49:405–418

Francis RM, Peacock M, Storer JH, Davies AEJ, Brown WB, Nordin BEC (1983a) Calcium malabsorption in the elderly: the effect of treatment with oral 25 hydroxyvitamin D$_3$. Eur J Clin Invest 13:391–396

Francis RM, Peacock M, Storer J (1983b) The effect of anti-thyroid drugs on calcium metabolism in thyrotoxicosis. Met Bon Dis Rel Res

Francis RM, Peacock M, Barkworth SA (1984) Renal impairment and its effects on calcium metabolism in elderly women. Age Ageing 13:14–20

Francis RM, Peacock M (1987) Local action of oral 1,25(OH)$_2$D on calcium absorption in osteoporosis. Am J Clin Nutr 46:315–318

Frick A, Rumrich G, Ullrich KJ, Lassiter WE (1965) Microperfusion study of calcium transport in the proximal tubule of the rat kidney. Pfluegers Arch 286:109–117

Froeling PGAM, Bijvoet OLM (1974) Kidney-mediated effects of parathyroid hormone on extracellular homeostasis of calcium, phosphate and acid-base balance in man. Neth J Med 17:174–183

Frost HM (1973) The origin and nature of transients in human bone remodeling dynamics. In: Frame B, Parfitt AM, Duncan H (eds) Clinical aspects of metabolic bone disease. Excerpta Medica, Amsterdam, p 124

Gallagher JC, Young MM, Nordin BEC (1972) Effects of artificial menopause on plasma and urine calcium and phosphate. Clin Endocrinol 1:57–64

Godsell JW, Bartis WJ, Insogna KL, Broadus AE, Stewart AF (1986) Nephrogenous cyclic AMP, adenylate cyclase stimulating activity and humoral hypercalcaemia of malignancy. Rec Prog Horm Res 42:705–743

Goldfarb S, Westby GR, Goldberg M, Agus ZS (1977) Renal tubular effects of chronic phosphate depletion. J Clin Invest 59:770–779

Gordan DL, Suvanich S, Erviti V, Schwartz MA, Martinez CJ (1974) The serum calcium level and its significance in hyperthyroidism: a prospective study. Am J Med Sci 268:31–36

Goulding A, McChesney R (1977) Oestrogen–progesterone oral contraceptives and urinary calcium excretion. Clin Endocrinol 6:449–454

Gray RW, Wilz DR, Caldas AE, Lemann J Jr (1977) The importance of phosphate in regulating plasma 1,25(OH)$_2$ vitamin D levels in humans: studies in healthy subjects in calcium stone formers and patients with primary hyperparathyroidism. J Clin Endocrinol Metab 45:299–306

Gray SP, Morris JEW, Brooks CJ (1973) Renal handling of calcium, magnesium, inorganic phosphate and hydrogen ions during prolonged exposure to elevated carbon dioxide concentrations. Clin Sci Mol Med 45:751

Greenberg AJ, McNamara H, McCrory WW (1966) Metabolic balance studies in primary renal tubular acidosis. Effects of acidosis on external calcium and phosphorus balances. J Pediatr 69:610–618

Habener JF, Rosenblatt M, Potts JT Jr (1984) Parathyroid hormone: biochemical aspects of biosynthesis, secretion, action and metabolism. Physiol Rev 64:985–1053

Hallson PC, Rose GA (1979) Uromucoids and urinary stone formation. Lancet I:1000–1002

Haussler MR, McCain TA (1977) Vitamin D metabolism and action. N Engl J Med 297:974–983, 1041–1050

Haymovitz A, Horwith M (1964) The miscible calcium pool in metabolic bone disease – in particular acromegaly. J Clin Endocrinol 24:4–14

Heaton FW, Hodgkinson A (1963) External factors affecting diurnal variation in electrolyte excretion with particular reference to calcium and magnesium. Clin Chim Acta 8:246–254

Herbert LA, Lemann J, Petersen JR, Lennon EJ (1966) Studies of the mechanism by which phosphate infusion lowers serum calcium concentration. J Clin Invest 45:1886–1894

Heyburn PJ, Peacock M (1977) The management of hypoparathyroidism with 1-alpha-hydroxy vitamin D$_3$. Clin Endocrinol 7 [Suppl]:209S–214S

Heyburn PJ, Robertson WG, Peacock M (1982) Phosphate treatment of recurrent calcium stone disease. Nephron 32:314–319

Higgins BA, Nassim JR, Collins J, Hilb A (1964) The effect of bendrofluazide on urine calcium excretion. Clin Sci 27:457–462

Hodgkinson A, Heaton FW (1965) The effect of food ingestion on the urinary excretion of calcium and magnesium. Clin Chim Acta 11:354–362

Hodgkinson A, Pyrah LN (1958) The urinary excretion of calcium and inorganic phosphate in 344 patients with calcium stone of renal origin. Br J Surg 46:10–18

Hodgkinson A, Marshall RW, Cochran M (1971) Diurnal variations in calcium phosphate and calcium oxalate activity products in normal and stone-forming urines. Isr J Med Sci 7, 11:1230–1234

Jamison RL, Frey NR, Lacy FB (1974) Calcium reabsorption in the thin loop of Henle. Am J Physiol 227:745–751

Kanis JA, Russell RGG (1977) Rate of reversal of hypercalcaemia and hypercalciuria induced by vitamin D and its 1-alpha-hydroxylated derivatives. Br Med J i:78–81

Kaye M, Silverman M (1965) Calcium metabolism in chronic renal failure. J Lab Clin Med 66:535–548

Kerr C, Loken HF, Glendening MB, Gordan GS, Page EW (1962) Calcium and phosphorus dynamics in pregnancy. Am J Obstet Gynecol 83:2–8

King JS, Jackson R, Ashe B (1964) Relation of sodium intake to urinary calcium excretion. Invest Urol 1:555–560

Kinne R, Keljo D, Ginaj P, Murer H (1978) The energy source of glucose and Ca transport in the renal proximal tubule. In: Vogel HG, Ullrich KJ (eds) New aspects of renal function. Excerpta Medica, Amsterdam, p 41

Kleeman CR, Bernstein D, Rockney R, Dowling JT, Maxwell MH (1961) Studies on the renal clearance of diffusible calcium and the role of the parathyroid glands in its regulation. Yale J Biol Med 34:1–30

Kleeman CR, Bohannan J, Bernstein D, Ling S, Maxwell MH (1964) Effect of variations in sodium intake on calcium excretion in normal humans. Proc Soc Exp Biol Med 115:29–32

Kleeman CR, Ling S, Bernstein D, Maxwell MH, Chapman L (1966). The effect of independent changes in glomerular filtration (GFR) and sodium (Na$^+$) excretion on the renal excretion of calcium (Ca^{++}) and magnesium (Mg^{++}) in acutely hypercalcaemic dogs. J Clin Invest 45:1032–1033

Knapp EL (1947) Factors influencing the urinary excretion of calcium in normal persons. J Clin Invest 26:182–202

Knebel L, Tschope W, Ritz E (1985) A one day cellulose phosphate (CP) test discriminates nonabsorptive from absorptive hypercalciuria. In: Schwille PO, Smith LH, Robertson WG, Vahlensieck W (eds) Urolithiasis. Plenum Press, New York, pp 303–306

Kretsinger RH (1980) Structure and function of calcium modulated proteins. CRC Crit Rev Biochem 8:119–174

Lamberg BA, Kuhlback B (1959) Effect of chlorothiazide and hydrochlorothiazide on the excretion of calcium in the urine. Scand J Clin Lab Invest 11:351–357

Lassiter WE, Gottschalk CW, Mylle M (1963) Micropuncture study of renal tubular reabsorption of calcium in normal rodents. Am J Physiol 104:771–775

Leitch I, Aitken FC (1959) The estimation of calcium requirement: a re-examination. Nutr Abstr Rev 29:393–411

Lemann J Jr, Litzow JR, Lennon EJ (1967) Studies of the mechanism by which chronic metabolic acidosis augments urinary calcium excretion in man. J Clin Invest 46:1318–1328

Lemann J Jr, Piering WF, Lennon EJ (1970) Studies of the acute effects of aldosterone and cortisol on the interrelationship between renal sodium, calcium and magnesium excretion in normal man. Nephron 7:117–130

Lemann J Jr, Adams ND, Gray RW (1979) Urinary calcium excretion in human beings. N Engl J Med 301:535–541

Levin ME, Boissean VC, Avioli LV (1976) Effects of diabetes mellitus on bone mass in juvenile and adult-onset diabetes. N Engl J Med 294:241–245

Lichtwitz A, De Seze S, Parlier R, Hioco D, Bordier P (1960) L'hypocalciurie glomerulaire. Bull Soc Med Hop (Paris) 76:98–119

Lindheimer MD, Katz AI (1981) The renal response to pregnancy. In: Brenner BM, Rector FC (eds) The kidney. Saunders, Philadelphia, pp 1732–1815

Linkswiler HM, Joyce CL, Anand CR (1974) Calcium retention of young adult males as affected by level of protein and of calcium intake. Trans NY Acad Sci 36:333–340

Liu SH, Chu HI (1943) Studies of calcium and phosphorus metabolism with special reference to pathogenesis and effect of dehydroxytachysterol (AT10) and iron. Medicine 22:103–161

Lotz M, Ney R, Bartter FC (1964) Osteomalacia and debility resulting from phosphorus depletion. Trans Assoc Am Physicians 77:281–295

Loutit JF, Papworth DG (1965) Diurnal variation in urinary excretion of calcium and strontium. Proc R Soc Lond [Biol] 162:458–472

McCance RA, Widdowson EM, Lehmann H (1942) The effect of protein intake on the absorption of calcium and magnesium. Biochem J 36:686–691

Malm OJ (1953) On phosphates and phosphoric acid as dietary factors in the calcium balance of man. Scand J Clin Lab Invest 5:75–84

Margen S, Chu JY, Kaufman NA, Calloway DH (1974) Studies in calcium metabolism. 1. The calciuric effect of dietary protein. Am J Clin Nutr 27:584–589

Marshall DH (1976) Calcium and phosphate kinetics. In: Nordin BEC (ed) Calcium, phosphate and magnesium metabolism. Churchill Livingstone, Edinburgh, pp 257–297

Marshall DH, Peacock M, Nordin BEC (1972) Plasma calcium homeostasis. In: Hioco DJ (ed) Rein et calcium. Sandoz Editions, Rueil-Malmaison, pp 15–25

Marshall DH, Nordin BEC, Speed R (1976) Calcium, phosphorus and magnesium requirement. Proc Nutr Soc 35:163–173

Marshall RW (1976) Plasma fractions. In: Nordin BEC (ed) Calcium, phosphate and magnesium metabolism. Churchill Livingstone, Edinburgh, pp 162–185

Marshall RW, Hodgkinson A (1983) Calculation of plasma ionised calcium from total calcium, proteins and pH: comparison with measured values. Clin Chim Acta 127:305–310

Massry SG (1982) Renal handling of calcium. In: Bronner F, Coburn JW (eds) Disorders of mineral metabolism, vol 2. Academic Press, New York, pp 189–235

Massry SG, Coburn JW, Chapman LW, Kleeman CR (1968) Role of serum Ca, parathyroid hormone, and NaCl infusion on renal Ca and Na clearances. Am J Physiol 214:1403–1409

Massry SG, Ahumada JJ, Coburn JW, Kleeman CR (1970) Effect of MgCl$_2$ infusion on urinary Ca and Na during reduction in their filtered loads. Am J Physiol 219:881–885

Mawer EB, Backhouse J, Davies M, Hill LF, Taylor CM (1976) Metabolic fate of administered 1,25-dihydroxycholecalciferol in controls and in patients with hypoparathyroidism. Lancet I:1203–1206

Meierhofer WJ, Gray RW, Cheung HS, Lemann J Jr (1983) Bone resorption stimulated by elevated serum 1,25(OH)$_2$ vitamin D concentrations in healthy men. Kidney Int 24:555–560

Melvin KEW, Tashjian AH Jr (1968) The syndrome of excessive thyrocalcitonin produced by medullary carcinoma of the thyroid. Proc Natl Acad Sci USA 59:1216–1222

Mioni G, D'Angelo A, Ossi E, Bertaglia E, Marcon G, Maschio G (1971) The renal handling of calcium in normal subjects and in renal disease. Eur J Clin Biol Res 16:881–887

Modlin M (1966) The interrelation of urinary calcium and sodium in normal adults. Invest Urol 4:180–189

Morgensen CE (1976) Renal function changes in diabetes. Diabetes 25:872–879

Murayama Y, Morel F, Le Grimellec C (1972) Phosphate, calcium and magnesium transfers in proximal tubules and loops of Henle, as measured by single nephron microperfusion experiments in the rat. Pfluegers Arch 333:1–16

Nordin BEC (1976) Plasma calcium and plasma magnesium homeostasis. In: Nordin BEC (ed) Calcium, phosphate and magnesium metabolism. Churchill Livingstone, Edinburgh, pp 186–216

Nordin BEC, Peacock M (1969) Role of the kidney in regulation of plasma calcium. Lancet II:1280–1283

Nordin BEC, Hodgkinson A, Peacock M (1967) The measurement and the meaning of urinary calcium. Clin Orthop 52:293–322

Nordin BEC, Peacock M, Wilkinson R (1972a) Hypercalciuria and renal stone disease. Clin Endocrinol Metabol 1:169–183

Nordin BEC, Peacock M, Wilkinson RW (1972b) Relative importance of gut, bone and kidney in the regulation of serum calcium. In: Talmage RV, Belanger (eds) Calcium, parathyroid hormone and the calcitonins. Excerpta Medica, Amsterdam, pp 236–272

Nordin BEC, Marshall DH, Francis RM, Crilly RG (1982) The effects of sex steroids and corticosteroid hormones on bone. J Steroid Biochem 15:171–174

O'Brien MM, Uhlemann I, McIntosh HW (1967) Urinary pyrophosphate in normal subjects and in stone-formers. Can Med Assoc J 96:100–103

O'Connor WJ, Summerville RA (1976) The effects of a meal of meat on glomerular filtration rate in dogs at normal urine flows. J Physiol 256:81–91

Paillard F, Ardaillou R, Malendin H, Fillastre JP, Prier S (1972) Renal effects of salmon calcitonin in man. J Lab Clin Med 80:200–216

Pak CYC, Okata M, Lawrence FC, Snyder W (1974) The hypercalciurias: causes, parathyroid function and diagnostic criteria. J Clin Invest 54:387–400

Parfitt AM (1969a) The acute effects of mersalyl, chlorothiazide and mannitol on the renal excretion of calcium and other ions in man. Clin Sci 36:267–282

Parfitt AM (1969b) Chlorothiazide-induced hypercalcaemia in juvenile osteoporosis and hyperparathyroidism. N Engl J Med 281:55–59

Parfitt AM (1972) The interactions of thiazide diuretics with parathyroid hormone and vitamin D. Studies in patients with hypoparathyroidism. J Clin Invest 51:1879–1888

Parfitt AM, Higgins BA, Nassim JR, Collins JA, Hilb A (1964) Metabolic studies in patients with hypercalciuria. Clin Sci 27:463–482

Parry ES, Lister IS (1975) Sunlight and hypercalciuria. Lancet I:1063–1065

Payne RB, Little AJ, Williams RB, Milner JR (1973) Interpretations of serum calcium in patients with abnormal serum proteins. Br Med J iv:643–646

Peacock M (1975) Stone and bone disease in primary hyperparathyroidism and their relationship to the action of parathyroid hormone on calcium absorption. In: Talmage RV, Owen M (eds) Calcium regulating hormones. Excerpta Medica, Amsterdam, pp 78–81

Peacock M (1980) Hypercalcaemia and calcium homeostasis. Metabol Bone Dis Rel Res 2:143–150

Peacock M (1982) The mechanisms of hypercalciuria are unnecessary for treatment of recurrent renal calcium stone formers. Contrib Nephrol 33:152–162

Peacock M (1984) Osteomalacia and rickets. In: Nordin BEC (ed) Metabolic bone and stone disease, 2nd edn. Churchill Livingstone, Edinburgh, pp 71–111

Peacock M, Marshall DH (1986) Assessment of calcium and magnesium status in humans. Group of European Nutritionists workshop on nutritional status assessment methodology for individuals and population groups. Perugia, Italy (in press)

Peacock M, Nordin BEC (1968) Tubular reabsorption of calcium in normals and hypercalciuric subjects. J Clin Pathol 21:353–358

Peacock M, Robertson WG (1988) Urinary calcium stone disease. In: De Groot L et al. (eds) Endocrinology, 2nd edn. Grune and Stratton, New York (in press)

Peacock M, Knowles F, Nordin BEC (1968) Effect of calcium administration and deprivation on serum and urine calcium in stone-forming and control subjects. Br Med J ii:729–731

Peacock M, Robertson WG, Nordin BEC (1969) Relation between serum and urinary calcium with particular reference to parathyroid activity. Lancet I:384–386

Peacock M, Hodgkinson A, Nordin BEC (1967) Importance of dietary calcium in the definition of hypercalciuria. Br Med J iii:469–471

Peacock M, Marshall RW, Robertson WG, Varnavides C (1976) Renal stone disease in primary hyperparathyroidism and idiopathic stone-formers, etiology and treatment. In: Finlayson B,

Thomas WC (eds) Colloquium on renal lithiasis. Gainsville University Press of Florida, pp 339–355

Peacock M, Aaron JE, Walker GS, Davison AM (1977) Bone disease and hyperparathyroidism in chronic renal failure: the effect of 1-alpha-hydroxy vitamin D_3. Clin Endocrinol 7:73S–81S

Peacock M, Taylor GA, Brown WB (1980) Plasma 1,25(OH)$_2$ vitamin D measured by radioimmunoassay and cytosol radioreceptor assay in normal subjects and patients with primary hyperparathyroidism and renal failure. Clin Chim Acta 101:93–101

Peacock M, Francis RM, Selby PL et al. (1982) 25OHD$_3$ test: normal subjects; idiopathic urinary calcium stone formers; female osteoporotics and vitamin D deficient osteomalacic patients. In: Norman AW, Schaefer K, Herrath DV, Grigoleit HG (eds) Vitamin D. Chemical, biochemical and clinical endocrinology of calcium metabolism. Walter de Gruyter, Berlin, pp 1057–1059

Peacock M, Francis RM, Selby PL (1983) Vitamin D and osteoporosis. In: Dixon AStJ, Russell RGG, Stamp TCB (eds) Osteoporosis. A multi-disciplinary problem. Academic Press and Royal Society of Medicine London, pp 245–254 (Royal Society of Medicine international congress and symposium series, no. 55)

Pedersen KO (1969) Determinations of calcium fractions of serum. 1. The separation of protein-bound and protein-free fractions by means of a simplified ultrafiltration technique. Scand J Clin Lab Invest 24:69–87

Pedersen KO (1970) Determination of calcium fractions of serum. 3. Ionized calcium in ultrafiltrates of normal serum and examination of pertinent variables. Scand J Clin Lab Invest 25:223–230

Phillips NJ, Cooke JNC (1967) The relation between urinary calcium and sodium in patients with idiopathic hypercalciuria. Lancet I:1354–1357

Pitts RF (1944) The effects of infusing glycine and of varying the dietary protein intake on renal hemodynamics in the dog. Am J Physiol 142:355–365

Porter RH, Cox BG, Heaney D, Hostetter TH, Stinebaugh PJ, Suki WN (1978) Treatment of hypoparathyroid patients with chlorothalidone. N Engl J Med 298:577–581

Pullman TN, Alving AS, Dern RJ, Landowne M (1954) The influence of dietary protein intake on specific renal functions in normal man. J Lab Clin Med 44:320–332

Quamme GA, Wong NLM, Sutton RAL, Dirks JH (1975) Interrelationship of chlorothiazide and parathyroid hormone: a micropuncture study. Am J Physiol 229:200–205

Quamme GA, O'Callaghan T, Wong NLM, Sutton RAL, Dirks JH (1976) Hypercalciuria in the phosphate depleted dog: a micropuncture study. Proc Am Soc Nephrol 9:6

Ralston SH, Fogelman I, Gardner MD, Dryburgh RA, Cowan RA, Boyle IT (1984) Hypercalcaemia of malignancy: evidence for an non-parathyroid humoral agent with an effect on tubular handling of calcium. Clin Sci 66:187–191

Rao PN, Prendeville V, Buxton A, Moss DG, Blacklock NJ (1982) Dietary management of urinary risk factors in renal stone formers. Br J Urol 54:578–583

Rasmussen H, Barrett PQ (1984) Calcium messenger system: an integrated view. Physiol Rev 64:938–985

Robertson WG (1969) Measurement of ionised calcium in biological fluids. Clin Chim Acta 24:149–157

Robertson WG (1976) Urinary excretion. In: Nordin BEC (ed) Calcium, phosphate and magnesium metabolism. Churchill Livingstone, Edinburgh, pp 113–161

Robertson WG (1985) Dietary factors important in calcium stone-formation. In: Schwille PO, Smith LH, Robertson WG, Vahlensieck W (eds) Urolithiasis and related clinical research. Plenum Press, New York, pp 61–68

Robertson WG, Marshall RW (1979) Calcium measurements in serum and plasma – total and ionised. CRC Crit Rev Clin Lab Sci 11:271–304

Robertson WG, Morgan DB (1972) The distribution of urinary calcium excretions in normal persons and stone-formers. Clin Chim Acta 37:503–508

Robertson WG, Peacock M (1968) New techniques for the separation and measurement of the calcium fractions of normal human serum. Clin Chim Acta 20:315–326

Robertson WG, Peacock M (1980) The cause of idiopathic calcium stone disease: hypercalciuria or hyperoxaluria. Nephron 26:105–110

Robertson WG, Peacock M (1985) The origin of metabolic abnormalities in primary calcium stone disease – natural or unnatural selection? In: Schwille PO, Smith LH, Robertson WG, Vahlensieck W (eds) Urolithiasis and related clinical research. Plenum Press, New York, pp 287–290

Robertson WG, Peacock M, Nordin BEC (1968) Activity products in stone-forming and non-stone-forming urine. Clin Sci 34:579–594

Robertson WG, Peacock M, Nordin BEC (1969) Calcium crystalluria in recurrent renal-stone formers. Lancet II:21–24

Robertson WG, Gallagher JC, Marshall DH, Peacock M, Nordin BEC (1974) Seasonal variations in urinary excretion of calcium. Br Med J iv:436–437

Robertson WG, Peacock M, Marshall RW, Speed R, Nordin BEC (1975) Seasonal variations in the composition of urine in relation to calcium stone-formation. Clin Sci 49:597–602

Robertson WG, Peacock M, Marshall RW, Marshall DH, Nordin BEC (1976) Saturation inhibition index as a measure of the risk of calcium oxalate stone formation in the urinary tract. N Engl J Med 294:249–252

Robertson WG, Peacock M, Heyburn PJ, Marshall DH, Clark PB (1978) Risk factors in calcium stone disease of the urinary tract. Br J Urol 50:449–454

Robertson WG, Peacock M, Hodgkinson A (1979a) Dietary changes and the incidence of urinary calculi in the UK between 1958 and 1976. J Chronic Dis 32:469–476

Robertson WG, Heyburn PJ, Peacock M, Hanes FA, Swaminathan R (1979b) The effect of high animal protein intake on the risk of calcium stone formation in the urinary tract. Clin Sci 57:285–288

Robertson WG, Peacock M, Heyburn PJ, Hanes FA (1980) Epidemiological risk factors in calcium stone disease. Scand J Urol Nephrol 53:15–28

Selby PL, Peacock M (1986) Ethinyl oestradiol and norethindrone in the treatment of primary hyperparathyroidism in postmenopausal women. N Engl J Med 314:1481–1485

Selby PL, Peacock M, Bambach CP (1984a) Hypomagnesaemia after small bowel resection – treatment with 1-alpha-hydroxylated vitamin D metabolites. Br J Surg 71:334–337

Selby PL, Peacock M, Marshall D (1984b) Hypercalcaemia. Management. Br J Hosp Med 31:186–197

Selby PL, Peacock M, Barkworth SA, Brown WB, Taylor GA (1985) Early effects of ethinyloestradiol and norethisterone treatment in post-menopausal women on bone resorption and calcium regulating hormones. Clin Sci 69:265–271

Shah PJR, Williams G, Green NA (1980) Idiopathic hypercalciuria: its control with unprocessed bran. Br J Urol 52:426–429

Sherman HC (1920) Calcium requirement in man. J Biol Chem 44:21–27

Sherwood LM, Potts JT Jr, Care AD, Mayer GP, Aurbach GD (1966) Evaluation by radioimmunoassay of factors controlling the secretion of parathyroid hormone. Nature 209:52–57

Sotornik I, Schuck O, Stribrna J (1969) Influence of diuretics on renal calcium excretion. Experientia (Basel) 25:591–592

Stamp TCB, Round JM (1974) Seasonal changes in human plasma levels of 25-hydroxycholecalciferol. Nature 147:563–565

Stanbury SW, Lumb GA (1962) Metabolic studies of renal osteodystrophy. 1. Ca, P and N metabolism in rickets, osteomalacia and hyperparathyroidism complicating chronic uraemia and in the osteomalacia of the adult Fanconi syndrome. Medicine 41:1–34

Sutton RAL, Dirks JH (1981) Renal handling of calcium phosphate and magnesium. In: Brenner BM, Rector FC (ed) The kidney, 2nd edn. Saunders, Philadelphia, pp 551–611

Sutton RAL, Wong NLM, Dirks JH (1975) 25-hydroxy vitamin D (25OHD$_3$): enhancement of distal tubular calcium reabsorption in the dog. Proc Am Soc Nephrol 8:8

Sutton RAL, Wong NLM, Dirks JH (1979) Effects of metabolic acidosis and alkalosis on sodium and calcium transport in the dog kidney. Kidney Int 15:520–533

Tan CM, Raman A, Sinnathyray JA (1972) Serum ionic calcium levels during pregnancy. J Obstet Gynaecol Br Commonw 79:694–697

Taylor A, Windhager EE (1979) Possible role of cytosolic calcium and Na–Ca exchange in regulation of transepithelial sodium transport. Am J Physiol 236:F505

Thorn NA (1960) The effect of anti-diuretic hormone on renal excretion of calcium in dogs. Dan Med Bull 7:110–112

Tjellsen L, Christiansen C, Humner L, Larsen NE (1983) Unchanged indices of bone turnover despite fluctuations in 1,25-hydroxy vitamin D during the menstrual cycle. Acta Endocrinol 102:476–480

Ullrich KJ, Rumrick G, Kloss S (1976) Acute Ca^{2+} reabsorption in the proximal tubule of the rat kidney. Pfluegers Arch 364:223–228

Walser M (1961) Calcium clearance as a function of sodium clearance in the dog. Am J Physiol 200:1099–1104

Walser M (1969) Renal excretion of alkaline earths. In: Comar CL, Bronner F (eds) Mineral metabolism, vol 3. Academic Press, New York, pp 235–320

Wasserman RH, Fullmer CS (1982) Vitamin D-induced calcium-binding protein. In: Cheung WY (ed) Calcium and cell function, vol II. Academic Press, New York, pp 175–216

Wasserman RH, Taylor AN (1969) Some aspects of the intestinal absorption of calcium with special reference to vitamin D. In: Comar CL, Bronner F (eds) Mineral metabolism – an advanced treatise. Academic Press, New York, pp 320–403

Wesson LG (1962) Magnesium, calcium and phosphate excretion during osmotic diuresis in the dog. J Lab Clin Med 60:422–433

Widrow SH, Levinsky NG (1962) The effect of parathyroid extract on renal tubular calcium reabsorption in the dog. J Clin Invest 41:2151–2159

Wilkinson R (1976) Absorption of calcium, phosphorus and magnesium. In: Nordin BEC (ed) Calcium, phosphate and magnesium metabolism. Churchill Livingstone, Edinburgh, pp 36–112

Wiske PS, Wentworth SM, Norton JA, Epstein S, Johnston CC (1982) Evaluation of bone mass and growth in young diabetics. Metabolism 31:848–854

Yendt ER, Gagne RJA, Cohanim M (1966) The effect of thiazides in idiopathic hypercalciuria. Am J Med Sci 251:449–460

Chapter 7

Calcified Tissues: Chemistry and Biochemistry

A. L. Boskey

Introduction: Composition of the Calcified Tissues

Over 90% of the calcium in the human body is deposited as calcium phosphate (hydroxyapatite) within the "hard tissues" of the skeleton and the teeth. Hydroxyapatite is also frequently found pathologically in metastatic and dystrophic deposits (Dieppe and Calvert 1983). The extracellular matrices of the physiologically calcified tissues (calcified cartilage, ligamentous and tendinous insertions, the mineralized matrix of bone itself, as well as the dentine, cementum, pulp and enamel of teeth) contain from 20% to 90% calcium phosphate. The presence of the mineral within these tissues provides them with strength and rigidity, allowing them to function properly. The mineral content is generally expressed either as the proportion of mineral found per unit weight of dry tissue (ash content) or as the proportion of mineral in the intact organ. These parameters may be quite different; for example, in the osteoporotic bone the mineral content is markedly reduced due to the increased porosity (decreased density) of the bone, while the distribution of mineral in the tissue of that bone (ash content) may not be altered. In general, the extent of mineralization depends upon tissue function, site within the tissue, age and species. For example, the ash content of embryonic chick bones increases during development from 13% to 51% (Pellegrino and Biltz 1972). In general, the ash content of most bone in mature animals is between 50% and 70%. However, the bones of the ear are 80%–90% mineral (Bonar and Lees 1984). The hydroxyapatite content of the extracellular matrices of some representative mammalian hard tissues is summarized in Table 7.1.

All the calcified tissues consist of hydroxyapatite mineral, proteins, water, lipids and cells. With the exception of enamel, whose matrix proteins are enamelins and amelogenins (Termine et al. 1979), the principal matrix protein upon which hydroxyapatite is deposited is collagen (Miller 1984). In addition to collagen, the calcified matrices also contain 2%–7% non-collagenous proteins (Table 7.2). Most of these non-collagenous proteins are synthesized by the connective tissue cells (Butler 1984) but some, such as albumin and the

Table 7.1. Ash content of representative hard tissues containing hydroxyapatite

Tissue	Age	Mineral content (% ash)	Reference
Embryonic chick bone	9 d	13	a
	12 d	26	a
	15 d	38	a
	18 d	45	a
	21 d	51	a
Chick bone	6 mo	64	a
Rat diaphyseal bone	11 d	43	b
	8 wk	65	c
	14 wk	70.1	c
	22 wk	72.2	c
Mature human enamel	. .	95–97	d
Bovine dentine	6 mo	75	d

a, Pellegrino and Biltz (1972); b, Boskey and Marks (1985); c, Burnell et al. (1980); d, Driessens (1982).

Table 7.2. Major non-collagenous proteins in bone, cartilage and dentine (all as % of non-collagenous proteins) (Recalculated from Butler 1984; Conn and Termine 1985; Fisher et al. 1983a, b; Linde 1984; Smith et al. 1985; Termine et al. 1981a, b; Thomas and Leaver 1975)

Bone	Calcified cartilage	Dentine
Osteonectin (20%–25%)	Proteoglycans (33%–39%)	Phosphoproteins (50%–55%)
Blood-derived proteins (20%–25%)	Chondrocalcin (5%–30%)	Acidic glycoproteins (20%–25%)
Gla proteins (13%–15%)	Acidic glycoproteins (7%–8%)	Proteoglycans (5%–10%)
Phosphoproteins (10%–13%)		Gla proteins (3%–8%)
Acidic glycoproteins (7%–8%)		
Proteoglycans (7%–9%)		

immunoglobulins, are not synthesized by these cells, and accumulate in the mineralized matrices because of their affinity for hydroxyapatite (Smith et al. 1985).

The average size and perfection of the hydroxyapatite crystals in the mineralized tissues also differ with tissue type and animal age (Table 7.3). Chemical analysis of the mineral isolated from the calcified matrices by ashing or acid hydrolysis indicates that ions other than calcium, phosphate and hydroxide are contained in these tissues (Table 7.4). Bone and dentine contain fairly large proportions of carbonate, and are often referred to as carbonate-apatites (Young and Brown 1982). All the biological apatites are considered "imperfect" due to the presence of foreign ions within their lattices and/or on their surfaces (Posner et al. 1980). Some of these ions, such as magnesium, iron, sodium and chloride, are predominantly adsorbed on to the surface of the hydroxyapatite crystallites, whereas others such as strontium, fluoride and carbonate are incorporated as part of the mineral lattice (Eanes and Termine 1983). The presence of these ions may markedly alter the properties of the calcified tissues: strontium and fluoride in enamel for example, decrease the solubility of the hydroxyapatite and thus are thought to be anti-caries agents capable of preventing the dissolution of tooth enamel (Driessens 1982). Fluoride is also being used therapeutically to stabilize bone mineral crystals in osteoporotic patients (Riggs et al. 1980). Aluminium (Ellis et al. 1979; Thomas and Meyer 1984) and iron (Huser et al. 1971) accumulate at the mineral surface in association with increased ingestion and/or imperfect clearance of these ions.

Table 7.3. Common features of calcification in mammalian tissues (Adapted from Anderson 1976; Franzen et al. 1982; Schenk et al. 1982; Bernard and Pease 1969; Carrino et al. 1984; Almuddaris and Dougherty 1979; Garant and Cho 1984; Robinson et al. 1983; Warshawsky 1984)

	Calcified cartilage	Bone	Dentine	Enamel
Cell type	Chondrocyte	Osteoblast	Odontoblast	Ameloblast
Site of initial mineral	Matrix vesicles	Collagen vesicles	Collagen vesicles	Dentine–enamel junction
Major matrix protein	Types II, I and X collagen	Type I collagen	Type I collagen	Enamelins and amelogenins
Mean mineral (ash) content	40%–50%	50%–70%	70%–75%	92%–96%
Length of mineral crystals	150–200 Å	200–300 Å	250–350 Å	1000–2000 Å
Characteristics of major non-collagenous proteins	Anionic	Anionic	Anionic	Anionic/ hydrophobic
Proteoglycans	Large	Small	Small	

Table 7.4. Elemental composition of the mineral within calcified tissues (weight % ash) (Based on unpublished data, and reports by Knuuttila et al. 1985 and Driessens 1982)

	Ca	P	Mg	CO_3	F	Na
Bone						
Human cortical	32–34	14–15	3.3	5.5	0–0.2	1.5–2.0
Human calcified cartilage	33–37	14–16	3–7	3–8	—	2–5
Teeth						
Human enamel	34–39	16–18	0.2–0.6	2–5	0.01–0.3	0.2–0.9
Human dentine	34–39	17–19	1.0–1.3	6–7	—	0.2–0.6

Chemical analyses of the hydroxyapatite crystals also reveal that the ratio of Ca : P : OH is different from the 5 : 3 : 1 ratio expected for stoichiometric hydroxyapatite.

Comparative Histogenesis of the Calcified Tissues

The cells involved in formation of the cartilage, bone, dentine and enamel matrices (Table 7.3), though morphologically and functionally unique, have in common the ability to regulate the calcification process by the synthesis of an appropriate extracellular matrix and the regulation of the flux of mineral ions into that matrix (Schenk et al. 1982).

The spatial relationship of the cells in the calcified matrix is different in each of the tissue types. In bone the osteoblasts, connected to one another by long processes, are totally surrounded by mineralized matrix. In dentine, the odontoblasts move away from the calcifying matrix so that only the highly polarized extensions of the cells extend into the calcifying matrix. Mature enamel is essentially acellular.

The chemical composition of the extracellular matrices produced by the different cell types is also quite different (Table 7.3). Bone contains predominantly type I collagen and the predominant non-collagenous proteins are osteonectin (Termine et al. 1981b) and osteocalcin (Price et al. 1976; Hauschka et al. 1983). Calcifying cartilage contains a high proportion of proteoglycans and predominantly type II collagen (Franzen et al. 1982). Proteoglycans are extremely large molecules containing carboxylated and sulphated glycosaminoglycan chains covalently bound to a protein core. In cartilage these anionic proteoglycan monomers associate non-covalently with hyaluronic acid to form high molecular weight, space-filling aggregates. Proteoglycans in calcifying cartilage are smaller and less highly aggregated than those in the non-calcifying cartilages (Buckwalter 1983), but the mineralized cartilagenous matrix retains a relatively high proportion of glycosaminoglycans relative to the other mineralized tissues (Poole et al. 1982). Dentine, similar to bone, contains predominantly type I collagen, but in this tissue the major (Maier et al. 1985) non-collagenous proteins are the phosphoproteins (Veis 1984; Tsay and Veis 1985).

Despite all these differences, there are several common histological features of early mineralization shared by these tissues. These can be seen by comparing the appearance of each of the newly calcified tissues (Hunziker et al. 1984). In the epiphyseal growth plate during the process of endochondral ossification (Holtrop 1972), cartilage becomes calcified and is replaced by bone. Initial mineral deposition in the cartilage occurs at discrete sites on membrane-bound bodies (matrix vesicles) in the extracellular matrix (Anderson 1969; Ali 1976). Initial mineral deposits are diffuse and not oriented. Mineral crystals proliferate, and mineralization proceeds filling the longitudinal, but not the transverse, septa. Changes in enzymatic activities (Vaananen and Korhonen 1978), especially of those enzymes capable of hydrolysing phosphate esters (Wuthier and Register 1984; Hsu 1983) and proteolytic enzymes (Woessner and Howell 1983), occur prior to or concomitant with the appearance of mineral. Following vascular invasion (Trueta 1973) lamellar bone is formed by osteoblasts directly on the surface of the pre-existing mineralized cartilage. The osteoblasts secrete a type I collagen matrix with very little proteoglycan (Sugahara et al. 1981; Carrino et al. 1984), and few extracellular matrix vesicles.

Similarly, during dentinogenesis (Garant and Cho 1984), initial mineralization of the first layer of mantle dentine occurs in association with extracellular matrix vesicles, while the next layer of mineral (circumpulpal dentine) forms apparently without vesicles. Thus mineralization occurs initially at discrete sites. The mineral deposition then progresses in a highly oriented fashion. However, the predentine is not mineralized, and the enzymes involved in modification of the predentine matrix (TenCate 1967) are similar to those involved in endochondral ossification.

Even in enamel mineralization, which does not occur on a collagenous matrix, there are many similarities. The enamel matrix consists of two groups of proteins, enamelins and amelogenins (Termine et al. 1979). There are no collagens, no proteoglycans and no matrix vesicles in the enamel matrix. Since the enamelins cannot be extracted from the mineralized tissue without decalcification, it has been suggested that enamelins provide the matrix site for mineral deposition (similar to the template provided by the collagen in bone and dentine) while amelogenins are more involved in control of mineral size and orientation (Warshawsky 1984). In contrast to the other calcified tissues the mature enamel matrix contains little organic matrix (less than 3%) and few cells. However, the hydroxyapatite crystals in enamel do form in a highly organized fashion (lying in prisms perpendicular to the junction between dentine and enamel), and this is also the case in the other calcified tissues.

In all these tissues the deposition of mineral is a dynamic process, controlled by phenotypic cells. These dynamic events can be observed in histological sections of young animals (Hunziker et al. 1984), in bone and dentine organ cultures (Wigglesworth 1968; Osdoby and Caplan 1981; Tenenbaum and Heersche 1982; Finkelman and Butler 1985) and in chondrocyte (Binderman et al. 1979; Osdoby and Caplan 1979) and osteoblast cell cultures (Binderman et al. 1974; Escarot-Charrier et al. 1983; Sudo et al. 1983).

Calcification Mechanisms

In Vitro Formation of Hydroxyapatite

The current picture of the chemical events involved in the formation of hydroxyapatite in biological tissues comes for the most part from solution studies of hydroxyapatite deposition (Posner and Betts 1981). The events involved in the formation of hydroxyapatite in solution are the same as those involved in any crystal formation and crystal growth process. Formation of an insoluble crystalline precipitate involves two energy-requiring steps: (a) nucleation and (b) crystal growth and ripening. The process of nucleation in which the first stable crystal embryos (nuclei) are formed is the thermodynamically least favourable stage (Garside 1982). During the nucleation stage many ions and/or ion clusters must collide simultaneously and in the correct orientation in order to form the first stable-insoluble crystals. The probability of forming such insoluble nuclei may be increased by increasing the collision frequency, by increasing the solution supersaturation (i.e. raising the concentration so that the ion product exceeds the solubility product to a greater extent), decreasing the solution viscosity and/or increasing the temperature.

The energy requirement for nucleation can be reduced by providing heterogeneous nuclei (dust or foreign particles). In fact, most crystallization occurs by heterogeneous rather than homogeneous nucleation (de novo nucleation). A special class of heterogeneous nucleators are the epitaxial nucleators (Irving 1981) which have surfaces which structurally resemble the precipitating phase.

Crystal growth and proliferation occur following nucleation via the addition of ions or ion clusters to growth centres on the nuclei. Ions may add preferentially to one or more surfaces causing oriented growth. Small crystals may break off the surface of larger crystals (secondary nucleation) causing an exponential increase in the number of nuclei. The rate of crystal growth is a function of the number and properties of nuclei, the solution supersaturation and so on. Crystal growth, in contrast to nucleation, can continue, however, at fairly low supersaturations. In fact, crystal growth generally does not stop, since the smaller, more imperfect crystals dissolve while the larger crystals continue to develop (Ostwald ripening).

Crystal growth may be inhibited by the addition of materials which bind either to the crystal surface or to ions in solution. Complexing ions in solution both decreases the effective supersaturation and alters the ability of the ions to fit into growth centres. Inhibitors which bind to the surface block or "poison" growth sites, preventing dissolution (and Ostwald ripening) as well as growth.

The formation of hydroxyapatite from solution has been extensively studied (for reviews see Posner and Betts 1981; Brown and Young 1982; Eanes 1984). Since calcium phosphate phases other than hydroxyapatite are formed more readily (having lower energy requirements) and are, in addition, much more insoluble, it has been suggested that calcium phosphate precursor formation precedes hydroxyapatite deposition in solution (Feenstra and DeBruyn 1981). Such precursors would either later be transformed to hydroxyapatite or would serve as epitaxial nucleators for hydroxyapatite. Precursor phases (Boskey 1984) identified during hydroxyapatite formation in solution include an x-ray amorphous tricalcium phosphate, octacalcium phosphate and brushite. The first two can convert in solution to hydroxyapatite. In fact, recent studies show that amorphous calcium phosphate converts under some circumstances to octacalcium phosphate, prior to its conversion to hydroxyapatite (Tung and Brown 1983; Chen 1985). Octacalcium phosphate (Nelson and McLean 1984) and brushite (Francis and Webb 1971) can provide a well-matched epitaxial surface for hydroxyapatite formation. The conditions favouring the formation of each of these phases are quite variable and there are in vitro conditions of low supersaturation where hydroxyapatite forms without the formation of a detectable intermediate phase (Boskey and Posner 1976).

Conversion of precursor phases and growth and ripening of hydroxyapatite in vitro can be altered by the addition of ions or macromolecules that affect the stability of the precursor or nucleator. Examples of such effectors and their actions are summarized in Table 7.5. The in vitro actions of these effectors may explain, in part, their in vivo actions in the presence of cells and matrix.

In Vivo Calcification

Physiological formation of hydroxyapatite (for more detailed reviews see Boskey 1981; Glimcher 1981; Wuthier 1982; Eanes 1984) involves a similar set of nucleation and growth events. The situation is much more complex in vivo than in vitro due to the presence of cells and extracellular matrix which are involved in providing calcified tissues of differing physical characteristics. In vitro studies using organ and cell cultures as well as analyses of the temporal changes in the

Table 7.5. Agents affecting hydroxyapatite formation in vitro

Agent	Effect on					
	ACP	OCP	Brushite	HA (de novo)	HA (growth)	Reference
Aluminium	+	...	...	−	−	a
Magnesium	+	+	+	−	=	b, c
Citrate	+	...	...	...	−	c, d
Fluoride	−	+/−	...	...	+	e
Pyrophosphate	+	+	...	−	−	f
Diphosphonates	+	+	+	...	−	f, g
Polycarboxylic acids	...	...	+	...	−	h
Phosphoproteins	−	...	...	+	−	i, j
Proteoglycans	+	...	...	−	−	k, l
Osteocalcin	...	...	+	=	−	m, n
Salivary proteins	...	...	+	...	−	o
Phospholipids	+	...	...	+	−	p, q

ACP, amorphous calcium phosphate; OCP, octacalcium phosphate; HA, hydroxyapatite (de novo formation refers to direct formation from solution, growth refers to seeded growth).
+ indicates the precursor is stabilized or formation and growth is favoured.
− indicates the precursor is less stable, and formation and growth is inhibited.
= indicates there is no effect, ... indicates effect has not been determined.
+/− indicates the effect is concentration dependent.
a, Blumenthal and Posner 1984; b, Blumenthal et al. 1977; c, Christoffersen and Christoffersen 1981a, b; d, Meyer and Thomas 1982; e, Young and Brown 1982; f, Meyer and Fleisch 1984; g, Francis 1979; h, Amjad 1984, i, Nawrot et al. 1976; j, Termine et al. 1980; k, Blumenthal et al. 1979; l, Chen and Boskey 1985; m, Price et al. 1976; n, Boskey et al. 1985; o, VanDyke et al. 1979; p, Boskey and Posner 1977; q, Wuthier and Eanes 1975.

chemical composition of calcifying tissues have begun to provide insight into the sequence of events involved in physiological calcification.

Physiological tissue mineralization requires (a) an extracellular matrix suitable for calcification, (b) a mechanism for initial deposition and (c) a mechanism for proliferation and growth of hydroxyapatite crystals.

The Extracellular Matrix

In addition to collagen there are present in lesser amounts, a large number of highly charged anionic non-collagenous proteins including phosphoproteins (Glimcher 1984; Veis 1984), glycoproteins (Fisher et al. 1983a), proteoglycans (Franzen et al. 1982; Fisher et al. 1983b), gamma-carboxy glutamic acid containing proteins (Price et al. 1976; Hauschka et al. 1983) and proteolipids (Boyan 1984). One feature of all of these anionic macromolecules is their ability to sequester calcium, perhaps thereby regulating the free ionic calcium concentration in the matrix. They may also, because of their negative charge, exclude phosphate from the matrix, thereby also controlling the supersaturation (CaxP product) and inhibiting hydroxyapatite formation (Chen et al. 1984). Many of these matrix proteins are modified enzymatically prior to, or concomitant with, matrix calcification, thereby increasing the CaxP product of the tissue. Modification of these proteins may also make them into nucleators or less effective inhibitors of hydroxyapatite formation. It is of interest to note that the enamelins and amelogenins, which are the major proteins secreted by the ameloblast, are also highly anionic.

Reconstituted type I collagen was the first calcified tissue component shown to be able to cause hydroxyapatite formation both in vitro (Glimcher and Krane 1968) and in vivo (Mergenhagen et al. 1960). Although the major collagens in bone, dentine, skin, ligament and tendon are identical in amino acid composition and structural properties (type I) (Miller 1984), recent studies by Mechanic et al. (1984) indicate that the collagen cross-linking (the chemical bonds stabilizing aggregates of collagen molecules) in mineralized tissues is different from that in non-mineralized tissue. However it is not known whether these differences are present prior to mineral deposition.

Recently type I collagen from which non-collagenous proteins had been extracted was shown to be incapable of supporting hydroxyapatite formation (Termine et al. 1981). Therefore it is believed that some other extracellular matrix component(s) must be present for hydroxyapatite nucleation. The only other calcified tissue matrix components shown to be capable of inducing hydroxyapatite formation when implanted in vivo (Raggio et al. 1986) are the proteolipids (Boyan 1984) and their component complexed acidic phospholipid phosphate complexes (Boskey and Posner 1975; Boyan-Salyers and Boskey 1980). Whether collagenous or non-collagenous structures serve as the sites for the first deposition of mineral remains to be determined. Nonetheless, collagen serves a structural role, providing a template for mineral alignment.

In growth and articular cartilage (Anderson 1969; Ali 1976), mantle dentine (Almuddaris 1979), ligamentous and tendinous insertions (Yamada 1976) and in lamellar bone (Bernard and Pease 1969) extracellular matrix vesicles have been identified at the site of initial mineral deposition. These extracellular membrane-bound bodies, enriched in the enzymes alkaline phosphatase, ATP-pyrophosphohydrolase and several neutral proteases, are derived from the cells by some yet to be determined process (Wuthier 1982). The matrix vesicles have several possible functions in the mineralization process: provision of a protected limited environment in which mineral ions can accumulate and form the first mineral; storage and accumulation of ions for precursor formation; and storage of enzymes required for modification of the matrix prior to the onset of calcification.

The matrix vesicle membranes, as well as the membranes of the cells of the calcifying connective tissues, contain calcium acidic phospholipid phosphate complexes as components of specific membrane proteolipids (Boyan 1984). These lipid macromolecules cause hydroxyapatite deposition in vitro (Boskey and Posner 1977) and in vivo (Raggio et al. 1986), hence it is not surprising that mineral crystal deposition has been reported to be associated with cell membranes in bone and in dentine (Dougherty 1978).

The only other matrix macromolecule that has been shown to promote in vitro calcification (in the presence of collagen) is the bone-specific glycoprotein, osteonectin (Termine et al. 1981a, b). Dentine and bone phosphoproteins (Termine et al. 1980), enamelins and amelogenins (Doi et al. 1984), bone and dentine gla protein (osteocalcin) (Boskey et al. 1985) and cartilage proteogly-cans (Cuervo et al. 1973; Blumenthal et al. 1979; Chen et al. 1984; Chen and Boskey 1985) all inhibit hydroxyapatite formation and growth in vitro. There are several other non-collagenous proteins which accumulate at the mineralization front, either because they play a yet to be determined role in mineralization or because they are adsorbed to the mineral surface. The functions of chondrocalcin (recently shown to be a type II collagen propeptide) (Poole et al. 1984), bone sialoprotein (Fisher et al. 1983b) and several less abundant glycoproteins remain to be elucidated. Further, recent studies suggest that several of the anionic macromolecules may function both as nucleators and inhibitors varying in function with concentration, extent of phosphorylation and interaction with other matrix molecules. For example, dentine phosphophoryn can both promote initial hydroxyapatite deposition (Nawrot et al. 1976) and inhibit hydroxyapatite growth (Termine et al. 1981b). It is conceivable that one of the functions of alkaline phosphatase, an enzyme long associated with mineralization (Robson 1923) and recently shown to function at physiological pH as a phosphoprotein phosphatase (Burch et al. 1985), is to control the properties of such matrix molecules by regulating both the extent of protein phosphorylation and the phosphate ion concentration at the site of mineral deposition.

The Initial Deposition of Mineral

This requires an elevation of local calcium and phosphate concentrations. This may occur as a result of the release of calcium and/or phosphate already bound to matrix molecules or as a result of an influx of additional ions. In the epiphyseal growth plate, chondrocyte mitochondria accumulate calcium just prior to the onset of calcification, and loose this calcium by the time calcification commences (Brighton and Hunt 1976). Although the total amount of calcium lost by the mitochondria accounts for less than 2% of the calcified cartilage matrix calcium, the release of calcium may cause a local increase in calcium phosphate ion product sufficient to facilitate initial mineral deposition. Accumulation of these ions within or on the surface of extracellular matrix vesicles would allow initial mineral deposition to occur in these extracellular bodies.

The nature of the initial mineral phase deposited in the calcified tissues is unknown, although as reviewed elsewhere (Boskey 1984; Eanes 1984) each of the calcium phosphate precursor phases formed in vitro, as well as others, have been identified in biological deposits. Which of these phases truly exists in the tissues and which are artefacts of preparation remains to be determined. If transient precursors exist in vivo, the mechanism of their transformation to hydroxyapatite, if different from that in vitro, also remains unknown.

Growth and Proliferation of Hydroxyapatite Mineral Crystals

Once the initial hydroxyapatite crystals are present in the tissue they must grow and fill the interstices of the matrix in an oriented fashion. Growth in this sense refers both to the accretion of additional small nuclei (by the process of secondary nucleation) and to the increase in size of initial mineral cystals. As seen from Table 7.3 the size of the crystals in the physiologically calcified tissues appears to be tightly controlled.

Collagen plays a key role in regulating the orientation of hydroxyapatite crystals, and may also be important in controlling crystal size, since crystals in non-collagenous dystrophic deposits (Boskey and Bullough 1984) and those in enamel are appreciably larger than those deposited in a collagenous matrix. Although there is some debate as to whether the mineral fills the collagen fibrils or is deposited on the surface of the collagen fibrils, it is well accepted that most of the growth of hydroxyapatite crystals occurs along the c-axis (needle direction) which in turn is aligned parallel to the fibril axis of the collagen (Glimcher 1981).

The non-collagenous proteins probably also play a role in regulating hydroxyapatite orientation and growth. Many have a high affinity for both calcium and hydroxyapatite, and several have been shown to regulate hydroxyapatite growth in solution (Table 7.4). Definitive proof that these molecules play an active role in controlling in vivo hydroxyapatite growth (rather than a passive role due to their accumulation on the surface of pre-existing mineral) awaits study of animals in which these molecules are absent or non-functional.

Remodelling

Bones are in a dynamic state, and are constantly being remodelled and reformed. The calcified cartilage segment of bone is almost completely resorbed and replaced by newly formed bone during the process of endochondral ossification (Schenk et al. 1982). The ligamentous and tendinous insertions are similarly remodelled. Turnover of cancellous and cortical bone occurs both in response to mechanical stress (according to Wolff's law, bone remodels so as to maximize its mechanical properties) and in response to the need for calcium homeostasis. In general, the more recently deposited, more highly vascularized bone is remodelled in preference to the more highly mineralized tissue.

According to the principles of physical chemistry, smaller crystals are appreciably more soluble than larger ones. The increased solubility of the smaller crystals is the basis of the Ostwald ripening phenomenon, and also explains, in part, why the newest mineral is resorbed first.

Bones of immature animals have a broader distribution of crystallite sizes than those of more mature animals (Boskey and Marks 1985). This distribution is sharpened in favour of larger crystals with maturity. Crystals in osteoporotic bone have a size distribution that is skewed even more in favour of larger crystals (Thompson et al. 1983), since this condition is characterized by increased resorption and/or decreased formation. In contrast, the bones of osteopetrotic rats, which lack functional resorbing cells (and hence resorb neither their calcified cartilage nor their bone), have smaller crystals than age-matched healthy animals (Boskey and Marks 1985).

Although it is not clear precisely what agents are released from osteoclasts (in bone) and chondroclasts (in cartilage) to facilitate removal of the mineral, the environment adjacent to the osteoclasts' ruffled border (the area of the bone-removing cell seen in apposition to the mineralized matrix) has an acidic pH. This low pH, in a solution of sufficient volume and buffering capacity, would be acidic enough to demineralize the entire bone. However, the area of bone which is exposed to this environment, and hence the area which is decalcified, is quite limited.

Detailed studies of the dissolution of hydroxyapatite mineral have generally investigated enamel crystals or large, synthetic hydroxyapatite crystals prepared at high temperatures as models of dental caries formation. Such studies have suggested the existence of two distinct sites of different solubility in the apatite crystals: one less soluble site along longitudinal channels within the crystal, and one more soluble site on the crystal surface (Higguchi et al. 1984). However, the general applicability of the two-site model has not yet been established and there are several other models predicting other dissolution patterns (Christofferson and Christofferson 1981a, b).

Solubilization of crystals generally begins with surface dissolution. In this way, impurities and defects adsorbed on the surface can be removed and replaced by lattice ions. Surface adsorption can also alter the solubility of crystals, thus enamel mineral and synthetic hydroxyapatite are rendered less soluble by adsorption of salivary proteins (VanDyke et al. 1979) and polycarboxylic acids (Amjad 1984). It is likely that in bone and cartilage, surface adsorption of matrix proteins or serum proteins can also alter the solubility of mineral crystals. Perhaps the function of some of the lysosomal enzymes released along with collagenase (Galloway et al. 1983) during resorption is to remove some of these protective proteins.

Some of the non-collagenous proteins that adsorb to the mineral in bone (e.g. osteocalcin) may also provide recognition signals for osteoclasts. Bone removed from osteocalcin-deficient animals and implanted in muscle of normal animals is not resorbed as efficiently as implants of normal bone (Lian et al. 1984), thus osteocalcin may be serving as a chemotactic agent for macrophages. Contradicting this observation is the report that in osteocalcin-deficient animals bone resorption associated with repair appears to be normal (Price et al. 1981) although there is excessive growth plate mineralization in these animals (Price et al. 1982).

After the mineral is removed, or during removal, the organic matrix of bone is degraded enzymatically. Collagenase (Sakamoto and Sakamoto 1984) and lysosomal and non-lysosomal proteolytic enzymes (Galloway et al. 1983) are responsible for further matrix resorption. Resorption is followed by matrix synthesis and calcification of the newly synthesized matrix as described previously. The processes of resorption and reformation are generally believed to be coupled (Farley and Baylink 1982; Urist et al. 1983) by factors released by one cell activating the other.

Conclusions

The development of sophisticated chemical and biological techniques has markedly expanded understanding of the events involved in formation of calcified tissues. The chemical composition of the calcified tissues has been precisely defined, and studies using techniques of molecular biology, cell culture and immunocytochemistry are beginning to define the functions of minor as well as major calcified tissue components. Such technologies hold promise of answering the unsolved questions: the nature and site of initial mineral deposition, the precise role of each of the calcified tissue matrix components in mineralization, the nature of the factors coupling resorption to formation and the ways in which normal and abnormal calcification can be controlled by outside intervention. Perhaps the most important of the questions about calcification is what allows mineral to form where it is needed, and prevents its formation where it is unwanted.

Acknowledgements. Dr. Boskey's research described in this chapter was supported by NIH grant DE 04141. The author greatly appreciates the many helpful discussions she had with Dr. Peter G. Bullough during the preparation of this manuscript.

References

Ali SY (1976) Analysis of matrix vesicles and their role in the calcification of epiphyseal cartilage. Fed Proc 35:135–142

Almuddaris MF, Dougherty WF (1979) The association of amorphous mineral deposits with the plasma membrane of pre- and young odontoblasts and their relationship to the origin of dentinal matrix vesicles in rat incisor teeth. Am J Anat 155:223–244

Amjad Z (1984) The influence of mellitic acid on the crystal growth of hydroxyapatite. In: Misra DN (ed) Adsorption on and surface chemistry of hydroxyapatite. Plenum Press, New York, pp 1–12

Anderson HC (1969) Vesicles associated with calcification in the matrix of epiphyseal cartilage. J Cell Biol 41:59–72

Anderson HC (1976) Matrix vesicle calcification. Fed Proc 35:105–108

Bauminger E, Ofer S, Gedalia I, Horowitz G, Mayer I (1985) Iron uptake by teeth and bones: a Mössbauer effect study. Calcif Tissue Int 37:386–389

Bernard GW, Pease DC (1969) An electron microscopic study of initial intramembranous osteogenesis. Am J Anat 125:271–290

Binderman I, Duksin D, Harrell A, Ephraim K, Sachs L (1974) Formation of bone tissue in culture from isolated bone cells. J Cell Biol 61:427–439

Binderman I, Green RM, Pennypacker JP (1979) Calcification of differentiating skeletal mesenchyme in vitro. Science 206:222–225

Blumenthal NC, Posner AS (1984) In vitro model of aluminum osteomalacia: inhibition of hydroxyapatite formation and growth. Calcif Tissue Int 36:439–441

Blumenthal NC, Betts F, Posner AS (1977) Stabilization of amorphous calcium phosphate by Mg and ATP. Calcif Tissue Res 23:245–250

Blumenthal NC, Posner, AS, Silverman LD, Rosenberg LC (1979) The effect of proteoglycans on in vitro hydroxyapatite formation. Calcif Tissue Res 27:75–82

Bonar LC, Lees S (1984) Structure and composition of very heavily mineralized bone of mammalian tympanic bullae. In: Butler WT (ed) The chemistry and biology of mineralized tissues. Ebsco Medica Inc., Birmingham, Alabama, p 419

Boskey AL (1981) Current concepts of the biochemistry and physiology of calcification. Clin Orthop 157:165–196

Boskey AL (1984) Overview of cellular elements and macromolecules implicated in the initiation of mineralization. In: Butler WT (ed) The chemistry and biology of mineralized tissues. Ebsco Medica Inc., Birmingham, Alabama pp 335–343

Boskey AL, Bullough PG (1984) Cartilage calcification: normal and aberrant. SEM Inc., AMF O'Hare (Chicago), pp 943–952

Boskey AL, Marks SC (1985) Mineral and matrix alterations in the bones of incisors-absent (ia/ia) osteopetrotic rats. Calcif Tissue Res 37:287–292

Boskey AL, Posner AS (1975) Extraction of a calcium–phospholipid–phosphate complex from bone. Calcif Tissue Res 19:273–283

Boskey AL, Posner AS (1976) The formation of hydroxyapatite at low supersaturations. J Phys Chem 80:40–46

Boskey AL, Posner AS (1977) The role of synthetic and bone extracted Ca–phospholipid–PO$_4$ complexes in hydroxyapatite formation. Calcif Tissue Res 23:251–258

Boskey AL, Wians FH Jr, Hauschka PV (1985) The effect of osteocalcin on in vitro lipid-induced hydroxyapatite formation and seeded hydroxyapatite growth. Calcif Tissue Int 37:57–62

Boyan BD (1984) Proteolipid-dependent calcification In: Butler WT (ed) The chemistry and biology of mineralized tissues. Ebsco Medica Inc., Birmingham, Alabama, pp 125–131

Boyan-Salyers BD, Boskey AL (1980) Relationship between proteolipids and calcium–phospholipid–phosphate complexes in *Bacterionema matruchotti* calcification. Calcif Tissue Int 30:167–174

Brighton CT, Hunt RM (1976) Histochemical localization of calcium in growth plate mitochondria and matrix vesicles. Fed Proc 35:143–147

Brown WE, Young RA (1982) Structures of biological minerals. In: Nancollas GW (ed) Biological mineralization and demineralization. Springer-Verlag, Berlin Heidelberg New York, pp 101–141

Buckwalter JA (1983) Proteoglycan structure calcifying cartilage. Clin Orthop 172:207–232

Burch WM, Warner G, Wuthier RE (1985) Phosphotyrosine and phosphoprotein phosphatase activity of alkaline phosphatase in mineralizing cartilage. Metabolism 34:169–175

Burnell JM, Teubner EJ, Miller AG (1980) Normal maturational changes in bone matrix, mineral, and crystal size in the rat. Calcif Tissue Int 31:13–19

Butler WT (1984) Matrix macromolecules of bone and dentin. Coll Rel Res 4:297–307

Carrino DA, Weitzhandler M, Caplan A (1984) Proteoglycans synthesized during the cartilage to bone transition. In: Butler WT (ed) The chemistry and biology of mineralized tissues. Ebsco Medica Inc. Birmingham, Alabama pp 197–208

Chen CC, Boskey AL (1985) Mechanisms of proteoglycan inhibition of hydroxyapatite growth. Calcif Tissue Int 37:395–400

Chen CC, Boskey AL, Rosenberg LC (1984) The inhibitory effects of cartilage proteoglycans on hydroxyapatite growth. Calcif Tissue Int 36:285–290

Chen P-T (1985) Octacalcium phosphate formation in vitro: implications for bone formation. Calcif Tissue Int 37:91–94

Christoffersen J, Christoffersen MR (1981a) Kinetics of dissolution of calcium hydroxyapatite. 4. The effect of some biologically important inhibitors. J Cryst Growth 53:42–54

Christoffersen J, Christoffersen MR (1981b) Kinetics of dissolution of calcium hydroxyapatite 5. J Cryst Growth 53:42–54

Conn KM, Termine JD (1985) Matrix protein profiles in calf bone development. Bone 6:33–36

Cuervo LA, Pita JC, Howell DS (1973) Inhibition of calcium phosphate mineral growth by proteoglycan aggregate fractions in a synthetic lymph. Calcif Tissue Res 13:1–10

Dieppe P, Calvert P (1983) Crystals and joint disease. Chapman and Hall, London

Doi Y, Eanes ED, Shimokawa H, Termine JD (1984) Inhibition of seeded growth of enamel apatite crystals by amelogenin and enamelin proteins in vitro. J Dent Res 63:98–103

Dougherty WG (1978) The occurrence of amorphous mineral deposits in association with the plasma membrane of active osteoblasts in rat and mouse alveolar bone. Metab Bone Dis Rel Res 1:119–123

Driessens FCM (1982) Mineral aspects of dentistry. In: Meyers HM (ed) Monographs on oral science, vol 10. Karger, Basel

Eanes ED (1984) Dynamic aspects of apatite phases of mineralized tissues. Model studies. In: Butler WT (ed) The chemistry and biology of mineralized tissues. Ebsco Medica Inc., Birmingham, Alabama, pp 213–220

Eanes ED, Termine JR (1983) Calcium in mineralized tissues. In: Spiro TG (ed) Calcium in biology. John Wiley, New York, pp 203–233

Ellis HA, McCarthy HP, Herrington J (1979) Bone aluminum in hemodialyzed patients and in rats injected with aluminum chloride: relationship to impaired bone mineralization. J Clin Pathol 32:832–844

Escarot-Charrier B, Glorieux FH, van der Rest M, Pereira G (1983) Osteoblasts isolated from mouse calvaria initiate matrix mineralization in culture. J Cell Biol 96:639–643

Farley JR, Baylink DJ (1982) Purification of a skeletal growth factor from human bone. Biochemistry 21:3502–3507

Feenstra TP, DeBruyn PL (1981) The ostwald rule of stages in precipitation from highly supersaturated solutions: a model and its application to the formation of the nonstoichiometric amorphous calcium phosphate precursor phase. J Colloid Interface Sci 84:66–72

Finkelman RD, Butler WT (1985) Appearance of dentine γ-carboxyglutamic acid-containing proteins in developing rat molars in vitro. J Dent Res 64:1008–1015

Fisher LW, Termine JD, Deiter SW et al. (1983a) Proteoglycans of developing bone. J Biol Chem 258:6588–6594

Fisher LW, Whitson SW, Avioli LV, Termine JD (1983b) Matrix sialoprotein of developing bone. J Biol Chem 258:12723–12727

Francis MD (1979) The inhibition of calcium hydroxyapatite crystal growth by polyphosphonates and polyphosphates. Calcif Tissue Res 3:151–162

Francis MD, Webb NC (1971) Hydroxyapatite formation from a hydrated calcium monohydrate phosphate precursor. Calcif Tissue Res 6:335–342

Franzen A, Heinegard D, Olsson S-E (1982) Proteoglycans and calcification of cartilage in the femoral head epiphysis of the immature rat. J Bone Joint Surg [Am] 64: 558–566

Galloway WA, Murphy G, Sandy JD, Gavrilovic J, Cawston TE, Reynolds JJ (1983) Purification and characterization of a rabbit bone metalloproteinase that degrades proteoglycans and other connective tissue components. Biochem J 209:741–752

Garant PA, Cho MI (1984) Ultrastructure of the odontoblast. In: Butler WT (ed) The chemistry and biology of mineralized tissues. Ebsco Medica Inc., Birmingham, Alabama, pp 22–32

Garside J (1982) Nucleation. In: Nancollas GH (ed) Biological mineralization and demineralization. Springer-Verlag, Berlin Heidelberg New York, pp 23–35

Glimcher MJ (1981) On the form and function of bone: from molecules to organ. Wolff's law revisited, 1981. In: Veis A (ed) The chemistry and biology of mineralized connective tissues. Elsevier, Amsterdam, pp 616–673

Glimcher MJ (1984) Recent studies of the mineral phase in bone and its possible linkage to the organic matrix by protein-bound phosphate bonds. Philos Trans R Soc Lond [Biol] 304:479–508

Glimcher MJ, Krane SM (1968) The organization and structure of bone, and the mechanism of calcification. In: Gould BS (ed) Treatise on collagen, vol 2, Biology of collagen, part B. Academic Press, New York, pp 68–251

Hauschka PV, Frenkel J, DeMuth R, Grundberg CM (1983) Presence of osteocalcin and related higher molecular weight 4-carboxyglutamic acid-containing proteins in developing bone. J Biol Chem 258:176–182

Higguchi WI, Cesar EY, Cho PW, Fox JL (1984) Powder suspension method for critically re-examining the two-site model for hydroxyapatite dissolution kinetics. J Pharm Sci 73:146–153

Holtrop ME (1972) The ultrastructure of the epiphyseal plate. 2. The hypertrophic chondrocyte. Calcif Tissue Res 9:140–151

Hsu HT (1983) Purification and partial characterization of ATP phosphohydrolase from fetal bovine epiphyseal cartilage. J Biol Chem 258:3463–3468

Hunziker EB, Herrmann KW, Schenk RK, Mueller M, Moor H (1984) Cartilage ultrastructure after high pressure freezing, freeze substitution, and low temperature embedding. 1. Chondrocyte ultrastructure – implications for the theories of mineralization and vascular invasion. J Cell Biol 98:267–276

Huser HJ, Eichenberg P, Cottier H (1971) Incorporation of iron into osteoid tissue and bone. Schweiz Med Wochenschr 101:1815

Irving JT (1981) Epitaxy down the ages. In: Veis A (ed) The chemistry and biology of mineralized connective tissues. Elsevier, Amsterdam, pp 253–266

Knuuttila M, Lappalainen R, Alakuijala P, Lammi S (1985) Statistical evidence for the relation between citrate and carbonate in human cortical bone. Calcif Tissue Int 37:363–366

Lian JB, Tassinari M, Glowacki J (1984) Resorption of implanted bone prepared from normal and warfarin-treated rats. J Clin Invest 73:1223–1226

Linde A (1984) Dynamic aspects of dentinogenesis. In: Butler WT (ed) The chemistry and biology of mineralized tissues. Ebsco Medica Inc., Birmingham, Alabama, pp 344–355

Maier GD, Evans JS, Veis A (1985) Characterization of the primary gene product of rat incisor α-phosphophoryn. Biochemistry 24:6370–6375

Mechanic GL, Banes AJ, Yamauchi MHM (1984) Possible collagen structural control of mineralization. In: Butler WT (ed) The chemistry and biology of mineralized tissues. Ebsco Medica Inc., Birmingham, Alabama, pp 98–102

Mergenhagen SE, Martin GR, Rizzo AA, Wright DN, Scott DB (1960) Calcification in vivo of implanted collagen. Biochim Biophys Acta 43:563–565

Meyer JL, Fleisch H (1984) Determination of calcium phosphate inhibitor activity. Critical assessment of methodology. Miner Electrolyte Metab 10:249–258

Meyer JL, Thomas WC (1982) Trace metal–citric acid complexes as inhibitors of calcification and crystal growth. J Urol 128:1372–1375

Miller EJ (1984) Recent information on the chemistry of the collagens. In: Butler WT (ed) The chemistry and biology of mineralized tissues. Ebsco Medica Inc., Birmingham, Alabama, pp 80–93

Nawrot CF, Campbell DJ, Schroeder JK, van Valkenberg M (1976) Dental phosphoprotein-induced formation of hydroxyapatite during in vitro synthesis of amorphous calcium phosphate. Biochemistry 15:3445–3449

Nelson DGA, McLean JD (1984) High-resolution electron microscopy of octacalcium phosphate and its hydrolysis products. Calcif Tissue Int 36:219–222

Osdoby P, Caplan AI (1979) Osteogenesis in cultures of limb mesenchymal cells. Dev Biol 73:84–102

Osdoby P, Caplan AI (1981) First bone formation in the developing chick limb. Dev Biol 86:145–156

Pellegrino ED, Biltz RM (1972) Mineralization in the chick embryo. 1. Monohydrogen phosphate and carbonate relationships during maturation of the bone crystal complex. Calcif Tissue Res 10:128–235

Poole AR, Reddi AH, Rosenberg LC (1982) Persistence of calcified cartilage proteoglycan and link protein during matrix-induced endochondral bone development – an immunofluorescent study. Dev Biol 89:532–539

Poole AR, Pidoux I, Reiner A, Choi H, Rosenberg LC (1984) Association of an extracellular protein (chondrocalcin) with the calcification of cartilage in endochondral bone formation. J Cell Biol 98:54–65

Posner AS, Betts F (1981) Molecular control of tissue mineralization. In: Veis A (ed) The chemistry and biology of mineralized connective tissues. Elsevier, Amsterdam , pp 257–266

Posner AS, Betts F, Blumenthal NC (1980) Formation and structure of synthetic and bone hydroxyapatites. Prog Crystal Growth Charact 3:49–64

Price PA, Williamson MK (1981) Effects of warfarin on bone. J Biol Chem 256:12754–12759

Price PA, Otsuka AS, Poser JW, Kristaponis J, Raman N (1976) Characterization of a γ-carboxyglutamic acid-containing protein from bone. Proc Natl Acad Sci USA 73:1447–1451

Price PA, Lothringer JW, Baukel SA, Reddi AH (1981) Developmental appearance of the vitamin K-dependent protein of bone during calcification. J Biol Chem 256:3781–3784

Price PA, Williamson MK, Haba T, Dell RB, Jee WS (1982) Excessive mineralization with growth plate closure in rats on chronic warfarin treatment. Proc Natl Acad Sci USA 79:7734–7738

Raggio CL, Boskey AL, Boyan BD (1986) In-vivo hydroxyapatite formation induced by lipids. J Bone Min Res 1:409–415

Riggs BL, Hodgson SF, Hoffman DL, Kelly PJ, Johnson KA, Taves D (1980) Treatment of primary osteoporosis with fluoride and calcium. JAMA 243:446–449

Robinson C, Weatherell JA, Hohlin HJ (1983) Formation and mineralization of dental enamel. Trends Biochem Sci 8:284–287

Robson R (1923) The possible significance of hexose phosphonic esters in ossification. Biochem J 17:286–293

Sakamoto S, Sakamoto M (1984) Isolation and characterization of collagenase synthesized by mouse bone cells in culture. Biomed Res 5:39–46

Schenk RK, Hunziker E, Herrmann W (1982) Structural properties of cells related to tissue mineralization. In: Nancollas GH (ed) Biological mineralization and demineralization, Springer-Verlag, Berlin Heidelberg New York, pp 143–160

Smith AJ, Matthews JB, Wilson C, Sewell HF (1985) Plasma proteins in human cortical bone: in vitro binding studies. Calcif Tissue Int 37:208–210

Sudo H, Kodama H-A, Amagai Y, Yamamoto S, Kasai S (1983) In vitro differentiation and calcification in a new clonal osteogenic-cell line derived from newborn mouse calvaria. J Cell Biol 96:191–198

Sugahara K, Ho P-L, Dorfman A (1981) Chemical and immunological characterization of proteoglycans of embryonic chick calvaria. Dev Biol 85:180–189

TenCate AR (1967) A histochemical study of the human ondontoblast. Arch Oral Biol 12:963–969

Tenenbaum HC, Heersche JNM (1982) Differentiation of osteoblasts and formation of mineralized bone in vitro. Calcif Tissue Int 34:76–79

Termine JD, Torchia DA, Conn KM (1979) Enamel matrix: structural proteins. J Dent Res 58B:773–778

Termine JD, Eanes ED, Conn KM (1980) Phosphoprotein modulation of apatite crystallization. Calcif Tissue Int 31:247–251

Termine JD, Belcourt AB, Conn KM, Kleinman HK (1981a) Mineral and collagen binding proteins of fetal calf bone. J Biol Chem 256:10403–10408

Termine JD, Kleinman HK, Whitson SW, Conn KM, McGarvey ML, Martin GR (1981b) Osteonectin, a bone-specific protein linking mineral to collagen. Cell 26:99–105

Thomas T, Leaver AG (1975) Identification and estimation of plasma proteins in human dentin. Arch Oral Biol 20:217–218

Thomas WC, Meyer JC (1984) Aluminum induced osteomalacia: an explanation. Am J Nephrol 4:201–213

Thompson DD, Posner AS, Laughlin WS, Blumenthal NC (1983) Comparison of bone apatite in osteoporotic and normal eskimos. Calcif Tissue Int 35:392–393

Trueta J (1973) The role of the vessels in osteogenesis. J Bone Joint Surg [Br] 45:402–418

Tsay T-G, Veis A (1985) Preparation, detection, and characterization of an antibody to rat incisor α-phosphophoryn. Biochemistry 24:6363–6369

Tung MS, Brown WE (1983) An intermediate state in hydrolysis of amorphous calcium phosphate. Calcif Tissue Int 35:783–780

Urist MR, DeLange RJ, Finerman GAM (1983) Bone cell differentiation and growth factors. Science 220:680–686

Vaananen K, Korhonen LK (1978) Histochemistry of epiphyseal plate. Cell Mol Biol 23:105–111

VanDyke TE, Levine MJ, Herzberg MC, Ellison SA, Hay DI (1979) Isolation of low molecular weight glycoprotein inhibitor from the extra-parotid saliva of Macaque monkeys. Arch Oral Biol 24:85–89

Veis A (1984) Phosphoprotein of dentin and bone: do they have a role in matrix mineralization. In: Butler WT (ed) The chemistry and biology of mineralized tissues. Ebsco Medica Inc., Birmingham, Alabama, pp 170–176

Warshawsky H (1984) Ultrastructural studies in amelogenesis. In: Butler WT (ed) The chemistry and biology of mineralized tissues. Ebsco Medica Inc., Birmingham, Alabama, pp 33–46

Wigglesworth DJ (1968) Formation and mineralization of enamel and dentin by rat tooth germs in vitro. Exp Cell Res 49:211–215

Woessner JF Jr, Howell DA (1983) Hydrolytic enzymes in cartilage. In: Maroudas A, Hollborrow EJ (eds) Studies in joint disease, vol 2. Pitman, London, pp 106–152

Wuthier RE (1982) A review of the primary mechanism of endochondral calcification with special emphasis on the role of cells, mitochondria and matrix vesicles. Clin Orthop 169:219–242

Wuthier RE, Eanes ED (1975) Effect of phospholipids on the transformation of amorphous calcium phosphate to hydroxyapatite in vitro. Calcif Tissue Res 19:197–201

Wuthier RE, Register TC (1984) Role of alkaline phosphatase, a polyfunctional enzyme, in mineralizing tissues. In: Butler WT (ed) The chemistry and biology of mineralized tissues. Ebsco Medica Inc., Birmingham, Alabama, pp 113–124

Yamada M (1976) Ultrastructural and cytochemical studies on the calcification of the tendon–bone joint. Arch Histol Jpn 39:347–378

Young RA, Brown WE (1982) Structures of biological minerals. In: Nancollas GH (ed) Biological mineralization and demineralization. Springer-Verlag, Berlin Heidelberg New York, pp 101–141

Chapter 8

Calcified Tissues: Cellular Dynamics

F. Melsen and L. Mosekilde

Introduction

Bone is a living tissue with properties of growth and renewal like most other types of tissue in the human organism. It has multiple functions among which support, weight-bearing and participation in calcium homeostasis are the most important. To understand these properties a certain knowledge of primary bone formation, postnatal growth (modelling) and the continuous renewal of adult bone throughout life (remodelling) (Frost 1969) is necessary. This chapter deals with these subjects and with certain disorders involving bone tissue which may compromise its biomechanical and/or calcium homeostatic properties.

Primary Bone Formation

In the embryonic state, primary bone formation is due either to intramembranous or to endochondral ossification. The clavicle and the majority of bones in the skull are formed by intramembranous ossification, which means that the bone formation is initiated in membranous mesenchymal plates. In these plates some of the mesenchymal cells differentiate to osteoblasts with subsequent formation of organic bone matrix (osteoid). After a period of maturation the osteoid starts to calcify. As the matrix grows, previously active osteoblasts are incorporated as osteocytes and the bone starts to form either cancellous structures, called primitive spongiosa, or more dense structures, called primitive compacta. This type of bone is characterized by a random orientation of the collagen fibres in the matrix (woven bone) that will persist until the internal reorganization, the remodelling, takes places.

The remaining bones in the central and peripheral skeleton are formed by endochondral ossification, which means that bone is laid down within the material of pre-existing hyaline cartilage. The primary ossification centres are to be found in tubular bones, most often in the mid-diaphyses. Through growth of the chondrocytes, which are cartilage cells embedded in the matrix, plate-like structures are formed on which calcification occurs. The primitive peri-chondrium around the cartilage material now becomes periosteum, which has osteogenic properties in its deeper layers. Osteogenic cells will form bone matrix and subsequently mineralized bone in concentric layers around the periphery of the tubular cortical bone. Then by osteoclastic resorption, vessels and osteogenetic cells will penetrate the concentric layers of bone from the periosteum into the calcified cartilage matrix. On the surface of this matrix, primary ossification takes place through differentiation of the osteogenetic cells to osteoblasts followed by matrix formation and calcification.

Postnatal Development of Bone

Modelling

After their primary formation the bones undergo continuous modelling processes which lead to the final gross structures. In long bones these processes include longitudinal growth, transverse growth and adaptation of shape. The longitudinal growth occurs in the epiphyseal growing plate due to cartilage proliferation and formation of primitive bone trabeculae. The transverse growth takes place at the periosteal surface. At the same time the marrow cavity expands due to bone resorption at the endosteal surface. The adaptation of individual bones, to growing processes, functional needs and prevailing mechanical forces, which involves a highly organized regulation or local bone resorption and formation, is extremely complex and not yet fully understood. Very little is known about the humoral or local factors which regulate normal bone growth, except that growth hormone and thyroid hormone(s) have some role to play.

Internal Reorganization

Remodelling

Bone is continuously renewed (remodelled) throughout life (Frost 1969) in order to maintain viability of the cells (especially the osteocytes) and biomechanical competence. Bone remodelling takes place at localized sites in cortical and trabecular bone. The typical remodelling sequence in cortical bone is shown in Fig. 8.1. Following activation, a group of osteoclasts form a cutting cone, which erodes a canal through existing bone (Parfitt 1976). The osteoclasts also resorb bone centrifugally at given sites. When resorption at a site is completed, after about 1 month, the osteoblasts start forming new concentric layers of lamellar bone, the closing cone. The formation of the new structural unit will at each site take about 3 months.

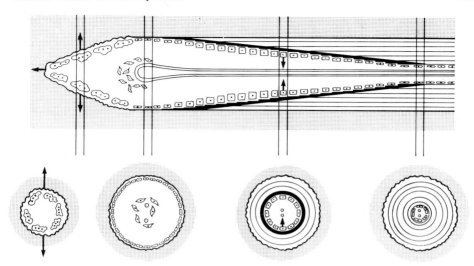

Fig. 8.1. Typical remodelling sequence in cortical bone. *Upper panel*, morphological changes with time at a given site undergoing remodelling, from the osteoclastic cutting cone to the osteoblastic closing cone. *Lower panel*, typical cross-sectional changes.

The amount of bone which is turned over by a remodelling process is termed a BSU (basic structural unit) or an osteon (Jaworsky 1971). In cortical bone these osteons in cross-section are nearly circular structures with concentric lamellae surrounding a central canal (Fig. 8.2a). The group of osteoclasts and osteoblasts which is responsible for a remodelling process at a given site has been termed a BMU (basic multicellular unit) (Frost 1969).

During the last few years it has become obvious that trabecular bone is renewed in the same way as cortical bone. A BSU of trabecular bone in cross-section is typically crescent-shaped with parallel lamellae (Kragstrup et al. 1982). It is outlined by the trabecular bone surface and a cement line (Fig. 8.2b). Renewal of trabecular bone, like renewal of cortical bone, occurs by osteoclastic resorption followed by osteoblastic formation (Eriksen 1980). Following activation, osteoclasts erode from the bone surface to a depth of about 15 μm, after which they are replaced by mononuclear cells with resorptive capacity. The final depth following resorption is about 65 μm (Eriksen et al. 1984b). The total resorptive period lasts about 40 days of which the first 8 days are dominated by osteoclast activity. After about 9 days, the osteoblasts start forming osteoid, which following an initial 15-day lag time is converted to mineralized lamellar bone (Eriksen et al. 1984a). The osteoblasts function at the same site for about 145 days, after which they are converted to flat, inactive-looking lining cells or surface osteocytes (Fig. 8.3).

The remodelling processes occur on three anatomically recognizable surfaces or envelopes (Frost 1969): the periosteal, Haversian and endosteal surfaces. Hormonal or local stimuli may differentially affect the remodelling processes at these three sites, and the proportion of resorption to formation will change at each site with increasing age (Garn et al. 1967).

Activation

Activation (Frost 1969) is the process by which osteoclast precursors are transformed to osteoclasts and later attracted to the bone surface, where they start resorbing bone. A given site on the trabecular bone surface undergoes remodelling about once every 2nd year (Eriksen et al. 1986a). This is referred to as the activation frequency.

It is probable that prostaglandins or lymphokines produced by surface cells or ajoining lymphocytes attract or signal the osteoclasts or their precursors to the bone surface. Osteoclasts do not have PTH receptors and do not respond to PTH in vitro unless osteoblast-like cells are present. It has therefore been postulated that PTH stimulates activation and osteoclast recruitment indirectly through other surface cells and lymphocytes. 1,25-dihydroxyvitamin D, on the other hand, seems to stimulate the transformation of mononuclear cells, probably of the monocyte-macrophage line, to multinucleated osteoclasts. Calcitonin acutely reduces the activity of the osteoclasts by a direct effect on the cells. This model for bone activation, which is shown in a very schematic form in Fig. 8.4, explains the well-known synergistic effect of PTH and 1,25-dihydroxy-vitamin D on osteoclast activity and number.

Coupling

After the resorptive cells have removed a certain amount of bone they are replaced by osteoblasts, which re-form the missing bone. Hence, during remodelling a complete coupling exists between bone resorption and bone formation (Parfitt 1982). This concept has been supported by histological and histomorphometric investigations of bone tissue, and a putative coupling factor has been described. Furthermore, the positive relationships found between biochemical, histomorphometric and calcium kinetic parameters of bone resorption and bone formation provide further evidence for the above concept (Eriksen 1986). The coupling ensures the integrity of the bone mass in spite of the disruptive process of bone remodelling. Decoupling, where resorption is not followed by re-formation or where bone formation starts without previous resorption, may however occasionally occur at the periosteal or endosteal surfaces. Imbalance between the amounts of bone resorbed and formed during a complete remodelling process should not be termed decoupling.

Mechanisms Causing Variations in Bone Mass

In spite of the described coupling between bone resorption and formation, changes in bone mass can occur due to different mechanisms (Parfitt 1976;

Fig. 8.2. **a** Several generations of osteons (basic structural units) in cortical bone. **b** Typical crescent-shaped trabecular bone packet (basic structural unit). **c** A normal osteoid seam (*red*) on mineralized trabecular bone (*blue*). **d** Pronounced osteomalacia with increased width and extent of osteoid (*red*). **e** Tunnel-like osteoclastic resorption in renal osteodystrophy. **f** Deposition of aluminium (*red lines*) at the osteoid/mineralized bone interface in a patient on chronic dialysis.

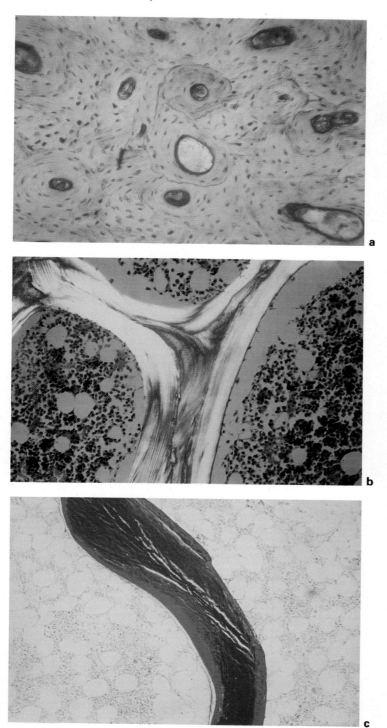

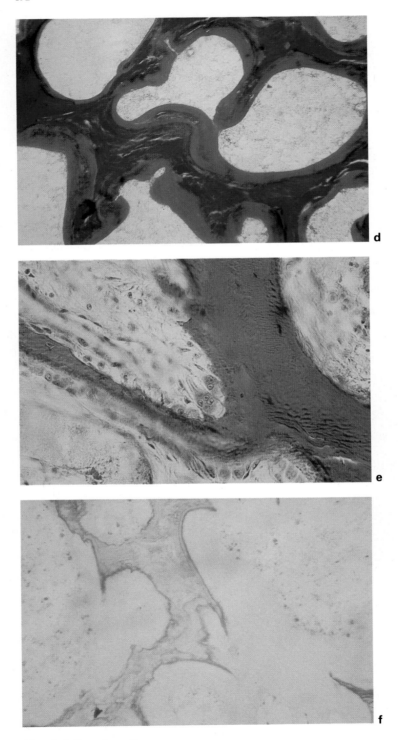

Fig. 8.2 (*continued*)

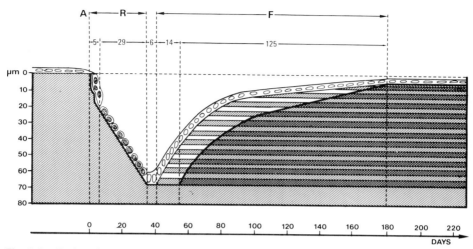

Fig. 8.3. Trabecular bone remodelling sequence reconstructed from 20 normal iliac crest bone biopsies obtained after intravital tetracycline double labelling. *A*, activation; *R*, resorption; *F*, formation (Eriksen et al. 1984a, b).

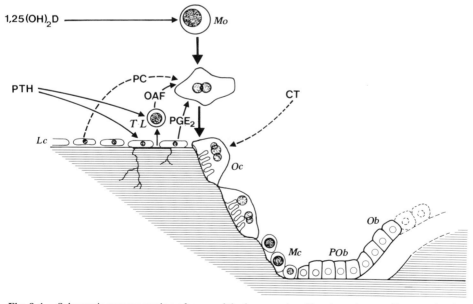

Fig. 8.4. Schematic representation of some of the hormonal and local regulators of bone activation. *Mo*, monocyte; *Oc* osteoclast; *Mc*, mononuclear cell; *POb*, preosteoblast-like cell; *Ob*, osteoblast; *Lc*, lining cell; *T-L*, T lymphocyte; *PGE₂*, prostaglandin E₂; *OAF*, osteoclast activating factor; *PC*, prostacycline; *CT*, calcitonin; *PTH*, parathyroid hormone; *1,25(OH)₂D*, 1,25-dihydroxyvitamin D. Stimulating effect: →. Inhibiting effect: ---→.

NORMAL TURNOVER   HIGH TURNOVER

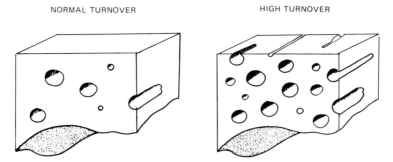

Fig. 8.5. Reversible bone mass changes due to variation in cortical bone porosity in normal and high bone turnover states. An increased activation frequency results in an increased number of remodelling sites where resorption has not yet been followed by formation. A decrease in activation frequency to a normal bone turnover state will result in a reduction in porosity because the created holes will be refilled by the later osteoblastic bone formation.

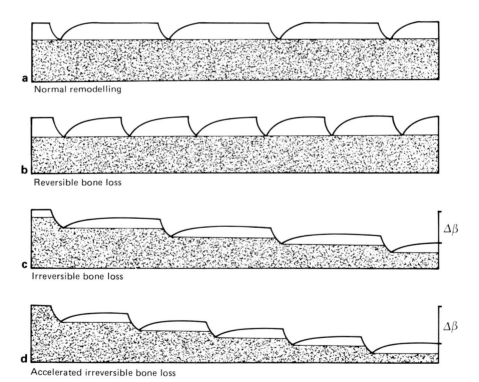

Fig. 8.6. Different types of bone loss. **a** Remodelling with no imbalance between resorption and formation, resulting in preservation of bone mass. **b** Increased formation rate of new remodelling cycles (activation frequency) resulting in an expansion in the remodelling space and reversible loss of bone. **c** Bone remodelling with a negative net balance per remodelling cycle resulting in an irreversible loss of bone. **d** Net irreversible bone loss per remodelling cycle in combination with an enhanced activation frequency, resulting in an accelerated irreversible bone loss.

Parfitt 1984). These mechanisms are often influenced by the activation frequency or the bone turnover. It is important to distinguish between reversible and irreversible bone changes. A reversible bone loss induced by an enhanced bone turnover is followed by an equivalent bone gain after normalization of the bone turnover, whereas an irreversible bone loss will not be followed by a gain in bone mass.

Figure 8.5 illustrates the effect of an increased activation frequency on cortical bone. Due to the increased number of newly activated remodelling sites, where bone is resorbed but not yet re-formed, the bone shows an increased porosity. This porosity is of course followed by mobilization of bone mineral and a decrease in bone mass. Following normalization of the activation frequency, resorption will decrease and the holes will then be refilled with bone. This is an example of a reversible bone loss due to an expansion of the remodelling space. Such changes occur in hyperparathyroidism (Eriksen et al. 1986b) and during antler growth in deers (Parfitt 1976).

A negative balance between the amount of bone resorbed and later formed will create an irreversible bone loss. This occurs in osteoporosis (Eriksen 1986) and during glucocorticoid excess. The negative balance at tissue and organ level will be accelerated if the activation frequency of new remodelling cycles is enhanced (accelerated irreversible bone loss). Furthermore, the irreversible bone loss will in this case be combined with a reversible bone loss. This type of bone loss is found in hyperthyroidism (Melsen and Mosekilde 1977; Eriksen et al. 1986a) and probably associated with the menopause. The different types of bone loss mentioned above are illustrated in Fig. 8.6.

In trabecular bone, another mechanism, trabecular perforation, also results in irreversible loss of bone. This mechanism, which is related to the activation frequency, is probably of major importance in the normal age-related loss of trabecular bone (Parfitt 1984). The average resorption depth in trabecular bone is about 65 μm but can be as great as several hundred μm (Eriksen et al. 1984b). The average trabecular thickness is about 150 μm but can range from close to zero to several hundred μm. The possibility exists therefore, that a deep resorption cavity may affect a narrow trabecular structure and thereby perforate that structure. This will remove the structural basis for the following formative period (uncoupling) and the final result will be a hole in the trabecular network, a loss of bone and bone mineral and a reduced strength of the remaining bone. The likelihood of trabecular perforation depends on the activation frequency, the resorption depth and the trabecular thickness. Figure 8.7 illustrates the effect of repeated trabecular perforation on trabecular bone mass and structure. The strong interconnected trabecular lattice found among young individuals (Fig. 8.7a) is successively disrupted (Figs. 8.7b and c) and finally converted to the more or less isolated structures of trabecular bone often found among elderly osteoporotic subjects (Fig. 8.7d).

Bone Structure and Bone Strength

Non-load-bearing trabecular bone in younger individuals is constructed of uniform trabecular plates forming a honeycomb structure which combines maximum strength with minimum bone mass. The architecture of load-bearing trabecular bone in the same individuals is characterized by thick plates and

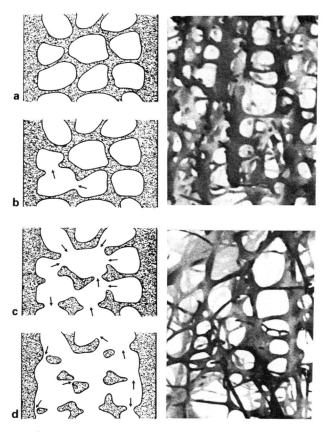

↑ : New perforations

Fig. 8.7. *Left.* Different stages in the age-dependent loss of trabecular bone and disintegration of trabecular structure. **a** Intact trabecular structure in a young individual. **b** First perforations of trabecular structures. **c** Progress of disintegration of trabecular structure due to new perforations. **d** Complete loss of interconnected trabecular architecture. *Right.* Difference in resulting three-dimensional trabecular structure in a vertebral body from a young (*upper panel*) and an old individual (*lower panel*).

columns orientated in the direction of the compressive forces, sustained by thinner trabecular struts in the non-load-bearing plane (anisotrophy). The trabecular elements in the young are completely interconnected giving maximum support. This trabecular framework changes with age however. The thick load-bearing plates are successively perforated during remodelling processes and converted into columns and the thinner non-load-bearing trabeculae are disconnected and often disappear. The biomechanical consequences of this loss of continuity in the three-dimensional lattice are pronounced (Mosekilde and Mosekilde 1986). As the thinner sustaining trabeculae disappear the slenderness ratio (i.e. the ratio of the length between supporting struts to the length of the load-bearing columns) of the remaining trabeculae increases. When this ratio reaches a critical value, elastic buckling and bending forces will dominate instead of compressive forces (Euler buckling). These mechanisms explain why bone strength decreases with age even after being normalized for bone mass (Mosekilde and Mosekilde 1986).

During adulthood the skeleton shows a generalized tendency towards slow appositional growth leading to an increased diameter of tubular bones (Garn et al. 1967) and an expansion of the marrow cavity due to a positive balance at the periosteal envelope and a negative balance on the endosteal envelope. Similar changes have been observed in irregular weight-bearing bones where they might to some extent reduce the effect on bone strength of the reduction in trabecular bone mass with age (Mosekilde and Mosekilde 1986).

Osteoid and its Determinants

Osteoid, or unmineralized bone matrix, is laid down by osteoblasts during bone formation. In trabecular bone it appears as osteoid seams (Fig. 8.2c), which cover about 20% of the trabecular bone surface. That the osteoid seams have a certain thickness, normally about 9–10 μm, indicates that there is a time lag between the formation of osteoid and its mineralization (Melsen and Mosekilde 1981). This lag time, which is also referred to as the osteoid maturation period (although it is unknown to what extent a maturation process is going on), is 20–25 days in normal individuals (Melsen and Mosekilde 1981). The amount of osteoid may be altered in different metabolic bone diseases due to changes in the surface extent or in the mean thickness of the osteoid seams. It appears that the surface extent of osteoid depends on the number of new seams initiated in unit time and the duration of the formative period. Due to the coupling between bone formation and bone resorption the formation rate of osteoid seams is equivalent to the birth rate of new BMU or remodelling cycles, that is, to the activation frequency. The duration of the bone-formation period is, as previously stated, about 4–5 months in normal individuals but may vary considerably in metabolic bone diseases (Eriksen 1986). The mean osteoid thickness is proportional to the linear formation rate of osteoid and the mineralization lag time. In steady-state situations, the formation rate of osteoid equals the bone mineralization rate, which in normal individuals is about 0.45 μm/day (Melsen and Mosekilde 1981).

Figure 8.8 shows in a schematic way the different variables which influence the amount of osteoid (Melsen and Mosekilde 1981). All the variables can be estimated from bone biopsies, carried out after tetracycline double labelling, using quantitative histomorphometry. The amount of osteoid present may be within normal levels in spite of large variations in its determinants. In hyperthyroidism, for example (Melsen and Mosekilde 1977), the amount of osteoid present is normal because a slight increase in the surface extent of osteoid is balanced by a slight decrease in the osteoid seam thickness. The activation frequency is markedly increased but the effect on osteoid surfaces is reduced by a shortened bone formation period (Eriksen et al. 1985). Also, the osteoid formation rate is increased but the mineralization lag time is shortened to such an extent that the final result is a slight decrease in osteoid thickness. In frank osteomalacia the increase in osteoid surface extent is caused by a marked prolongation of the formative period. The increase in osteoid thickness is caused by prolongation of the mineralization lag time (up to 2000–3000 days) even

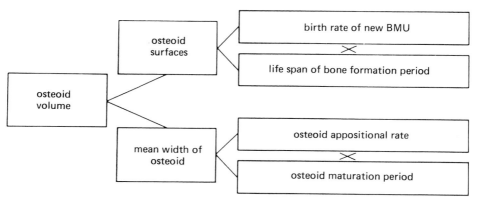

Fig. 8.8. Causes of changes in the amount of osteoid in trabecular bone. (For further explanation see text.)

though the osteoid formation rate is markedly reduced (Melsen and Mosekilde 1981).

Resorptive Parameters

The surface extent of resorption, which covers about 4%–5% of normal trabecular bone (Melsen and Mosekilde 1981), depends on the frequency with which new resorptive sites are activated (activation frequency) and the life span of the resorptive sites. The average resorption period is 45 days but may vary considerably in metabolic bone diseases and with age (Eriksen 1986). In primary hyperparathyroidism the surface extent is mainly increased due to an enhanced activation frequency (Eriksen et al. 1986b). In hyper- and hypothyroidism the extent of resorption is only moderately affected due to opposite effects of activation frequency and life span of resorptive sites in both diseases (see below). Until recently resorption could only be evaluated by its extent and the occurrence of osteoclasts. The introduction of a stereological method for estimating final resorption depth (Eriksen et al. 1984b), however, has made it possible to measure resorption rate in single bone biopsies after tetracycline double labelling. Furthermore, simultaneous determination of final resorption depth and the thickness of the completed packets (structural units) has made it possible to estimate the actual balance between bone resorption and bone formation during the remodelling process (Eriksen 1986).

Bone and Calcium Homeostasis

Serum calcium levels are normally kept within narrow limits. Changes in the remodelling processes described above very seldom lead to marked changes in serum calcium due to the fundamental coupling between bone resorption and formation. An exception is the slight hypercalcaemia often seen in hyperthyroidism (Mosekilde and Christensen 1977), caused mainly by a marked increase in the activation frequency leading to a reversible bone loss and to a lesser degree by a negative balance in the remodelling cycle (Eriksen et al. 1985, 1986a). In hyperparathyroidism, on the other hand, in which there is often marked hypercalcaemia, no major continuous loss of bone takes place (Charles et al. 1986; Eriksen et al. 1986b) due to the steady state of the enhanced activation frequency and a maintained balance between resorption and formation. Mechanisms other than changes in bone remodelling, therefore, are needed in order to explain the elevated serum calcium in primary hyperparathyroidism.

Quiescent surfaces, where neither resorption nor formation take place, are covered by flat fibroblast-like osteoblasts (lining cells), which communicate with deeper osteocytes and form a functional barrier between extra- and intraosseous extracellular fluids. An altered blood–bone equilibrium at these surfaces and the increased tubular renal reabsorption of calcium form two PTH-regulated threshold systems which can explain the often very constant hypercalcaemia observed in patients with established primary hyperparathyroidism (Parfitt 1982; Talmage 1969).

Metabolic States Mainly Characterized by Altered Bone Turnover

Hypothyroidism

In patients with hypothyroidism, serum levels of calcium, phosphorus and immunoreactive PTH (iPTH) are usually normal although serum calcium levels may show a larger variability than in normal individuals. Serum 1,25-dihydroxy-vitamin D, intestinal calcium absorption and the renal excretions of calcium may be increased, whereas serum levels of alkaline phosphatase and osteocalcin and renal hydroxyproline excretion are reduced (Charles et al. 1985; Eriksen et al. 1986a), indicating reduced bone turnover at organ level. This is confirmed by [47]Ca-kinetic studies (Charles et al. 1985). The amount of trabecular bone is normal or slightly increased (Mosekilde and Melsen 1978a). The formation rate of new remodelling cycles (activation frequency) is reduced with a marked prolongation of bone resorption and formation periods (Eriksen et al. 1986a) and therefore of the total remodelling cycle.

The balance between the amount of bone resorbed and formed per remodelling cycle is positive but the effect on bone mass is minimal due to the markedly reduced formation rate of new remodelling cycles (Fig. 8.9). The osteoblasts and osteoclasts are less active than normal (Mosekilde and Melsen 1978a). The surface extent of osteoid and the resorptive surfaces are normal due to the combined effect of a reduced activation frequency and prolonged formative and resorptive periods. The mean osteoid seam thickness is slightly reduced because of a marked reduction in the osteoid formation rate and in spite of a prolongation of the mineralization lag time.

Thus lack of thyroid hormones can be seen to induce a low turnover state with a prolongation of all phases in the remodelling cycle. This may explain the larger variability in serum calcium levels since the buffer capacity of the skeleton may

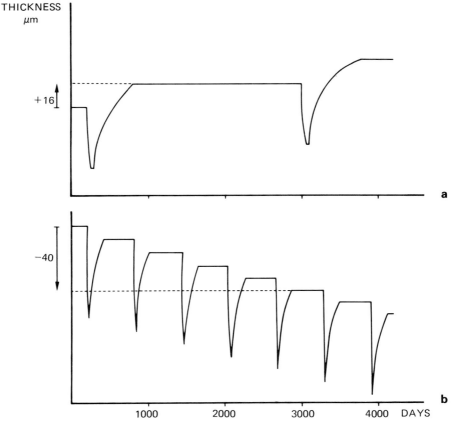

Fig. 8.9. Reconstructed remodelling sequence in hypothyroid (**a**) and hyperthyroid (**b**) patients. The activation frequency of a given site on the trabecular bone surface is indicated by the fast resorptive phase followed by a more slow formation phase. The balance per remodelling cycle has been included in the overall change in bone thickness. After 3000 days the hypothyroid patient has gained about 17 μm in thickness while the hyperthyroid patient has lost 48 μm.

be reduced. The enhanced renal calcium excretion may be explained by the increased serum level of $1,25(OH)_2D$ which enhances intestinal calcium absorption.

Hyperthyroidism

In patients with hyperthyroidism, mean serum levels and renal excretions of calcium and phosphorus are often increased. The hypercalcaemia, however, can be masked by a reduction in serum albumin. Serum iPTH is usually reduced and levels are inversely correlated with serum calcium levels (Mosekilde and Christensen 1977). Serum 25-hydroxyvitamin D is normal, whereas serum 1,25-

dihydroxyvitamin D is reduced and serum 24,25-dihydroxyvitamin D is increased. The fractional intestinal calcium absorption is reduced in accordance with the altered vitamin D metabolism. Serum levels of alkaline phosphatase and osteocalcin and renal hydroxyproline excretions are increased indicating an enhanced bone turnover at organ level (Charles et al. 1985; Eriksen et al. 1985). The average bone mineral content is reduced. In accordance with this, [47]Ca-kinetics have shown increased mineralization and resorption rates with a negative calcium balance (Charles et al. 1985). The biochemical changes correlate with the degree of hyperthyroidism.

Bone histomorphometry has disclosed an increased activation of new remodelling cycles in trabecular bone with a shortening of bone resorption and formation periods (Eriksen et al. 1985). Osteclasts and osteoblasts are more active than normal. The balance between the amount of bone resorbed and formed per remodelling cycle is negative (Eriksen et al. 1985). Trabecular bone mass is reduced (Melsen and Mosekilde 1977) due to an expansion of the remodelling space and an accelerated irreversible bone loss. In cortical bone the enhanced activation frequency leads to an increased porosity (Melsen and Mosekilde 1977). The extent of osteoid surfaces slightly increases because of the enhanced activation frequency and in spite of the shortened bone formation period. The mean osteoid seam width is reduced because of a reduced mineralization lag time and in spite of a slightly enhanced osteoid formation rate. The resorption surface is increased because of the enhanced activation frequency. The size of the osteocytic lacunae is normal (Mosekilde and Melsen 1978b).

Excess thyroid hormone stimulates bone turnover leading to a reversible bone loss (expansion of remodelling space) and an accelerated irreversible bone loss with an associated risk of trabecular perforations (Fig. 8.9). Bone mineral is mobilized leading to hypercalcaemia, suppressed parathyroid secretion, hyperphosphataemia and enhanced renal excretion rates of calcium and phosphorus. The altered vitamin D metabolism is probably caused by a reduced activity of the renal 1-α-hydroxylase secondary to hyperphosphataemia, hypercalcaemia and reduced parathyroid function. The increased CaxP product found in hyperthyroidism may shorten the mineralization lag time by inducing mineralization at an earlier stage of the formation of osteoid. The reduction in bone mineral content may be explained mainly by the bone loss. A lower mean age of the remaining bone due to the enhanced remodelling activity can, however, contribute since young BSU are less mineralized than the old.

Primary Hyperparathyroidism

In patients with primary hyperparathyroidism, serum levels and renal excretions of calcium are increased whereas serum phosphorus levels are reduced. Serum iPTH levels are increased and usually correlate with the biochemical changes. If renal function is normal, serum 1,25-dihydroxyvitamin D is usually increased, in spite of a normal serum 25-hydroxyvitamin D, leading to enhanced intestinal calcium absorption. The average serum levels of alkaline phosphatase and osteocalcin are elevated, as is renal excretion of hydroxyproline (Charles et al. 1986; Eriksen et al. 1986b). In accordance with these observations, combined [47]Ca-kinetic and balance studies show increased mineralization and resorption rates at organ level with a normal or slightly negative calcium balance (Charles et al. 1986). The bone mineral content is reduced.

Bone histomorphometry has revealed an increased activation of new remodelling cycles with near normal bone formation and resorption periods (Eriksen et al. 1986b). The osteoblasts and osteoclasts are less active than normal. The balance between the amount of bone resorbed and later formed per remodelling cycle is normal in accordance with the normal amount of trabecular bone found in slight hyperparathyroidism. In cortical bone the enhanced activation frequency may lead to an increased porosity. The extent of osteoid and resorptive surfaces is increased because of the high activation frequency (Mosekilde and Melsen 1978b). The mean osteoid seam width and the average mineralization lag time is normal. As previously discussed, the size of the periosteocytic lacunae is increased suggesting that the osteocytes participate in the abnormal serum calcium regulation (Talmage 1969).

Excess parathyroid hormone induces a moderate increase in bone turnover probably followed by an initial reversible bone loss due to an enhanced remodelling space. Apparently the bone loss does not continue since the amount of trabecular bone is normal in spite of what is often a long-standing disease. This is in agreement with the normal balance between bone resorption and formation per remodelling cycle. The reduced bone mineral content can easily be explained by the slight initial bone loss and the reduction in mean bone age due to enhanced remodelling activity. Hence loss of bone mineral is not a major cause of the established hypercalcaemia in hyperparathyroidism. Increased osteocytic osteolysis, however, may be of importance for maintaining the hypercalcaemia, since it creates a functional barrier between extraosseous and intraosseous fluid (Talmage 1969). Furthermore, an enhanced tubular reabsorption of calcium directly induced by the excess of parathyroid hormone is of importance.

Metabolic States Mainly Characterized by Disturbed Osteoid Mineralization

Osteomalacia

Diagnosis of osteomalacia has previously been based on the presence of an excess of osteoid. However, according to the osteoid model described above, an increase in the amount of osteoid may be due to several pathophysiological mechanisms. Following the introduction of routine intravital double labelling with tetracycline it is now generally accepted that osteomalacia is a condition characterized by increased osteoid seam thickness caused by a prolonged mineralization lag time.

The exact mechanism responsible for the defective mineralization is unknown. Sufficient amounts of calcium and phosphorus at the mineralization site are important and the hypophosphataemia in x-linked hypophosphataemic rickets may directly explain the mineralization defect found in this disease. Lack of alkaline phosphatase in congenital hypophosphatasia may lead to osteomalacia because of the consequent reduced ability to accumulate phosphate at the mineralization fronts. Whether 1,25-dihydroxyvitamin D acts directly on the mineralization process or by increasing the CaxP product is at present unknown.

In malabsorptive states and malnutrition, gradual bone changes take place. Initially malabsorption of calcium with often very slight hypocalcaemia lead to a secondary hyperparathyroidism. Serum 25-hydroxyvitamin D is low but serum 1,25-dihydroxyvitamin D may be normal or even slightly increased due to the stimulatory effect of increased serum PTH and hypophosphataemia on the renal 1-α-hydroxylase. Serum levels of alkaline phosphatase and renal hydroxyproline excretions are usually increased. In accordance with these biochemical findings, the typical bone changes consist of an enhanced activation frequency with an increase in the surface extent of osteoid and resorption (Melsen and Mosekilde 1981). At this stage the mineralization process is normal. In more advanced cases serum calcium and serum 1,25-dihydroxyvitamin D are low in spite of the secondary hyperparathyroidism and the mineralization process is compromised. The bone changes are now due to a combination of secondary hyperparathyroidism and a defective mineralization which leads to an increased thickness of the osteoid seams and an often clearly prolonged mineralization lag time. The osteoid volume is now increased due both to an increase in osteoid thickness and to an increase in osteoid surface extent.

In frank osteomalacia, bone turnover declines and bone surfaces become inactive and covered with accumulating osteoid due to a pronounced mineralization defect (Fig. 8.2d). The reduction in activation frequency may partly be explained by the lack of ability of the osteoclasts to penetrate osteoid and resorb bone. Serum alkaline phosphatase may still be elevated in marked osteomalacia in spite of a reduced bone turnover. The explanation for this finding is unknown but enhanced production by individual osteoblasts to overcome the lack of phosphate at the mineralization front may be of importance. The amount of trabecular bone is usually normal even in pronounced osteomalacia, but the amount of mineralized bone is reduced. The increase in bone turnover in the initial stages and the accumulation of non-mineralized bone in the later stages lead to a decrease in the amount of bone mineral.

Hence, osteomalacia is characterized by a progressive development from one clinical and histopathological state to the next ending with the classical picture of marked osteoid accumulation, mineralization defect and inactive cell populations. The osteomalacic conditions can only be classified and separated from other hyperosteoidoses by measuring the dynamic variables after tetracycline double labelling.

Bone Disease in Chronic Renal Failure

In renal failure the chronic loss of nephrons leads to severe disturbances in bone and calcium metabolism (Nielsen et al. 1980). The decrease in glomerular filtration rate (GFR) compromises the renal excretion of phosphorus resulting in hyperphosphataemia. Renal production of 1,25-dihydroxyvitamin D from 25-hydroxyvitamin D is reduced due to the hyperphosphataemia as well as the damage to the renal tubules, which inhibit the renal 1-α-hyroxylase. At GFRs below 50 ml/min serum iPTH levels increase secondary to reduced renal production of 1,25-dihydroxyvitamin D, intestinal malabsorption of calcium and increased resistance of bone to PTH. Using C-terminal radioimmunoassays for estimating serum iPTH, some of the increase may, however, be shown to be caused by renal retention of inactive *C*-terminal PTH fragments. With this complexity in calcium, phosphorus and vitamin D metabolism it is not surprising that the concomitant bone changes are extremely varied, ranging from those seen in patients with secondary hyperparathyroidism to those seen in patients with extreme osteomalacia. In the same way the amount of bone may vary between, for example, osteopenia and osteosclerosis.

Renal Osteodystrophy

In mild renal failure, the most frequent finding affecting bone is secondary hyperparathyroidism with its increased bone resorption and bone formation surfaces (Nielsen et al. 1980). This disease may progress to a very severe state with numerous large multinucleated osteoclasts situated in larger resorption lacunae, sometimes creating tunnel-like resorption defects with consequent disturbance of the bony structure (Fig. 8.2e). Marrow fibrosis is often pronounced in relation to the resorptive sites. In these cases the osteoid seams are typically increased in surface extent whereas the osteoid seam thickness, the osteoblastic activity and the mineralization of osteoid are normal.

A mineralization defect may develop in some cases of pronounced and long-lasting renal failure. This osteomalacic variant is characterized by wide osteoid seams with reduced mineralization activity and prolonged mineralization lag time. Even in severe mineralization defects the expression of secondary hyperparathyroidism is often more marked than in nutritional osteomalacia. In a few cases the osteomalacia completely dominates the bone with all surfaces covered with wide osteoid seams which don't show tetracycline uptake (Fig. 8.2d).

Dialysis Osteodystrophy

During chronic dialysis treatment the bone changes are very much the same as in renal osteodystrophy (Nilsson et al. 1985). However, accumulation of aluminium from the dialysate and oral phosphate-binders may affect bone mineralization and lead to frank osteomalacia. The frequency of this complication depends on the amount of aluminium in the dialysate. The aluminium is deposited at the mineralization fronts, where it can be demonstrated by special histological staining methods (Fig. 8.2f). Treatment with desferroximin, which strongly binds aluminium, may prevent or cure this type of often very severe osteomalacia.

Post-transplant Bone Changes

Severe osteopenia with spontaneous fractures is a frequent complication following renal transplantation (Melsen and Nielsen 1977). This may to a certain degree be due to the pre-existing bone disease described above. The most important aetiological factor however is the post-transplant combined immunosuppressive treatment which leads to bone changes indistinguishable from those induced by corticosteroids alone.

Metabolic States Mainly Characterized by a Reduction in Bone Mass

Senile Osteoporosis

In our terminology osteopenia designates a reduced amount of bone whereas osteoporosis is the clinical state characterized by atraumatic fractures caused by a reduction in bone mass (and therefore bone strength). The typical osteoporotic fractures occur in areas of the skeleton mainly dominated by trabecular bone. The amount of trabecular bone decreases with age in both sexes (Melsen and Mosekilde 1981). At age 70, about half the amount of trabecular bone found at age 30 is left. Bone strength is significantly related to bone mass or bone mineral density. However, during adult life trabecular bone strength decreases to about 20% of its initial level even after normalization for bone mass (Mosekilde and Mosekilde 1986). This indicates that bone strength depends not only on bone mass, but also on bone structure.

With increasing age the thickness of bone resorbed per remodelling cycle decreases slightly (Eriksen 1986) whereas the thickness of bone reformed per cycle decreases more markedly (Kragstrup et al. 1982). This gives rise to a more pronounced negative net balance per remodelling cycle with increasing age and a continuous reduction in the amount of trabecular bone. Furthermore, perforations of thinner trabecular structures by osteoclastic resorption will lead to disruption of the strong interconnected trabecular lattice, resulting in more isolated trabecular structures with reduced strength (Fig. 8.7). The average trabecular thickness does not change very much with age, which may be explained by the opposite effects of the negative net balance per remodelling cycle (Eriksen 1986), which reduces trabecular thickness, and the trabecular perforations, which mainly remove the thin trabeculae. Also the thickness of cortical bone decreases with age due to a more rapid rate of expansion of the marrow cavity (endosteal envelope) than of periosteal growth (Garn et al. 1967). The skeletal bone loss takes place without significant changes in the surface-related static histomorphometric parameters, i.e. extent of resorption surfaces, and extent, width and total amount of osteoid surfaces (Melsen and Mosekilde 1981). Furthermore, serum levels of alkaline phosphatase and osteocalcin and renal hydroxyproline excretion show no consistent changes. Due to reduced renal production of 1,25-hydroxyvitamin D and malabsorption of calcium and to occasional nutritional vitamin D deficiency among the elderly, the osteopenia may in this population be complicated by a mineralization defect, which reacts favourably to treatment with vitamin D and calcium.

With the progressive decrease in bone mass after the age of 30 all individuals will experience osteoporotic fractures if they live long enough. At present, however, it is not known why some individuals develop frank osteoporosis during a normal lifespan while others maintain sufficient skeletal strength to withstand daily external trauma. Variation among individuals in the maximum skeletal mass after cessation of growth and in the rate of bone loss during adult life are probably both of importance.

Postmenopausal Osteoporosis

Serum levels of alkaline phosphatase and osteocalcin and renal hydroxyproline and calcium excretion increase in females after the menopause, suggesting an increase in bone turnover with an accelerated bone loss. Oestrogen treatment reduces the biochemical markers of bone turnover and increases bone mineral content. However, at present there is no histomorphometric evidence for these apparent changes in bone turnover. An increase in bone turnover after the menopause would induce a reversible bone loss due to expansion of the remodelling space. Furthermore, if the individual is in a negative net balance of bone per remodelling cycle, the enhanced activation frequency would induce an accelerated irreversible bone loss. Finally, an enhanced formation rate of new remodelling cycles would increase the risk of trabecular perforation. This risk would be further increased by increased final resorption depth in the perimenopausal period (Eriksen 1986). The described mechanisms may explain the increased risk of osteoporotic fractures in females after the menopause and the potential effect of oestrogen in osteoporosis prophylaxis.

Other Osteopenic States

Several chronic diseased states are followed by osteopenia or frank osteoporosis, for example, rheumatoid arthritis, chronic inflammatory bowel diseases, chronic inflammatory lung diseases, diabetes mellitus and Cushing's syndrome. From a clinical point of view, the most important state is glucocorticoid-induced osteopenia. The mechanisms by which these diseases cause osteopenia are not known. In steroid-induced osteopenia, activation, bone formation and mineralization are much reduced as is shown by tetracycline double labelling. Hence, the most likely explanation for the decrease in bone mass in this state is a marked negative bone balance per remodelling cycle caused by a reduction in the thickness of bone formed. Histomorphometric evidence for this view is however lacking.

Practical Implications of Iliac Crest Bone Biopsy

Among all the techniques used in the evaluation of metabolic bone diseases, iliac crest bone biopsy after tetracycline double labelling is the only procedure that provides information on resorption depth and formation thickness of structural units, single cell activity in the various phases of the remodelling cycle, activation frequency, duration of resorption and formation and the occurrence of reversible and irreversible bone changes (Melsen and Mosekilde 1981; Eriksen 1986). Bone histomorphometry, therefore, is a valuable scientific tool in the evaluation of the pathophysiological background of metabolic bone diseases and of the reaction of bone to a pharmacological challenge, i.e. treatment of osteopenic states. In the daily management of single patients, however, its practical use is fairly limited, due to the biological variation in the measured variables and the errors involved.

However, bone biopsy is a very essential diagnostic tool in patients at risk from mineralization defects. Slight to moderate nutritional or malabsorption osteomalacia can only be diagnosed with certainty using bone biopsies, which also provide the best control for the therapeutic effect of vitamin D or its metabolites. Further, bone biopsy may serve as a valuable indication of the success of the treatment of the various types of renal osteodystrophy and aluminium-induced osteomalacia. Finally, bone biopsy is essential in the final diagnosis of more rare inherited or acquired bone diseases such as hypophospha-taemic rickets and hypophosphatasia, osteopoikilosis, osteopetrosis, osteogenesis imperfecta and Paget's disease. The biopsy procedure and its few complications have been described in detail elsewhere (Melsen and Mosekilde 1981).

References

Charles P, Poser JW, Mosekilde L, Jensen FT (1985) Estimation of bone turnover evaluated by [47]Ca kinetics: efficiency of serum bone gamma-carboxyglutamic acid-containing protein, serum alkaline phosphatase, and urinary hydroxyproline excretion. J Clin Invest 76:2254–2258

Charles P, Mosekilde L, Jensen FT (1986) Primary hyperparathyroidism: evaluated by [47]Ca kinetics, calcium balance and serum bone Gla-protein. Eur J Clin Invest 16:277–283

Eriksen EF (1986) Normal and pathological remodeling of human trabecular bone: three-dimensional reconstruction of the remodeling sequence in normals and in metabolic bone disease. Endocr Rev 7:379–408

Eriksen EF, Gundersen HJG, Melsen F, Mosekilde L (1984a) Reconstruction of the formative site in iliac trabecular bone in 20 normal individuals employing a kinetic model for matrix and mineral apposition. Metab Bone Dis Rel Res 5:243–252

Eriksen EF, Melsen F, Mosekilde L (1984b) Reconstruction of the resorptive site in iliac trabecular bone: a kinetic model for bone resorption in 20 normal individuals. Metab Bone Dis Rel Res 5:235–242

Eriksen EF, Mosekilde L, Melsen F (1985) Trabecular bone remodeling and bone balance in hyperthyroidism. Bone 6:421–428

Eriksen EF, Mosekilde L, Melsen F (1986a) Kinetics of trabecular bone resorption and formation in hypothyroidism: evidence for a positive balance per remodeling cycle. Bone 7:101–108

Eriksen EF, Mosekilde L, Melsen F (1986b) Trabecular bone remodeling and balance in primary hyperparathyroidism. Bone 7:213–222

Frost HM (1969) Tetracycline-based histological analysis of bone remodeling. Calcif Tissue Res 3:211–237

Garn SM, Rohmann CG, Wagner B, Ascoli W (1967) Continuing bone growth throughout life: a general phenomenon. Am J Phys Anthropol 26:313–317

Jaworsky ZF (1971) Some morphologic and dynamic aspects of remodelling on the endosteal–cortical and trabecular surfaces. In: Menczel J, Harell A (eds) Calcified tissue: structural, functional and metabolic aspects. Academic Press, New York, pp 159–160

Kragstrup J, Gundersen HJG, Melsen F, Mosekilde L (1982) Estimation of the three-dimensional wall thickness of completed remodeling sites in iliac trabecular bone. Metab Bone Dis Rel Res 4:113–119

Melsen F, Mosekilde L (1977) Morphometric and dynamic studies of bone changes in hyperthyroid-ism. Acta Pathol Microbiol Scand [A] 85:141–150

Melsen F, Mosekilde L (1981) The role of bone biopsy in metabolic bone disease. Orthop Clin North Am 12:571–602

Melsen F, Nielsen HE (1977) Osteonecrosis following renal allotransplantation. Acta Pathol Microbiol Scand [A] 85:99–104

Mosekilde L, Christensen MS (1977) Decreased parathyroid function in hyperthyroidism: interrelationships between serum parathyroid hormone, calcium–phosphorus metabolism and thyroid function. Acta Endocrinol 84:566–575

Mosekilde L, Melsen F (1978a) Morphometric and dynamic studies of bone changes in hypothyroidism. Acta Pathol Microbiol Scand [A] 86:56–62

Mosekilde L, Melsen F (1978b) A tetracycline based histomorphometric evaluation of bone resorption and bone turnover in hyperthyroidism and hyperparathyroidism. Acta Med Scand 204:97–102

Mosekilde L, Mosekilde L (1986) Normal vertebral body size and compressive strength: relations to age and to vertebral and iliac trabecular bone compressive strength. Bone 7:207–212

Nielsen HE, Melsen F, Christensen MS (1980) Interrelationships between calcium–phosphorus metabolism, serum parathyroid hormone and bone histomorphometry in non-dialyzed and dialyzed patients with chronic renal failure. Mineral Electrolyte Metab 4:113–122

Nilsson P, Melsen F, Malmaeus J, Danielson BG, Mosekilde L (1985) Relationships between calcium and phosphorus homeostasis, parathyroid hormone levels, bone aluminium, and bone histomorphometry in patients on maintenance hemodialysis. Bone 6:21–27

Parfitt AM (1976) The actions of parathyroid hormone on bone: relation to bone remodeling and turnover, calcium homeostasis, and metabolic bone disease. 2. PTH and osteoblasts, the relationship between bone turnover and bone loss, and the state of bone in primary hyperparathyroidism. Metabolism 25:1033–1087

Parfitt AM (1982) The coupling of bone formation and bone resorption: a critical analysis of the concept and of its relevance to the pathogenesis of osteoporosis. Metab Bone Dis Rel Res 4:1–6

Parfitt AM (1984) Age related structural changes in trabecular and cortical bone. Cellular mechanisms and biomechanical consequences. Calcif Tissue Int 36:123–128

Talmage RV (1969) Calcium homeostasis–calcium transport and parathyroid action: the effects of parathyroid hormone on the movement of calcium between bone and fluid. Clin Orthop 67:210–224

Chapter 9

Calcified Tissues: Structure–Function Relationships

J. Dequeker

Introduction

Bone tissue is the most complex of all the building materials of the body. It probably was the last to have appeared in the course of evolution. Its unique physical and chemical attributes mirror the diversity of its functions.

The most obvious function, of course, is its supportive one; strong, hard bones make useful limbs. A second function is that of protecting vital soft tissues such as the brain, spinal cord, heart and bone marrow. Thirdly, bone is involved in calcium homeostasis. Bone also plays an important role as a trap for a variety of blood-borne ions, such as lead, fluoride and strontium, which may exchange with calcium ions or otherwise become incorporated in the apatite crystal lattice or bound to the organic matrix of bone.

Macroanatomy

Bone is an ideal supporting material by virtue of its remarkable strength. It is a two-phase material, consisting of two contrasting substances: fibrous protein collagen (which is strong in tension) and the mineral apatite (which is strong in compression). The crystallites of apatite (or calcium phosphate) are exceedingly small and are aligned along the collagen fibrils. The way in which these two substances are actually bound together is still not properly understood. If the mineral matter is removed by acids (ethylenediaminetetraacetate or EDTA), the result is a rubbery bone which is so flexible that it can be tied in a knot (Fig. 9.1). If the organic matter is destroyed, a brittle bone results. In some way the mineral matter locks in the protein so that this organic matter is able to survive for millions of years. Presumably the close packing of the crystals seals off the organic matter from natural destructive agents.

Fig. 9.1. Bones become pliable when the mineral matter is removed.

Although bone is ideally suited to be a structural material, like any building material it has to be put together in the proper manner. To be effective, its internal architecture must be appropriate for the function it has to perform. For a given volume of material to support a weight, it is more efficient that it should be organized in the form of a cylinder rather than as a solid block. This is normal engineering practice and this is how the long bones of the limbs are constructed.

The solid, compact part of limb bones is the outer cortex of the midshafts. At the ends of the bones, where they approach the joints, the situation is entirely different and the bones are of a loose, spongy texture. When examined closely, this spongy-looking tissue is, in fact, well organized, showing a delicate internal architecture. There are a series of narrow beams, or trabeculae, joined by minor cross struts, which together form an intricate three-dimensional scaffolding. The main trabeculae line up along the major axes of force to which the bone is subjected during development. When the weight of the body is transmitted from one bone to another, as at the knee or hip joints during walking, for example, the forces are distributed through this tracery of trabeculae to the compact cortex of the outer parts of the shafts (Fig. 9.2). This complexity at the ends of the bones is necessary because both movement and weight have to be transmitted with the bones in different positions. In whatever manner the limbs are positioned, within reason, the forces can still be effectively distributed through the trabeculae (Halstead and Middleton 1972).

At the macroscopic level there are thus two major types of bone: compact or cortical bone and trabecular or cancellous bone. The former is seen in the diaphyses of long bones and the surfaces of flat bones. There is also a thin cortical shell at the epiphyses and metaphyses of long bones. Trabecular bone is limited to the epiphyseal and metaphyseal regions of long bones and is present within the cortical coverings in the smaller flat and short bones.

The cortical bone appears dense macroscopically and cortical bone is histologically characterized as Haversian bone. The Haversian systems or secondary osteons approximate cylindrical structures with concentric lamellae. The osteons in the diaphyses of the long bones are oriented in the direction of the long axis of the bone. Figure 9.3 is a schematic representation of cortical bone tissue.

Cancellous bone is structurally an open cell form, continuous with the inner surface of the cortical shell, and is a three-dimensional lattice composed of bone

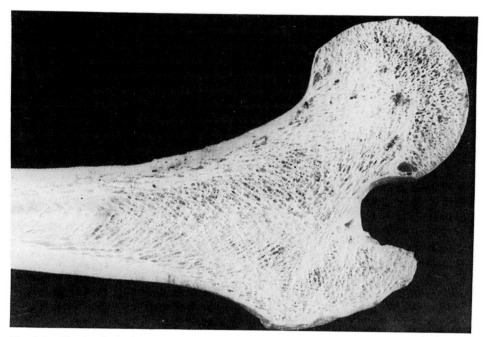

Fig. 9.2. The head of a human thigh bone cut to reveal the fine trabeculae which distribute the forces applied to the bone.

plates and bone columns. The mechanical properties and the role in mineral metabolism of both types of bone architecture have to be distinguished.

The overall shape of a bone and its internal architecture are a reflection of the forces acting upon it. During the life of any individual, the forces acting on the

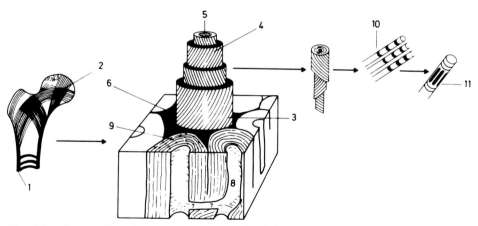

Fig. 9.3. Organization of bone at macroscopic level (1, cortical bone; 2, trabecular bone), at microscopic level (3, osteon; 4, osteon lamella; 5 and 8, Haversian canal; 6, interstitial lamella; 7, Volkmann canal; 9, osteocyte) and at ultrastructural level (10, collagen fibrils; 11, hydroxyapatite crystals).

skeleton do not remain constant. If the organism is to function properly, it must be capable of altering the shape of its bones, albeit slightly, to accommodate changing requirements. This does indeed happen and is obvious if a limb is broken. If correctly set, the limb will heal and eventually the new bone will be as good as the old.

So far we have discussed the nature of bone itself and its basic structure. We have mentioned the forces to which bones are subjected, but these will vary enormously depending on the particular species studied. One of the most obvious facts about different species is that they differ in size. The size of an animal, for example, reflects to a very large extent the nature and overall shape of its internal skeleton. The form of the skeleton will also be determined by the species' life style. The different proportions of the bones of the limbs tell us, for example, whether the creature is a fast runner or a powerful digger of holes.

These examples emphasize that bone is a living tissue capable of regeneration. For a bone to change its shape in response to changing conditions, as well as being capable of producing new bone, it must at the same time be able to destroy bone. Whereas there are special cells which produce bone, the osteoblasts, there are also cells, the osteoclasts, whose function is to destroy bone.

In the normal course of events, the remodelling that goes on in our bones is imperceptible. Bone is constantly being resorbed and renewed. In adult life there is a delicate equilibrium between these two processes which only begins to break down in old age, when the rate of resorption is greater than the rate of redeposition. The bones become more porous, resulting in the condition known as osteoporosis. They are then less able to withstand the strains normally placed on them, and become more liable to fracture.

Age-Related Changes in Bone Mass

In old age, changes occur which are easily detected in the dead body, and one of the principal of these is found in the bones which become "thin in their shell and spongy in their texture" (Astley Cooper 1824).

Non-invasive Methods for Measuring Bone Mass

The observation of Astley Cooper on the effect of ageing on bone was confirmed in dried skeletons by Trotter and Peterson (1955) and Trotter et al. (1960), who found that the weight of the skeleton decreases with age as does the physical density (weight/volume) of individual bones.

Despite these striking observations, it was only from 1960 onwards that technical tools became available for the quantitative evaluation of bone in vivo. Since then, several non-invasive methods of quantifying bone mass at the appendicular and axial skeleton have been developed. In the 1960s radiogram-metry of cortical bone, in the 1970s single photon absorptiometry of the forearm and in the 1980s dual photon absorptiometry and quantitative computed tomography of the spine became firmly established as vital processes for the investigation of age-related bone loss and metabolic bone diseases (Dequeker and Johnston 1982).

Determination of bone quantity based on the radiodensity of routine skeletal roentgenograms is very imprecise. Estimates, especially of the spinal column, are less than ideal due to variation in film exposure and development, the varying quantity and quality of overlying soft tissue (Doyle et al. 1967) and the different interpretations of different observers (Bland et al. 1969). Moreover, roentgenologic estimation of bone quantity is quite insensitive; as much as 30% of bone tissue can be removed before the loss is detectable (Lachmann 1955).

Singh and colleagues (1970) noted that bony trabeculae in the upper femur, which is subjected to the greatest tensile and compressive stresses, are the last to disappear in involutional osteopenia. These workers developed a six-point scale (Singh index) to semiquantitate the loss of femoral trabeculae (Fig. 9.4). The authors concluded that this index would distinguish those patients with vertebral crush fractures or femoral neck fractures from age–sex-matched controls. Others have found that the method is difficult to reproduce and is not highly prognostic of subsequent vertebral compression (Khairi et al. 1976). In our hands, the Singh index was of limited use in the assessment of total bone mass but was of value in screening subjects liable to femoral neck fractures (Dequeker et al. 1974).

A method for assessing cortical bone thickness, termed radiogrammetry, introduced by Virtama and Mahönen (1960) in Finland, and Barnett and Nordin (1960) in Great Britain, and 2 years later by Meema (1962) in Canada, was used extensively by Garn (1972) and Dequeker (1972, 1976) in population surveys and in the diagnosis of specific diseases. Although the method is applicable to

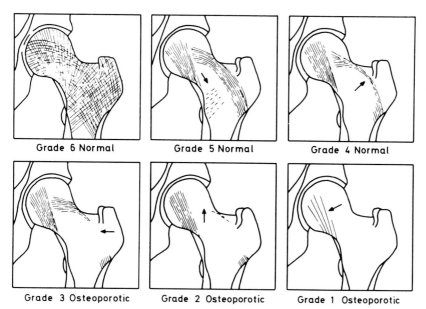

Fig. 9.4. Bone trabecular patterns of the upper end of the femur and corresponding grades. Arrows indicate progressive radiographic disappearance of trabecular groups (Singh et al. 1970).

any bone, the site most frequently measured is the midpoint of the second metacarpal. Precision calipers are used to define the periosteal and endosteal margins from which estimates of cortical thickness and area may be derived. At present, this method is the only means of quantitating bone loss or gain at the periosteal and endosteal bone surfaces. However, it does not detect intracortical porosity and it does not measure trabecular bone.

Photon beam absorptiometry, introduced by Cameron and Sörenson (1963), has generally replaced radiographic photodensitometry as the method of choice for the evaluation of bone mass. The technique involves the emission of photons from a finely collimated monochromatic ^{125}I source for peripheral cortical bone evaluation and a dichromatic ^{153}Gd source for the evaluation of axial trabecular bone mineral content, and a collimated detector. Photon absorption is directly related to the mineral content of the bone that is being scanned.

Commercial instruments are now available that scan the bone areas, usually the radius for peripheral bone and the lumbar spine for axial trabecular bone, and provide a direct read-out of bone mineral content in grams per cm or cm^2 of bone scanned. Photon absorptiometry readings correlate with mineral content at the site being measured to within 4% (Mazess 1971), as well as with total body calcium in a normal population (Manzke et al. 1975).

The measured bone mineral content is not always equivalent to the amount of bony tissue. In osteomalacia, osteitis fibrosa and other high turnover bone diseases, a low bone mineral content may be found without low bone volume. However, in most situations, bone mineral content (BMC) represents bone mass.

Single (monochromatic) photon absorptiometry (SPA) was first established for measurement of cortical bone at the midshaft of the radius, where there are no major problems with surrounding soft tissues and fat. For the measurement of axial bone mass in the spine, the development of dual (dichromatic) photon absorptiometry (DPA) was necessary in order to correct for individual variations in soft tissue and fat. The development of dual photon absorptiometry was of particular importance because it is now possible to measure bone mass, especially trabecular bone mass of the spine and femoral neck, at the sites where most of the fractures occur and where bone turnover is highest.

Considerable effort has been expended in recent years on applying quantitative computed tomography (QCT) to bone mineral measurements. This has been of particular importance and is the most advanced technique in this field at present. The usefulness of computed tomography for measurement of bone lies in its ability to provide a quantitative image and, thereby, measure trabecular, cortical or integral bone, centrally or peripherally (Cann and Genant 1980). Despite methodological problems, such as relocation and fat interference and the cost of the instrumentation, quantitative computed tomography, is being used on an increasing scale for bone mineral measurements.

Population Studies

There are now data on thousands of subjects obtained using the precise and accurate tools described above. The ageing changes in compact bone are reasonably clear (Dequeker 1975; Mazess 1982). Cross-sectional data on bone mass in normal adult populations reported by different authors show similar patterns. There is an increase in bone mass up to the age of 30–40 years, followed by a progressive decrease, which is much more pronounced in women (Fig. 9.5), and faster after the age of 50 years in women. After age 50 this loss is approximately 10% per decade in women and 5% per decade in men (Newton-John and Morgan 1968). Males exceed females in the absolute amount of their bone (cortical area, bone mineral content) throughout the age range 15–80 years; this difference is approximately 30% in the fourth decade. Bone loss with age is not confined to modern population samples. Similar decreases in bone mass have been reported in prehistoric populations (Dewey et al. 1969; Van Gerven et al. 1969; Perigian 1973). The uniformly reported cross-sectional data on bone loss with age have been confirmed by longitudinal data which preclude differential sampling and selective mortality (Garn et al. 1967; Adams et al. 1970; Dequeker et al. 1972).

Reports on ageing changes in trabecular bone are less consistent (Mazess 1982). Most of the evidence (Riggs et al. 1981), including our own observation in the iliac crest (Dequeker et al. 1971) and in the lumbar spine (Geusens et al. 1986), suggests that the decrease in trabecular density begins during young adulthood (20–40 years) in both sexes and amounts to approximately 6% per decade in women and 3% per decade in men.

However, although the onset of loss in the females antedates the menopause, the rate of trabecular bone loss is distinctly accelerated after the menopause (Dequeker and Geusens 1985). Longitudinal studies confirm that trabecular bone loss does indeed occur after oophorectomy (Cann et al. 1980), and at

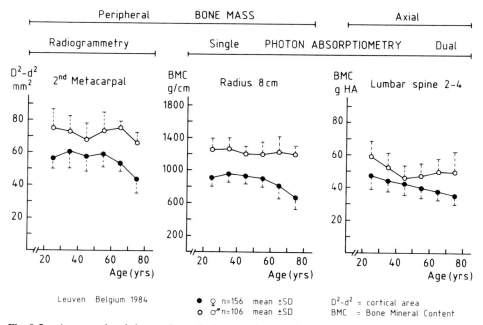

Fig. 9.5. Age-associated changes in cortical area at the second metacarpal, in bone mineral content at the radius midshaft and in bone mineral content at the lumbar spine.

natural menopause (Krölner and Pors Nielsen 1982) at a rate several times greater than that occurring either earlier or later. The loss of bone after the menopause was twice as fast in the spine as in peripheral cortical bone over a period of one decade, -15% and -7% respectively (Dequeker and Geusens 1985b).

In order to reveal real menopausal effects, age at menopause has to be taken into account. When we divided our female population between 40 and 59 years of age into those past the menopause (no menstruation in the previous 6 months) and those still menstruating, a clear and significant effect of the menopause could be seen. Our data confirm the suggestion of Richelson et al. (1984) that the bulk of loss of vertebral bone mineral content is related to oestrogen deficiency and that in the peripheral bones the diminution is mainly related to ageing.

Population studies reveal not only major differences in bone mass between men and women and differential changes in the bone mineral content of the appendicular and axial skeletons with ageing and menopause, but also a marked biological variation in bone mass of $\pm15\%-25\%$ in specific age–sex groupings. A slight increase in variance with increasing age is sometimes found. The heterogeneity of skeletal size has an influence on the variance since there is a highly significant correlation between body length and compact and trabecular absolute bone mass in the third decade in men and women. This positive correlation with skeletal size no longer obtains in the older age groups, indicating that bone loss is not uniform with ageing and that some people lose more bone than others (Table 9.1) (Dequeker et al. 1987a). Introducing skeletal size corrections for bone length, bone width and bone area, and expressing the results as densities, considerably reduces the variation between the sexes and within each age–sex group.

Table 9.1. Decade-specific correlation coefficients of bone mass on skeletal size in women

Age groups	Cortical area D^2-d^2 on metacarpal length	Radius 8 cm BMC/cm and width	L_2–L_3–L_4 BMC and body length
20–29	0.225	0.832***	0.737**
30–39	0.537***	0.566**	0.658**
40–49	0.501***	0.067	0.225
50–59	0.244	0.233	0.225
60–69	0.413	0.260	0.094
70–79	0.052	−0.036	−0.010

* $p<0.05$; ** $p<0.01$; *** $p<0.001$

The quantity of bone in the skeleton is influenced by many factors, among them age, sex, menopause and body size. Each factor contributes to the variation of bone mass found in populations of all ages and such variation increases the difficulty of separating normal from abnormal. Statistical techniques have been developed for removing partially the effects of these variables (Cohn et al. 1976; Dequeker et al. 1980a). This process is sometimes called "normalization". However, as will be shown later on, this normalization may obscure the most important indicator of fracture risk, the absolute bone mass.

Peak Bone Mass

Because of the physiological loss of bone which occurs with age, peak bone mass is the single most important factor that determines how much bone can be lost (due to age or disease) before a critically low bone mass is reached and fractures occur.

Figure 9.6 shows peak bone mass and age- and menopause-related changes in peripheral and axial bone mass as found in our population study (Geusens et al. 1986). From this figure it is clear that maximum bone mass in the axial, mainly trabecular, skeleton is reached in the third decade, while maximum cortical bone mass is reached in the fourth and fifth decades, as judged by cortical area in the metacarpal and bone mineral content in the radius midshaft. The age at which bone growth and modelling achieve maximum bone mass differs according to the bone site and between the sexes. In men, maximum bone mass was achieved at all sites at the age of 25 years. Parfitt (1983) reports that peak adult bone mass, representing the summated contributions of growth (90%) and consolidation (10%), is reached at about age 35 for cortical bone and earlier for trabecular bone.

The factors that determine peak bone mass result from genetic, mechanical, nutritional and hormonal forces. The latter three are probably important as modulators for the achievement of maximum genetic potential.

Genetic Factors

Epidemiological studies utilizing the above screening techniques, such as metacarpal cortical thickness and bone mineral content, have shown significant differences in cortical bone mass between populations from different countries (Garn et al. 1969; Mazess and Mather 1974, 1975; Mazess 1978; Dequeker et al. 1980a; Mazess and Christiansen 1982).

Because of racial heterogeneity, differences in the level of physical activity, the composition of the diet and other factors, it is difficult to dissociate environmental influences from hereditary factors as determinants of peak bone mass. Among the most convincing studies that suggest the importance of genetic factors in the pathogenesis of osteoporosis are comparisons between white and black populations in the United States. American Blacks have a greater bone mass at skeletal maturity than do Whites (Garn 1972). In addition, the rate of bone loss after the age of 40 in Blacks, both males and females, is lower than in age-matched white subjects. Differences in geographic origin, economic background and dietary composition do not account for these observations.

There is other evidence for genetic influences on bone mass. Smith et al. (1973) and Möller et al. (1978) found in studies of cortical bone mass in monozygotic and dizygotic twins that there was a genetic effect on cortical bone mass in adults. In juvenile twins however, Smith et al. (1973) could not demonstrate conclusively a genetic effect on cortical bone mass. In a study of cortical bone mass in twins, we demonstrated a significant genetic effect also on trabecular bone development in twins younger than 25 years old, but a genetic determinant of axial bone development could not be conclusively demonstrated in adult twins (Dequeker et al. 1987b).

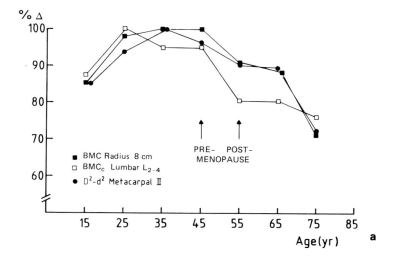

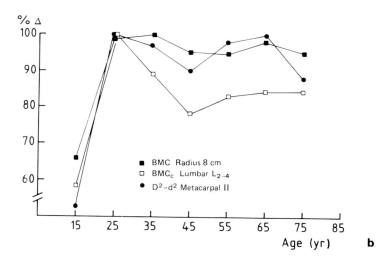

Fig. 9.6. Peak bone mass and age-related changes in women (**a**) and men (**b**).

Mechanical Factors

The role of physical activity in the development and maintenance of bone mass has been well documented in a variety of studies in humans and experimental animals (Howell 1917; Allison and Brooks 1921; Armstrong 1946; Saville and Whyte 1969; Pennock et al. 1972).

Mechanical loading is responsible for variation in mass around the genetically determined level. Increased loading results in increased bone mass and decreased loading results in decreased mass. As a general rule, within any individual there tends to be a constant relationship between muscle mass and skeletal mass. Both of course reflect and respond to mechanical work, so this relationship is not surprising. It is quite striking, however, to note that despite a 30%–40% decline in bone mass with age in women, total body calcium and total body potassium (reflecting cell mass, largely muscles) maintain an essentially constant ratio (Cohn et al. 1976; Heaney 1983).

A relationship between muscle weight and bone mass in humans was shown by Doyle et al. (1970) in a necropsy study of the ash weight of the third lumbar vertebral body and the weight of the left psoas. Other examples of this relationship are seen in the hypertrophy of both bone and muscle in the dominant arm of tennis players (Huddleston et al. 1980) and in the legs of ballet dancers (Nilsson et al. 1980) and the atrophy of both bone and muscle that occurs on prolonged bed rest during space flight (Rambaut and Goode 1985) and in paralytic syndromes such as poliomyelitis (Roh et al. 1973b). In a study on the relationship between bone mass and upper extremity strength in 106 eighteen-year-old boys, a significant correlation between arm pull strength and cortical area was found (Dequeker and Van Tendeloo 1982). This study suggests that static strength is one of the determinants of bone gain before maturity.

Nutritional Factors

Garn (1972), studying bone mass at the metacarpal bone in large numbers of individuals from different economic and social classes and nationalities, found that in general the more affluent of both black and white races have larger bones and greater skeletal weight. The relationship between per capita income and the mass of skeleton reflects the direct availability of both calories and nutrients to those of higher incomes. The difference in skeletal mass between those receiving top incomes and those receiving slightly below average incomes is in the order of 9%–10%. Rates of bone formation and bone resorption appear to be proportional within these economic brackets, however, so that the major difference is in the mass of bone rather than in the proportion of compact bone in the total (anatomical) bone envelope. Nonetheless, being smaller in both size and volume the bones of the poor should be weaker and more subject to fracture.

That dietary calcium may contribute to the bone mass acquired during the period of skeletal growth is suggested by the report of Matkovic et al. (1979). Two populations in Yugoslavia of similar ethnic origin and roughly comparable living conditions and a similar pattern of physical exercise were studied. In one population, calcium intake was relatively high (approximately 1000 mg/day) and in the other calcium intake was relatively low (approximately 500 mg/day); this difference was related to differences in the consumption of milk and dairy products. It was found that bone mass was greater (about 5% in women and 10% in men) and proximal femur fracture rate lower in the population with the higher calcium intake. However, the differences in bone mass between the two populations were detected by age 30 and did not increase with age. Therefore, it is possible that the higher fracture rate and the lower bone mass of the population on the lower calcium intake were related to their having attained a lower bone mass during growth.

To permit normal bone growth, calcium intake during childhood should meet the requirements of net calcium retention and obligatory calcium loss (Parfitt

1983). Mean net calcium retention for the entire growth period is in the range 100–200 mg/day, with a peak retention during the prepubertal growth spurt of 300–400 mg/day (Garn 1970; Nordin et al. 1979). Both in utero and during the 1st year of life, modest shortages in the supply of calcium or vitamin D, not severe enough to cause rickets, reduce the rate of increase of cortical bone mass (Raman et al. 1978; Hillman 1983; Tsang 1983). It is not known whether the deficit can be regained in later childhood or whether it persists to reduce peak adult bone mass.

Hormonal Factors

From conception until epiphyseal closure, there is a progressive increase in trabecular bone due to endochondral ossification and in cortical bone due to net periosteal and endosteal apposition. Bone mass is controlled by the remodelling process in which microscopic cavities are eroded on bone surfaces by osteoclasts and subsequently refilled, completely or incompletely, by osteoblasts (Parfitt 1982). The cycle of remodelling is under hormonal control and is initiated by the activation of a quiescent bone surface; a cycle normally lasts 3–4 months. The frequency of activation which is increased by PTH and decreased by calcitonin, determines the rate of bone turnover. Net gains or losses of calcium over a few hours, as in the normal circadian changes in balance, are mediated by a short-term storage and release mechanism that does not involve structural change in bone (Parfitt 1981).

The effect of hormonal change on the bone surfaces during life can best be appreciated on x-rays of the tubular bone of the hand. During infancy, puberty, adolescence and ageing, important changes occur at the periosteal and endosteal surface, as has been demonstrated radiogrammetrically by Garn in a population of 12 290 participants in the United States (Fig. 9.7). Periosteal appositional growth is not constant: there is a phase of childhood apposition, a period of adolescent gain and a continuing bone expansion after the third decade, more so in males than in females. The changes observed at the endosteal surface are even more impressive: there is a resorptive phase in infancy, a childhood appositional phase, a marked pubertal resorptive phase, an adolescence through adulthood appositional phase and finally the adult resorptive phase.

Between the two bone surfaces described, cortical area (and therefore cortical volume) increases greatly from infancy to adulthood, and then undergoes a precipitous decrease, often with peak rates of loss in the fifth and sixth decades. The different phases of bone apposition and resorption at the different bone surfaces coincide with changes in the hormonal status of the individual during life.

Bone gain and bone loss both have in part a hormonal basis. Adolescent bone gain, both subperiosteal and endosteal, follows the overt evidence of sexual maturation and gametogenesis. Bone gained at the inner surface is clearly under the influence of sex hormones, since in Turner's syndrome, an anoestrogenic condition, this bone gain at the endosteal surface is not observed while periosteal apposition is normal (Dequeker 1971). The apposition of bone from middle adolescence to the end of the fourth decade, especially in females, resembles the medullary bone formation under oestrogenic and androgenic

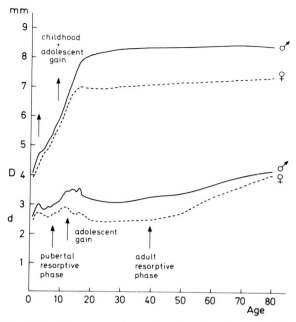

Fig. 9.7. Changes at the periosteal and endosteal surfaces in 12 290 white male and female participants in the "Ten-State Nutrition Survey, USA, 1968–1970". (Adapted from Garn 1972)

control in birds (Taylor and Belanger 1969). During the pre-laying period, the marrow cavities of many of the bones of female birds are invaded by a system of secondary bone which acts as a labile store for calcium for the calcification of the egg shell. In this context it is also of interest that multiparity is associated with increased bone mass rather than with depleted bone mass (Smith 1967; Garn 1970; Daniell 1976). Heaney and Skillmann (1971) demonstrated very active calcium storage mechanisms during human pregnancy, probably resulting in the development of calcium reserves considerably in excess of foetal skeletal demands.

Periosteal bone gain has been shown to be under the influence of growth hormone (Dequeker 1971; Garn et al. 1971; Dequeker et al. 1980b). Parathyroid hormone also has a positive effect on the periosteal site but a negative effect on the endosteal site in postmenopausal women (Dequeker 1970). Normal hormonal development is obviously necessary for full expression of the genetically determined peak adult bone mass.

Thus the mass of the total skeleton is determined by a hierarchy and interaction of factors; variations in mass around the level determined by a genetic effect are due to mechanical loading, the internal hormonal environment and the availability of key nutrients.

After cessation of growth there is a period of consolidation lasting about 15 years during which the cortical porosity which increased during the adolescent growth spurt declines slowly to a nadir, and the cortices become slightly thicker because of very slow net apposition on both periosteal and endosteal surfaces (Parfitt 1983).

A few years after peak adult bone mass is attained, age-related bone loss begins as mentioned above, a universal phenomenon in human biology not seen in animal biology, that occurs regardless of sex, race, occupation, economic development, geographical location, historical epoch or dietary habits.

Age-Related Fracture and Osteoporosis

Definition of Osteoporosis

The classical definition of osteoporosis by Albright and Reifenstein (1948), "too little bone, but what bone there is, is normal", has been widely accepted. This fundamental definition has never seriously been challenged and a clear distinction has been made between osteoporosis and osteomalacia. In osteomalacia the ratio of mineral to organic matrix is low, indicating a mineralization defect which in general is due to a fault in vitamin D metabolism. However, Albright's definition does not define "too little bone". If this means too little by the standard for adults in the prime of life as Nordin (1983) advocates, many people without symptoms have osteoporosis simply because the physiological changes of ageing have affected their bones.

The term osteopenia is now often used to designate a loss of absolute bone volume without fracture, whereas osteoporosis is taken to imply mechanical failure of the skeleton in addition to osteopenia. The fractures in osteoporosis, generally involve the spine, but can also occur in the hip and wrist, and are due to minor trauma in persons (or body regions) with reduced skeletal mass, and not simply to the reduction in skeletal mass itself. This dichotomy of terms was previously not necessary because patients first came to medical attention with fractures, and because heretofore it was difficult reliably to detect decreased skeletal mass per se in someone without fractures. Thus, for most physicians, fracture itself was and is the principal diagnostic evidence of osteoporosis. Nevertheless, the fact of fracture and the fact of decreased skeletal mass are distinct and ought not to be confused.

Osteoporosis at present is not considered to be a disease but a syndrome, revealed by a fracture from minimal trauma in a person with reduced skeletal mass. Analogous syndromes include congestive heart failure, renal insufficiency, hepatic failure, mental deficiency, arterial insufficiency and respiratory insufficiency. All these syndromes have a common physiopathology, namely insufficient organ reserves to meet intercurrent minor stress.

Because of the confusion in the use of the term osteoporosis, the National Institute of Health (USA) Consensus Conference on Osteoporosis in 1984 formulated the following definition for osteoporosis. Primary osteoporosis is an age-related disorder characterized by decreased bone mass and by increased susceptibility to fracture in the absence of other recognizable causes of bone loss. Secondary osteoporosis is then the condition where the osteoporosis and fracture are the result of excessive bone resorption in specific endocrinopathies, as for example hyperparathyroidism and hyperthyroidism, or of excessive and prolonged use of drugs which interfere with bone metabolism, as for example corticosteroids.

Epidemiology

Osteoporosis is the most common of the diseases that affect bones. At all ages the incidence is higher in women than in men, and in the USA it is higher in Whites than in Blacks. One out of four white women will develop osteoporosis after the menopause. The principal fractures associated with osteoporosis are those of the distal forearm, the vertebrae and the femoral neck. There is a steep rise in the distal forearm fracture rate in women from about the time of menopause, but little change with age in men. Wedging of the vertebrae is present in about 60% of elderly women, but only 5%–10% develop two or more crush fractures. The rise in the incidence of femoral neck fractures with age in both sexes is steeper than can be accounted for by loss of bone alone; an important contributory factor is the increasing incidence of falls with advancing age, particularly in women, due to the increasing prevalence of cardiovascular and cerebrovascular disease with age, to impaired vision and to other disabilities associated with ageing. Nonetheless, falls alone do not explain the fractures; loss of bone is a vital contributory factor.

All these fractures are interrelated, i.e. the patients who have a wrist fracture incur a significantly increased risk of vertebral and/or femoral neck fracture, patients with vertebral fractures are at increased risk from wrist and femoral neck fractures and patients with femoral neck fractures have had significantly more wrist fractures and vertebral fractures than age-matched controls. The cumulative prevalence of all these fractures in women is about 7% by age 60 and about 25% by age 80 (Nordin 1983) (Fig. 9.8).

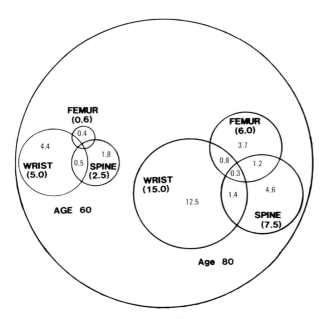

Fig. 9.8. Fracture prevalence (%) in women age 60 and age 80. The outer circle represents the whole population; the inner circles represent the fracture populations drawn to scale. The overlapping areas represent cases with more than one fracture. The number denotes the percentages of the total population who have sustained the different fractures. (Nordin 1983)

In our modern society with an increasingly geriatric population, the number of individuals predisposed to hip, femoral, forearm or vertebral fractures will steadily increase and become a formidable public health problem. Fractures in aged individuals require prolonged periods of hospitalization, often resulting in significant inactivity and morbidity. The mortality associated with hip fractures is as high as 17%, which is about 20 times that expected (Lewis 1981). Several recent studies from Great Britain have reported a rising incidence of fractures of the proximal femur in both sexes (Lewis 1981; Boyce and Vessey 1985). This rise is more than just a consequence of demographic change. The raised incidence might be due to the emergence of one or more new risk factors – alcohol consumption rises steadily with age, as does the use of drugs, particularly psychotropic drugs – or it may reflect an increase in the prevalence or severity of existing factors. Various explanations, such as poor diet during the 1914–1918 War (Nilsson and Obrant 1978) or reduced mobility (Lewis 1981) have been suggested. At present there is not enough evidence to choose between any of these theories. The more general theory is that the more elderly there are in our society (compared with 40 years ago) the more infirm it is and the more likely it is to possess some of the known risk factors.

Heterogeneity of Osteoporosis

Osteoporosis Fracture Patterns

The age and sex distribution of each type of osteoporotic fracture differs. Riggs and Melton (1983) have suggested that two distinct types of involutional osteoporosis exist: *Type I* postmenopausal osteoporosis occurring in a small subset (5%–10%) of women within 15–20 years after the menopause, involving predominantly trabecular bone loss and manifested chiefly by vertebral fractures; and *Type II* senile osteoporosis occurring in both sexes after age 75, involving proportionate trabecular and cortical bone loss, and manifested by vertebral and/or hip fracture.

Primary osteoporosis occurs not only in the older age groups; for no apparent reason, osteoporosis is also seen (although rarely) in children, in young adults and in pregnancy. The term "juvenile osteoporosis" is applied to individuals who present with fractures and decreased bone mass in childhood or adolescence in the absence of another contributing systemic disorder. The term "idiopathic osteoporosis" is used to define the disease in those individuals in whom the disease becomes manifest between the ages of 20 and 45. Severe osteoporosis in young women has also been described in association with pregnancy and lactation (Nordin and Roper 1955; Dent and Friedman 1965; Smith et al. 1985).

Biological Heterogeneity

The heterogeneity of osteoporosis is not only characterized by this apparent difference in fracture sites and in age and sex distribution, but also by differences in bone remodelling – at the time of bone failure (Meunier et al. 1979; White et al. 1982; Johnston et al. 1985).

Even in postmenopausal women, osteoporosis is a heterogeneous disease. Osteoporosis may be the end result of different (pathological) mechanisms occurring during life on top of the physiological bone loss which occurs with ageing, e.g. oestrogen/progestogen deficiency, reduction in intestinal calcium absorption, renal calcium loss, immobilization, alcohol consumption and imbalances in vitamin D metabolites, parathyroid hormone, calcitonin, growth hormone, thyroid hormone and endogenous corticosteroids.

Osteoporosis is generally regarded as a disorder which is not accompanied by pathognomonic biochemical alterations. Most studies show a considerable overlap with normal in all the various measurements made. Small subsets with increased or decreased values in one or another parameter are often recorded, indicating heterogeneity within osteoporosis. We found that in symptomatic vertebral crush fracture cases, 39% had a high fasting calcium/creatinine ratio, 84% a high fasting hydroxyproline/creatinine ratio, 21% a high (for their age) 24-hour calcium excretion of more than 200 mg and 12% a low (less than 50 mg/ 24 hours) calcium excretion (Dequeker et al. 1984a). Some patients at the time of fracture have indications of a high or low bone turnover. Osteocalcin (bone Gla protein, BGP), a specific marker for bone formation and turnover, has been found to be increased in a number of patients with postmenopausal osteoporosis, indicating that overall bone turnover is increased and that an absolute decrease in bone formation is not the major cause of bone loss (Delmas et al. 1983).

Many aspects of calcium metabolism change with age and thus age-related change will correlate with bone mineral loss; causal effects in osteoporosis have been argued a priori for many of these changes (e.g. parathyroid hormone, calcitonin and 1,25-dihydroxyvitamin D). It is therefore necessary in any study to include age–sex-matched controls who do not have and are unlikely to develop osteoporosis. The now well-established evidence for an inverse relationship between osteoporosis and osteoarthritis suggests the use of patients with osteoarthritis as proper controls in studies trying to elucidate pathogenetic effects other than ageing in osteoporosis (Dequeker 1985). No significant differences in blood parameters such as serum calcium, phosphorus, alkaline phosphatase, creatinine or total protein have been found between osteoporosis and osteoarthritis cases (Geusens et al. 1983). A significantly higher excretion of calcium but not of hydroxyproline has been found in patients with osteoporosis, indicating a negative calcium balance in osteoporotics (Geusens et al. 1983).

The only difference in hormonal status between osteoporosis and osteoarthritis cases was in the growth hormone level in the fasting state and after an insulin stimulation test, osteoarthritis cases having a higher level of growth hormone than osteoporosis cases (Dequeker et al. 1982). Since growth hormone is the most important anabolic hormone, this finding might be of particular importance. Despite the heterogeneous results, investigation at the time of fracture might be of importance for the choice of therapy.

In femoral neck fracture cases, there seems to be some evidence that in a subset of cases vitamin D metabolism is disturbed with reduced 25-hydroxyvitamin D levels, decreased kidney function and slightly raised PTH levels (Lips et al. 1983). Osteomalacia occurs in patients with hip fracture but its incidence has been overestimated (Parfitt et al. 1982).

Bone Mass and Risk of Fractures

"If hypertension is a silent killer, osteoporosis is a silent thief. It insidiously robs the skeleton of its banked resources, often for decades, before the bone is weak enough to sustain a spontaneous fracture" (Kaplan 1983).

Bone Mass and Strength

Newton-John and Morgan (1970) and Morgan (1983), in their reviews of ageing osteopenia, made assumptions about the interrelationships of bone loss, bone strength and fracture. They suggested that the loss of bone with age was an inevitable universal consequence of senescence and so were fractures. The relationship between bone mass and compressive strength or breaking strength of vertebrae and long bones has been examined under laboratory conditions (Yamada 1970; Lindahl 1976; Kazarian and Graves 1977). The compressive strength of the bodies of vertebrae is about equal in the lower cervical and upper thoracic vertebrae, then increases gradually to another plateau in the region of lumbar vertebra 2 to lumbar vertebra 4. The mass of a lumbar vertebra increases from L1 to L4, but its areal density (g/cm^2) remains unchanged. Peak strength of vertebrae is reached in the third decade, after which there is a loss of compressive strength with age, more pronounced in women. The rate of loss of compressive strength with age is faster in the vertebrae than in the proximal tibia. Clinical and epidemiological studies support these observations.

It is clear that 80% of the variance in the compressive strength of trabecular bone and 90% of the variance in compact bone is due to the mass of the material tested (Bell et al. 1967; Rockoff et al. 1969; Dalen et al. 1976; Nokso-Kovisto et al. 1976; Smith and Smith 1976; Hansson et al. 1980). Changes in bone strength with age generally parallel changes of bone quantity but are usually more accentuated (Carter and Hayes 1976; Wall et al. 1979; Martens 1985). The decreased strength does not reflect a decreased relative mineral content of dry, defatted bone in either compact (Woodard 1964) or trabecular (Dequeker and Merlevede 1971) bone.

Accordingly, the changes in bone mass that occur with age have been taken to be indicative of decreased bone strength and increasing susceptibility to fracture. This suggests that the total mass of trabecular or cortical bone rather than density needs to be measured in order to evaluate fracture risk. However, in longitudinal studies, density indices are preferable in order to eliminate the effects of relocation errors. Furthermore measurements on trabecular and compact bone must be considered separately as well as together.

The above studies suggest that the bone mass is, indeed, an important factor in the genesis of fracture. It is, perhaps, the most important, but other factors must also contribute.

Bone Quality and Strength

Structural continuity and anatomical orientation in trabecular bone, as well as the amount of bone present, determine its strength. So discontinuity due to perforation of trabeculae, as may occur in rapid bone loss, may produce greater loss of strength than the same extent of bone loss due to thinning alone (Pugh et al. 1973; Parfitt 1984).

The lower mineral content in trabecular bone substance than in compact bone (60% versus 68%), particularly in women (Baker and Angel 1965; Mbuyi-Muamba and Dequeker 1987), may influence bone strength and susceptibility to fracture (Vose and Kubala 1959; Currey 1969; Mazess 1982). Therefore it may require a smaller change in bone mass in trabecular areas than in compact areas to produce decreases of strength (Mazess 1982). Microfractures (fatigue fractures), which seem to accumulate in trabecular bone with age (Vernon-Roberts and Pirie 1973), may also contribute to fracturing with minimal changes of bone mass. The number of such fractures increases as bone density decreases (Todd et al. 1972; Freeman et al. 1974). As a consequence of these factors, it cannot be assumed that ageing osteopenia in trabecular and compact bone produces equivalent risks of fracture.

From measurements in vivo in subjects with and without fracture, similar conclusions can be drawn. In most studies, individuals with fractures have lower bone mass than those without. However, considerable overlap between those with fractures and sex–age-matched control groups has been found when bone mass is measured in the radius of patients with and without vertebral collapse (Smith et al. 1972; Dequeker et al. 1974; Riggs et al. 1981); in the femoral neck of patients with and without fracture of the proximal femur (Vose and Lockwood 1965; Riggs et al. 1982a; Bohr and Schaadt 1983; Aitken 1984; Johnston et al. 1985); in the vertebrae of subjects with and without vertebral crush fractures (Arnold 1965; Riggs et al. 1981; Krölner and Nielsen 1982; Dequeker et al. 1984b) and in the opposite radius of patients with wrist (Colles') fracture (Nilsson and Westlin 1974).

An explanation could be that the overlap occurs partially because the methods are insufficiently sensitive (for example, because they use density indices rather than absolute bone mass indices), or that the measurements were made at locations other than the fracture sites. A more likely explanation is that other unmeasured variables contribute to the pathogenesis of the fracture. Factors which contribute to the genesis of fractures may be extrinsic or intrinsic to bone. Intrinsic factors include heterogeneity in bone remodelling and repair and a certain degree of osteomalacia.

Risk of Trauma and Fracture

Most fractures are associated with some degree of trauma which is usually related to falling. In a longitudinal study of falls occurring in an ambulatory institutionalized population over age 65, the annual fall rate was 668 per 1000. A rising frequency was found after age 75. Women had a higher fall rate than men, and the severity of their injuries increased with age (Gryfen et al. 1977). Overstall et al. (1977) examined 243 elderly people who had fallen and measured sway in the group with an ataxiameter. Sway increased with age and was higher in women.

These studies argue that changes in the central nervous system or muscle function may also be a major contributing factor in the pathogenesis of fracture, especially of femoral neck fractures, particularly among those who already have low bone mass.

Thus, the risk of age-related fractures can be expressed as follows:

$$\text{Risk of fracture} = \frac{\text{Degree of trauma}}{\text{Bone mass}}$$

The risk factors for age-related fractures then can be arbitrarily divided into two kinds: (a) those that increase the risk of trauma, in terms of severity, frequency or both; and (b) those that cause a reduction in bone mass.

Identification of Those at Risk for Fractures

If osteoporosis is to be prevented, the first step is to identify in good time those at high risk. The effect of oestrogen deficiency on bone loss is well documented (Albright and Reifenstein 1948; Dequeker 1972; Lindsay et al. 1977).

Patients at high risk of osteoporosis are women with premature oestrogen deficiency, surgical or natural, who clearly require oestrogen replacement treatment. Knowledge of factors other than oestrogen deficiency likely to cause or aggravate age- and sex-related bone loss is of clinical value for prediction and counselling.

Clinical Criteria

There seems to be general agreement that small, thin, fair, light-skinned women, especially of Northern European extraction, appear to have significantly higher fracture risk as do spinsters or mothers of only one or two children, those with strong family histories of hip and spine fracture and possibly those who, in addition to the foregoing, are heavy smokers and/or alcohol abusers (Heaney 1981).

Of clinical interest in this context is the observation that primary osteoporosis and generalized osteoarthritis are rarely seen together, although both are common in the older age groups. Actual data suggest that real differences between people with the two conditions exist (Dequeker 1985). Women with osteoarthritis have been shown to score highly on anthropometric indices of body weight, skinfold thickness and muscle girth and strength, while women with osteoporosis scored below average on these indices (Dequeker et al. 1982b). No significant differences in serum PTH, 25-hydroxy- or 1,25-dihydroxyvitamin D metabolite levels could be established between osteoporosis and osteoarthritis cases (Geusens et al. 1983), but a significantly raised fasting growth hormone level was found in osteoarthritis and not in osteoporosis (Dequeker et al. 1983). Are these differences of pathogenic importance for the development of the two diseases? Are the excess of body weight, excess of fat and growth hormone or factors related to these associated with reduced bone resorption or increased bone formation in osteoarthritis?

In previous studies (Roh et al. 1973b, 1974a, b), the observation by Foss and Byers (1972) that patients with osteoarthritis had a higher bone mass than those with osteoporosis matched for age and sex was confirmed. An explanation for this difference could be that patients with osteoarthritis, with their excess of fat, have an increased peripheral aromatization of androstenedione relative to oestrone, with a reduced bone resorption as a result (Frumar et al. 1980), in addition to the anabolic effect of growth hormone.

Since the first clinical signs of generalized osteoarthritis are readily observable in the hand joints (Heberden and Bouchard nodes) in the perimenopausal period, these changes can be useful for identifying a group of women who are probably not at risk of developing symptomatic osteoporosis and thus do not need preventive therapy.

Factors that aggravate age-related bone loss or superimpose another bone-losing mechanism on age-related loss include the following: immobilization and prolonged bed-rest, thyrotoxicosis, glucocorticoid therapy, malabsorption syndromes, excessive use of aluminium-containing antacids and miscellaneous disease states ranging from renal failure to disseminated collagen diseases.

Objective Criteria

Objective identification of the fracture risk group is a goal of great practical significance since treatment of a group larger than necessary would be difficult, expensive and in certain instances fraught with needless risk.

Refinements in bone mass measurements of a sort readily available in medical practice, such as radiogrammetry and single and dual photon absorptiometry, would aid in identification of preosteoporotic women with low bone mass. (It remains to be seen whether these women are those most at risk.)

These methods allow reasonably accurate quantification of cortical bone mass in the forearm and have been very useful in epidemiological research and in follow-up measurements on groups of people (Dequeker and Johnston 1982). However, some fractures in osteoporosis involve areas of predominantly trabecular bone in the spine and there is now abundant evidence that forearm bone mass measurements discriminate poorly between cases with vertebral crush fractures and normals but discriminate better between femoral neck fracture cases and normals because in this case a predominantly cortical bone area is involved (Dequeker et al. 1975).

For screening purposes and follow-up it is now essential to employ two or more measurements of bone at different sites, including the more metabolically active trabecular bone in the spine and the more compact cortical bone at the periphery.

Two new fundamentally different techniques, dual photon absorptiometry (DPA) and quantitative computed tomography (QCT), now make it possible to measure bone mass in areas affected by osteoporosis. The newer methods allow correction for soft tissue absorption and thus allow bone measurements in such areas as the lumbar spine and proximal femur. Quantitative computed tomography measures the density of a defined volume of trabecular bone within the vertebral body and allows exclusion of cortical bone in the laminae, pedicles and vertebral body envelope. Dual photon absorptiometry measures the mass of the whole vertebral segment and so includes both cortical and trabecular components. The latter technique (DPA), incorporating total cortical and trabecular bone mineral content into a single measurement, may more accurately predict strength than techniques that assess either component independently. Furthermore, the radiation exposure during dual photon absorptiometry measurement is minimal in contrast to the much larger exposure when QCT is used.

Sensitive, accurate and clinically meaningful assays of bone mass in selected body regions will soon be more widely available. The main role of such measurements will be to detect people already at high risk of fracture. At present, a person who has a bone mass measurement 20% below the average of an age–sex-matched control population, can be considered to have a low bone mass and to be a candidate for preventive treatment.

If measurements of the radius are used for screening it should be realized that low radius bone mineral content (BMC) indicates a high probability of concomitant low total skeletal calcium and low spinal bone mineral but a normal value does not exclude spinal osteoporosis. Radius BMC may be of more value in predicting bone loss at the hip than in the spine (Johnston 1983; Wahner et al. 1984). Bone mineral measurement of trabecular bone in the spine will give a better indication of those who are at risk from vertebral crush fracture.

There is accumulating evidence that there is a threshold value of BMC for fracture in the vertebrae. This value corresponds approximately to a BMC value two standard deviations below the mean of adult peak bone mass (Riggs et al. 1981; Krölner and Pors Nielsen 1982; Dequeker et al. 1984b).

Many studies express results as the mass of bone per unit area or volume (apparent density) for purposes of comparison between individuals but the strength of the bone is (as discussed above) attributable to its total mass (and internal architecture) rather than its density. There are minor differences between the sexes in the apparent density of trabecular bone segments. The differences between the sexes in fracture pattern, like that in strength, must reflect bone size and mass more than density. Even among women, it is those with smaller bones who develop fractures and not necessarily those with the lowest bone density.

This suggests that total mass indices rather than density (or relative) indices need to be used in order to evaluate fracture risk. In studies over time however, density indices may be preferable in order to average out relocation shifts (Mazess 1982).

As noted above, no method can be expected to predict fracture risk infallibly, since factors other than low bone mass play a role in fracture pathogenesis.

Treatment of Osteoporosis

Prevention

Few would disagree that prevention is better than cure, but if we are to ameliorate this aspect of ageing, it will mean giving "treatment" to asymptomatic "normal" people and the necessity for the therapy to be absolutely safe will be critical.

Potential hazards and inconvenience associated with such therapy might, however, be reasonable for a "high risk" group of patients.

Prevention refers here primarily to the slowing or cessation of the age-related bone loss in postmenopausal women. Prevention of fracture is of course the only goal of practical importance, but no controlled prospective studies bearing on this outcome have been carried out, and for logistic reasons may never be carried out. Rich et al. (1966) estimated that demonstration of a 40% reduction in fracture frequency would require 1400 patients followed over a 3-year period. Hence we are left with a fairly reasonable presumption, but a presumption nonetheless, that prevention of bone loss is a necessary condition for prevention of bone fracture.

Oestrogens

There is now evidence to suggest that preventing the accelerated cortical bone loss in postmenopausal women is both practical and effective. Several double-blind, placebo-controlled studies using oestrogen as a prophylactic have been reported (Lindsay et al. 1976; Horsman et al. 1977; Recker et al. 1977; Christiansen et al. 1980). These studies have shown significant protection against age-related bone loss, i.e. a reduction in rate of loss to undetectable levels and even in some cases an increase in peripheral bone mass of 1.2% per year (Fig.

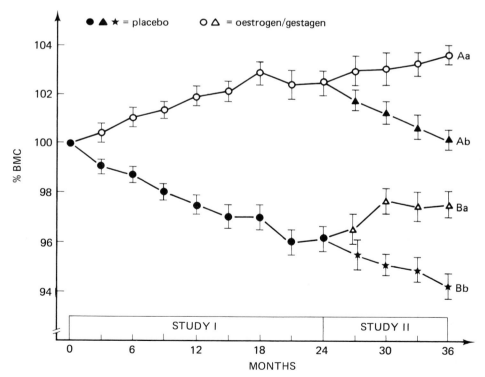

Fig. 9.9. Bone mineral content (BMC) as a function of time and treatment (with oestrogen/gestagen) in women soon after the menopause. (Christiansen et al. 1981; pp. 459–461)

9.9). Hutchinson et al. (1979), using case control methods, were able to demonstrate, apparently for the first time, protection against hip and radius fractures in women given postmenopausal oestrogen therapy.

Side-effects associated with prolonged oestrogen treatment are well known and need no further mention. There is new evidence that cyclic administration of oestrogens with an added progestogen may reduce the risk of breast and endometrial cancers (Thom et al. 1979; Gambrell 1982). How long preventive oestrogen therapy has to be given for is still a matter of discussion and in most cases it has to be governed by individual judgement. At least 5 years', but preferably 10 to 15 years' treatment would take the person at risk beyond the natural fracture time (Dequeker et al. 1984a).

Calcium

There is evidence that simple calcium supplementation may slow down bone loss in postmenopausal women but to a lesser degree than does oestrogen therapy (Recker et al. 1977; Nordin et al. 1980). Longitudinal studies of subjects with deliberately elevated calcium intake have shown that high intakes of about 1.5 g of calcium per day will inhibit age-related bone loss in postmenopausal white women. As women get older, they need more calcium but tend to take less. Mean calcium requirements estimated by the balance method are higher in women than in men and increase with age in both sexes and after the menopause in women, mainly because of reduced absorption efficiency (Heaney et al. 1982). Since calcium is the only nutrient for which there is epidemiological evidence of a relationship to fracture rate and since it is cheap, easy to administer and has no harmful effect, it should be considered in any case with a low bone mass (Parfitt 1983).

Exercise

The value of physical activity in preserving bone mass is emphasized by several studies which showed that patients lost calcium while they were confined to bed-rest and that the loss did not reverse until they were back on their feet. Mechanical stress not only preserves bone mass but also seems to enhance bone formation (Dequeker and Van Tendeloo 1982).

The goal of a prophylactic exercise programme should be to achieve the highest possible peak bone mass before old age and to maintain that peak as long as possible. Such a preventive programme might account for as much as 10%–20% additional bone mass later on, and that difference might be "crucial" in preventing injury and fractures due to falls in the later years.

Inhibition of involutional bone loss by exercise has been reported by Aloia et al. (1978), by Smith et al. (1984) and by Krölner and Nielsen (1982), in small groups over a restricted time period.

However, excessive exercise is not to be recommended since it is now well established that exercise-induced menstrual dysfunction develops in 10%–40% of female runners and that abnormally low density of vertebral bone has been found in young athletes with amenorrhoea (Drinkwater et al. 1984; Lindberg et al. 1984).

Curative Treatment for Osteoporosis

In patients with fractures, oestrogen and calcium treatment has been considered but without much enthusiasm, since it is only palliative. However, recent data have shown that untreated patients with osteoporosis lose bone at a higher rate than do age-matched controls (Riggs et al. 1976) and oestrogen administration markedly inhibits this bone loss (Nordin et al. 1980). Risk of further fracture can be decreased as well (Riggs et al. 1982b).

Elderly patients, sometimes severely disabled from many fractures, usually find that the management of vaginal bleeding 10 to 15 years after the menopause and follow-up visits to the gynaecologist, cause enough discomfort and inconvenience to outweigh the benefits derived from inhibiting bone loss. On the other hand, younger women who have had no more than one or two fractures may yet derive enough benefit from administration of oestrogens to make it worthwhile.

Anabolic steroids have been tested in patients with osteoporosis. They have effects on bones similar to those of oestrogens (Dequeker et al. 1984a; Geusens and Dequeker 1986). Bone loss has been seen to stop during 2 or more years of observation and a slight gain in bone may occur (Lindsay et al. 1976; Chesnut et al. 1977, 1983). There seems to be minimal risk associated with the use of anabolic steroids. Androgens were once used in the treatment of this condition but their use has since been abandoned because of their unwanted virilizing effect.

There appears to be a differential response by parts of the skeleton to fluoride treatment for osteoporosis; some parts gain bone and others lose it. The more central parts (vertebral bodies, pelvis), which are rich in trabecular bone, accumulate new bone (Briançon and Meunier 1981; Harrison et al. 1981); the more peripheral parts (distal forearm, femur, metacarpal) lose compact bone or at least show no gain (Dambacher et al. 1978; Franke 1978; Ringe et al. 1978; Riggs et al. 1980). So what might be considered as advantageous for one part of the skeleton could be associated with deleterious effects for others.

Although there is some concern about the quality of bone that may be produced during fluoride treatment, studies have indicated actual reduction in fracture incidence (Reutter and Olah 1978; Riggs et al. 1980). Riggs and co-workers (1982b) reported that the combination of calcium, oestrogens and fluoride virtually abolished fractures after the first year of treatment.

Although fluoride therapy can be useful in vertebral crush fracture syndromes, fluoride cannot be recommended for general use and definitely not for treatment of femoral neck fracture cases.

The value of calcitonin and 1,25-vitamin D as therapeutic agents for established osteoporosis is not clear yet. There are indications however that these hormones might have a differential effect on bone with a more marked effect on trabecular bone than on peripheral bone (Gallagher 1983; Dequeker et al. 1984a).

In the near future it is expected that because of better knowledge of the physiopathology of osteoporosis and a better monitoring of trabecular and cortical bone mass changes, better and more effective treatment regimens will be disclosed. At present, treatment of osteoporosis remains controversial and unsatisfactory (Kanis 1984).

Acknowledgements. The author is indebted to Drs. P. Geusens, A. Verstraeten and G. Gevers for reviewing this chapter and to Prof. B.E.C. Nordin (Fig. 9.8), Prof. C. Christiansen (Fig. 9.9), and Dr. M. Martens (Figs. 9.2, 9.3) for giving permission to reproduce their original figures.

References

Adams P, Davies GT, Sweetnan P (1970) Osteoporosis and the effects of ageing and bone mass in elderly men and women. Q J Med 39: 601–615

Aitken JM (1984) Relevance of osteoporosis in women with fracture of the femoral neck. Br Med J 288:597–601

Albright F, Reifenstein EC (1948) The parathyroid glands and metabolic bone disease. Williams and Wilkins, Baltimore

Allison N, Brooks B (1921) Bone atrophy. An experimental and clinical study of the changes in bone which result from non-use. Surg Gynecol Obstet 33:250–260

Aloia JF, Cohn SH, Ostuni JA, Cane R, Ellis K (1978) Prevention of involutional bone loss by exercise. Ann Intern Med 89:356–358

Armstrong WD (1946) Bone growth in paralysed limbs. Proc Soc Exp Biol Med 61:358–362

Arnold JS (1965) Amount and quality of trabecular bone in osteoporotic vertebral fractures. Clin Endocrinol Metabol 2:221–238

Baker PT, Angel JL (1965) Old age changes in bone density: sex, and race factors in the United States. Hum Biol 37:104–121

Barnett E, Nordin BEC (1960) The radiological diagnosis of osteoporosis. A new approach. Clin Radiol 11:166–174

Bell GH, Dunbar O, Beck JS (1967) Variations in strength of vertebrae with age and their relation to osteoporosis. Calcif Tissue Res 1:75–86

Bland JH, Soule AB, van Buskirk FW, Brown E, Clayton RV (1969) A study of inter- and intra-observer error in reading plain roentgenograms of the hands. Am J Roentgenol 105:853–859

Bohr H, Schaadt O (1983) Bone mineral content of femoral bone and the lumbar spine measured in women with fracture of the femoral neck by dual photon absorptiometry. Clin Orthop Rel Res 179:240–245

Boyce WJ, Vessey MP (1985) Rising incidence of fracture of the proximal femur. Lancet I:150–151

Briançon D, Meunier PJ (1981) Treatment of osteoporosis with fluoride, calcium and vitamin D. Orthop Clin North Am 12:629–648

Cameron JR, Sörenson J (1963) Measurement of bone mineral in vivo: an improved method. Science 142:230–232

Cann CE, Genant HK (1980) Precise measurement of vertebral mineral content using computed tomography. J Comput Assist Tomogr 4:493–500

Cann CE, Genant HK, Ettinger B, Gordon GS (1980) Spinal mineral loss in oophorectomized women. JAMA 244:2056–2059

Carter DR, Hayes WC (1976) Bone compressive strength: the influence of density and strain rate. Science 194:1174–1176

Chesnut CH, Nelp WB, Baylink DJ, Denney JD (1977) Effect of methandrostenolone on post-menopausal bone wasting as assessed by changes in total bone mineral mass. Metabolism 26:267–277

Chesnut CH, Ivey JL, Gruber HE et al. (1983) Stenozolol in postmenopausal osteoporosis: therapeutic efficacy and possible mechanism of action. Metabolism 32:571–580

Christiansen C, Christensen MS, McNair P, Hagen C, Stocklund K, Transbol I (1980) Prevention of early postmenopausal bone loss: controlled 2-year study in 315 normal females. Eur J Clin Invest 10:273–279

Christiansen C, Christensen MS, Transbol I (1981) Bone mass in postmenopausal women after withdrawal of oestrogen/gestagen replacement therapy. Lancet I:459–461

Cohn SH (1982) Techniques of determining the efficacy of treatment of osteoporosis. Calcif Tissue Int 34:433–438

Cohn SH, Vaswani A, Zanzi I, Aloia JF, Roginsky MS, Ellis KJ (1976) Changes in body chemical composition with age measured by total-body neutron activation. Metabolism 25:85–95

Cooper A (1824) A treatise on dislocations, and on fractures of the joints, 4th edn. Longman, Hurst, Rees, Orme, Brown and Green, London

Currey JD (1969) The mechanical consequences of variation in the mineral content of bone. J Biomech 2:1

Dalén N, Hellstrom LG, Jacobson B (1976) Bone mineral content and mechanical strength of the femoral neck. Acta Orthop Scand 47:503–508

Dambacher MA, Lauffenberger T, Lammle B, Haas HG (1978) Long term effects of sodium fluoride on osteoporosis. In: Courvoisier B, Donath A, Baud CA (eds) Fluoride and bone. Hans Huber, Bern, pp 238–241

Daniell HW (1976) Osteoporosis and the slender smoker. Arch Intern Med 136:298–304

Delmas PD, Wahner HW, Mann KG, Riggs BL (1983) Assessment of bone turnover in postmenopausal osteoporosis by measurement of serum bone Gla-protein. J Lab Clin Med 102:470–476

Dent OE, Friedman M (1965) Pregnancy and idiopathic osteoporosis. Q J Med 34:341–357

Dequeker J (1970) Parathyroid activity and postmenopausal osteoporosis. Lancet II:211–212

Dequeker J (1971) Periosteal and endosteal surface remodelling in pathologic conditions. Invest Radiol 6:260–265

Dequeker J (1972) Bone loss in normal and pathological conditions. University Press, Leuven

Dequeker J (1975) Bone and ageing. Ann Rheum Dis 34:100–115

Dequeker J (1976) Quantitative radiology: radiogrammetry of cortical bone. Br J Radiol 49:912–920

Dequeker J (1985) The relationship between osteoporosis and osteoarthritis. Clin Rheum Dis 11:271–296

Dequeker J, Geusens P (1985a) Anabolic steroids and osteoporosis. Acta Endocrinol [Suppl] 271:45–52

Dequeker J, Geusens P (1985b) Contributions of ageing and estrogen deficiency to postmenopausal bone loss. N Engl J Med 313:453

Dequeker J, Johnston CC (1982) Non-invasive bone measurements: methodological problems. (Proceedings of the workshop on non-invasive bone measurements, Knokke, Belgium, 1981). IRL Press, Oxford

Dequeker J, Merlevede W (1971) Collagen content and collagen extractability pattern of adult human bone according to age, sex and degree of porosity. Biochim Biophys Acta 244:410–424

Dequeker J, Van Tendeloo G (1982) Metacarpal bone mass and upper-extremity strength in 18-year-old boys. Invest Radiol 17:427–429

Dequeker J, Remans J, Franssens R, Waes J (1971) Ageing patterns of trabecular and cortical bone and their relationship. Calcif Tissue Res 7:23–30

Dequeker J, Roh YS, Van Dessel D, Gautama K, Burssens A (1973) Bone mineral estimation in vivo by photon absorptiometry. Influence of skeletal size and its value for detecting osteoporosis. J Belge Rheumatol Méd Phys 28:293–301

Dequeker J, Gautama K, Roh YS (1974) Femoral trabecular patterns in asymptomatic spinal osteoporosis and femoral neck fracture. Clin Radiol 25:243–246

Dequeker J, Burssens A, Creytens G, Bouillon R (1975) Ageing of bone. Its relation to osteoporosis in postmenopausal women. In: van Keep PA, Lauritzen C (eds) Estrogens in the post-menopause. Karger, Basel, pp 116–130 (Frontiers of Hormone Research, vol 3)

Dequeker J, Wielandts L, Nijs J, Ringe JD (1980a) Evaluation of bone mineral content data using isowidth and percentile curves. In: Mazess RB (ed) Proceedings of the fourth international conference on bone measurement. NIH Publication 080–1938:69–80

Dequeker J, Geusens P, De Proft G, Nijs J (1980b) Bone mass and soft-tissue measurements in acromegaly. In: Mazess RB (ed) Proceedings of the fourth international conference on bone measurements. NIH Publication 080–1938:147–153

Dequeker J, Burssens A, Bouillon R (1982) Dynamics of growth hormone secretion in patients with osteoporosis and in patients with osteoarthrosis. Horm Res 16:353–356

Dequeker J, Goris P, Uytterhoeven R (1983) Osteoporosis and osteoarthritis (osteoarthrosis): anthropometric distinctions. JAMA 249:1448–1451

Dequeker J, Geusens P, Verstraeten A (1984a) Osteoporosis and its treatment. In: Van Herendael H, Van Herendael B, Riphagen FE, Goessens L, Van der Pas H (eds) The climacteric: an update. MTP Press, Lancaster, pp 157–179

Dequeker J, Geusens P, Wielandts L, Nijs J, Verstraeten A (1984b) Lumbar BMC skeletal size nomogram. In: Christiansen C, Arnaud CD, Nordin BEC, Parfitt AM, Peck WA, Riggs BL (eds) Osteoporosis. Proceedings of the international symposium on osteoporosis, Copenhagen 1984, pp 341–344

Dequeker J, Geusens P, Verstraeten A, Nijs J (1987a) Experience with dual photon absorptiometry of the lumbar spine compared to information gained at the peripheral skeleton. In: Mazess RB (ed) (to be published)

Dequeker J, Nijs J, Verstraeten A, Geusens P, Gevers G (1987b) Genetic determinants of bone mineral content at the axial and peripheral skeleton: a twin study. Bone 8:207–209

Dewey JR, Armelagos GJ, Bartley MH (1969) Femoral cortical involution in three Nubian archaeological populations. Hum Biol 41:13–28

Doyle F, Gutteridge DH, Joplin GF, Russel F (1967) An assessment of radiological criteria used in the study of spinal osteoporosis. Br J Radiol 40:241–250

Doyle F, Brown J, Lachange C (1970) Relation between bone mass and muscle weight. Lancet I:311–313

Drinkwater BL, Nilson K, Chesnut CH, Brenner WJ, Shainholtz S, Southworth MB (1984) Bone mineral content of amenorrheic and eumenorrheic athletes. N Engl J Med 311:277–281

Foss MUL, Byers PD (1972) Bone density, osteoarthrosis of the hip and fracture of the upper end of the femur. Ann Rheum Dis 31:259–264

Franke J (1978) Our experience in the treatment of osteoporosis with relative low sodium-fluoride doses. In: Courvoisier B, Donath A, Baud CA (eds) Fluoride and bone. Hans Huber, Bern, pp 256–262

Freeman MAR, Todd RC, Price CJ (1974) The role of fatigue fracture in the pathogenesis of senile femoral neck fracture. J Bone Joint Surg [Br] 56:698–702

Frumar AM, Meldrum DR, Geola F et al. (1980) Relationship of fasting urinary calcium to circulating estrogen and body weight in postmenopausal women. J Clin Endocrinol Metab 50:70–75

Gallagher JC (1983) The use of calcitriol (1.25 dihydroxyvitamin D) in osteoporosis. In: Parfitt AM, Frame B (eds) Proceedings of the symposium on clinical disorders of bone and mineral metabolism, Detroit, May 1983. Excerpta Medica, Amsterdam, pp 364–367

Gambrell RD (1982) Role of hormones in the etiology and prevention of endometrial and breast cancer. Acta Obstet Gynecol Scand [Suppl] 106:37–46

Garn SM (1970) The earlier gain and later loss of cortical bone in nutritional perspective. CC Thomas, Springfield, Illinois

Garn SM (1972) The course of bone gain and the phases of bone loss. Orthop Clin North Am 3:503–520

Garn SM, Rohmann CG, Wagner B, Ascoli W (1967) Continuing bone growth throughout life: a general phenomenon. Am J Phys Anthrop 26:313–318

Garn SM, Rohmann CG, Wagner B, Damla GH, Ascoli W (1969) Population similarities in the onset and rate of adult endosteal bone loss. Clin Orthop 65:51–60

Garn SM, Poznanski AK, Nagy JM (1971) Bone measurement in the differential diagnosis of osteopenia and osteoporosis. Radiology 100:509–518

Geusens P, Dequeker J (1986) Longterm effect of nandrolone decanoate, 1-alpha-hydroxyvitamin D3 or intermittent calcium infusion therapy on bone mineral content, bone remodeling and fraction rate in symptomatic osteoporosis. A double-blind study. Bone Miner 1:347–357

Geusens P, Dequeker J, Verstraeten A (1983) Age-related blood changes in hip osteoarthritis patients: a possible indicator of bone quality. Ann Rheum Dis 42:112–113

Geusens P, Dequeker J, Verstraeten A, Nijs J (1986) Age-, sex- and menopause-related changes of vertebral and peripheral bone: a population study using dual and single photon absorptiometry and radiogrammetry. J Nucl Med 27:1540–1549

Gryfen CI, Arnies A, Ashley MJ (1977) A longitudinal study of falls in an elderly population. 1. Incidence and morbidity. Age Ageing 6:201–210

Halstead B, Middleton J (1972) Bare bones, an explanation in art and science. Oliver and Boyd, Edinburgh

Hansson T, Ross B, Nachemson A (1980) The bone mineral content and ultimate compressive strength of lumbar vertebrae. Spine 5:46–55

Harrison JE, McNeill KG, Sturtridge WC, Bayley TA, Murray TM, William C, Tan C, Fornasier V (1981) Three-year changes in bone mineral mass of postmenopausal osteoporotic patients based on neutron activation analysis of the central third of the skeleton. J Clin Endocrinol Metab 52:751–758

Heaney RP (1981) Osteoporosis. In: Bronner F, Cobun JW (eds) Disorders of mineral metabolism. Academic Press, New York, pp 91–117

Heaney RP (1983) Prevention of age-related osteoporosis in women. In: Avioli LV (ed) The osteoporotic syndrome. Grune and Stratton, New York, pp 123–144

Heaney RP, Skillman TG (1971) Calcium metabolism in normal human pregnancy. J Clin Endocrinol Metab 33:661–670

Heaney RP, Gallagher JC, Johnston CC, Neer R, Parfitt AM (1982) Calcium nutrition and bone health in the elderly. Am J Clin Nutr 36:986–1013

Hillman LS (1983) Neonatal osteoporosis – diagnosis and management. In: Frame B, Potts J Jr (eds) Clinical disorders of bone and mineral metabolism. Excerpta Medica, Amsterdam, pp 427–430 (International Congress Series vol 617)

Horsman A, Gallagher JC, Simpson M, Nordin BEC (1977) Prospective trial of oestrogen and calcium in postmenopausal women. Br Med J ii:789–792

Howell JA (1917) An experimental study of the effect of stress and strain on bone development. Anat Rec 13:235–252

Huddleston AL, Rockwell D, Kulund DN, Harrison RB (1980) Bone mass in lifetime tennis athletes. JAMA 244:1107–1109

Hutchinson TA, Polansky SM, Feinstein AR (1979) Post-menopausal oestrogens protect against fractures of hip and distal radius. A case-control study. Lancet II:705–709

Johnston CC Jr (1983) Noninvasive methods for quantitating appendicular bone mass. In: Avioli V (ed) The osteoporotic syndrome: detection, prevention and treatment. Grune and Stratton, New York, pp 73–84

Johnston CC, Norton J, Khairi MRA et al. (1985) Heterogeneity of fracture syndromes in post-menopausal women. J Clin Endocrinol Metab 61:551–556

Kanis JA (1984) Treatment of osteoporotic fracture. Lancet I:27–33

Kaplan FS (1983) Osteoporosis. Ciba-Geigy, West Caldwell, New Jersey, p 5 (Ciba clinical symposia no. 35)

Kazarian L, Graves GA (1977) Compressive strength characteristics of the human vertebral centrum. Spine 2:1–14

Khairi MRA, Cronin JH, Robb JA, Smith DM, Yu PL, Johnston CC (1976) Femoral trabecular pattern index and bone mineral content measurement by photon absorption in senile osteoporosis. J Bone Joint Surg [Am] 58:221–226

Krölner B, Pors Nielsen S (1982) Bone mineral content of the lumbar spine in normal and osteoporotic women: cross-sectional and longitudinal studies. Clin Sci 62:329–336

Krölner B, Toft B, Pors Nielsen S, Tondevold E (1983) Physical exercise as prophylaxis against involutional vertebral bone loss: a controlled trial. Clin Sci 64:541–546

Lachmann E (1955) Osteoporosis: the potentialities and limitations of its roentgenologic diagnosis. Am J Roentgenol 74:712–715 (editorial)

Lewis AF (1981) Fracture of the neck of the femur: changing incidence. Br Med J 283:1217–1220

Lindahl O (1976) Mechanical properties of dried defatted spongy bone. Acta Orthop Scand 47:11–19

Lindberg JS, Fears WB, Hunt MM, Powell MR, Boll D, Wade CE (1984) Exercise-induced amenorrhea and bone density. Ann Intern Med 101:647–648

Lindsay R, Aitken JM, Anderson JB, Hart DM, MacDonald EBC, Clarke AC (1976) Long-term prevention of postmenopausal osteoporosis by oestrogen. Lancet I:1038–1041

Lindsay R, Coutts JTR, Hart DM (1977) The effect of endogenous oestrogen on plasma and urinary calcium and phosphate in oophorectomized women. Clin Endocrinol 6:87–93

Lindsay R, Hart DM, Kraszewski A (1980) Prospective double-blind trial of synthetic steroid (Org.OD14) for preventing postmenopausal osteoporosis. Br Med J i:1207–1209

Lips R, Jongen HJM, van Ginkel FC et al. (1983) Vitamin D deficiency and hip fractures: cause or incidence. In: Frame B, Pott JT Jr (eds) Clinical disorders of bone and mineral metabolism. Excerpta Medica, Amsterdam, pp 204–207 (International Congress Series, vol 617)

Manzke E, Chestnut CH III, Wergedal JE, Baylink DJ, Nelp WB (1975) Relationship between local and total bone mass in osteoporosis. Metabolism 24:605–615

Martens M (1985) Mechanical properties of bone. PhD Thesis, Faculty of Medicine, University of Leuven, Belgium

Matkovic V, Kostial K, Simonovic I, Brodarec A, Nordin BEC (1979) Bone status and fracture rates in two regions of Yugoslavia. Am J Clin Nutr 32:540–549

Mazess RB (1971) Estimation of bone and skeletal weight by direct photon absorptiometry. Invest Radiol 6:52–60

Mazess RB (1978) Bone mineral in Vilcabamba, Ecuador. Am J Roentgenol 120:671–674

Mazess RB (1982) On ageing bone loss. Clin Orthop 165:239–252

Mazess RB, Christiansen C (1982) A comparison on bone mineral results from Denmark and the US. Hum Biol 54:343–354

Mazess RB, Mather WE (1974) Bone mineral content of North Alaskan Eskimos. Am J Clin Nutr 27:916–925

Mazess RB, Mather WE (1975) Bone mineral content in Canadian Eskimos. Hum Biol 47:45–63

Mbuyi-Muamba JM, Dequeker J (1987) Biochemical anatomy of human bone: comparative study of compact and spongy bone in femur, rib and iliac crest. Acta Anat 124:184–187

Meema HE (1962) The occurrence of cortical bone atrophy in old age and in osteoporosis. J Can Assoc Radiol 13:27–32

Meunier PJ, Courpron P, Edouard C et al. (1979) Bone histomorphometry in osteoporotic states. In:Barzel US (ed) Osteoporosis II. Grune and Stratton, New York, pp 27–47

Möller M, Horsman A, Harvald B, Hauge M, Henningsen K, Nordin BEC (1978) Metacarpal morphometry in monozygotic and dizygotic twins. Calc Tissue Res 25:197–201

Morgan DB (1983) The epidemiology of osteoporosis. In: Dixon AStJ, Russell RGP, Stamp JCB (eds) Osteoporosis, a multi-disciplinary problem. Academic Press, London, pp 81–87 (Royal Society of Medicine international congress and symposium series, vol 55)

Newton-John HF, Morgan DB (1968) Osteoporosis: disease or senescence? Lancet I:232–233

Newton-John HF, Morgan DB (1970) The loss of bone with age, osteoporosis, and fractures. Clin Orthop 71:229–252

Nilsson BE, Obrant KJ (1978) Secular tendencies of the incidence of fracture of the upper end of the femur. Acta Orthop Scand 49:389–391

Nilsson BE, Westlin EE (1974) The bone mineral content in the forearm of women with Colles' fracture. Acta Orthop Scand 45:836–844

Nilsson BE, Anderson SM, Havdmys TV, Westlin NE (1980) Bone mineral content in ballet dancers and weight lifters. In: Mazess RA (ed) Proceedings of the fourth international conference on bone measurement. NIH Publication 080–1938:81–86

Nokso-Kovisto VM, Alhava EM, Olkkonen H (1976) Measurement of cancellous bone strength. Correlations with mineral density, aging and disease. Ann Clin Res 8:399–402

Nordin BEC (1983) Osteoporosis with particular reference to the menopause. In: Avioli LV (ed) The osteoporotic syndrome. Grune and Stratton, New York, pp 13–43

Nordin BEC, Roper A (1955) Postpregnancy osteoporosis. A syndrome. Lancet I:431–434

Nordin BEC, Horsman A, Marshall DH, Simpson M, Waterhouse GM (1979) Calcium requirements and calcium therapy. Clin Orthop 140:216–239

Nordin BEC, Horsman A, Crilly RG, Marshall DH, Simpson M (1980) Treatment of spinal osteoporosis in postmenopausal women. Br Med J 280:451–454

Overstall OW, Exton-Smith AM, Hums FJ, Johnson AL (1977) Falls in the elderly related to postural imbalance. Br Med J i:261–264

Parfitt AM (1981) Integration of skeletal and mineral homeostasis. In: De Luca HF, Frost H, Jee W, Johnston CC, Parfitt AM (eds) Osteoporosis: recent advances in pathogenesis and treatment. Baltimore University Press, pp 115–126

Parfitt AM (1982) The coupling of bone resorption to bone formation: a critical analysis of the concept and its relevance to the pathogenesis of osteoporosis. Metab Bone Dis Rel Res 4:1–6

Parfitt AM (1983) Dietary risk factors for age-related bone loss and fractures. Lancet I:1181–1185

Parfitt AM (1984) Age-related structural changes in trabecular and cortical bone: cellular mechanism and biomechanical consequences. Calc Tissue Int 36:S123–128

Parfitt AM, Chir B, Gallagher JC et al. (1982) Vitamin D and bone health in the elderly. Am J Clin Nutr 36:1014–1031

Pennock JM, Kahn DN, Clark MB, Easter GV, Doyle FH (1972) Hypoplasia of bone induced by immobilization. Br J Radiol 45:641–646

Perigian AJ (1973) The antiquity of age-associated bone demineralization in man. J Am Geriatr Soc 21:100–105

Pugh JW, Rose RM, Radin EL (1973) Elastic and viscoelastic properties of trabecular bone: dependence on structure. J Biomech 6:475–485

Raman L, Rajalaksksni K, Krishnamachari KAVR, Sastry JG (1978) Effect of calcium supplementation to undernourished mothers during pregnancy on the bone density of the neonates. Am J Clin Nutr 31:466–469

Rambaut PC, Goode AW (1985) Skeletal changes during space flight. Lancet II:1050–1052

Recker RR, Saville PD, Heaney RP (1977) Effect of oestrogens and calcium carbonate on bone loss in postmenopausal women. Ann Intern Med 87:649–655

Reutter FW, Olah AJ (1978) Bone biopsy findings and clinical observations in long-term treatment of osteoporosis with sodium fluoride and vitamin D$_3$. In: Courvoisier B, Donath A, Baud CA (eds) Fluoride and bone. Hans Huber, Bern, pp 249–255

Rich C, Bernstein DS, Gates S et al. (1966) Factors involved in an objective study of the efficacy of treatment of osteoporosis. Clin Orthop Relat Res 45:63–66

Richelson LS, Wahner HW, Melton LJ III, Riggs BL (1984) Relative contributions of aging and estrogen deficiency to menopausal bone loss. N Engl J Med 311:1273–1275

Riggs BL, Melton LJ (1983) Evidence for two distinctive syndromes of involutional osteoporosis. Am J Med 75:899–901

Riggs BL, Jowsey J, Kelly PJ, Hoffman DL, Arnaud CD (1976) Effects of oral therapy with calcium and vitamin D in primary osteoporosis. J Clin Endocrinol Metab 42:1139–1144

Riggs BL, Hodgson SF, Hoffman DL, Kelley PJ, Johnson KA, Taves D (1980) Treatment of primary osteoporosis with fluoride and calcium. JAMA 243:446–449

Riggs BL, Wahner HW, Dunn WL, Mazess RB, Offord KP, Melton LJ III (1981) Differential changes in bone mineral density of the appendicular and axial skeleton with aging. J Clin Invest 67:328–335

Riggs BL, Wahner HW, Seeman E, et al. (1982a) Changes in bone mineral density of the proximal femur and spine with aging. J Clin Invest 70:716–723

Riggs BL, Seeman E, Hodgson SF, Taves DR, O'Fallon WM (1982b) Effects of the fluoride/calcium regimen on vertebral fracture occurrence in postmenopausal osteoporosis. N Engl J Med 306:446–450

Ringe JD, Kruse HP, Kuhlencordt F (1978) Long-term treatment of primary osteoporosis by sodium fluoride. In: Courvoisier B, Donath A, Baud CA (eds) Fluoride and bone. Hans Huber, Bern, pp 228–232

Rockoff SD, Sweet E, Bleustein J (1969) The relative contribution of trabecular and cortical bone to the strength of human lumbar vertebrae. Calcif Tissue Res 3:163–175

Roh YS, Dequeker J, Mulier JC (1973a) Osteoarthrosis at the hand skeleton in primary osteoarthrosis of the hip and in normal controls. Clin Orthop Rel Res 90:90–94

Roh YS, Dequeker J, Mulier JC (1973b) Cortical bone remodelling and bone mass in poliomyelitis. Acta Orthop Belg 39:758–771

Roh YS, Dequeker J, Mulier JC (1974a) Bone mass in osteoarthrosis, measured in vivo by photon absorption. J Bone Joint Surg [Am] 56:587–591

Roh YS, Dequeker J, Mulier JC (1974b) Trabecular pattern of the upper end of the femur in primary osteoarthrosis and in symptomatic osteoporosis. Belg Tijdschr Radiol 57:89–94

Saville PD, Whyte MP (1969) Muscle and bone hypertrophy. Clin Orthop 65:81–88

Singh M, Nagrath AR, Maini PS (1970) Changes in trabecular pattern of the upper end of the femur as an index of osteoporosis. J Bone Jt Surg 52:457–467

Smith CB, Smith DA (1976) Relations between age, mineral density and mechanical properties of human femoral compacta. Acta Orthop Scand 47:496–502

Smith DM, Johnston CC Jr, Yu PL (1972) In vivo measurement of bone mass: its use in demineralization states such as osteoporosis. JAMA 319:321–329

Smith DM, Nance WC, Kang KW, Christian JC, Johnston CC Jr (1973) Genetic factors in determining bone mass. J Clin Invest 52:2800–2808

Smith EL, Smith PE, Enniger C, Shea MM (1984) Bone involutions decrease in exercising middle-aged women. Calcif Tissue Int 36(1):129–138

Smith RW (1967) Dietary and hormonal factors in bone loss. Fed Proc 26:1737–1746

Smith R, Wineads CG, Stevenson JC, Woods CG, Wordsworth BP (1985) Osteoporosis of pregnancy. Lancet I:1178–1180

Taylor TG, Belanger LF (1969) The mechanism of bone resorption in laying hens. Calc Tissue Res 4:162–173

Thom MH, White PJ, Williams RM et al. (1979) Prevention of endometrial disease in climacteric women receiving oestrogen therapy. Lancet II:455–457

Todd RC, Freeman MAR, Pirie CJ (1972) Isolated trabecular fatigue fractures in the femoral head. J Bone Joint Surg [Br] 54:723–728

Trotter M, Peterson RR (1955) Ash weight of human skeletons in percent of their dry-fat-free weight. Anat Rec 123:341–358

Trotter M, Broman GF, Peterson RM (1960) Densities of bone of white and negro skeletons. J Bone Joint Surg [Am] 42:50–58

Tsang RC (1983) The quandary of vitamin D in the newborn infant. Lancet I:1370–1372

Van Gerven DP, Armelages GJ, Bartly MH (1969) Roentgenographic and direct measurement of cortical involution in a prehistoric Mississipian population. Am J Phys Anthrop 31:23–38

Vernon-Roberts B, Pirie CJ (1973) Healing trabecular microfractures in the bodies of lumbar vertebrae. Ann Rheum Dis 32:406–412

Virtama P, Mahönen H (1960) Thickness of the cortical layer as an estimate of mineral content of human finger bones. Br J Radiol 33:60–62

Vose GP, Kubala AL Jr (1959) Bone strength: its relationship to X-ray determined ash content. Hum Biol 31:261–270

Vose GP, Lockwood RM (1965) Femoral neck fracturing. Its relationship to radiographic bone density. J Gerontol 20:300–350

Wahner HW, Dunn WL, Riggs BL (1984) Assessment of bone mineral: 1. J Nucl Med 25:1134–1141

Wall JC, Chaterji SK, Jeffery JW (1979) Age-related changes in the density and tensile strength of human femoral cortical bone. Calcif Tissue Int 27:105–108

White MP, Bergfeld MA, Murphy NA, Avioli LV, Teitelbaum SL (1982) Postmenopausal osteoporosis: a heterogenous disorder as assessed by histomorphometric analysis of iliac crest bone from untreated patients. Am J Med 72:193–202

Woodard HQ (1964) The composition of human cortical bone: effect of age and some abnormalities. Clin Orthop 37:187–193

Yamada H (1970) Strength of biological materials. Williams and Wilkins, Baltimore

Chapter 10

Calcium in Extracellular Fluid: Homeostasis

G. Schaafsma

Introduction

Calcium in Extracellular Fluid

The total amount of extracellular calcium in the body is less than 1% of the total body calcium and equals the sum of the calcium in the plasma and in the interstitial fluid (Table 10.1). The normal mean plasma calcium concentration is 10 mg/100 ml and as the plasma accounts for approximately 5% of body weight, the total amount of calcium in the plasma of a 70-kg man is about 350 mg. The normal mean concentration of (ultrafiltrable) calcium in the interstitial fluid is about 5.6 mg/100 ml. Interstitial fluid accounts for approximately 18% of body weight. The total amount of calcium in the interstitial fluid of a 70-kg man is therefore about 700 mg.

Table 10.1. Distribution of calcium in extracellular fluids in a 70-kg man

	Volume (l)	Calcium concentration (mg/l)	Calcium content (mg)
Plasma	3.5	100	350
Interstitial fluid	12.5	56	700

Calcium in Plasma

Calcium in plasma occurs in three different forms: (a) as ionic or free calcium, (b) as complexed calcium or (c) as calcium bound to proteins.

Approximately 50% of the plasma calcium is in the free or ionic form (Table 10.2). This fraction is physiologically active and is under strict hormonal control. The mean normal plasma ionized calcium concentration is 4.7 mg/100 ml. The intra-individual variation in the plasma ionized calcium concentration is probably not more than 0.1 mg/100 ml which is close to the detection limits of most analytical techniques (Marshall 1976; Moore 1970).

Table 10.2.   Distribution of calcium in plasma[a]

	Normal range (mg/100 ml)	Distribution (approximate) (%)
Total calcium	9.2–11.0	100
Ionized	4.4– 5.1	50
Complexed	0.6– 1.2	10
Bound to proteins	3.7– 4.3[b]	40

[a] Data adapted from Marshall (1976) and Henry et al. (1974).
[b] Based on the normal range of total plasma proteins.

The concentration of complexed calcium in plasma varies between 0.6 and 1.2 mg/100 ml, as computed from the concentration of ligands (mainly citrate, phosphate and bicarbonate) and binding constants at physiological pH. Together with ionized calcium, this fraction contributes to the ultrafiltrable calcium in interstitial fluid and glomerular filtrate.

About 40% of calcium in plasma is bound to proteins. About 80% of that calcium is bound to albumin, the rest to globulins (Moore 1970). No physiological role has been identified for the complexed and protein-bound fractions; the physiologically important extracellular calcium homeostasis is concerned only with ionized calcium.

The ionized calcium concentration can be measured with an ion-selective electrode. Alternatively, it can be computed from the total calcium concentration by taking into account the concentrations of plasma proteins and the pH (Marshall and Hodgkinson 1983). However, since an excellent internal correlation exists between total calcium and ionized calcium in plasma, if computed over the full range of plasma calcium values (Marshall 1976), total plasma calcium is often used as an index of calcium homeostasis. This is not strictly valid if the plasma protein concentration or plasma pH are abnormal.

Plasma Calcium Homeostasis

General Outline

The plasma calcium concentration is maintained within narrow limits by the complex and integrated hormonal regulation of intestinal calcium absorption, urinary calcium excretion and bone formation and resorption. This homeostasis is controlled mainly by parathyroid hormone (PTH). If the plasma ionized calcium concentration falls below the parathyroid gland setpoint of about 4.5 mg/100 ml, PTH is secreted and restores the ionized calcium concentration by acting in three ways:

1. On bone to stimulate calcium resorption.

2. On the kidneys to increase distal tubular calcium reabsorption, thereby decreasing urinary calcium excretion.
3. On the kidneys to enhance the synthesis of calcitriol (1,25-dihydroxyvitamin D), the active metabolite of vitamin D.

Calcitriol stimulates intestinal calcium absorption, tubular calcium reabsorption and calcium resorption from bone. An outline of the actions of PTH and calcitriol on bone, kidney and intestine is given in Fig. 10.1.

The second calciotropic hormone, calcitonin, is secreted from the parafollicular cells of the thyroid gland in response to hypercalcaemia. In animals it has a hypocalcaemic action, inhibiting bone resorption and renal tubular reabsorption of calcium, and thus antagonizing the actions of PTH. Its role in humans has been questioned (Kalu and Foster 1976). According to Stevenson et al. (1979), calcitonin might be of some importance in antagonizing the bone resporptive actions of calcitriol and PTH during periods of calcium stress, such as pregnancy and lactation.

Figure 10.2 outlines calcium metabolism in adults. If the daily diet contains 1000 mg calcium, about 300 mg of this (or 30%) is absorbed from the intestinal tract. This is called true absorption. About 150 mg calcium is secreted into the gut with digestive juices and desquamation of epithelial cells and is thought to enter into the same gut pool as dietary calcium. About 45 mg (30%) of this endogenous calcium is reabsorbed. Thus faecal calcium consists of 700 mg non-absorbed dietary calcium and 105 mg non-reabsorbed endogenous calcium, making a total of 805 mg. Net absorption, defined as intake minus faecal output,

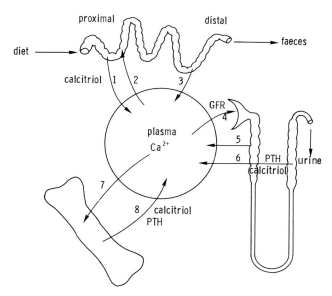

Fig. 10.1. Outline of calcium fluxes and of the actions of PTH and calcitriol on (active) calcium transport from intestine, bone and kidney: *1*, calcitriol-dependent "active" calcium absorption; *2*, endogenous calcium secretion; *3*, "passive" calcium absorption; *4*, glomerular filtration rate (GFR) and filtered calcium load; *5*, proximal tubular calcium reabsorption (passive); *6*, distal tubular calcium reabsorption (active); *7*, bone formation (passive); *8*, bone resorption (active).

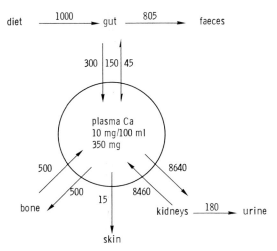

Fig. 10.2. Outline of calcium metabolism (mg/day) under equilibrium conditions (input = output) in adults.

is 195 mg, or almost 20%. About 15 mg of calcium is lost through the skin. Bone turnover (formation and resorption) involves about 250–500 mg calcium per day. If the glomerular filtration rate is 100 ml/min and 60% of the plasma calcium is ultrafiltrable, the filtered load corresponds to 8640 mg/day, 98% of which is reabsorbed by the tubular cells, leaving a urinary calcium excretion of 180 mg.

In the following sections the discussion on the regulation of the extracellular calcium concentration will focus on the role of the target tissues involved, i.e. kidneys, bone and intestine.

Role of the Kidneys

The important role of the kidneys in the maintenance of the extracellular calcium concentration can be illustrated by a simple tank model (Fig. 10.3). In this model, the fluid level is equal to the extracellular calcium concentration and the kidneys are represented by the outlet tube, its size reflecting the glomerular filtration rate and its height the tubular reabsorption of calcium. In Fig. 10.3 the input of calcium is equal to the sum of the (net) influxes of calcium from bone and intestine and under conditions of calcium balance this input is equal to the output. The simple tank model in Fig. 10.3 predicts that above the threshold level the output is a linear function of the extracellular calcium concentration with a slope which is dependent on the glomerular filtration rate (Fig. 10.4). Indeed it has been demonstrated experimentally by means of calcium infusion in subjects with different glomerular filtration rates that at any given plasma calcium concentration urinary calcium excretion is higher in those with the higher glomerular filtration rate (Fig. 10.5).

The urinary excretion of calcium is equal to the difference between the filtered calcium load and calcium reabsorbed in the tubules. Thus the amount of calcium reabsorbed can be computed from measurements of the filtered load and

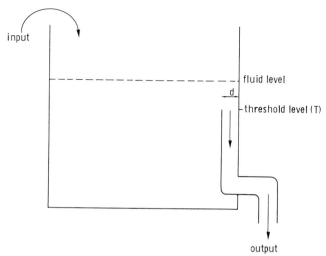

Fig. 10.3. Simple tank model of plasma calcium homeostasis. Taken from Nordin (1976) in a modified form; *d* represents the glomerular filtration rate and *T* the threshold or maximum tubular reabsorption.

calcium excretion. The different quantities can be expressed in mg per unit of time but this requires the measurement of the glomerular filtration rate; this can be estimated from the endogenous creatinine clearance but it is a laborious method liable to collection errors. A very convenient solution is to express the urinary calcium excretion in mg/100 ml glomerular filtrate. This quantity can be obtained simply by multiplying the calcium/creatinine ratio in the urine by the plasma creatinine concentration (all in mg/100 ml). This measurement has the advantage that when it is used to calculate calcium reabsorption, the filtered calcium load is expressed directly as the plasma ultrafiltrable calcium concentration (mg/100 ml).

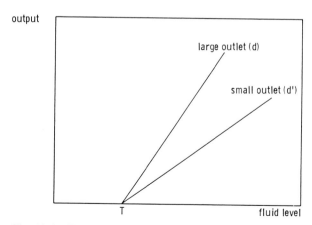

Fig. 10.4. Regression of output on fluid level in the tank model of plasma calcium homeostasis. Outlet *d* (*d'*) represents the glomerular filtration rate. *T* is the threshold level of Fig. 10.3.

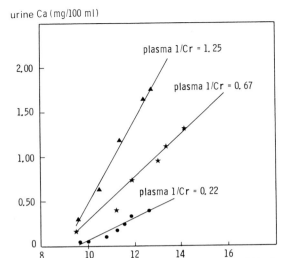

Fig. 10.5. The relationship between plasma and urine calcium (calcium/creatinine ratio) during calcium infusions in three subjects with different glomerular filtration rates (expressed as the inverse of the creatinine concentration in plasma). (Adapted from Nordin 1976)

The relation between urinary calcium excretion and plasma calcium concentration has been studied in normal subjects loaded with oral calcium or infused with a calcium solution. Regression of urinary calcium excretion (in mg/100 ml glomerular filtrate) on plasma calcium resulted in a relationship which is apparently linear (Fig. 10.6) as was predicted by the simple model of Fig. 10.3. The theoretical plasma calcium threshold, obtained by extrapolation to zero excretion, is about 9 mg/100 ml and the slope is about 0.3. Since about 60% of the plasma calcium is ultrafiltrable, these figures correspond to an ultrafiltrable calcium of 5.4 mg/100 ml (= 60% × 0.9) and a slope of 0.5 (= 0.3/60%).

The linear relationship between plasma calcium and urinary calcium excretion and the existence of a threshold have been explained by Mioni et al. (1971) who postulated a simple diffusion process for tubular reabsorption which is exclusively dependent on the calcium concentration in the glomerular filtrate and an active reabsorption process which accounts for the threshold. By the use of calcium infusions in hypocalcaemic subjects with osteomalacia and in normal subjects, Peacock et al. (1969) were able to investigate further the relationship between plasma and urinary calcium in the splay area of Fig. 10.6 (plasma calcium below 9–10 mg/100 ml). From their results it appears that a saturable active reabsorption mechanism for calcium exists in the tubules without a diffusion component and the calculated maximum reabsorptive capacity (Tm) for calcium was 8.05 mg/100 ml glomerular filtrate (Fig. 10.7). This value corresponds to a total plasma calcium concentration of 13.5 mg/100 ml.

On the basis of microperfusion studies, in vivo as well as in vitro, the fractions of calcium that are reabsorbed in the different parts of the tubules have been estimated by Raymakers (1978). The calcium transport in the proximal tubule and in the loop of Henle is passive. In the proximal tubule about 70% of the

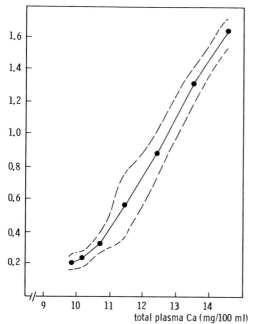

Fig. 10.6. The relationship between plasma calcium and calcium excretion (mg/100 ml glomerular filtrate) in healthy subjects after oral calcium supplementation or during calcium infusion. Constructed on the basis of data from Peacock et al. (1969). The *dotted lines* indicate the 95% confidence limits.

filtered load of calcium is reabsorbed and this reabsorption is dependent on the concentration gradient produced by the reabsorption of salt and water. Of the calcium that reaches the loop of Henle (20% of the filtered calcium load) 70% is

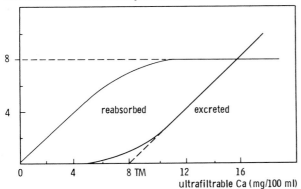

Fig. 10.7. The relationship between filtered, excreted and reabsorbed calcium from calcium infusion data in healthy and osteomalacic subjects. (Nordin 1976)

reabsorbed and this transport depends largely on the active transport of chloride in the ascending loop. In the distal tubule, where 10%–15% of the filtered calcium load is available, there is active reabsorption of calcium. The process is influenced by PTH, calcitriol, acid–base balance and the plasma concentrations of calcium and phosporus. It appears from microperfusion studies that about 90% of the filtered load is reabsorbed by diffusion. This figure does not accord with the slope of the line in Fig. 10.6, which suggests that in normal subjects infused or orally loaded with calcium, only about 50% of the filtered calcium is reabsorbed by this process. According to Raymakers (1978), this difference can be explained by the effects of calcium infusion as such on extracellular calcium, PTH secretion and sodium reabsorption, which change tubular function. In addition, the anion infused with calcium may also decrease reabsorption of calcium in the tubules.

Role of Bone

Because of the obligatory endogenous losses of calcium (via kidneys, intestine and skin), a daily net influx of calcium into the extracellular fluid compartment is necessary. Normally this net influx comes from intestinal calcium absorption, but if the dietary supply of calcium is insufficient or if there is calcium malabsorption, calcium is resorbed from bone to maintain the plasma calcium concentration at the expense of the skeleton.

Bone is a dynamic tissue that is formed and broken down continuously by the action of bone cells (osteoblasts and osteoclasts). This process, which is mediated mainly by PTH and is called bone remodelling, certainly contributes to extracellular calcium homeostasis. The influx of calcium from the intestine is not constant and it may not be sufficient to maintain the plasma calcium concentration during periods between meals. Within minutes after a decrease of plasma calcium, PTH is secreted into the circulation and mobilizes calcium from bone by its effect on calcium transport in the cellular membrane which separates the bone from the extracellular fluid. There is a concentration gradient across this membrane with a slightly lower calcium concentration in the bone fluid compartment. If the secretion of PTH is prolonged, the hormone also stimulates the formation of osteoclasts from monocytic precursors and activates the osteoclasts to supply calcium to the extracellular fluid by bone resorption (Gaillard 1961). Parathyroid hormone is assisted in its action by calcitriol (Garabedian et al. 1974).

In growing individuals, the role played by bone in extracellular calcium homeostasis is limited because of the net uptake of calcium from the circulation, required for bone growth. If the dietary supply of calcium is inadequate for a prolonged period of time, bone growth is retarded but the plasma calcium is maintained as close as possible to normal.

Role of Intestine

The intestinal absorption of calcium is described in detail elsewhere in this monograph. The process involves an "active" saturable calcitriol-dependent component in the proximal small intestine and passive transport in the ileum

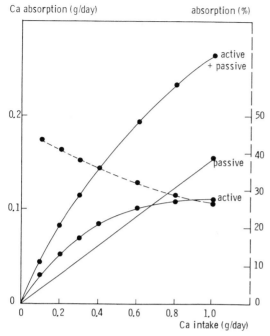

Fig. 10.8.   True intestinal calcium absorption (y) as a function of calcium intake (x), constructed on the basis of the formula $y = 0.154x + 0.312 \times e^{-1.0539x}$, developed by Heaney et al. (1975) from data in normal subjects. The first and second terms of this formula represent passive and active calcium transport respectively. The *dotted line* indicates true absorption, expressed as a percentage of intake.

(see Fig. 10.8). The active component of the intestinal calcium transport system contributes to the regulation of plasma calcium, particularly if the supply of dietary calcium is limited.

Effects of Non-nutritional Factors on Plasma Calcium Homeostasis

Age and Sex

The plasma ionized calcium concentration changes little throughout life and a clear-cut sex difference in this concentration does not exist (Marshall 1976). According to Keating et al. (1969) total plasma calcium in adult males, but not in females, slightly but significantly decreases with age at an average rate of 0.007 mg/100 ml per year. Due to loss of oestrogen production and an increase of bone resorption, plasma calcium increases in women during the early postmenopausal years (Riggs et al. 1969, 1972).

The loss of bone tissue is an important feature of the ageing process. It starts in the fifth decade of life and proceeds about twice as fast in females as in males. In females the average loss is about 12% of the bone mass per decade (Garn et al. 1967), but is higher during the early postmenopausal years. According to Cohn et al. (1976), on the basis of measurements of total body calcium, the loss of calcium from bone can be characterized by two components, one with a rate of 0.37% per year and the other, a more rapid one, which starts in the menopause, with a rate of 1.08% per year. In a longitudinal study of 250 postmenopausal women, we have observed an average rate of loss of calcium from bone of 1.9% per year as measured by photon absorptiometry on the distal third of the radius (Van Beresteyn et al., unpublished results). The loss of calcium from the skeleton corresponds to a net influx of calcium into the extracellular fluid of about 20–60 mg per day, and this quantity is excreted in the urine.

Another important age-related phenomenon is a decrease in intestinal calcium absorption. It is well documented in man as well as in animals (Allen 1982) and is attributable to a decline of the active component of the intestinal calcium transport system (Wilkinson 1971; Horst et al. 1978; Armbrecht et al. 1979); this would be compatible with a decreased production of calcitriol in elderly people due to the normal decline of renal function. This decline amounts on average to 30% in persons from 30 to 80 years, as measured by the creatinine clearance (Rowe et al. 1976).

To make skeletal growth and mineralization possible, about 1200 g of calcium has to be retained between birth and maturity. Until the age of about 30 years, when peak bone mass is reached, there is a net positive flux of calcium into bone which amounts to about 150–200 mg/day during early life (first 6 months) and up to 400 mg/day (in boys) during the rapid growth phase of adolescence (American Academy of Pediatrics 1978). During these periods of increased calcium requirement, the raised plasma levels of the calcitropic hormones (PTH and calcitriol) stimulate bone turnover and intestinal calcium absorption. Urinary calcium excretion in children averages 2.9 mg/kg/day and it does not change significantly with age if expressed in this way (Von Misselwitz and Kauf 1981).

Renal Function

Because of the important role of the kidneys in plasma calcium homeostasis, it is not surprising that renal failure affects homeostasis. Indeed, a complex disorder, called renal osteodystrophy, occurs in uraemic patients suffering from end-stage renal failure when renal function falls below about 30% (glomerular filtration rate less than 40 ml/min) (Kumar 1979). The resulting bone lesions are mainly osteitis fibrosa and osteomalacia, the latter being more common in areas where vitamin D intake and/or exposure to ultraviolet light are marginal.

The most important factor in the development of renal osteodystrophy is secondary hyperparathyroidism. It is the first detectable abnormality in very early renal failure (glomerular filtration rate 70–80 ml/min). According to Bricker et al. (1969) and Slatopolsky and Bricker (1973), a decline in glomerular filtration rate leads to a small increase of plasma phosphate. The rise in plasma phosphate decreases plasma ionized calcium and this triggers the secretion of PTH, which in turn decreases tubular reabsorption of phosphate and increases

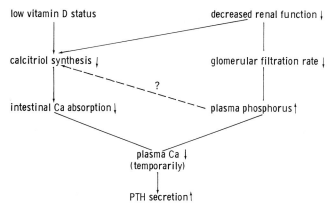

Fig. 10.9. Outline of the pathogenesis of secondary hyperparathyroidism. Other contributing factors are diets low in calcium and/or high in phosphorus.

phosphate output per remaining nephron. In this way a new steady state is achieved but at the price of a higher rate of PTH secretion. In the later stages of renal failure, the body can no longer cope with the intake of phosphorus and hyperphosphataemia develops. In this condition the production of calcitriol is severely reduced (Eisman et al. 1976) and this causes intestinal malabsorption of calcium, hypocalcaemia and bone loss. Figure 10.9 shows in outline the pathogenesis of secondary hyperparathyroidism.

In young infants (up to 3 months of age) kidney function has not yet developed fully and tubular reabsorption of phosphate is high. This may lead to hyperphosphataemia if the milk formula contains more than 35 mg phosphorus per 100 ml (ESPGAN 1977). The hyperphosphataemia causes an increased flux of calcium into the bone and hypocalcaemia. This hypocalcaemia is responsible for the classical picture of cow's milk tetany in young infants.

Pregnancy and Lactation

At birth a human baby contains 25–35 g of calcium, most of which is delivered to the foetus during the last trimester of pregnancy. During a 6-month lactation period, about 50 g of calcium is secreted into the breast milk. These calcium fluxes are, at least in part, brought about by increased circulating levels of the calciotropic hormones PTH and calcitriol (Kumar et al. 1979), which mobilize calcium from the maternal skeleton and increase active intestinal calcium transport. Heaney and Skillman (1971) reported an increased intestinal calcium absorption of 150 mg/day in women from the 20th week of pregnancy, which returned to normal by 3 months postpartum. No information is available about the effect of lactation on intestinal calcium absorption in women. In the rat, calcium absorption is raised during pregnancy and lactation, even in the absence of vitamin D, and this may be brought about by prolactin (Halloran and DeLuca 1980). Provided the diet contains adequate calcium, lactation does not appear to decrease bone density (Chan et al. 1982).

Plasma ionized calcium decreases slightly (by 0.20 mg/100 ml) during the third trimester of pregnancy (Tan et al. 1972) and does not appear to be affected by lactation (Hillman et al. 1981). No data are available indicating that human pregnancy and lactation change the urinary excretion of calcium.

Effects of Nutritional Factors on Plasma Calcium Homeostasis

Calcium

It might be anticipated, on the basis of the regulation of the plasma calcium concentration, that dietary calcium deficiency would produce hypocalcaemia, hyperparathyroidism, increased plasma calcitriol concentrations, increased bone resorption, decreased bone mineralization and ultimately osteoporosis. All of these effects have been clearly demonstrated in experimental animals (Ambrus et al. 1978; Rader et al. 1979). The increased secretion of PTH and calcitriol raises fractional intestinal calcium absorption, decreases urinary and faecal calcium excretion and enhances bone turnover (Table 10.3). It also increases intestinal phosphate absorption and urinary phosphate excretion (Table 10.3), and the increase in plasma phosphate may contribute to the hypocalcaemia. However, signs of calcium deficiency in humans are rare, possibly because skeletal growth in the human is so much slower than in most experimental animals. Only in very young children has a calcium deficiency syndrome been reported (Pettifor et al. 1979). The clinical and biochemical signs resembled those of rickets and developed in patients with a calcium intake of about 125 mg/day. It is still a matter of debate as to what extent human osteoporosis is the result of dietary calcium deficiency.

Table 10.3. Effects of dietary calcium, phosphorus and vitamin D deficiency in rats on weight gain, plasma calcium, calcium metabolism, bone ash and urinary hydroxyproline[a] (mean values±SE)

| | Diets[b] | | | |
	Control (n=6)	Ca deficient (n=6)	P deficient (n=6)	Vitamin D deficient (n=6)
Weight gain (g/3 weeks)	101 ± 7[c]	95 ±10	76 ±10*	55 ± 8*
Plasma Ca (mg/100 ml)	11.3± 0.3	9.2± 0.2*	12.7± 0.4*	6.3± 0.3*
Ca intake (mg/day)	65.3± 3.6	14.0± 1.1*	58.4± 4.2*	43.8± 4.4*
Ca in faeces (mg/day)	19.2± 2.1	0.1± 0.0*	7.5± 6.1*	26.6± 2.4*
Ca in urine (mg/day)	3.1± 1.3	0.1± 0.0*	24.8± 6.6*	0.2± 0.1*
Ca retention (mg/day)	43.2± 4.2	13.9± 1.1*	26.1± 2.6*	17.0± 2.2*
Ca absorption (% intake)	70.5± 4.2	99.2± 0.4*	87.1±10.5*	39.2± 1.9*
Femur ash (mg)	81 ±11	32 ± 3*	38 ± 5*	38 ± 8*
Urinary hypro (mg/g creat.)	245 ±23	445 ±61*	302 ±26*	432 ±81*

* Significantly different from control group ($p \leq 0.05$).
[a] Schaafsma and Dekker, unpublished results.
[b] Male Wistar rats were fed the different diets for 3 weeks after weaning; rats fed the vitamin D-deficient diet were born from vitamin D-deficient mothers; control diet: 0.5% Ca, 0.3% P; Ca-deficient diet: 0.1% Ca, 0.3% P; P-deficient diet: 0.5% Ca, 0.15% P; vitamin D-deficient diet: 0.5% Ca, 0.3% P.

Because intestinal calcium transport is in part a simple diffusion process (Fig. 10.8) which is not saturable, urinary calcium is raised when calcium intake is increased. The slope of urinary on dietary calcium has a gradient of about 6% (Heaney et al. 1975).

Phosphorus

Dietary phosphorus deficiency increases the renal synthesis of calcitriol independently of PTH (Tanaka and DeLuca 1973). Phosphorus-deficient rats have increased circulating levels of calcitriol and decreased levels of PTH (Rader et al. 1979). This tends to increase the intestinal absorption of phosphorus and to reduce its urinary excretion (Table 10.4) but it also increases intestinal absorption of calcium and leads to hypercalcaemia and hypercalciuria (Table 10.3).

Dietary phosphorus reduces the urinary excretion of calcium by increasing tubular reabsorption of calcium in the kidneys, an effect which has been attributed to PTH (Yuen and Draper 1983). From the available human data it appears that when the phosphorus requirement is reached, an increase of 500 mg in daily dietary phosphorus reduces urinary calcium by about 15–40 mg/day.

Experiments in adult animals have shown that high-phosphorus diets cause secondary hyperparathyroidism and increased bone resorption, leading to bone loss (Draper and Bell 1979). The hyperparathyroidism results from the removal of calcium from plasma, its secretion into the gut and its excretion in the faeces (Draper and Bell 1979). The stimulating effect of oral phosphorus on parathyroid function has been confirmed in the human (Reiss et al. 1970), but there is no direct evidence that high-phosphorus diets cause bone loss in man. Spencer et al. (1984) have demonstrated in humans that increasing the phosphorus intake from 800 to 2000 mg/day at levels of calcium intake of 200, 800 or 2000 mg/day, slightly increases faecal calcium (by about 60 mg/day), but this increase is more than offset by the decrease in urinary calcium.

Vitamin D

Vitamin D is required for transporting dietary calcium (and phosphorus) to the bone and therefore for bone mineralization. The effects of vitamin D deficiency in rats are shown by Tables 10.3 and 10.4. The intestinal absorption of calcium and phosphorus is reduced, plasma and urinary calcium are low, bone mineralization is decreased and bone turnover is enhanced. The effect of vitamin D on bone mineralization may be an indirect one due to the contribution of calcitriol to the maintenance of adequate concentrations of calcium and phosphate in the extracellular fluid (Underwood and DeLuca 1984; Weinstein et al. 1984).

Protein

Early in this century, Sherman (1920) reported that a man who was in calcium balance on a diet of 390 mg calcium daily, lost body calcium when meat was

Table 10.4. Effects of dietary calcium, phosphorus and vitamin D deficiency in rats on plasma phosphate and phosphate metabolism[a] (mean values±SE)

	Diets[b]			
	Control (n=6)	Ca deficient (n=6)	P deficient (n=6)	Vitamin D deficient (n=6)
Plasma P (mg/100 ml)	9.8±0.2[c]	9.8±0.3	6.6±0.5*	9.8±0.9
P intake (mg/day)	43.8±2.4	39.1±3.1	19.6±1.4	29.6±3.0*
P in faeces (mg/day)	4.9±0.6	0.8±0.2*	1.0±0.2*	9.0±0.7*
P in urine (mg/day)	8.2±1.8	23.9±2.3*	<0.1*	4.1±2.2*
P retention (mg/day)	30.7±1.4	14.5±1.5*	18.6±1.3*	16.5±1.3*
P absorption (% intake)	88.6±1.8	97.9±0.4*	94.8±1.2*	69.2±4.0*

* Significantly different from control group ($p \leq 0.05$).
[a] Schaafsma and Dekker, unpublished results.
[b] See footnote to Table 10.3.

added to his diet. Later, McCance et al. (1942) observed that a rise in the protein content of the diet of three healthy subjects resulted in an increased intestinal absorption and urinary excretion of calcium and magnesium. It was suggested that the primary action of dietary protein was that it increased intestinal absorption.

More recent studies in young adult males by Linkswiler and co-workers (1974) have shown that, regardless of the level of calcium intake, urinary calcium increases when the protein content of the diet is raised. When the calcium content of the diet was low (500 mg/d), intestinal absorption of calcium was unaffected by the protein content of the diet, and calcium retention became negative if the protein content of the diet was raised from 47 to either 95 or 142 g/d. At higher levels of calcium intake, protein increased intestinal calcium absorption, the maximum effect being reached at a level of 95 g of protein. Almost all balances were negative at the highest level of protein intake. Kim and Linkswiler (1979) have shown that the calciuretic effect of dietary protein is caused by a change in the renal handling of calcium: the glomerular filtration rate and the filtered calcium load are increased and the tubular reabsorption of calcium is decreased. The former effect would result from increased renal arterial pressure, induced by protein stimulation of a hormonal factor (glomerulopressin) (Hirsch 1985), while the latter effect has been reported to be related to increased urinary excretion of sulphate, originating from the oxidation of methionine and cystine (Whiting and Draper 1980) and to increased secretion of insulin (Allen et al. 1981).

The observations described above were obtained in experiments with humans who were given purified proteins for relatively short periods (10 days), care being taken to ensure that other dietary components known to affect calcium metabolism (particularly phosphorus) were kept constant. This probably explains the different results reported by Spencer et al. (1978). They found in a longer study (60 days) that increasing the protein intake from 1 to 2 g/kg of body weight had no effect on urinary calcium when calcium intake was either 200 or 800 mg daily and only had a temporary effect when calcium intake was 1100 or 2000 mg per day. It was assumed that the calciuretic effect of protein was counterbalanced by a concomitant increase in the intake of phosphorus. It is now clear that dietary protein raises the requirements for both calcium and phosphorus (Schuette and Linkswiler 1982). However, it has been calculated (Heaney et al. 1982; Heaney and Recker 1982) that the decrease in urinary calcium caused by dietary phosphorus is not large enough to neutralize all of the protein effect at the phosphorus/protein ratios found in most natural protein sources, and from the available data it can be concluded that increasing the protein content of the diet from natural sources by 20 g/day increases urinary calcium by about 20 mg/day.

Another factor which has to be considered in protein-induced hypercalciuria is diet acidity. When the protein intake of postmenopausal women was raised from 50 to 110 g/day, with calcium and phosphorus intakes held constant at 700 and 1080 mg/day respectively, urinary calcium doubled and renal acid excretion increased three- to fourfold (Lutz and Linkswiler 1981). It appeared in a later study (Lutz 1984) that ingestion of sodium bicarbonate alkalinized the urine and reduced the increase in urinary calcium associated with a higher protein intake. Moreover, it has been demonstrated in rats that the hypercalciuric effect of protein was reduced when dietary phosphorus was given in a less acidic form (Petito and Evans 1984). The effect of diet acidity on urinary calcium has been attributed to decreased tubular reabsorption of calcium in the kidney (Lemann et al. 1979).

It is not clear whether plasma levels of calcitriol and PTH change in response to an increased protein intake. Conflicting results have been reported in this regard (Coe et al. 1975; Adams et al. 1979; Licata 1981).

Lactose

It is well known that lactose stimulates intestinal calcium absorption in both experimental animals and humans (Ali and Evans 1973). The effect is independent of vitamin D and is exerted in the ileum on the diffusion component of the intestinal calcium transport system (Pansu et al. 1979). In vitamin D-deficient hypocalcaemic rats, lactose raises the plasma calcium concentration, increases bone mineralization, decreases bone turnover and almost normalizes bone histology (Table 10.5).

Sodium

It has been recognized for more than 20 years that the renal clearance of calcium is closely linked to that of sodium (Kleeman et al. 1964; Phillips and Cooke 1967). For this reason the normal range of urinary calcium excretion is a function of sodium intake. The link between calcium and sodium excretion is explained by the fact that calcium reabsorption along a large part of the nephron is dependent on the transport of sodium and chloride. During volume expansion, the reabsorption of sodium and chloride is reduced and this leads to a reduction of calcium reabsorption. It can be calculated from the data of Sabto et al. (1984) that in normal adults an increase in sodium excretion of 1 g/d (2.54 g salt) is associated with an increase in urinary calcium of on average 26 mg/d. Castenmiller et al. (1985) observed an average increase in urinary calcium of 20 mg/g creatinine when the diets of young healthy males were supplemented with 3.5 g sodium per day while other dietary factors were kept constant.

Table 10.5. Effects of dietary lactose and calcium on plasma calcium and bone in vitamin D-deficient rats[a] (mean values ± SE)

	Diets[b]			
	Control ($n=8$)	D^- ($n=6$)	D^-Ca ($n=6$)	D^-Ca^+ Lact. ($n=6$)
Plasma Ca (mg/100 ml)	10.4± 0.4[c]	6.1± 0.5*	7.0± 0.8*	9.8± 0.7*
Urinary hypro (mg/g creat.)	104 ±17	213 ±21*	162 ± 7*	142 ±31*
Femur chemistry				
dry weight (mg)	282 ±52	204 ±24*	209 ±24*	234 ±24*
ash (mg)	178 ±32	118 ±17*	125 ±18*	146 ±14*
ash (% dry weight)	63.3± 1.6	57.7± 2.3*	59.5± 1.6*	62.5± 1.4

* Significantly different from control group ($p \leqslant 0.05$).
[a] Schaafsma et al. in preparation.
[b] Male Wistar rats (all born from vitamin D-deficient mothers) were fed the different diets for 6 weeks after weaning: control diet, 1 IU vitamin D_3/g, 0.5% Ca, 0.3% P; D^-: idem control diet, minus vitamin D; D^-Ca: idem D^-, 1.5% Ca, 0.9% P; $D^-Ca^+Lact.$: idem D^-Ca, 15% lactose.
[c] Mean value ± SD.

In normal subjects, sodium-induced hypercalciuria is accompanied by increased calcitriol production and intestinal calcium absorption but this adaptation does not occur in patients with hypoparathyroidism, suggesting PTH mediation (Breslau et al. 1982). An increased release of PTH is consistent with a reduced plasma ionized calcium concentration following salt loading, as has been reported in the rat (Pernot et al. 1979).

Other Nutritional Factors

Caffeine increases the urinary excretion of calcium in a dose-dependent way (Heaney and Recker 1982; Massey and Wise 1984). In 12 healthy women, urinary calcium increased by on average 8 and 26 mg in the 3 hours after the intake of 150 and 300 mg caffeine respectively (Massey and Wise 1984). The mechanism underlying the effect of caffeine on urinary calcium is not known but since one cup of coffee contains 100–150 mg of caffeine, it is an effect which should not be ignored.

Some dietary factors, mainly phytic acid from plant products, uronic acids from cereals, oxalic acids from certain vegetables and long-chain fatty acids, may reduce calcium absorption by complexing calcium in the intestine (Allen 1982; Kelsey 1985; Kies 1985). It is not certain whether this is a significant factor in calcium homeostasis with the amounts of these compounds usually present in Western diets.

References

Adams ND, Gray RW, Lemann J (1979) The calciuria of increased fixed acid production in humans: evidence against a role of parathyroid hormone and 1,25(OH)$_2$-vitamin D. Calcif Tissue Int 28:233–238

Ali R, Evans JL (1973) Lactose and calcium metabolism: a review. J Agr Univers, Puerto Rico, 57:149–164

Allen LH (1982) Calcium bioavailability and absorption: a review. Am J Clin Nutr 35:783–808

Allen LH, Block GD, Wood RY, Bryce GF (1981) The role of insulin and parathyroid hormone in the protein-induced calciuria of man. Nutr Res 1:3–11

Ambrus JL, Robin JC, Kelly RS, Mannley N, Thomas CC (1978) Studies on osteoporosis. 1. Experimental models. Effect of age, sex, genetic background, diet, steroid and heparin treatment on calcium metabolism of mice. Res Commun Chem Pathol Pharmacol 22:3–14

American Academy of Pediatrics, Committee on Nutrition (1978) Calcium requirements in infancy and childhood. Pediatrics 62:826–834

Armbrecht HJ, Zenser TV, Bruns ME, Davis BB (1979) Effect of age on intestinal calcium absorption and adaptation to dietary calcium. Am J Physiol 236:E769–774

Breslau NA, McGuire JL, Zerwekh JE, Pak CYC (1982) The role of dietary sodium on renal excretion and intestinal absorption of calcium and on vitamin D metabolism. J Clin Endocrinol Metab 55:369–373

Bricker NS, Slatopolsky E, Reiss E, Avioli LV (1969) Calcium, phosphorus and bone in renal disease and transplantation. Arch Intern Med 123:543–553

Castenmiller JJM, Mensink RP, van der Heijden L et al. (1985) The effect of dietary sodium on urinary calcium and potassium excretion in normotensive men with different calcium intakes. Am J Clin Nutr 41:52–60

Chan GM, Roberts CC, Folland D, Jackson R (1982) Growth and bone mineralisation of normal breast-fed infants and the effects of lactation on maternal bone mineral status. Am J Clin Nutr 36:438–443

Coe FL, Firpo JJ, Hollandsworth DL, Segil L, Canterbury JM, Reiss E (1975) Effect of acute and chronic metabolic acidosis on serum immunoreactive parathyroid hormone in man. Kidney Int 8:262–273

Cohn SH, Vaswani A, Zazi I, Ellis KJ (1976) Effect of aging on bone mass in adult women. Am J Physiol 230:143–148

Draper HH, Bell RR (1979) Nutrition and osteoporosis. In: Draper HH (ed) Advances in nutritional research. Plenum, New York, pp 79–106

Eisman JA, Hamstra J, Kream BE, DeLuca HF (1976) 1,25-Dihydroxyvitamin D in biological fluids. A simplified and sensitive assay. Science 193:1021–1023

ESPGAN Committee on Nutrition (1977) Guidelines on infant nutrition. 1. Recommendations for the composition of an adapted formula. Acta Paediatr Scand [Suppl] 262

Gaillard PJ (1961) Parathyroid and bone in tissue culture. In: Greep RO, Talmage RV (eds) The parathyroids. CC Thomas, Springfield, Illinois, pp 20–45

Garabedian M, Tanaka Y, Holick MF, DeLuca HF (1974) Response of intestinal calcium transport and bone calcium mobilization to 1,25-dihydroxy-vitamin D in thyroparathyroidectomized rats. Endocrinology 94:1022–1027

Garn SM, Rohmann CG, Wagner B (1967) Bone loss as a general phenomenon in man. Fed Proc 26:1729–1736

Halloran BP, DeLuca HF (1980) Calcium transport in small intestine during pregnancy and lactation. Am J Physiol 239:E64–68

Heaney RP, Recker RR (1982) Effects of nitrogen, phosphorus and caffeine on calcium balance in women. J Lab Clin Med 99:46–55

Heaney RP, Skillman TG (1971) Calcium metabolism in normal human pregnancy. J Clin Endocrinol Metab 33:661–670

Heaney RP, Saville PD, Recker RR (1975) Calcium absorption as a function of calcium intake. J Lab Clin Med 85:881–890

Heaney AP, Gallagher JC, Johnston CC et al. (1982) Calcium nutrition and bone health in the elderly. Am J Clin Nutr 36:986–1013

Henry RJ, Cannon DC, Winkelman JW (1974) Clinical chemistry, principles and techniques. Harper and Row, New York

Hillman L, Sateesha S, Haussler M, Weist W, Slatopolsky E, Haddad J (1981) Control of mineral homeostasis during lactation: interrelationships of 25-hydroxyvitamin D, 24,25-dihydroxyvitamin D, 1,25-dihydroxyvitamin D, parathyroid hormone, calcitonin, prolactin and estradiol. Am J Obstet Gynecol 139:471–476

Hirsch DJ (1985) Limited-protein diet: a means of delaying the progression of chronic renal disease? Can Med Ass J 132:913–917

Horst RW, DeLuca HF, Jorgenson N (1978) The effect of age on calcium absorption and accumulation of 1,25-$(OH)_2$-D_3 in intestinal mucosa of rats. Metab Bone Dis Rel Res 1:29–33

Kalu DN, Foster GV (1976) Calcitonin, its physiological roles. NY State J Med 76:230–233

Keating FR, Jones JD, Elveback LR, Randall RV (1969) The relation of age and sex to distribution of values in healthy adults of serum calcium, inorganic phosphorus, magnesium, alkaline phosphatase, total proteins, albumin, and blood urea. J Lab Clin Med 73:825–834

Kelsey JL (1985) Effect of oxalic acid on calcium bio-availability. In: Kies C (ed) Nutritional bio-availability of calcium. American Chemical Society, Washington DC, pp 105–116 (ACS Symposium Series 275)

Kies C (1985) Effect of dietary fat and fiber on calcium bio-availability. In: Kies C (ed) Nutritional bio-availability of calcium. American Chemical Society, Washington DC, pp 175–187 (ACS Symposium Series 275)

Kim Y, Linkswiler HM (1979) Effect of level of protein intake on calcium metabolism and on parathyroid and renal function in the adult human male. J Nutr 109:1399–1404

Kleeman CR, Bohannan J, Bernstein D, Ling S, Maxwell MH (1964) Effect of variations in sodium intake on calcium excretion in normal humans. Proc Soc Exp Biol Med 115:29–32

Kumar R (1979) Renal osteodystrophy: a complex disorder. J Lab Clin Med 93:895–898

Kumar R, Cohen WR, Silva P, Epstein FH (1979) Elevated 1,25-dihydroxyvitamin D plasma levels in normal human pregnancy and lactation. J Clin Invest 63:342–344

Lemann J, Adams ND, Gray RW (1979) Urinary calcium excretion in human beings. N Engl J Med 301:535–541

Licata AA (1981) Acute effects of increased meat protein on urinary electrolytes and cyclic adenosine monophosphate and serum parathyroid hormone. Am J Clin Nutr 34:1779–1784

Linkswiler HM, Joyce CL, Anand CR (1974) Calcium retention of young adult males as affected by level of protein and calcium intake. Trans NY Acad Sci 36:333–340

Lutz J (1984) Calcium balance and acid-base status of women as affected by increased protein intake and by sodium bicarbonate ingestion. Am J Clin Nutr 39:281–288

Lutz J, Linkswiler HM (1981) Calcium metabolism in postmenopausal and osteoporotic women consuming two levels of dietary protein. Am J Clin Nutr 34:2178–2186

Marshall RW (1976) Plasma fractions. In: Nordin BEC (ed) Calcium, phosphate and magnesium metabolism. Churchill Livingstone, Edinburgh, pp 162–185

Marshall RW, Hodgkinson A (1983) Calculation of plasma ionised calcium from total calcium, proteins and pH: comparison with measured values. Clin Chim Acta 127:305–310

Massey LK, Wise KJ (1984) The effect of dietary caffeine on urinary excretion of calcium, magnesium, sodium and potassium in healthy young females. Nutr Res 4:43–50

McCance RA, Widdowson EM, Lehmann H (1942) The effect of protein intake on the absorption of calcium and magnesium. Biochem J 36:686–691

Mioni G, D'Angelo A, Ossi E, Bertaglia E, Mareon G, Maschio G (1971) The renal handling of calcium in normal subjects and in renal disease. Eur J Clin Biol Res 16:881–887

Moore EW (1970) Ionized calcium in normal serum, ultrafiltrates and whole blood, determined by ion-exchange electrodes. J Clin Invest 49:318–334

Nordin BEC (1976) Plasma calcium and plasma magnesium homeostasis. In: Nordin BEC (ed) Calcium, phosphate and magnesium metabolism. Churchill Livingstone, Edinburgh, pp 186–216

Pansu D, Bellaton C, Bronner F (1979) Effect of lactose on duodenal calcium-binding protein and calcium absorption. J Nutr 109:508–512

Peacock M, Robertson WG, Nordin BEC (1969) Relation between serum and urinary calcium with particular reference to parathyroid hormone. Lancet I:384–386

Pernot F, Berthelot A, Schleiffer R, Gairard A (1979) Ionized serum calcium, urinary cAMP and immunoreactive PTH after DOCA + NaCl treatment in the rat. Electrolyte Metab 2:258

Petito SL, Evans J (1984) Calcium status of the growing rat as affected by diet acidity from ammonium chloride, phosphate and protein. J Nutr 114:1049–1059

Pettifor JM, Ross P, Moodley G et al. (1979) Calcium deficiency in rural black children in South Africa: a comparison between rural and urban communities. Am J Clin Nutr 32:2477–2483

Phillips MJ, Cooke JNC (1967) Relation between urinary calcium and sodium in patients with idiopathic hypercalciuria. Lancet I:1354–1357

Rader JE, Baylink DJ, Hughes MR, Sagilian EF, Haussler MA (1979) Calcium and phosphorus deficiency in rats: effects on PTH and 1,25-dihydroxyvitamin D_3. Am J Physiol 236:E118–122

Raymakers JA (1978) De calciumuitscheiding door de nieren. Thesis, State University of Utrecht, The Netherlands

Reiss E, Canterbury JM, Bercovitz MA, Kaplan EL (1970) The role of phosphate in the secretion of parathyroid hormone in man. J Clin Invest 49:2146–2149

Riggs BL, Jowsey J, Kelly PJ, Jones JD, Maher FT (1969) Effect of sex hormones on bone in primary osteoporosis. J Clin Invest 48:1065–1073

Riggs BL, Jowsey J, Goldsmith RS, Kelly PJ, Hoffman DL, Arnaud CD (1972) Short- and long-term effects of estrogen and synthetic anabolic hormone in postmenopausal osteoporosis. J Clin Invest 51:1659–1663

Rowe JW, Andres R, Jordan D, Norris AH, Shock NW (1976) The effect of age on creatinine clearance in men: a cross-sectional and longitudinal study. J Gerontol 31:155–163

Sabto J, Powell MY, Breidahl MJ, Guz FW (1984) Influence of urinary sodium on calcium excretion in normal individuals. A redefinition of hypercalciuria. Med J Aust 140:354–356

Schuette SA, Linkswiler HM (1982) Effects on Ca and P metabolism in humans by adding meat, meat plus milk or purified proteins plus Ca and P to a low protein diet. J Nutr 112:338–349

Sherman HC (1920) Calcium requirement of maintenance in man. J Biol Chem 44:21–27

Slatopolsky E, Bricker NS (1973) The role of phosphorus restriction in the prevention of secondary hyperparathyroidism in chronic renal failure. Kidney Int 4:141–145

Spencer H, Kramer L, Osis D, Norris C (1978) Effect of a high protein (meat) intake on calcium metabolism in man. Am J Clin Nutr 31:2167–2180

Spencer H, Kramer L, Osis D (1984) Effect of calcium on phosphorus metabolism in man. Am J Clin Nutr 40:219–225

Stevenson JC, Hillyard CJ, MacIntyre I, Cooper H, Whitehead MI (1979) A physiological role for calcitonin: protection of the maternal skeleton. Lancet II:769–770

Tan CM, Raman A, Sinnathyray TA (1972) Serum ionic calcium levels during pregnancy. J Obstet Gynaecol Br Commonw 79:694–697

Tanaka Y, DeLuca HF (1973) The control of 25-hydroxyvitamin D metabolism by inorganic phosphorus. Arch Biochem Biophys 154:566–574

Underwood JL, DeLuca HF (1984) Vitamin D is not directly necessary for bone growth and mineralization. Am J Physiol 246:E493–498

Von Misselwitz J, Kauf E (1981) Die Kalzium Ausscheidung im Urin bei gesunden Kindern als Grundlage für die Suche nach einer Hyperkalziurie bei Kalziumnephrolithiasis. Kinderarzliche Praxis 49:64–71

Weinstein RS, Underwood JL, Hutson MS, DeLuca HF (1984) Bone histomorphometry in vitamin D deficient rats infused with calcium and phosphorus. Am J Physiol 246:E499–505

Whiting SJ, Draper HH (1980) The role of sulphate in the calciuria of high protein diets in adult rats. J Nutr 110:212–222

Wilkinson R (1971) Studies on calcium absorption by small intestine of rat and man. Thesis, University of Leeds, UK

Yuen DE, Draper HH (1983) Long-term effects of excess protein and phosphorus on bone homeostasis in adult mice. J Nutr 113:1374–1380

Chapter 11

Calcium as an Intracellular Regulator

A. K. Campbell

Intracellular Calcium and Cell Behaviour

The survival and reproduction of human life requires calcium, not only outside the cell for maintenance of skeletal and tissue architecture, and for key extracellular enzymes such as those involved in blood clotting or digestion, but also inside cells controlling their behaviour (Campbell 1983). The evolutionary success of eukaryotic cells has been critically dependent on the development of mechanisms which enable the whole organism to regulate its behaviour in response to both internal and external stimuli. Signals from the brain can provoke release of neurotransmitters from nerve terminals leading to muscle movement. Individual cell movement occurs in neutrophils, the first cells to appear at a site of infection. The daily fasting–feeding cycle requires regulation of intermediary metabolism by substrates and hormones, in order to channel nutrients through the pathways of lipid or carbohydrate oxidation necessary for supplying the energy needs of individual tissues and the body as a whole (Campbell and Hales 1976).

Calcium is required by four enzymes in the blood-clotting process, and for binding the C1 complex of the "classical" complement pathway to an antibody–antigen complex. Removal of calcium stops both clotting and complement activation. Yet calcium is not a regulator in a physiological sense. Changes in plasma calcium neither provoke nor significantly alter these two events. The role of calcium is thus "passive". In complete contrast, calcium inside the cell plays an "active" role. Here changes in free calcium caused by physiological stimuli provoke a change in behaviour of the cell.

Whereas the free calcium concentration outside the cell is in the mM range, it is a change in free calcium concentration in the μM range within cells which activates the mechanisms responsible for phenomena such as cell movement, secretion and alterations in the direction of intermediary metabolism. Increases in intracellular calcium occur during egg fertilization and during cell division. Disturbances in free and bound calcium within the cell also play an important role in the mechanism underlying cell injury caused, for example, by components of the immune system, viruses and toxins. An understanding of the molecular basis of the control of intracellular calcium is thus vital if we are to comprehend fully the pathogenesis of major diseases such as heart disease, the arthritides, multiple sclerosis and diabetes. This will also enable new regimes to be developed for therapeutic intervention or disease prevention.

What then is the evidence that a small absolute change in a cation, calcium, can lead to such dramatic changes in cell behaviour? How are these intracellular changes in calcium regulated physiologically, and disturbed pathologically or pharmacologically? How does calcium work, and what is its interaction with other intracellular regulators such as cyclic nucleotides and diacylglycerol? Before examining these particular questions it is necessary first to define more precisely the cell biology of cell activation, together with the balance of calcium within the resting cell which provides the springboard enabling the cell to transform to a different state. Attention will be focused initially on physiological examples of cell activation before consideration is given to the pathological significance of changes in intracellular calcium.

The Cell Biology of Intracellular Calcium

A Framework for Cell Regulation

Events in cells responsible for changes in their behaviour, or that of the tissue to which they belong, are provoked by a primary stimulus. This originates usually from outside the cell, though in some cases, for example cell division, it may occur as a result of internal programming. These primary stimuli may be physical, for example touch, an electrical signal and absorption of light; or chemical, for example hormones, neurotransmitters or regulatory metabolites. Thus action potentials arriving at a nerve terminal provoke transmitter secretion, or in muscle cells stimulate contraction. Hormones, neurotransmitters and metabolites can activate endocrine or exocrine secretions, as well as alter intermediary metabolism. Growth factors and components of the immune system can provoke cell transformation and cell division, and contact between egg by sperm results in the activation of processes necessary for fertilization to occur. The primary stimuli, acting directly at the cell membrane or occasionally on a membrane within the cell, result in the production of a chemical signal which leads to activation of events somewhere else in the cell. Here we are concerned with phenomena where this chemical signal is intracellular calcium, or is one which leads to release of calcium inside the cell (Table 11.1). Some stimuli, however, may work primarily through cyclic AMP, cyclic GMP or diacylglycerol (Berridge 1976; Berridge and Irvine 1984), though even here there is invariably some interaction with intracellular calcium.

Table 11.1. Cell activation provoked by a rise in intracellular free Ca^{2+}

Phenomenon	Example	Primary stimulus	Secondary regulators
Cell movement	Muscle – skeletal	Acetylcholine	—
	– smooth	Noradrenaline	Eicosanoids (+,−)
	– cardiac	Action potential	Adrenaline (+)
	Amoeboid – neturophil chemotaxis	C5a	Eicosanoids (+,−)
	Ciliate – *Paramecium* reversal	Touch	—
Secretion from vesicles	Nerve terminal	Action potential	Adenosine (−)
	Endocrine – mast cell	Antigen on IgE	Adrenaline (−)
	Exocrine – pancreas	Cholecystokinin	?
Cell aggregation	Platelet	Thrombin	Eicosanoids (+,−)
Cell transformation	Lymphocyte	Antigen	T cells/macrophages
	Oocyte maturation – starfish	l Methyladenine	?
	Egg fertilization	Sperm	?
Intermediary metabolism	Glycogenolysis in liver	Adrenaline (α)	Insulin

These are only intended as examples to illustrate the wide range of phenomena provoked by intracellular Ca^{2+}.

The efficacy of the primary stimulus can be modified by secondary regulators (Table 11.1), which may alter the time of onset, the duration or the magnitude of the cell response. Some secondary regulators may potentiate, others inhibit, the effect of the primary stimulus, either by altering the affinity of the cell for the stimulus or the maximal capacity of the cell to respond. The question therefore arises; if the primary stimulus activates the cell through an increase in intracellular calcium, does the secondary regulator act on the calcium change, the mode of action of calcium or on some mechanism independent of calcium?

Two factors can complicate the investigation of the molecular basis of cell activation. Firstly the cell may respond to more than one stimulus. The questions then are why, and do they work through the same mechanism? Secondly more than one process may be activated in the same cell by the same stimulus. For example, activation of intermediary metabolism is usually necessary if cells are to have the necessary energy for activation, be it muscle contraction, secretion or cell division. In which case one might expect the two processes to be tightly coupled and therefore activated by the same intracellular signals. Alternatively one process may follow another, and thus they may be controlled by different intracellular signals. For example, neutrophils move down a concentration gradient of bacterial chemotactic factors or complement fragment C5a following an infection. Once at the site of infection phagocytosis and reactive oxygen metabolite production occur, provoked by C3b-coated bacteria, enabling the cell to engulf and kill the invading microorganisms. Likewise platelets can be activated to aggregate and secrete both by different stimuli or the same stimulus. A consequence of this is that adenosine diphosphate (ADP), secreted as a result of thrombin activation of the platelets, acts to precipitate a cascade, or chain reaction, in other platelets, thereby accelerating clot formation initiated by thrombin.

A significant characteristic of many phenomena involving activation of tissues, often ignored by those interested in molecular mechanisms, is the fact that they

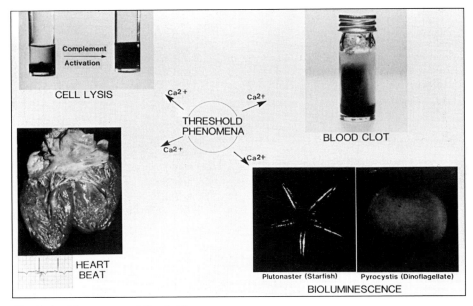

Fig. 11.1.   The concept of threshold.

occur through "threshold" responses in individual cells (Campbell 1983; Fig. 11.1). Thus a heart cell beats or it does not, a platelet aggregates with the clot or remains free, a cell in a luminous organism flashes or remains invisible. The relationship between the dose of stimulus and the time course and magnitude of the response of a cell population therefore depends on the time of onset, together with the duration of the phenomenon, in each cell. The number of cells "switched on" and their response will depend on the balance between the amounts of primary stimuli and secondary regulators present (Table 11.1). For example, adrenaline speeds up the response of an individual myocyte to the primary stimulus, the action potential, as well as increasing the strength of the contraction itself. Acetylcholine antagonizes both effects.

A similar principle based on individual cell "thresholds" can be applied to problems of cell injury induced by pathogens and the response of cells to drugs. The killing of cells by viruses or the terminal complex of the complement pathway requires a critical threshold point beyond which not only irreversible damage occurs but cell death rapidly ensues (Edwards et al. 1983). The time taken to reach this threshold varies from cell to cell. Since nucleated cells also have recovery mechanisms enabling them to remove potentially lethal attacking agents, via processes such as vesiculation and endocytosis (Campbell and Morgan 1985; Morgan et al. 1986a), the number of cells killed will depend on how many have succeeded in protecting themselves before the critical threshold point for cell death has been reached.

A rise in free calcium within the cell is ideally suited to act as the chemical switch responsible for initiating a threshold response, be it activation or cell death (Campbell 1987). However, somewhat paradoxically, calcium is also geared to activate protection mechanisms following an insult to a cell, thereby allowing any cell damage to be reversible rather than irreversible. To see how this situation can come about we must first examine how calcium is distributed within the cell, and how the cell is able to maintain itself in an environment containing calcium concentrations which would be hostile were they to exist inside the cell.

Cell Calcium Balance

In human cells, the total cell calcium can vary from as little as 0.02 mmol/kg cell water in erythrocytes with no organelles, to more than 5 or 10 mmol/kg cell water in cells such as muscle or platelets with large stores of intercellular calcium. Yet, whereas only some 50%–60% of the calcium in the fluid outside the cell is bound, in the cell more than 99.9% is bound, mainly within intracellular organelles (Fig. 11.2).

The cytoplasm of a resting, healthy cell contains about 10 μM calcium of which only about 0.1 μM is free calcium, the absolute value depending on the cell type and method of analysis used. The remainder is bound to proteins and small organic ligands such as citrate and ATP. How can the cell maintain such a low intracellular free calcium, when not only is the external free calcium some 10 000

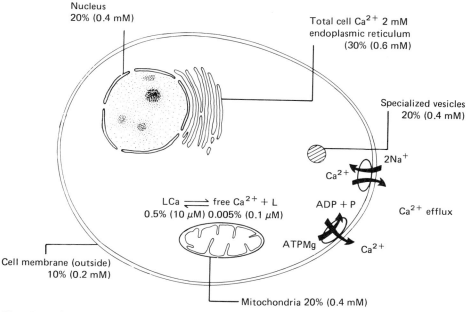

Fig. 11.2. Calcium balance of a typical eukaryotic cell. The figures are only intended to indicate orders of magnitude.

times greater than this but also there is a negative membrane potential of some 50–100 mV trying to pull calcium into the cell?

There are, of course, calcium pumps and transport mechanisms within organelles such as endoplasmic reticulum, mitochondria and vesicles enabling them to accumulate and store their calcium. However, these in themselves would not be sufficient to maintain cell calcium in a steady state. If the cell is not to go on accumulating calcium there must be an energy-dependent process continuously pumping calcium out of the cell. Three mechanisms have been identified in eukaryotic cells: a Ca^{2+} Mg-ATPase, a Na^+–Ca^{2+} exchange system (energetics require 2–3 Na^{2+} per Ca^{2+}) and a Ca^{2+}–H^+ exchange. The first appears to be the major mechanism in resting cells. The second comes into play, in cells which have it, e.g. heart muscle and nerve, during cell activation or injury when intracellular free calcium rises to 1–10 μM. Ca^{2+}–H^+ exchange may be found in intracellular organelles such as secretory vesicles or mitochondria but does not appear to be common at the plasma membrane. But how much energy do these Ca^{2+} pumps need?

Fortunately the passive permeability of the cell membrane to calcium is relatively low, sometimes several hundred times less than for Na^+ or K^+. This passive calcium entry occurs at a rate of between 0.01 and 0.2 pmol cm^2/s, equivalent to 30–600 nM/s (nmol/s/kg cell water) in a cell of radius 10 μ (e.g. liver). An equivalent amount of ATP is therefore required if the relationship is 1 ATP hydrolysed per calcium ion pumped out. However the total cell ATP is about 5–10 mM and its turnover may be as much as 0.2 mmol/s/kg H_2O, of which 30%–50% is used by the Na^+ pump responsible for maintaining the Na^+/K^+ gradient across the cell membrane. Thus the calcium pump probably consumes <0.3% of the resting cell's ATP production. A further consequence of this is that the calcium need have no counter ion since its effect on membrane potential will be negligible. After a rise in intracellular free calcium of 1–10 μM, which can occur during cell activation or cell injury, ATP utilization may rise 10-to 100-fold. The result is often a drop in cell ATP, which recovers once the free calcium has been restored to the resting level.

Physiological, Pathological and Pharmacological Consequences

In view of the enormous electrochemical gradient of calcium across the cell membrane, greater than for any of the other major ions in the cell, it is now clear why intracellular calcium is so ideally suited to acting as a "switch" for cell activation. A small increase in the permeability of the cell membrane, or a small fractional release of calcium from an internal store, will lead to a large fractional rise in cytoplasmic calcium, to a level some several fold greater than that of the non-activated cell. Furthermore damage to the cell membrane by a pathogen or the binding of a lipophilic drug is also inevitably going to lead to a large rise in cytoplasmic free calcium.

But how can we tell whether such rises, be they induced by naturally occurring stimuli, pathogens or drugs, are a cause as opposed to a consequence of cell attack? To answer this we must examine the experimental evidence which distinguishes between these alternatives, focusing on the key experiment, the direct measurement of free calcium in the living cell.

The Evidence for Calcium as an Intracellular Regulator

One Hundred Years of Experiments

Many people are often surprised to learn that Ca^{2+}, not cyclic AMP, was the first "second messenger" to be discovered, the idea being at least 70 years old (see Heilbrunn 1937; Campbell 1986b, 1987; Table 11.2). Sydney Ringer was the first physiologist to investigate systematically the need for individual inorganic ions in the fluid bathing cells and tissue for their normal function and survival. In 1882 he had apparently shown that calcium was *not* necessary for the beating of the frog heart. However, on realizing that his assistant had made up the solutions using London tap water, now known to contain Ca^{2+}, he repeated the experiment using distilled water. He was then able to show that calcium *was* required for the normal contraction of frog heart muscle (Ringer 1883), as well as for the development of fertilized eggs and tadpoles (Ringer and Sainsbury 1894). He, and other physiologists at the turn of the century, was able to show

Table 11.2. Important experiments establishing intracellular Ca^{2+} as a universal regulator

Date	Experiment	Workers
1808	Discovery of calcium as an element	Davy
1883	Extracellular Ca^{2+} required for frog heart contraction	Ringer
1894	Extracellular Ca^{2+} required for transmission of impulses from nerve to muscle	Locke
1928	Detection of a change in cytoplasmic Ca^{2+} in Amoeba	Pollack
1940	Ca^{2+} action on isolated protoplasm from muscle	Heilbrunn
1940	Ca^{2+} required for acetylcholine release	Harvey and MacIntosh
1942	Ca^{2+} activates myosin ATPase	Bailey, Needham
1943	Ca^{2+} on muscle protoplasm causes contraction	Kamata and Kinoshita
1947	Ca^{2+} injection into muscle causes contraction	Heilbrunn and Wiercinski
1953	Non-Na^+-dependent action potentials in crustacea	Fatt and Katz
1953	Phosphatidyl inositol turnover	Hokin and Hokin
1957	Discovery that cytoplasmic Ca^{2+} is very low ($<10~\mu M$)	Hodgkin and Keynes
1961	Ca^{2+} may trigger exocytosis	Douglas and Rubin
1962	ATP-dependent Ca^{2+} uptake system	Ebashi and Lipmann
1963	"Active tropomyosin" leads to the discovery of trophin C	Ebashi
1964	Ca^{2+} spikes in excitable cells	Hagiwara et al.
1967	Direct measurement of a Ca^{2+} transient with aequorin	Ridgway and Ashley
1967	Discovery of calmodulin	Cheung
1974	Visualization of Ca^{2+}-activated photoproteins	Morin and Reynolds
1975	Phosphatidyl inositol turnover involved in cell regulation	Michell
1975	Visualization of localized free Ca^{2+} changes in cytoplasm	Rose and Loewenstein; Taylor et al.
1981	Free Ca^{2+} measured in a small cell by dye fluorescence	Tsien
1983	Inositol trisphosphate releases intracellular Ca^{2+}	Berridge et al.

that calcium was important in maintaining cell adhesion and viability (Herbst 1900; Gray 1922), as well as being involved in the actions of adrenaline and digitalis on the heart (Mines, 1911; Loewi 1917, 1918; for reviews see Berliner 1933; Campbell 1983). It was also shown that removal of external Ca^{2+} could reduce rates of cell growth in culture, cause hyperexcitability of nerves and prevent transmission of signals from nerve to muscle by preventing secretion of transmitter at the neuromuscular junction (Locke 1894; Overton 1902; Houssay and Molinelli 1928; Harvey and McIntosh 1940).

The problem in interpreting the significance of these pioneering experiments was discovering what the Ca^{2+} was doing and where? Was Ca^{2+} simply playing a structural role in maintaining cell integrity? Some far-sighted physiologists working with adrenaline in the heart (Lawaczek 1928; Hermann 1932) and on amoebae, on pseudopod formation in amoebae (Pollack 1928) and on the parthogenetic stimulation of egg division (Heilbrunn and Wilbur 1937; Mazia 1937) believed that internal release of Ca^{2+} was required for cell activation.

By the 1940s evidence was growing that in skeletal muscle too a release of internal Ca^{2+} was the trigger for contraction, in spite of the fact that in most skeletal muscles removal of Ca^{2+} had no acute effect on contraction. Addition of Ca^{2+} to isolated muscle protoplasm (Heilbrunn 1940), or injection of Ca^{2+} into a fibre (Kamata and Kinoshita 1943; Heilbrunn and Wiercinski 1947) provoked contraction. Ca^{2+} was also found to be a potent activator of isolated myosin ATPase (Bailey 1942; Needham 1942).

Yet the "Ca^{2+} hypothesis", propagated so convincingly by Heilbrunn (1927, 1928, 1937, 1943, 1956), was slow to gain universal acceptance (Ebashi 1980). There were three particular problems. Firstly it was not generally realized how low was the free Ca^{2+} in living cells. Secondly, how could a simple cation, Ca^{2+}, acting at low concentrations, provide the energy for processes such as muscle contraction? Anyway the ATPase of purified actomyosin worked quite happily with ATP magnesium, but without Ca^{2+}. Thirdly, why did some so-called Ca^{2+}-activated cells work perfectly well without external Ca^{2+}, at least initially?

In 1957 Hodgkin and Keynes, working with the giant axon of the squid, estimated from the slow diffusion of $^{45}Ca^{2+}$ through the axoplasm, that the internal free Ca^{2+} must be less than 10 μM. Using the newly introduced calcium/EGTA buffers, Portzehl et al. (1964) measured the minimum free Ca^{2+} necessary to induce contraction of crab muscle and estimated the free Ca^{2+} in resting muscle to be less than 0.3 μM. The submicromolar concentration of free Ca^{2+} was finally measured directly in barnacle muscle by Ridgway and Ashley in 1967 using the Ca^{2+}-activated photoprotein aequorin as their intracellular Ca^{2+} indicator.

The second problem, concerning where the energy comes from, was solved by the discovery of Ca^{2+}-binding proteins (Table 11.3) which either depress or activate reactions and cellular processes whose energy comes from intermediary metabolism via ATP hydrolysis. In 1963 Ebashi discovered a "factor", a protein complex, in muscle homogenates which restored Ca^{2+} sensitivity to the actomyosin complex. This led to the isolation of the protein responsible, troponin C. The discovery of calmodulin, a high-affinity Ca^{2+}-binding protein found in all eukaryotic cells and having 45% amino acid homology with troponin C, is credited to Cheung (1967). However, it was not until the early 1970s that the Ca^{2+} sensitivity, significance and ubiquitous occurrence in eukaryotic cells was established (Cheung 1970; Kakiuchi and Yamazaki 1970; Smoake et al. 1974; Brostrom et al. 1975; Waisman et al. 1978; for reviews see Dedman 1986a, b). The realization that Ca^{2+} works through high-affinity binding proteins has resulted in the isolation of other, apparently structurally different, proteins associated with the cytoskeleton, vesicles and other intracellular structures. The discovery of the low intracellular free Ca^{2+} coupled with these high-affinity, Ca^{2+}-selective proteins dispelled much of the confusion generated by experiments carried out in the 1950s and 1960s which showed inhibition of 20 or 30 intracellular enzymes by mM Ca^{2+} concentrations. These are now realized to be virtually all physiologically irrelevant.

Table 11.3. Some calcium bindings associated with cell activation (Weeds 1982; Campbell 1983, 1986a, b; Dedman 1986a, b)

Protein	Approx. mol. wt.	Cell source
Troponin C	17 K	Skeletal and cardiac muscle
Leiotonin C subunit	20 K	Smooth muscle
Calmodulin	16.7 K	All eukaryotic cells
Actin-binding proteins (α-actinin, actinogelin, fragmin, gelsolin, villin, vinculin)	40–150 K	Many non-muscle cells, e.g. phagocytes, slime moulds, fibroblasts, liver
Calcimedins	33 K, 35 K, 67 K	Smooth muscle
Ca^{2+}-activated photoproteins	20 K	Luminous coelenterates Luminous radiolarians
Luciferin-binding proteins	18.5 K	Luminous anthozoans
Bacterial Ca^{2+}-binding proteins	33 K, 47 K, 60 K	*E. coli*

The third piece of the puzzle was solved with the identification of the mechanisms responsible for regulating intracellular Ca^{2+}: Ca^{2+} pumps responsible for maintaining the low intracellular free Ca^{2+} in the resting cell, Ca^{2+}-selective channels allowing intracellular Ca^{2+} to rise rapidly following cell activation and Ca^{2+} stores releasing Ca^{2+} following activation by a stimulus. In 1951 March identified a factor in muscle homogenates which modified myofibril contraction. Further studies on this factor showed it to be vesicular, removing Ca^{2+} in the presence of Mg^{2+} and ATP (Kumagai et al. 1955; Ebashi and Lipmann 1962; Hasselbach and Makinose 1963; Weber et al. 1964, 1966). Discovering that the release of Ca^{2+} from the sacroplasmic reticulum in muscle triggered contraction and that the removal of Ca^{2+} led to relaxation, led to the search for similar releasable endoplasmic reticulum stores of Ca^{2+} in other cells. The recent discovery of inositol trisphosphate-induced release of Ca^{2+} from such stores in non-muscle cells has now established the link between primary stimulus and the source of intracellular Ca^{2+} for cell activation (Streb et al. 1983; Berridge 1984).

The characterization of cell membrane Ca^{2+} pumps, a Ca^{2+} Mg-ATPase (Schatzmann 1966, 1975) and a Na^+-dependent Ca^{2+} extrusion mechanism (Baker and Blaustein 1968; Blaustein 1974), together with voltage-dependent and receptor-operated Ca^{2+} channels (Fatt and Katz 1953; Reuter 1983), has enabled the molecular mechanisms for controlling the efflux and influx of Ca^{2+} into the cell during, and following, cell activation to be identified.

Thus 100 years of experiments on cellular calcium have not only defined the distribution of Ca^{2+} within the cell, but also have identified molecules ideally suited for the regulation of its concentration and for mediating its action within the cell, and have explained how Ca^{2+} can act as an internal trigger of cell activation. How then should one now set about providing the necessary evidence in a system where the role of intracellular Ca^{2+} has yet to be established?

The Modern Experimental Approach

The complete description of the involvement of intracellular Ca^{2+} in the activation of a cell by a primary stimulus, together with its modification by secondary regulators, requires four intimately related phases (Fig. 11.3):

1. Definitive evidence that Ca^{2+} is the internal signal mediating the effect of the primary stimulus.
2. Identification of the source of the Ca^{2+}, together with the molecular basis of its control.
3. Characterization of the molecular basis of how Ca^{2+} activates the cell response.
4. Rationalization of the role of intracellular Ca^{2+} in the action of the primary and secondary regulators, its interaction with other internal signals such as cyclic nucleotides and diacylglycerol, and how this relates to the cell biology of the phenomenon under investigation.

In order to complete the first of these phases, four criteria must be satisfied.

1. The effect of the primary stimulus, or secondary regulator, must be dependent on Ca^{2+}, though not necessarily extracellular Ca^{2+}.
2. There must be a change in free Ca^{2+} somewhere in the cell during activation. Some stimuli may act by altering the affinity of proteins for Ca^{2+}, enabling cell activation to occur without the need for a significant change in intracellular free Ca^{2+} (Takai et al. 1979; Nishizuka 1984; Inouye et al. 1985). However, here Ca^{2+} can be regarded as playing a "passive" role in the activation process since it is not itself responsible for initiating anything, though it must be there for the physiological event to occur. This is an intracellular analogy to the role of Ca^{2+} in blood clotting examined earlier.
3. There must be a change, usually an increase, in Ca^{2+} bound to sites responsible for regulating the phenomenon.
4. There must be a change of flux of Ca^{2+} across either the cell membrane or the membrane of at least one of the organelles within the cell.

Indirect evidence that intracellular Ca^{2+} may be involved in cell activation relies on the acute and long-term effects of manipulation of extracellular Ca^{2+} using Ca^{2+} chelators such as EGTA, and manipulation of intracellular Ca^{2+} by microinjection of EGTA, Ca^{2+}-EGTA or addition of ionophores such as

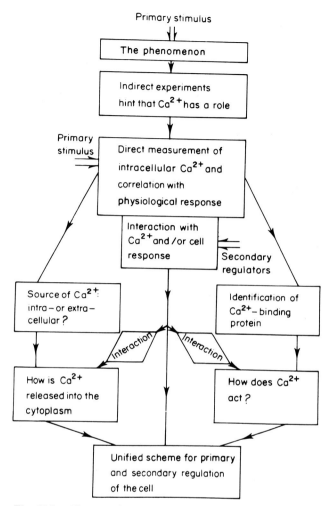

Fig. 11.3. The experimental approach to intracellular calcium (Campbell 1983).

A23187 or ionomycin which make the membrane permeable to Ca^{2+}. Ca^{2+} can also be "microinjected" at μM concentrations into the cell cytoplasm by using fusion of vesicles containing Ca^{2+}, iontophoresis from micropipettes or patched clamp electrodes, or permeabilization of the cell membrane with digitonin or high-voltage discharge. The aim is to mimic the response of the cell to the primary stimulus, to show Ca^{2+} is the only cation which will trigger the cell response and to define the concentration range necessary, expected to be up to about 10 μM free Ca^{2+}.

Attempts to support evidence obtained from such manipulation experiments have been based on measuring $^{45}Ca^{2+}$ influx and efflux and the use of Ca^{2+} "inhibitors".

Interpretation of Ca^{2+} flux data is very difficult without knowledge of the specific activity of the intracellular Ca^{2+} pools. Furthermore, a change in flux on its own is insufficient to show whether this is a cause or a consequence of cell activation, unless the flux change clearly occurs prior to the cellular event. Ca^{2+} "inhibitors" include local anaesthetics; Ca^{2+} channel blockers such as verapamil, D600 and nifedipine; calmodulin inhibitors such as the phenothiazine trifluoperazine; and intracellular organelle effectors such as oligomycin and ruthenium red for mitochondria, and caffeine or dantrolene for the endoplasmic reticulum. The specificity for Ca^{2+} of many of these effectors is questionable. For example, phenothiazines inhibit protein kinase C as well as calmodulin-activated protein kinases (Nishizuka 1984). Ca^{2+} channel blockers often effect both electrically excitable and non-excitable cells. In the latter, increased Ca^{2+} entry appears to occur via receptor-operated Ca^{2+} channels as opposed to those opened by a decrease in membrane potential. There are now many instances where these indirect experiments have led to spurious conclusions about the role of intracellular Ca^{2+} in cell activation. For example, phagocytosis in neutrophils induced by particles is inhibited by removal of external Ca^{2+} and by addition of trifluoperazine, and yet does not appear to involve a rise in intracellular free Ca^{2+} (Hallett and Campbell 1983a). Direct evidence requires the measurement of free Ca^{2+} in the living cell.

The Importance of Measuring Intracellular Free Ca^{2+}

Proof that a rise in free Ca^{2+} in the cell is necessary for cell activation can only be obtained by showing that the primary stimulus causes an increase in free Ca^{2+} prior to the onset of the physiological event, and that prevention of the Ca^{2+} rise also prevents the occurrence of cellular events. It is then possible to establish whether the secondary regulators act by altering the time course or magnitude of the intracellular free Ca^{2+}, or via some other mechanism.

There are five main reasons why measurement of intracellular free Ca^{2+} in intact, functioning, cells is now realized to be of such importance (Ashley and Campbell 1979; Thomas 1982; Campbell 1983; Scarpa 1985). Firstly, it provides direct links between the external signal, the internal signal and the onset and course of the physiological or pathological event. Secondly, by quantifying the free Ca^{2+} change, the range of affinity constants necessary for Ca^{2+}-binding proteins to be effective can be defined. Thirdly, the source and mechanism of control of intracellular Ca^{2+} changes can be identified. Fourthly, Ca^{2+} flux data can be properly interpreted. Fifthly, by careful correlation of the time course of Ca^{2+} changes with the onset of cell injury it is possible to show whether an increase in intracellular Ca^{2+} is a cause or consequence of cell damage, and to show whether therapeutic intervention works through modification of the Ca^{2+} rise.

We have already seen that methods based on measuring Ca^{2+} diffusion and Ca^{2+} buffering in the cell suggested that resting cells had a cytoplasmic free Ca^{2+} of about 0.1 μM. But how can this be measured directly? An indicator is needed, and it must be possible to incorporate it into the cell without damaging its response or altering the electrochemical gradient of Ca^{2+} across the cell membrane. Once this gradient is destroyed it is impossible to observe the natural change in intracellular free Ca^{2+} induced by the primary stimulus.

How to Measure Free Calcium in Living Cells

The Criteria

An ideal indicator of intracellular free Ca^{2+} should satisfy eight criteria:

1. *Signal*. Addition of Ca^{2+} to the indicator must cause a change in one of its properties which will produce a signal from within a cell detectable from outside. A change in the intensity, polarization or colour of light, or a similar change in another part of the electromagnetic spectrum would be easy to measure. Ideally the signal should be detectable at concentrations of indicator which would not significantly affect the Ca^{2+} balance of the cell.

2. *Specificity*. The change in signal must be caused only by Ca^{2+}. Effects of other biological ions (e.g. Na^+, K^+, H^+, Mg^{2+}, Zn^{2+} and other transition metals and anions) on the Ca^{2+} signal should be negligible. Lack of sensitivity to changes in pH is particularly important since intracellular pH can change in phenomena such as egg fertilization.

3. *Incorporation into living cells*. The indicator must be non-toxic, be capable of insertion into the cell without structural, functional or chemical damage, remain stable and evenly distributed in the cell compartment being studied throughout the experiment and not leak out of the cell.

4. *Sensitivity to Ca^{2+}*. The indicator should be sensitive to the entire range of free Ca^{2+} likely to be experienced in the resting, stimulated or damaged cells. Sensitivity to a range of 10 nM–100 μM would be adequate.

5. *Quantification*. It should be possible to calibrate the signal measured from within the cell and convert it to the absolute free Ca^{2+} concentration.

6. *Time course*. The indicator should be capable of responding, without distortion, to fast rises and falls in free Ca^{2+}, e.g. milliseconds in electrically excitable cells, and yet still be able to monitor changes in Ca^{2+} occurring over many minutes, or even hours, in dividing cells, found in blastulae following egg fertilization.

7. *Ca^{2+} distribution in cells*. The indicator should be capable of identifying where within the cytoplasm Ca^{2+} changes occur, and relate this to where the event provoked by the cell stimulus occurs. It should also be possible to monitor free Ca^{2+} changes within organelles, particularly the nucleus, mitochondria and endoplasmic reticulum.

8. *Availability*. The indicator should be readily available in sufficient quantities, not only for one's own experiments, but for others who may wish to confirm and extend them.

No Ca^{2+} indicators satisfy all of these criteria perfectly. Nevertheless several (Fig. 11.4) have provided important information about the control of intracellular free Ca^{2+} during cell activation and cell injury. Let us therefore see how they can detect Ca^{2+} at sub μM concentrations and how they can be used in living cells, remembering that familiar techniques such as atomic absorption and x-ray microprobe analysis are either insufficiently sensitive or incapable of distinguishing free, as opposed to total, Ca^{2+} in a single cell.

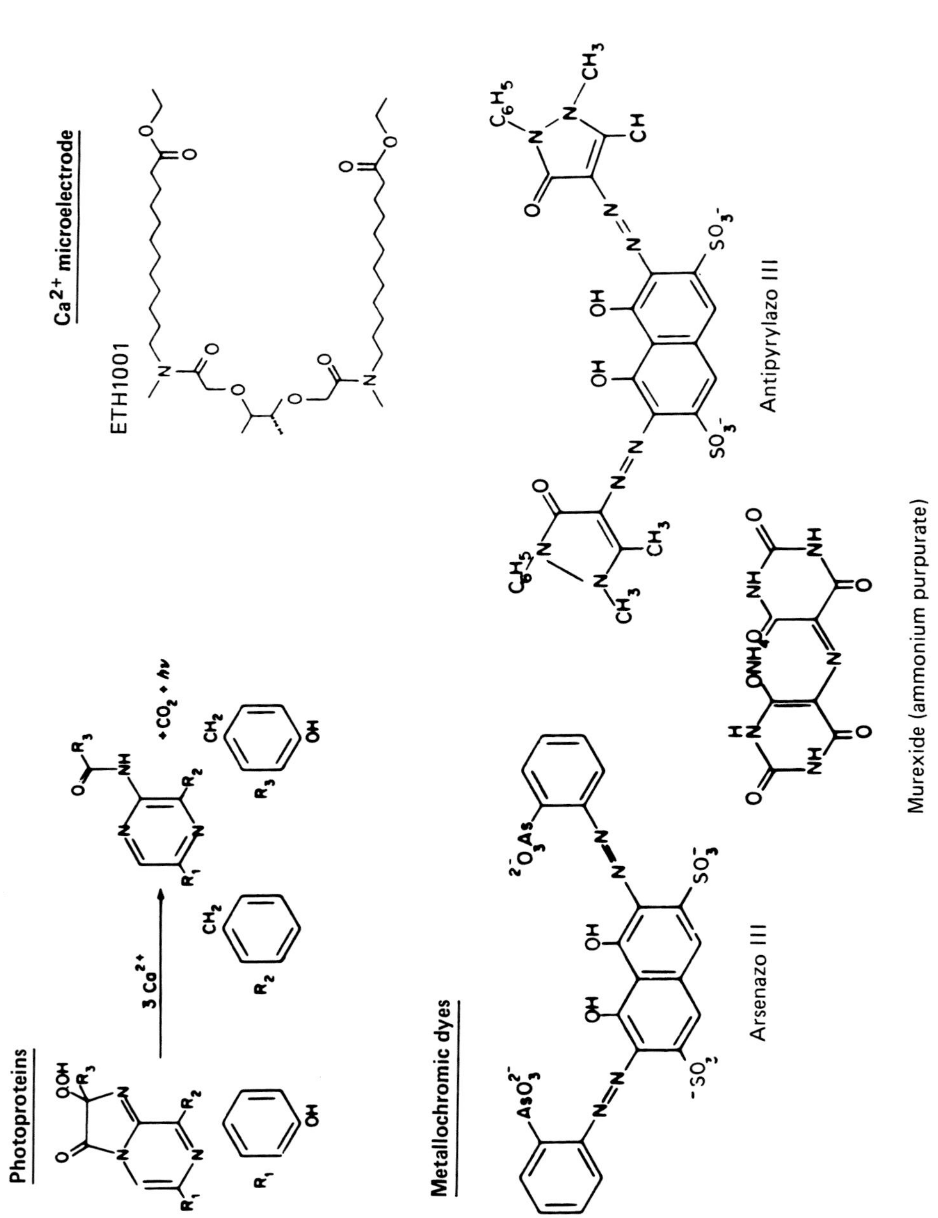

Ca²⁺ microelectrode

ETH1001

Antipyrylazo III

Murexide (ammonium purpurate)

Photoproteins

Metallochromic dyes

Arsenazo III

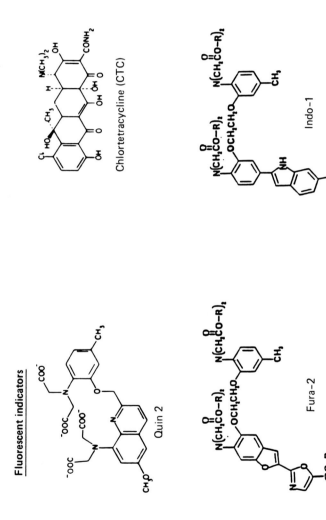

Fig. 11.4. Indicators of intracellular free Ca^{2+}.

The Indicators and How they Work

The first attempt to measure directly free Ca^{2+} in a living cell was carried out by Pollack in 1928 when he injected the dye alizarin sulphonate into an amoeba and observed a shower of red crystals close to the site of pseudopod formation. However it was a further 40 years before the Ca^{2+}-activated photoprotein aequorin was extracted from the luminous jelly fish *Aequorea forskalea*, which proved to be an intracellular Ca^{2+} indicator with the necessary specificity and sensitivity for measuring free Ca^{2+} in any eukaryotic cell (Ridgway and Ashley 1967; Ashley and Ridgway 1968; Blinks et al. 1976, 1982; Campbell et al. 1985a, b). Photoproteins, together with four other classes of Ca^{2+} indicator (Fig. 11.4), have since been developed and have been widely used to measure free Ca^{2+} in a variety of large and small cells. Measurement is based on changes in absorbance, fluorescence, chemiluminescence or electrical potential when the indicator binds Ca^{2+} (Ashley and Campbell 1979; Thomas 1982; Tsien 1983; Scarpa 1985). A group of indicators whose ^{19}F nuclear magnetic resonance (NMR) signal changes when they bind Ca^{2+} has also been developed (Fig. 11.4; Smith et al. 1983). The indicators of intracellular free Ca^{2+} now available are (for reviews see Blinks et al. 1982; Dormer et al. 1985; Scarpa 1985):

1. Absorbance – the metallochromic indicators murexide, antipyrylazo III and arsenazo III (Scarpa et al. 1978).
2. Fluorescence – the tetracarboxylic indicators quin2, fura-2 and indo-1 (Tsien 1980, 1981; Grynkiewicz et al. 1985).
3. Chemiluminescence – the Ca^{2+}-activated photoproteins aequorin and obelin (Blinks et al. 1976; Campbell et al. 1985b).
4. Nuclear magnetic resonance – ^{19}F derivatives of bis(o-amino phenoxy)-ethanetetraacetate (FBAPTA) (Smith et al. 1983).
5. Ca^{2+}-sensitive microelectrodes (Ammann et al. 1979; Ammann 1985).

Metallochromics

The metallochromic indicators change colour when they bind Ca^{2+}. For example arsenazo III turns from purple to blue in the presence of Ca^{2+}. Their relative sensitivities to Ca^{2+} are arsenazo III >antipyrylazo III >murexide, the latter being of no real value in cells. Arsenazo III has been the most widely used, having a reasonable absorption coefficient (25 000 M/cm), a response time apparently of only a few ms and capable of monitoring free Ca^{2+} over the range 50 nM–20 μM. Measurements are usually carried out at two wavelengths simultaneously, e.g. at 675 and 685 nm for arsenazo III, to reduce artefacts from light scattering. The selectivity for Ca^{2+} over Mg^{2+} of 4000 : 1 is only just acceptable. Cells which move or contract cause problems in signal interpretation. Furthermore concentrations of dye at which Ca^{2+} buffering occurs are necessary to obtain a sufficient signal. These indicators also leak out of cells during cell injury. A further factor complicating quantification is that the stoichiometry of arsenazo III : Ca^{2+} is not always a simple 1 : 1.

Fluors

In a field where literally thousands of papers are published each year it is rare to find the work of one individual standing out. Yet it was Roger Tsien who, in the late 1970s (Tsien 1980, 1983), ingeniously designed a family of fluorescent Ca^{2+} indicators. His studies with Timothy Rink established that these fluors had wide-ranging application to the study of intracellular Ca^{2+} in eukaryotic cells.

The calcium chelator, EGTA, has a high selectivity for Ca^{2+} over Mg^{2+}, but is not fluorescent. Its Ca^{2+}-binding capacity is also extremely sensitive to pH over the physiological range. Tsien solved both of these problems by using a methoxyquinolone coupled to a tetracarboxylic acid which he called quin2 (Fig. 11.5). When quin2 binds Ca^{2+} (1 : 1) its fluorescence increases about six times. It is not sensitive to pH in the physiological range since its highest pK_a is 6.5, nor is it seriously affected by physiological (mM) Mg^{2+} concentrations in cells, though its selectivity for Ca^{2+} is only 10 000 times better than for Mg^{2+}, not as good as EGTA. It has a pK for Ca^{2+} of about 0.1 μM so it is half saturated at the free Ca^{2+} concentration of an average cell at rest. Quin2 is suitable for measuring free Ca^{2+} changes over the range 10 nM–1 μM, or even up to perhaps 5 μM. It has also a fast response time to Ca^{2+} changes in

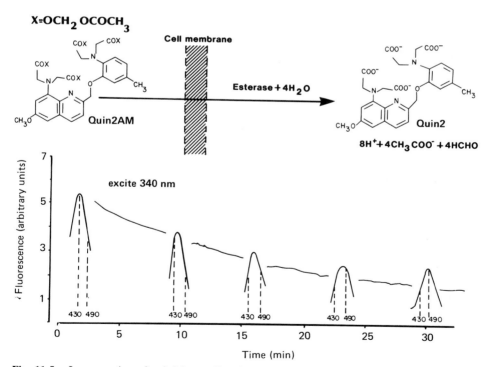

Fig. 11.5. Incorporation of quin2 into cells. The fluorescence emission maximum changes from emission at 430 nm to emission at 490 nm as the ester is converted to the free acid form inside the cell (Hallett and Lucas, unpublished).

the order of milliseconds. Quin2 is excited at 339 nm with fluorescence measured at 490 nm. If fluorescence of quin2 without $Ca^{2+} = F_{min}$ and fluorescence of quin2 all bound to $Ca^{2+} = F_{max}$ then

$$Ca^{2+} = K_d (F - F_{min})/(F_{max} - F_{min}) \tag{1}$$

where $K_d = Ca^{2+}$ dissociation constant and F = measured fluorescence.

However, as Tsien recognized from the outset, quin2 was not ideal. There were two main problems. Firstly, the excitation maximum for quin2 of 339 nm was very close to that for nicotinamide adenine dinucleotide phosphate, reduced (NADPH) in cells, which fluoresces at 460 nm. A wavelength separable from endogenous fluorescence, preferrably as a ratio, would be better. Secondly, quin2 is not a particularly good fluor having an absorption coefficient (ϵ) <5000, and a fluorescence quantum yield (ϕ_F) <0.15 (cf. a good fluor like fluorescein has $\epsilon \doteq 70\,000$, $\phi_F = 0.7$). The solution was to use a stilbene fluorophore instead of a quinoline (Grynkiewicz et al. 1985). The result was the synthesis of two fluors with some 30 times better fluorescence with Ca^{2+} than quin2 called fura-2 and indo-1 (Fig. 11.4). Ca^{2+} alters the excitation spectrum of fura-2 from a peak at 380 nm in the absence of Ca^{2+} to 340 nm when bound to Ca^{2+}. In contrast Ca^{2+} alters the emission maximum of indo-1 from 480 nm to 405 nm. Ca^{2+} has little effect on the excitation or emission peaks of quin2. Thus fura-2 and indo-1 are best quantified using a dual wavelength fluorimeter.

Finally a word about the fluor chlortetracycline (CTC) which has been used by several workers in an attempt to detect changes in intracellular Ca^{2+} (Caswell 1979). This fluor also increases its fluorescence when it binds Ca^{2+}. However it is lipophilic and when added to cells seems to remain in the membrane. Thus *decreases* in CTC fluorescence have often been interpreted as showing release of Ca^{2+} from the membrane. There is no convincing evidence that this interpretation is valid. The method is also non-quantitative, does not measure free Ca^{2+} directly and is susceptible to artefacts as a result of membrane perturbation.

Photoproteins

The photoproteins aequorin and obelin are proteins of molecular weight around 20 000, purified from the coelenterates *Aequorea forskalea* and *Obelia geniculata* respectively. They contain a prosthetic group (Fig. 11.4) covalently linked to the protein. Binding of three Ca^{2+} by the protein triggers an oxidative chemiluminescent reaction leading to the emission of blue light. When the proteins are saturated with Ca^{2+} all of the prosthetic group is consumed within a few seconds. However, the Kd^{Ca} under physiological conditions is about 10 μM so that in the resting cell the consumption rate is very low. With a good photomultiplier tube about 10 tipomol (10^{-20} mol) of photoprotein reacting per second can be detected. This means that the amount of photoprotein required inside cells to produce a detectable light signal does not significantly buffer the intracellular Ca^{2+}. This is in contrast to quin2 where concentrations in the range 0.1–1 mM are necessary in cells to distinguish quin2 fluorescence from the cell's autofluorescence. Even with fura-2, a concentration in the range

10–100 μM is necessary, some 100–1000 times the resting free Ca^{2+} concentration.

Photoproteins also cover a wider range of free Ca^{2+}, from 10 nM–100 μM, and provide an amplification of the Ca^{2+} signal because of the need to bind three Ca^{2+} in order to chemiluminesce. Their response time is about 3–10 ms, obelin being faster than aequorin. The rate constant of photoprotein decay is related to the absolute free Ca^{2+} concentration.

$$\text{light intensity } (I_t) = Q\text{PhP}_O \cdot \text{kapp} \cdot \exp(-\text{kapp})\, t \tag{2}$$

where Q = quantum yield, PhP_O = amount of photoprotein at time O, kapp = apparent rate constant at the particular Ca^{2+} concentration being studied. kapp can be measured in a cell from the relationship:

$$\text{kapp} = I_t/Q\text{PhP}_t \tag{3}$$

where PhP_t = amount of active photoprotein present at time t. The relationship between kapp and free Ca^{2+} is estimated in vitro by constructing a standard curve under ionic conditions equivalent to those in the cell.

NMR

The NMR indicators (Smith et al. 1983) have been synthesized by linking the natural isotope of fluorine, ^{19}F, detectable in very small amounts by NMR, to another of Tsien's compounds, BAPTA, designed originally as a pH-"insensitive" Ca^{2+} chelator for intracellular studies. Binding of Ca^{2+} to FBAPTA (Fig. 11.4) results in a shift in the NMR spectrum. The affinity for Ca^{2+} is similar to that of quin2. Free Ca^{2+} can therefore be estimated from:

$$[Ca^{2+}] = Kd\,(B/F) \tag{4}$$

where Kd = dissociation constant for Ca.FBAPTA, B and F = resonance areas from the NMR spectrum corresponding to Ca.FBAPTA an FBAPTA alone respectively.

In order to obtain a resolvable NMR spectrum, accumulation times of several minutes are usually required for a cell suspension at about 10^8/ml containing 0.1–1 mM intracellular FBAPTA.

Microelectrodes

In 1976 two Ca^{2+}-selective microelectrodes were described, one of a synthetic neutral carrier known as ETHlOOl (Fig. 11.4) (Oehme et al. 1976) and the other of an organic phosphate (Brown et al. 1976). The former have turned out to have superior properties for intracellular Ca^{2+} measurements. Their construction requires covering the tip, about 1 μm diameter, of a glass pipette (the electrode) with a membrane made from PVC containing the Ca^{2+} carrier ETAlOOl, a solvent o-nitrophenyl-octyl ether (o-NPOE) and either tetraphenylarsonium tetrakis (p-biphenyl) borate (TPASTBPB) or sodium tetraphenyl borate (NaTPB). The inner chamber of the electrode is filled with $CaCl_2$. Thus a potential is generated across the membrane when the electrode is immersed in a solution containing Ca^{2+}. Ideally the relationship between the potential and the log of the activity of Ca^{2+} should be linear, as predicted by the Nernst equation. However, at low free Ca^{2+} the relationship deviates from linearity because of interference from other ions. These microelectrodes are sufficiently selective for Ca^{2+} over Mg^{2+}, Na^+, K^+ and H^+ to be used inside cells. Their detection limit for Ca^{2+} is 10–100 nM. However, the resolution time is of the order of seconds, rather slow for quantifying changes in free Ca^{2+} in cells. An intracellular reference electrode is also needed to correct for changes in membrane potential.

How to Incorporate Indicators into Live Cells

Giant cells like squid axon and barnacle muscle are several cm long and can be impaled readily with electrodes and micropipettes enabling Ca^{2+} indicators to be injected into the cell. Much has been learned about the control of intracellular Ca^{2+} from the use of aequorin in these cells (Ashley and Ridgway 1968, 1970; Baker et al. 1971), as well as from the use of arsenazo III (Brinley et al. 1979); and more recently quin2 and fura-2 (Ashley et al. 1985a, b). Some mammalian cells can be impaled with microelectrodes and in skilled hands small numbers of single heart cells, fibroblasts and hepatocytes can be microinjected with photoproteins (Allen and Blinks 1978; Cobbold et al. 1983; Cobbold and Bourne 1985; Woods et al. 1986). However, cells such as human platelets and neutrophils are too small to microinject. Furthermore, whilst single-cell analysis is vital to a full understanding of intracellular Ca^{2+}, there is also the need to incorporate the Ca^{2+} indicator into perhaps a million or more cells in order to correlate intracellular free Ca^{2+} with the physiological response of the whole population. The Tsien fluors, the NMR indicators and the photoproteins do not penetrate healthy cell membranes. A number of ingenious methods have therefore been devised to incorporate them into the cell.

The tetracarboxylate fluors quin2, fura-2 and indo-1, and FBAPTA are added to the cells as the tetraacetoxymethyl ester. This derivative is lipophilic and thus penetrates the cell membrane. Fortunately in the cytoplasm there are esterases which hydrolyse the ester back to the tetracarboxylate which can now bind Ca^{2+} again, and being highly charged does not cross the cell membrane (Fig. 11.5). The conversion of the ester to free acid can be detected by a spectral shift. Thus incubation of cells with μM concentrations of the ester can lead to the accumulation of $0.1 - 1$ mM fluor in the cytoplasm, necessary for detection of the quin2 signal. Of course it may also be possible to inject a single small cell with the unesterified indicator using a micropipette or a patch clamp electrode.

Three methods (Fig. 11.5) have been developed to incorporate photoproteins into the cytoplasm of populations of mammalian cells. These are based on reversible permeation of the cell membrane (Campbell and Dormer 1978; Borle and Snowdowne 1982; Morgan and Morgan 1984; Johnson et al. 1985), vesicle–cell fusion (Hallett and Campbell 1982; Campbell and Hallett 1983) and release from micropinocytotic vesicles (Hallett and Campbell 1983a). These have enabled photoproteins to be incorporated into cells as small as platelets, neutrophils, macrophages, smooth muscle and tissue culture cells.

Whatever Ca^{2+} indicator or method of entrapment is used, valid interpretation of experimental data requires that all of the indicator must be in the cytoplasm. It must not leak out of the cells during the experiment and must be uniformly distributed within and between cells. Also the cells must be viable, being impermeable to trypan blue and having normal content of ATP and lactate dehydrogenase, as well as being able to respond normally to stimuli. A useful way of quantifying cytoplasmic indicator, and distinguishing it from any vesicular pools within the cell, is to activate complement on the cell surface with a cell-specific antibody. This leads to a rise in cytoplasmic free Ca^{2+} of about 10–30 μM within a few seconds, saturating any cytoplasmic indicator but not reacting with any vesicular pool (Campbell et al. 1979a, b; Hallett and Campbell 1983a; Campbell et al. 1985a, b).

Now that we have our indicator in the cytoplasm of viable cells, what of any consequence has been learnt to justify this effort?

Important Consequences of Measuring Intracellular Free Ca^{2+}

Since the late 1960s changes in free Ca^{2+} have been monitored in literally dozens of cells from more than six of the major phyla in the animal kingdom (Table 11.4, Fig. 11.6a–e), as well as several plant cells. Bacteria still pose a problem however. The animal cells include protozoa and slime moulds, echinoderm eggs, nerve and muscle cells from molluscs and arthropods, and a variety of electrically excitable and non-excitable cells from amphibians, fish and mammals. Up to 1980 most of the important new information about intracellular Ca^{2+} came from using Ca^{2+}-activated photoproteins, mainly aequorin. However, the invention of the tetracarboxylate fluors by Tsien has given large numbers of workers access to indicators which can be used in many small cells, and can be detected with apparatus already available in most laboratories. Early workers using aequorin had to construct their own chemiluminometers, though sensitive equipment is now commercially available. Many important discoveries would not have been made without these direct measurements in living cells, and there have been some surprises. Eight significant features can be identified.

Identification of Primary Stimuli Dependent or Independent of a Rise in Free Ca^{2+}

Many electrical and chemical stimuli have been identified which cause a rise in cytoplasmic free Ca^{2+} prior to cellular events and where prevention of the Ca^{2+} rise, by chelation of intracellular Ca^{2+}, by depletion of the intracellular Ca^{2+} stores or by removal of external Ca^{2+}, prevents the cell response. Photoproteins have shown the need for a rise in intracellular free Ca^{2+} in skeletal, cardiac and smooth muscle activated electrically or by neurotransmitter (Ashley and Ridgway 1970; Allen and Blinks 1978; Morgan and Morgan 1982, 1984); in exocytotic pancreatic cells activated by carbamyl-choline (Dormer 1984); in platelet aggregation and secretion activated by thrombin (Johnson et al. 1985); in oocyte maturation (Moreau and Guerrier 1979); in egg fertilization (Gilkey et al. 1978); and in activation by chemotactic factors of reactive oxygen metabolite production by neutrophils (Fig. 11.6b)

Table 11.4. Some examples of cells in which intracellular Ca^{2+} has been measured

Phylum	Species (common name and cell)	Cell volume (approx.)	Indicator	Reference
Animal kingdom				
Protozoa	*Chaos* (amoeba)	20 nl	Aequorin	Taylor et al. (1975)
Echinodermata	*Marthasterias* (star fish egg)	1.8 nl	Aequorin	Moreau and Guernier (1979)
Arthropoda	*Balanus* (barnacle muscle)	>12 μl	Aequorin	Ashley and Ridgway (1968, 1970)
			Fura-2	Ashley and Timmerman (1986), Ashley et al. (1985a,b)
			Microelectrode	Ashley et al. (1978)
	Chironomus (midge salivary gland)	0.5 μl	Aequorin	Rose and Loewenstein (1975)
Mollusca	*Loligo* (squid giant axon)	60 μl	Aequorin	Baker et al. (1971)
	Aplysia (sea hare)	1.2 μl	Arsenazo III	Gorman and Thomas (1978)
Vertebrate	*Xenopus* (toad egg)	1 μl	Aequorin	Baker and Warner (1972)
	Oryzia (fish egg)	0.4 μl	Aequorin	Gilkey et al. (1978)
	Homo (human platelet)	0.1 pl	Quin2	Rink et al. (1983)
	Homo (human neutrophil)	0.5 pl	Aequorin	Johnson et al. (1985)
			Quin2	Lew et al. (1985)
			Obelin	Hallett and Campbell (1982)
	Rattus (rat neutrophil)	0.5 pl	Fura-2	Tsien et al. (1985)
	Mus (mouse thymocyte)	0.5 pl	Obelin	Campbell and Dormer (1978)
	Columbia (pigeon erythrocyte)	0.5 pl	Aequorin	Woods et al. (1986)
	Rattus (rat hepatocyte)	4 pl		
Plant kingdom				
	Chara (alga)	1 μl	Aequorin	Ashley and Williamson (1981)
	Haemanthus (African blood lily)	>1 μl	Quin2	Keith et al. (1985)

These are but a small representative of the cells in which free Ca^{2+} has been measured directly, numbering now more than 50 cell types. For further references see Ashley and Campbell (1979); Blinks et al. (1982); Campbell (1983); Scarpa (1985).

(Hallett and Campbell 1982; Campbell and Hallett 1983a). Similarly quin2, and more recently fura-2, have demonstrated a role for intracellular free Ca^{2+} in platelet shape change and aggregation activated by ADP (Hallam and Rink 1985) (Fig. 11.6e); in smooth muscle activated by angiotensin II or vasopressin (Capponi et al. 1985); in liver glycogenolysis activated by adrenergic stimuli, angiotensin II or vasopressin (Williamson et al. 1985), by growth factors acting on 3T3 cells and by T cell mitogens on lymphocytes (Tsien et al. 1982); in neutrophil activation by chemotactic peptides (Lew et al. 1985a); and in the activation of many exocrine and endocrine secretory cells (Powers et al. 1985).

In some cells stimuli have been identified that do not require an increase in free Ca^{2+} but whose effects can be inhibited by the so-called "Ca^{2+} inhibitors" such as trifluoperazine (Hallett and Campbell 1985). Examples are found in platelets (Rink et al. 1982) and in the activation of reactive oxygen metabolite production in neutrophils by certain phagocytic stimuli from phorbol esters (Campbell and Hallett 1983a, b; Di Virgilio et al. 1984; Lew et al. 1985a). These are thought to work by activation of protein kinase C (Nishizuka 1984), which is itself activated by diacylglycerol or phorbol ester. The enzyme requires

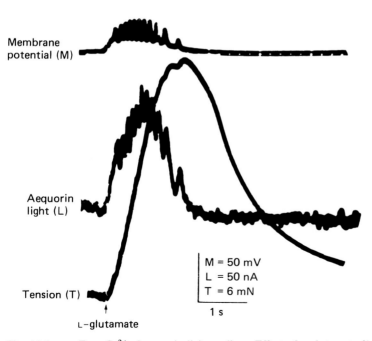

Membrane potential (M)

Aequorin light (L)

Tension (T)

M = 50 mV
L = 50 nA
T = 6 mN

1 s

L-glutamate

a

Fig. 11.6a–e. Free Ca^{2+} changes in living cells. **a** Effect of L-glutamate (1 mM) on an isolated muscle fibre from the barnacle *Balanus nubilus*, injected with aequorin (Ashley and Campbell 1978).

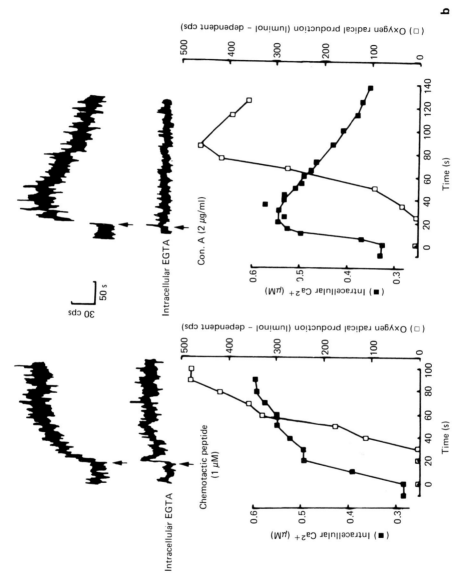

Fig. 11.6 (*continued*) **b** Effect of *N*-formyl met leu phe (chemotactic peptide) and concanavalin A on neutrophil–erythrocyte "ghost" hybrids containing obelin (Campbell and Hallett 1983).

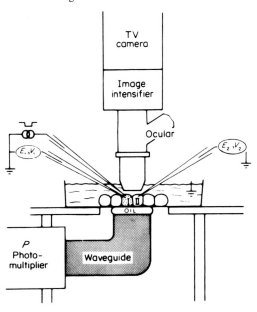

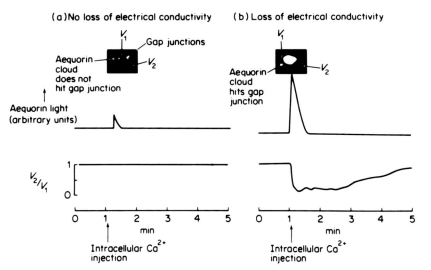

Fig. 11.6 (*continued*) **c** Visualization of free Ca^{2+} in salivary gland cells injected with aequorin, and correlation with gap junction conductance (Rose and Loewenstein 1975).

Ca^{2+}. However, since activation appears to increase the enzymes's affinity for Ca^{2+} it is activated maximally at resting free Ca^{2+} of around 0.1 μM, so no increase in free Ca^{2+} is necessary or has any effect on enzymatic activity. Fusagenic viruses cause a rise in intracellular free Ca^{2+}. However, it appears

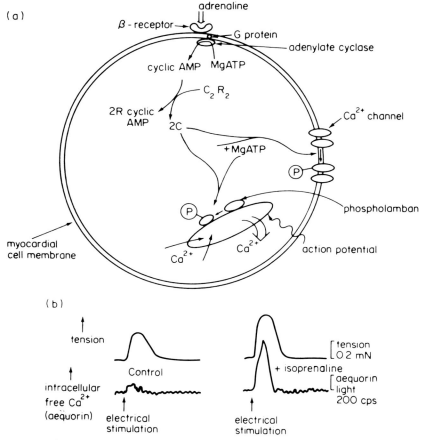

Fig. 11.6 (*continued*) **d** Effect of a β-adrenergic agonist on free Ca^{2+} transients, measured using aequorin in heart muscle (Allen and Blinks 1978 from Campbell 1983).

that this is not an absolute requirement for cell fusion, but rather increases the rate at which fusion occurs (Hallett and Campbell 1982).

Correlation of Cell Response with the Time Course and Absolute Free Ca^{2+} Rise

Comparisons of the absolute rises in free Ca^{2+} in neutrophils provoked by chemotactic peptide and the Ca^{2+} ionophore A23187 with reactive oxygen metabolite production, or of free Ca^{2+} in platelets activated by ADP or the Ca^{2+} ionophore ionomycin with shape change and aggregation (Hallam and Rink 1985), have shown that for a given cell response a lower intracellular free Ca^{2+} is required for the natural stimulus compared with the ionophore. This

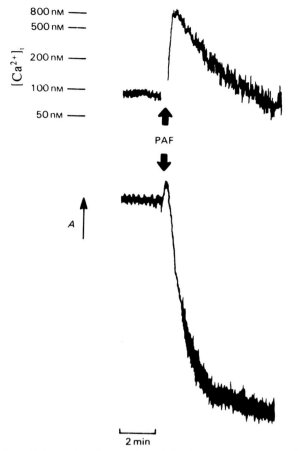

Fig. 11.6 (*continued*) **e**   Effect of platelet-aggregating factor (PAF) on free Ca^{2+}, measured using quin2, and aggregation (A decreasing) in human platelets (Hallam et al. 1984).

may indicate that the natural stimulus either produces some other "Ca^{2+} potentiating" factor or it produces only a local change as opposed to an increase throughout the whole cell cytoplasm.

Characterization of Secondary Regulators and Drugs

The positive ionotropic action of adrenaline on cardiac muscle is at least partly due to an increase in the magnitude and faster time course of the Ca^{2+} transient provoked by the action potential in the cell membrane (Allen and Blinks 1978) (Fig. 11.6d). A similar effect may occur in neutrophils where adenosine inhibits the action of stimuli dependent on, but not those independent of, a rise in intracellular free Ca^{2+} (Roberts et al. 1985).

Requirement of an Intracellular Ca^{2+} Store

Removal of external Ca^{2+} has little acute effect on the stimulation of skeletal muscle contraction, of platelet aggregation by ADP, of pancreatic exocrine secretion provoked by carbamyl choline or cholecystokinin or of egg fertilization by sperm. Yet by using an intracellular Ca^{2+} indicator it has been possible to show not only that the natural stimulus causes a rise in intracellular free Ca^{2+} but also that this rise still occurs after removal of extracellular Ca^{2+}. Only when the internal store is deliberately depleted, for example using no external Ca^{2+} plus a Ca^{2+} ionophore for several minutes, is there a massive reduction in the stimulated Ca^{2+} rise and in the cell response. In contrast removal of external Ca^{2+} virtually abolishes histamine secretion from mast cells induced by antigen binding to IgE receptors, or secretion of neurotransmitter from nerve terminals, and markedly inhibits chemotactic activation of neutrophils, electrically activated contraction of smooth muscle and activation of several endocrine cells. Here a reduction, by removal of external Ca^{2+}, of the Ca^{2+} transient can establish a need for both release from an intracellular Ca^{2+} store and a rapid influx of Ca^{2+} through the cell membrane.

Measurement of free Ca^{2+} in intact cells with quin2, fura-2 or Ca^{2+} release from digitonin-permeabilized cells detected with a Ca^{2+} electrode, has played a vital part in the identification of inositol trisphosphate as the signal from the cell membrane responsible for releasing Ca^{2+} from internal stores in many cells (Berridge 1984; see p. 297).

The Ranges of Intracellular Free Ca^{2+}

Estimation of free Ca^{2+} in resting cells has varied from 20 nM (Dipolo et al. 1976) to 430 nM (O'Doherty and Stark 1982) depending on the cell type and method used. The mean cytoplasmic free Ca^{2+} rises to 0.5–5 μM on stimulation of the cell, the level being related to the dose of stimulus. Transient, higher levels may occur locally, for example in egg fertilization (Gilkey et al. 1978). The rise in free Ca^{2+} may occur within milliseconds of the primary stimulus and be over in less than a second, as in a muscle twitch; or the onset may be delayed and remain high for many minutes or even up to an hour. Injury to the cell membrane, for example by the membrane attack complex of complement or viruses (Campbell et al. 1979a, b; Hallett et al. 1981), can lead to prolonged elevation in the pathological range of 5–30 μM. Above 50–100 μM intracellular free Ca^{2+} the cell appears to die very quickly.

The Importance of Studies on Single Cells and Intracellular Ca^{2+} Distribution

The pioneering studies of Reynolds (see review in 1978) using image intensification to visualize aequorin signals from within single cells have demonstrated how vital quantification of free Ca^{2+} in single cells and its distribution within the cell will be in the future. Using aequorin in fish eggs (Gilkey et al. 1978) or sea urchin eggs (Eisen and Reynolds 1985; Eisen et al. 1984), waves or clouds of Ca^{2+} can be seen moving across or through the cell following activation by sperm. The time course and location of the Ca^{2+} change can be correlated with cellular events at a particular place in the cell, for example membrane depolarization, increased NAD(P)H fluorescence or exocytotic cortical granule release. Visualization, using aequorin, of artificially induced Ca^{2+} changes in *Chironomus* salivary gland have shown that only when the Ca^{2+} reaches a gap junction is the electrical conductivity between two adjacent cells switched off (Rose and Loewenstein 1975, 1976) (Fig. 11.6c). More recently computerized imaging has been possible with quin2 (Keith et al. 1985), and fura-2 (Poinie et al. 1985; Williams et al. 1986). The former study showed Ca^{2+} gradients in the cytoplasm of a plant cell with elevation close to the mitotic spindle at anaphase. The latter study apparently showed a Ca^{2+} gradient across the nuclear membrane in cultured mammalian cells.

Single Cell Analysis and the "Threshold" Phenomenon

By measuring the mean Ca^{2+} concentration at which a phenomenon occurs in a cell population or a single cell it is possible to show that the level of Ca^{2+}, or its location, is necessary for a cellular event to occur. Thus, using aequorin in starfish oocytes, it was shown that meiosis stimulated by the maturation hormone 1 methyl adenine only occurred at 1 methyl adenine concentrations >0.2 μM, yet lower concentrations were still capable of causing a transient intracellular Ca^{2+}. Similarly in muscle, and in platelets (Rink and Hallam 1984), the "Ca^{2+} threshold" for cell activation has been defined. The occurrence of Ca^{2+} gradients in cells (Brownlee and Wood 1986; Tsien and Poinie 1986) provides an obvious mechanism for initiating a threshold. Only when the Ca^{2+} cloud hits the intracellular target will the cellular event occur.

Studies in single cells using aequorin in hepatocytes (Woods et al. 1986), or fura-2 in thymocytes (Tsien et al. 1985), have shown oscillations in free Ca^{2+} during hormone action or cell division. The intriguing possibility exists that bursts of Ca^{2+} transients in small cells may allow the threshold for cell activation to be maintained without the large mitochondrial Ca^{2+} uptake or cell efflux which would occur if the free Ca^{2+} rise was sustained. Thus, intracellular Ca^{2+}-dependent thresholds may control which cellular phenomenon is switched on, the nature of activation once the cell is activated as well as the number of cells switched on.

Intracellular Ca^{2+} and Cell Pathology

Correlation of intracellular free Ca^{2+} changes with the onset of cell damage, detected biochemically or morphologically, can establish Ca^{2+} as a cause or consequence of cell injury. For example blebing in individual myocytes induced by CN^- or 2-deoxyglucose was detectable well before any aequorin-monitored Ca^{2+} rise (Cobbold and Bourne 1985). Similarly the rise in free Ca^{2+} induced by the membrane attack complex of complement is at least partly responsible for the ability of nucleated cells to protect themselves against complement attack by removing the potentially lethal complex via vesiculation (Campbell and Morgan 1985; Morgan and Campbell 1985).

Which Intracellular Ca^{2+} Indicator is Best?

The simple answer to this question is none. All have their merits and problems. All have contributed something, in some cell or other, to our understanding of the role and control of intracellular Ca^{2+}. The two most valuable are the Tsien fluors and the Ca^{2+} activated photoproteins. The latter, though less easy to obtain and use, still have qualities which suggest that it will be some time before they are completely superseded by the tetracarboxylate fluors. Apoaequorin has been cloned (Inouye et al. 1985; Prasher et al. 1985), and can be reactivated using a synthetic prosthetic group (Shimomura and Johnson 1976; Campbell et al. 1981b). This will not only lead to the wider availability of the photoproteins, but also should enable their most serious disadvantage to be circumvented, that is, consumption.

Good points for the photoproteins aequorin and obelin are (a) the high sensitivity of the chemiluminescence analysis, so that they can be used under conditions where little or no buffering of intracellular Ca^{2+} occurs, (b) the amplification of the Ca^{2+} signal enables small localized changes in free Ca^{2+} to be detected, (c) the Ca^{2+} range is wide, 10 nM–100 μM, (d) the signal can be visualized by image intensification, (e) they are non-toxic and (f) they do not leak out of cells easily, a feature of particular importance in studying cell injury. Their disadvantages are (a) consumption of prosthetic group, particularly at high Ca^{2+} concentrations, (b) non-linearity and interference from H$^+$, Na$^+$, K$^+$ and Mg^{2+} complicating calibration curves, (c) the rather extreme methods necessary for incorporation into small cells and (d) the lack of easy availability at present.

The Tsien fluors are commercially available as free acids or the membrane permeant acetoxy methyl esters. Their signals are easy to quantify and visualize. They can be incorporated easily into most, though not all, mammalian cells. However, there are a number of potential problems which may not all be serious, but must be checked in each cell system. Not all cells have the esterase necessary for entrapment, e.g. mast cells. Large extracellular conversion can occur during loading. The formaldehyde released as a result of ester hydrolysis does not appear to be toxic. However, Ca^{2+} buffering by 0.1–1 mM intracellular quin2 may reduce or abolish the cell response. The more sensitive fura-2 may alleviate this problem, though it will still have to be used at concentrations which buffer some Ca^{2+}. Lack of Ca^{2+} signal amplification can result in missing a small, localized change in intracellular free Ca^{2+}. Their precise intracellular location needs closer examination. Certainly much is found in the cytoplasm. However, many cells have granules containing esterases which could accumulate the indicator, and fura-2, being more lipophilic than quin2, may diffuse across organelle membranes. Leakage or secretion may occur during cell activation, and will certainly occur in many types of cell injury. In these cases the "intracellular" signal must be corrected for any indicator outside or within granules exposed to high Ca^{2+}, e.g. by fusion, during cell activation or attack. A final worry is the effect of transition metals. Many eukaryotic cells contain μM–mM total Zn^{2+}. Zn^{2+} binds avidly to quin2 with a K_d of 26 pM, and to FBAPTA with a K_d of 8 nM. Zn^{2+}, unlike Ca^{2+}, has little effect on the fluorescence. Thus formation of zinc–quin2 will reduce the estimated free Ca^{2+}, and may account for some of the low levels, <50 nM, reported using this indicator. Transition metals are not good stimulators of photoprotein chemiluminescence.

The aim of this section has been to examine in some detail what has become, in the 1980s, the cornerstone for establishing and characterizing the role of free Ca^{2+} in living cells. We are now in a position to find out where the Ca^{2+} comes from, how this is regulated and what it does.

The Source of Intracellular Calcium for Cell Activation

From Outside or Inside?

In principle a rise in cytoplasmic free Ca^{2+} could occur through release from internal stores, increased permeability of the cell membrane or inhibition of the pumps responsible for efflux or uptake into intracellular organelles. Whilst it has been agreed that all three processes occur in many cells where a rise in intracellular Ca^{2+} is the trigger for cell activation, there have been two major controversies. Firstly, how much Ca^{2+} is released internally relative to that coming through the cell membrane? And secondly, is the endoplasmic reticulum or the mitochondrion the major source of releasable Ca^{2+} during cell activation? The conceptual breakthrough enabling these questions to be answered was the realization that there is an ordered sequence of events within the cell following cell activation. The two experimental breakthroughs have been the measurement of the time course of changes in intracellular free Ca^{2+} during this sequence, together with the discovery of the molecular basis of Ca^{2+} release from the endoplasmic reticulum.

How much Ca^{2+} actually has to move to initiate and sustain cell activation? The concentration of Ca^{2+}-binding proteins such as calmodulin found in all cells and troponin C found in muscle which are responsible for mediating the effect of the rise in internal free Ca^{2+} can be anything between 1 and 100 μM. Since they have four Ca^{2+}-binding sites, full activation could require up to 400 μM Ca^{2+}. In muscle this is the case. Even in small cells where the Ca^{2+}-binding protein involved is at much lower concentration, the *total* Ca^{2+} movement during activation will often be some 10 to a 100 times greater than the maximum rise in *free* Ca^{2+}. An unresolved puzzle is found in experiments using quin2. Free Ca^{2+} changes are apparently normal yet the cell has an additional 0.25–0.5 mM Ca^{2+} buffering capacity. Thus the total Ca^{2+} movement in these experiments must be some 4–10 times greater than in normal cells not containing quin2.

The current evidence, based on experiments removing external Ca^{2+}, depleting internal Ca^{2+} stores, or inhibiting release from them, is that the bulk of the intracellular Ca^{2+} for cell activation in most cells comes from an internal store, probably a portion of the endoplasmic reticulum. This is true not only in cells such as skeletal muscle, platelets, fertilized or maturing eggs and exocrine pancreas where removal of external Ca^{2+} has no immediate effect on internal Ca^{2+} transients or cell stimulation, but also in cardiac and smooth muscle, nerve terminals, endocrine cells, hepatocytes and neutrophils where the Ca^{2+} transient and cell response can be either abolished immediately or at least reduced, by removing external Ca^{2+}.

The sequence of events for Ca^{2+}, from the appearance of the stimulus to recovery of the cell after its disappearance, involves four phases. Two classes of cell type can be identified, the difference being that the order of the first two phases is reversed.

Class A, e.g. skeletal muscle, platelets, liver, exocrine pancreas

Phase 1: Rise in Cytoplasmic Ca^{2+}. Response of the cell membrane to the primary stimulus leads to the production of a second messenger or signal which releases Ca^{2+} from an internal store into the cytoplasm. As soon as Ca^{2+} rises in the right part of the cell, activation begins.

Phase 2: Increased Ca^{2+} Influx. In order to maintain the rise in free Ca^{2+} and sustain the cell response there is an increase in the permeability of the cell membrane to Ca^{2+} resulting in an increase in Ca^{2+} influx, together with an inhibition of the cell membrane Ca^{2+} pump(s) (Ca^{2+} Mg-ATPase and/or Ca^{2+}/Na^+ exchange). These two effects combine together to reduce loss of Ca^{2+} from the cell resulting from the rise in its concentration in the cytoplasm.

Phase 3: Reduction in Cytoplasmic Ca^{2+} (only in Cells where Stimulus is Sustained). In the continued presence of the stimulus, the cytoplasmic free Ca^{2+} may return to near resting levels, though the internal Ca^{2+} store may now be virtually depleted. Down-regulation of receptors may also occur, but the cell response reduces more slowly since covalent modifications of internal proteins maintain it.

Phase 4: Recovery of Cell Ca^{2+}. Removal of the stimulus allows the cell to recover. The increased Ca^{2+} influx remains, or may be further enhanced in some cells, to allow rapid replenishment of internal Ca^{2+} stores. There may also be a relocation of Ca^{2+} from the mitochondria, which has accumulated it during phases 1 and 2, back to the endoplasmic reticulum.

Thus in the twitch of a skeletal muscle, the action potential generated at the endplate by acetyl choline passes on an electrical signal to the sarcoplasmic reticulum which releases Ca^{2+} within milliseconds. As the Ca^{2+} is released it binds rapidly to troponin C, close to the site of release, which allows contraction to occur. Woodward (1949) showed an increased influx of $^{45}Ca^{2+}$ in a sustained contraction. Following repolarization, the Ca^{2+} is pumped back rapidly into the sarcoplasmic reticulum via a Ca^{2+}–Mg-ATPase. In liver the phases are similar but the time scale and molecular mechanisms are different from skeletal muscle. Binding of α-adrenergic stimuli, vasopressin or angiotensin to their respective membrane receptors leads to production, within 1–3 s, of a phospholipid-derived, soluble second messenger, inositol trisphosphate (p. 297). This in turn causes release of Ca^{2+} from an endoplasmic reticulum Ca^{2+} store leading to an elevation of free Ca^{2+}, detectable within 5 s and maximal within 6–10 s. The Ca^{2+} activates the calmodulin on phosphorylase kinase, which then activates phosphorylase by catalysing conversion of the "b", nonphosphorylated, to the "a", phosphorylated, form. Phosphorylase activation is detectable within 15 s and maximal within 30–60 s. The result is increased glucose release from glycogen. The internal Ca^{2+} pool is depleted within 3–4 minutes, though increased Ca^{2+} influx and inhibition of the cell membrane Ca^{2+}–Mg-ATPase have occurred prior to this. This sequence of Ca^{2+} events explains why $^{45}Ca^{2+}$ influx and efflux data have, in the past, been so difficult to understand, particularly when no method for monitoring cytoplasmic free Ca^{2+} was available. The initial elevation in cytoplasmic free Ca^{2+} will lead to increased $^{45}Ca^{2+}$ efflux if the store from which it originates is radioactive, but the sustaining of the free Ca^{2+} rise may then lead to a reduction in measured $^{45}Ca^{2+}$ efflux, particularly since the specific activity of the cytoplasmic pool will be reduced as a result of influx of non-radioactive Ca^{2+} from outside (see Campbell 1983).

Class B, e.g. heart muscle, smooth muscle, nerve terminals and neutrophils

Phase 1: Rise in Cytoplasmic Ca²⁺. Electrical or chemical primary signals at the cell membrane cause a rapid increase in its permeability to Ca^{2+} leading to a rise in intracellular free Ca^{2+} near to the inner surface. Second messengers also may be released into the cytoplasm from the cell membrane.

Phase 2: Release of Ca²⁺ from Internal Store. There is a large release of Ca^{2+} from an internal store leading to cell activation.

Phase 3: Increased Ca²⁺ Influx. Sustained stimulation may lead to inhibition of Ca^{2+} efflux.

Phase 4: Recovery of Cell Ca²⁺. Cell recovery after the removal of the stimulus requires replenishment of the intracellular Ca^{2+} store and closing of the activated influx pathway.

Thus, in heart muscle the action potential causes opening of voltage-sensitive Ca^{2+} channels and the resulting small rise in intracellular Ca^{2+} may itself cause release of Ca^{2+} from the endoplasmic reticulum which then stimulates the cell to contract. Rapid removal of Ca^{2+} back into the internal store, together with closing of the Ca^{2+} channels in the cell membrane, allow the myofibrils to relax thus completing the beat. Repolarization of the cell membrane makes the cell ready for the next action potential. In nerve terminals where the action potential conducted from the axon causes similar Ca^{2+} channels to open it is not clear whether all of the Ca^{2+} for provoking neurotransmitter release comes from outside, or whether there is also a rapidly releasable internal Ca^{2+} store as in cardiac and smooth muscle. Thus in neurosecretion, phase 2 may not be necessary, sufficient intracellular Ca^{2+} being made available through phase 1.

In view of the importance of release of Ca^{2+} from internal stores in both electrically excitable and non-excitable cells, the question arises, what triggers this Ca^{2+} to be released?

Release from Internal Stores

There has been much speculation since the late 1950s about the possible release into the cytoplasm of Ca^{2+} from endoplasmic reticulum, mitochondria or the inner surface of the plasma membrane, particularly in non-muscle cells. The nucleus, cytoplasmic Ca^{2+}-binding proteins or small organic ligands, and secretory vesicles known to contain a Ca^{2+} pump, also contain bound Ca^{2+} which could, in principle, be released into cytoplasm for cell activation. However, these three other candidates have received much less attention. The Ca^{2+} pump (Ca^{2+}/H^+ exchange) on secretory vesicles found for example in cholinergic nerve terminals, platelets or the adrenal medulla may be required either to get Ca^{2+} into the vesicle where it plays a structural role or as a mechanism for removing Ca^{2+} during the recovery of the cell from stimulation.

The reduction of membrane-bound chlortetracycline fluorescence in several activated cells led to the suggestion that Ca^{2+} might be released from the inner surface of the plasma membrane. The fact that inositol phospholipids can bind Ca^{2+} and are broken down during the activation of many cells (Michell 1975) provided a potential mechanism for such release. However, the total amount of Ca^{2+} on the inner surface of the plasma membrane, given the low affinity of these phospholipids for Ca^{2+} relative to proteins such as calmodulin, and the low free Ca^{2+} in the cytoplasm, makes it unlikely that there would be enough Ca^{2+} to elevate significantly the concentration of Ca^{2+} in the cytoplasm.

The first clues that mitochondria might take up Ca^{2+} came from experiments carried out in the 1940s and 1950s (see Campbell 1983 for references). By the mid-1950s it became clear that mitochondria have the capacity to take up large quantities of Ca^{2+} from the fluid surrounding them (Lehninger 1962; Chance 1965; Carafoli and Crompton 1976) using the electrochemical H^+ gradient generated by the respiratory chain, and at the expense of ATP synthesis. Mitochondrial Ca^{2+} uptake thus increases respiration. Unfortunately, up until the late 1970s nearly all the experiments on mitochondrial Ca^{2+} uptake used fluid Ca^{2+} concentrations in the range 50–200 μM, some 500–2000 times that surrounding mitochondria in the resting cell. In fact in the cell there is a "balance point" for mitochondrial Ca^{2+} of about 0.1–0.3 μM cytoplasmic free Ca^{2+} where mitochondrial Ca^{2+} uptake balances release. When the cytoplasmic free Ca^{2+} rises to 5 μM on cell activation the mitochondria try to buffer it, releasing again the Ca^{2+} taken up as the cell returns to its resting state. Polyamines like spermine may enhance cycling of Ca^{2+} through the mitochondria (Nicchitta and Williamson 1984). In spite of the discovery in vitro of a Na^+-activated Ca^{2+} release from the mitochondria of heart and squid axon, but not of liver, there is no convincing evidence that mitochondria release Ca^{2+} into the cytoplasm as the signal for cell activation. They do however act as an internal Ca^{2+} sink and buffer, restricting free Ca^{2+} changes to particular localities within the cytoplasm. Furthermore, several mitochondrial enzymes, such as the kinase and phosphatase of pyruvate dehydrogenase (Severson et al. 1976; Denton et al. 1980; Denton and McCormack 1985), can be regulated by μM Ca^{2+}, this being under the control of hormones such as insulin.

The principal source of Ca^{2+} for cell activation in all eukaryotes is now thought to be a special portion of the endoplasmic reticulum.

The sarcoplasmic reticulum in muscle was first observed in the light microscope by Verati in 1902, and in fine detail by Bennett and Porter in 1953. In 1951 March found a factor in muscle homogenates which modified contraction. The relaxing factor was subsequently shown to be vesicular fragments of the sarcoplasmic reticulum, requiring Mg-ATP^{2-} to remove Ca^{2+} from the surrounding fluid (Ebashi 1961). As much as 90% of the sarcoplasmic reticulum membrane protein may be the Ca^{2+}-activated Mg-ATPase, molecular weight about 102 000, capable of pumping two Ca^{2+} into the lamella of the reticulum for each ATP hydrolysed. The membrane contains so much ATPase that a single turnover of each enzyme molecule may be sufficient to pump all of the Ca^{2+} released for contraction back into the reticulum.

Several Ca^{2+}-binding proteins have been isolated from the sarcoplasmic reticulum, and are thought to be responsible for sequestering Ca^{2+} within the lamella (MacLennan et al. 1972). The most important of these Ca^{2+}-binding proteins is calsequestrin. One part of the sarcoplasmic reticulum is very close to (12–20 nm), but not touching, the T tubule of the muscle's plasma membrane. The rapid depolarization of the plasma membrane induced by the action potential leads to an approximately 100 mV depolarization of the reticulum membrane which, as a result, releases about 200 μM Ca^{2+} in a few milliseconds. The Ca^{2+} is pumped back within 100–200 ms into the reticulum, at a different site from where release occurred during the relaxation phase. As the free Ca^{2+} drops, Ca^{2+} dissociates from the troponin C. Parvalbumins, the high affinity Ca^{2+}-binding proteins found in the cytoplasm of many muscles, may act as a temporary sink for Ca^{2+} during both the contraction and relaxation phases.

A similar process occurs in heart muscle. However, in this case the small amount of Ca^{2+} entering the cell membrane through the voltage-sensitive Ca^{2+} channels appears to trigger release of Ca^{2+} from the reticulum. Furthermore the Ca^{2+}-pumping capacity of the reticulum in heart is increased by phosphorylation of the protein phospholamban, catalysed by cyclic AMP-dependent protein kinase (protein kinase A) when the myocytes are exposed to β-adrenergic agonists.

There are thus two mechanisms for stimulating Ca^{2+} release from sarcoplasmic reticulum – one electrical, found in skeletal muscle, the other Ca^{2+} induced, found in heart muscle (Martonosi 1984). There is a third, found in the endoplasmic reticulum of smooth muscle and non-muscle cells – inositol trisphosphate (Berridge 1984; Berridge and Irvine 1984).

In 1953 Hokin and Hokin discovered an increased turnover during cell activation of a group of minor phospholipids, accounting for only about 1% of the total in the membrane, the phosphatidyl inositides (Fig. 11.7). The pathway for these phospholipids involves the formation of phosphatidyl inositol from CDP-diacylglycerol and inositol, followed by the phosphorylation, using ATP, of two of the inositol hydroxyls to form phosphatidyl inositol 4,5-bisphosphate (PIP$_2$). Note that bis and tris designate phosphorylation at different sites as opposed to bi- and tri- where the phosphates are in a linear, linked chain. By the mid-1970s Michell (1975) and others had shown that the activation of smooth muscle cells, liver cells, secretory cells and other cells resulted in the activation, within a few seconds, of a phosphodiesterase with phospholipase C-like action. The result is the formation of two key intracellular regulators, diacylglycerol within the membrane and inositol 1,4,5-trisphosphate (IP$_3$) which is water

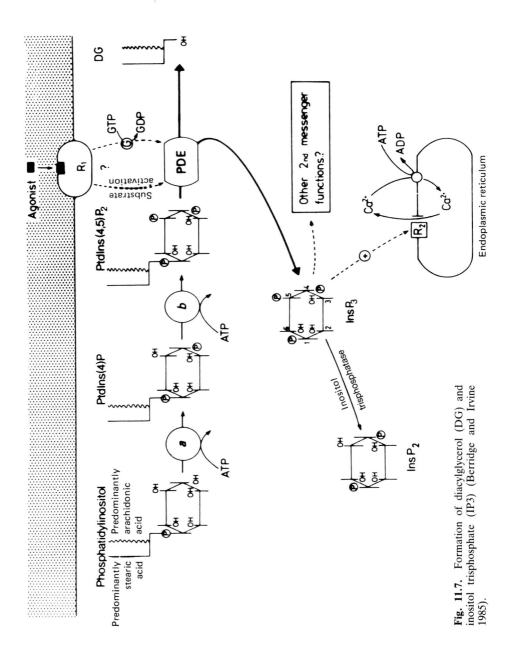

Fig. 11.7. Formation of diacylglycerol (DG) and inositol trisphosphate (IP3) (Berridge and Irvine 1985).

soluble and released into the cytoplasm in μM concentrations. Diacylglycerol increases the affinity of protein kinase C for Ca^{2+}, enabling it to be activated maximally at resting free Ca^{2+} (Nishizuka 1984). Inositol trisphosphate releases Ca^{2+} from an endoplasmic reticulum store.

Two key pieces of experimental evidence exist for the latter. Firstly, degradation of phosphatidyl inositol 4,5-bisphosphate following activation did not require, nor could it be caused by, an increase in intracellular free Ca^{2+} (Billah and Michell 1979). Secondly, addition of μM inositol trisphosphate to exocrine pancreatic cells permeabilized with digitonin resulted in a detectable increase in free Ca^{2+}, being released from a non-mitochondrial internal store (Streb et al. 1983, 1984).

Inositol 1,4,5-trisphosphate (IP_3) appears to be the second messenger, produced as a result of receptor occupancy, responsible for internal Ca^{2+} release both in smooth muscle cells and in non-muscle cells such as exo- and endocrine pancreas, hepatocytes activated by α-adrenergic agents, vasopressin or angiotensin II, platelets activated by thrombin or ADP, neutrophils activated by chemotactic agents and eggs fertilized by sperm (see Berridge 1984; Williamson et al. 1985 for references).

What then are the criteria necessary to show that IP_3 is the internal mediator of cell activation, causing a rise in cytoplasmic free Ca^{2+}?

1. The primary cell stimulus must act via a rise in cytoplasmic free Ca^{2+}, the Ca^{2+} for cell activation coming from a non-mitochondrial [uncoupler and inhibitor, e.g. CCCP (carboxy cyanide p-trifluoromethoxyphenyl hydrazone) and ruthenium red, insensitive] pool.

2. The primary stimulus should increase the turnover of phosphatidyl inositol (PI) lipids, detected using ^{32}P-, ^{14}C- or ^3H-labelled inositol, this increased turnover being detectable before the rise in cytoplasmic free Ca^{2+}. Typically in liver, PI turnover is detectable within 1–3 s of adding adrenaline or vasopressin, an increase in free Ca^{2+} being detectable within 6–10 s and phosphorylase activation occurring within 15 s.

3. The absolute IP_3 concentration should rise.

4. The increased PI turnover and IP_3 level itself should not be caused by or be dependent on a rise in cytoplasmic free Ca^{2+}, e.g. not mimicked by A23187 or ionomycin, or inhibited by intracellular EGTA.

5. Inhibition of phospholipase C by neomycin should inhibit IP_3 production and thereby reduce the rise in cytoplasmic free Ca^{2+}, whereas inhibition of IP_3 breakdown and inositol phosphatase, e.g. by Li^+, Mg^{2+} or 2,3-diphosphoglycerate, should prolong the Ca^{2+} rise.

6. Addition of IP_3 into the cell should initiate cell activation, and the concentration required should be in the μM range, though it may vary from 0.1–10 μM depending on the cell type.

Several of these criteria have been satisfied in many non-muscle cells. However, direct measurement of free IP_3 in living cells is now required, together with a full characterization of Ca^{2+} release from isolated vesicles. The endoplasmic reticulum from non-muscle cells has long been known to contain a Ca^{2+}-activated Mg-ATPase capable of pumping Ca^{2+} into it, and to have distinctive morphological features, e.g. in platelets, eggs and adipocytes. It has even been suggested that IP_3 is the releaser of Ca^{2+} in skeletal muscle, rather than direct electrical activation (Somlyo 1985, 1986).

Three questions now arise. Firstly, what is the precise nature of the endoplasmic reticulum Ca^{2+} store? Is it smooth, or rough as some have suggested? Secondly, what is the molecular basis of the activation of the phosphodiesterase responsible for IP_3 release? Does this involve a GTP-binding protein (G protein), one of the family responsible for hormone–receptor enzyme coupling in cells, as has been suggested? Thirdly, how does IP_3 cause release of Ca^{2+} from the endoplasmic reticulum? Is there a membrane receptor (Spat et al. 1986), perhaps a G protein?

A factor complicating these studies is the existence of other inositol phosphates, in particular inositol 1,3,4-trisphosphate, inositol 2,4,5-trisphosphate, inositol cyclic phosphates and higher phosphorylated forms such as IP_4 (inositol 1,3,4,5-tetrabisphosphate), phytic acid (IP_5) found in avian erythrocytes and di and tri phosphates of as yet unproven function, though IP_4 may open membrane Ca^{2+} channels (Irvine and Moor 1986).

A further problem is the fact that dose–response curves of agonist concentration versus cell response or IP_3 production do not always correlate. In fact sometimes a cell response is measurable when radioactive IP_3 formation is undetectable.

A crucial development will be a sensitive method for producing artificially, detecting and localizing free IP_3 in living cells, together with specific methods for preventing it rising or inhibiting its action in the living cell.

Ca^{2+} through the Cell Membrane

The passive movement of Ca^{2+} across the eukaryotic cell membrane is electrogenic, i.e. there is apparently no coupled counterion acting through either a symport (e.g. Ca^{2+} and HPO_4^{2-}) or an antiport (e.g. Ca^{2+} for Ca^{2+} or $2Na^+$). An increase in Ca^{2+} influx, leading to a rise in cytoplasmic free Ca^{2+}, could occur either by structural perturbation of the cell membrane or by increasing the solubility of the lipid bilayer to Ca^{2+}. The former could involve either a "pore" or "channel" selectively permeable to Ca^{2+}, or a breakdown on phospholipid bilayer, for example through the formation of miscelles. The latter would require the generation of an ionophore. Many such compounds have been proposed and searched for in eukaryotes. At present the evidence is firmly on the side of the opening of Ca^{2+} channels as the cause of increased influx of Ca^{2+} following exposure of a cell to its primary stimulus. Two sorts of channel have been identified, voltage-dependent and receptor-operated (Reuter 1983). The former occur in excitable cells activated electrically by action potentials or neurotransmitters, the latter in non-electrically excitable cells activated by hormones and neurotransmitters.

Voltage-dependent Ca^{2+} channels were discovered as a result of the work of Fatt and Katz (1953), who showed that action potentials in crab muscle still occurred in the absence of external Na^+. Further investigation showed that Ca^{2+}, not Na^+, was the major current carrier during electrical excitability (Hagiwara 1973). Ca^{2+} channels have since been found in many invertebrate and vertebrate excitable cells, including mammalian cardiac and smooth muscle and nerve terminals, working on their own or in conjunction with Na^+

channels. The use of the patch clamp technique has enabled the electrical properties of single Ca^{2+} channels to be defined (Stevens et al. 1982; Reuter 1983; Lux and Brown 1984). A mammalian heart cell may have about 5–15 Ca^{2+} channels per μ^2, equivalent to some 2000–10 000 Ca^{2+} channels over the whole surface of the cell. Although not completely selective for Ca^{2+} over other ions, they are some tenfold more selective than is the Na^+ channel for Na^+. The channels switch randomly between an open and a closed state. They appear to spend no time in an intermediate state. Depolarization of the membrane increases the probability of a channel opening. Thus

$$\text{mean current in any patch of membrane} = I_{Ca} = Npi \tag{5}$$

where N = number channels in the patch, p = probability of channel being open and i = single channel current.

I_{Ca} in cardiac muscle is increased by β-adrenergic agonists and reduced by muscarinic acetyl choline receptor occupancy. The former work by phosphorylation of the "channel" through activation of cyclic AMP-dependent protein kinase. This causes either an increase in the number of Ca^{2+} channels capable of opening during depolarization or an increase in the probability of a Ca^{2+} channel opening at a given membrane potential (Reuter 1983). The present model for the Ca^{2+} channel has two closed states (C) and one open state (O):

$$C_1 \underset{k_{-1}}{\overset{k_1}{\rightleftharpoons}} C_2 \underset{k_{-2}}{\overset{k_2}{\rightleftharpoons}} 0 \tag{6}$$

Cyclic AMP increases k_1 and k_2, thereby increasing the probability of a channel opening.

Unfortunately, unlike the highly specific tetrodotoxin of the Na^+ channel, there is no such highly selective compound for voltage-dependent Ca^{2+} channels. However, they can be blocked experimentally by several cations, including Mg^{2+}, Co^{2+}, Ni^{2+}, Mn^{2+}, Cd^{2+} and La^{3+}, as well as by several so-called Ca^{2+} antagonist drugs, the most important clinical ones being verapamil, dihydropyridines such as nifedipines, and diltiazem.

Little is known of the molecular basis of receptor-operated channels, nor are there any useful experimental compounds for blocking them specifically. The most likely candidate for linking receptor occupancy to increased Ca^{2+} permeability is a GTP-binding protein analogous to those responsible for linking activating or inhibitory receptors to adenylate cyclase (Greengard et al. 1984). GTP-binding proteins may also mediate inhibition of voltage-dependent calcium channels initiated by inhibitory neurotransmitters (Holz et al. 1986).

Recovery after Disappearance of the Cell Stimulus

Recovery of a cell following activation involves four components:

1. Removal of the stimulus from the cell membrane.
2. Reduction of the cytoplasmic free Ca^{2+} to a resting or sub-threshold level, enabling Ca^{2+} to dissociate from the Ca^{2+}-binding protein responsible for activating intracellular processes.

3. Replenishment of Ca^{2+} stores, together with relocation of Ca^{2+} from internal buffers or sinks.
4. Replenishment of the process being activated, together with recovery of cell ATP and normal metabolism.

In the case of receptor occupancy, the first of these components requires dissociation of the agonist from its receptor. This can occur simply by the agonist dropping off externally and diffusing away from the cell or by membrane recycling, a procedure whereby the bound receptor is removed into an endocytotic, clathrin-containing vesicle. This vesicle fuses with an acid-containing intracellular vesicle which removes the agonist, and the receptor is recycled. In contrast, in electrically excitable cells the signal for switching off is a repolarization of the cell membrane, often involving a Ca^{2+}-activated K^+ conductance (Meech 1979). Ca^{2+} is then pumped back into its endoplasmic reticulum store, and recycled back from the mitochondrial sink, a process activated by polyamines such as spermine (Nicchitta and Williamson 1984). Dissociation from the Ca^{2+}-binding proteins may be relatively fast. However, covalent modification, such as phosphorylation, of internal proteins may result in prolonged activation of the cell even when the intracellular free Ca^{2+} has returned to its resting level. An increased Ca^{2+} influx may also remain until the internal Ca^{2+} stores are fully replenished. Finally, once the cell ATP is back to normal and any cellular components lost during activation replenished, e.g. secretory production or glycogen, the cell is ready for another burst of activity.

How Calcium Acts

Direct Versus Indirect Action

Injection of Ca^{2+} into cells activates directly the process under the control of the primary and secondary regulators (Heilbrunn and Wiercinski 1947). However, direct, physiologically relevant, effects of Ca^{2+} on purified systems are much more difficult to demonstrate. In 1942 Bailey and Needham both showed that Ca^{2+} could activate the Mg-ATPase of myosin isolated from muscle. However, workers in the 1950s using purified actin and myosin were able to show activation of the Mg-ATPase without the need for Ca^{2+}. Furthermore there were many reports in the 1960s of Ca^{2+} inhibiting, rather than activating, enzymes in vitro. These included glycolytic enzymes, adenylate cyclase and Na^+ and K^+ Mg-ATPase. Also, some enzymes, such as phosphorylase kinase, were activated irreversibly by proteolysis (see Campbell 1983 for references). Most of these experiments are now known to be physiologically irrelevant since they employed Ca^{2+} concentrations in the mM range.

The Discovery of Ca^{2+}-Binding Proteins

The solution to the problem of how intracellular Ca^{2+} acts was solved when Ebashi discovered a protein complex conferring Ca^{2+} sensitivity on the actomyosin of muscle (Ebashi 1963; Ebashi and Kodama 1965). The complex was subsequently shown to consist of three proteins, named troponin I, troponin T and troponin C. The latter is a highly negatively charged protein with an acylated N terminus, and four Ca^{2+} sites (cardiac troponin C only has three). These Ca^{2+} sites have high affinity for Ca^{2+} and selectivity over Mg^{2+} when the intracellular free Ca^{2+} rises from 0.1 to 5 μM following muscle activation. The primary and three-dimensional structure of troponin C have now been elucidated (Collins 1976; Herzberg and James 1985). Binding of Ca^{2+} causes a conformational change altering the interaction of troponins I and T with the actin polymer, thereby derepressing the interaction of actin with myosin and allowing sliding of the filaments, and thus contraction to occur (Potter and Gergeley 1975; Ebashi 1980).

An alternative to derepression of an enzyme complex as a result of Ca^{2+} binding is direct activation of an enzyme, or process, by the Ca^{2+}-binding protein. A second breakthrough establishing this mechanism occurred with the discovery, in the late 1960s, of calmodulin (see Cheung 1980; Klee et al. 1980; Means et al. 1982; Klee and Newton 1985; Dedman 1986a, b for reviews). Although Cheung reported in 1967 that cyclic AMP phosphodiesterase was activatable by an endogenous "activator", the definitive demonstration of a separable protein activator was shown some 3 years later by Cheung (1970). Kakiuchi and co-workers (1970) showed that it was activatable by Ca^{2+}. The "activator" was subsequently indentified as a Ca^{2+}-binding protein (Teo and Wang 1973). This clearly distinguished direct activation via a Ca^{2+}-binding protein from activation via proteolysis, now known to occur by cleavage of the calmodulin-binding domain in several proteins (Dedman 1986a, b). By the mid to late 1970s calmodulin, as the activator became known, in the presence of μM Ca^{2+} was shown to activate many other important intracellular enzymes including adenylate cyclase (Brostrom et al. 1975), erythrocyte Ca^{2+}–Mg -ATPase (Gopinath and Vincenzi 1977; Jarrett and Penniston 1977), myosin light chain kinase (Drabrowska et al. 1978), a brain protein kinase (Schulman and Greengard 1978) and plant NAD^+ kinase (Anderson and Cormier 1978). The Ca^{2+} regulatory site of phosphorylase kinase was shown also to be calmodulin, tightly bound in this case to the enzyme complex and missed in the original studies (Cohen et al. 1978).

Calmodulin is now thought to be present in all eukaryotic cells at a concentration of around 1–20 μM. It is not found in prokaryotes, though these cells do have high affinity Ca^{2+}-binding proteins (Prasher et al. 1985). Calmodulin is found free in the cytoplasm, bound to soluble or membrane enzymes and bound to structures such as microtubules and actin. Antibodies to calmodulin labelled with fluorescein have enabled calmodulin to be localized in the cell following activation (Welsh et al. 1978). It is found for example in the mitotic spindle, moving during mitosis.

Calmodulins, since there is more than one protein in Nature, are highly negatively charged proteins ($pI = 3.4$), heat stable ($t_{\frac{1}{2}} = 7$ min at 100°C), with four Ca^{2+} sites, selective over Mg^{2+}, having K_d's for Ca^{2+} of 1–10 μM and exhibiting some cooperativity. The molecular weight is about 16 700 (148 amino acids) and the structure is highly unusual with a trimethyl lysine at position 115 (Watterson et al. 1980). The primary sequence is highly conserved over the eukaryotes, with just a few amino acid changes, mainly at the carboxyl end, between the mammalian, protozoan and plant calmodulins. They show some

40% homology with troponin C. Both calmodulin and troponin C belong to a family of Ca^{2+}-binding proteins, including also parvalbumins, with distinctive three-dimensional Ca^{2+} domains, named "E–F" hand regions by Kretsinger (1976). Oxygen, nitrogen and sulphur are the three main inorganic elements in proteins capable of coordinating with Ca^{2+}. Selectivity for Ca^{2+}, particularly over Mg^{2+}, has been achieved in evolution by using oxygen, both from the carboxyls of glutamate and aspartate residues and from the carbonyls of the peptide chain. Insertion of nitrogen residues into cation-binding sites results in loss of the selectivity between Ca^{2+} and Mg^{2+} (see Campbell 1983 for further details).

Important as the discovery and characterization of the calmodulins has been, there are several other high-affinity Ca^{2+}-binding proteins, of higher molecular weight (Table 11.4), several of which appear to interact with the cytoskeleton.

If the action of intracellular Ca^{2+} as a cell regulator is mediated by a Ca^{2+}-binding protein, how do these proteins work?

Mechanism of Ca^{2+}-Binding Proteins

To understand how Ca^{2+} binding to a specialized protein can result in activation of an event in the cell three questions must be answered:

1. What structural modification occurs in the protein when Ca^{2+} binds?
2. How does this lead to the necessary interaction with the target protein, and in particular does the Ca^{2+}-binding protein activate or derepress it?
3. What does the target protein do?

The three-dimensional structure of calmodulin has been elucidated (Schutt 1985). Both calmodulin and troponin C have a dumbell shape. The "balls" corresponding to each Ca^{2+} site are linked by an α-helical region. Binding of Ca^{2+} to calmodulin increases the α-helical content, exposing a hydrophobic domain which enables the protein either to bind to its target, for example cyclic AMP phosphodiesterase, or to alter its interaction with the protein to which it is already bound, for example phosphorylase kinase. The hydrophobic site of calmodulin binds drugs such as phenothiazines which inhibit the action of calmodulin, if it is not already bound to its target protein. Several calmodulin-binding proteins have been identified in eukaryotes, named caldesmons in smooth muscle (Marston and Lehman 1985).

A feature common in the activation of most non-skeletal muscle cells is an increase in covalent modification, e.g. phosphorylation or ADP ribosylation of intracellular proteins. Gel electrophoresis, followed by ^{32}P autoradiography, demonstrates that a plethora of as yet unidentified proteins are phosphorylated when cells are activated. In cells where actin–myosin interactions play a key role in cell activation, the phosphorylation of a protein with molecular weight 27 000, the myosin light chain, by a kinase activated by Ca^{2+}–calmodulin, seems to be important.

A number of problems still remain, however:

1. It is not clear how protein phosphorylation, or Ca^{2+}-binding protein interaction with its target, leads to cell activation, particularly when this involves a threshold in the cell.

2. Whilst Ca^{2+}–calmodulin activates many enzymes in vitro, the evidence that this occurs in vivo is often poor.

3. There is a calmodulin paradox. Ca^{2+}–calmodulin in vitro can activate opposing enzymes, e.g. adenylate cyclase and cyclic AMP phosphodiesterase, or kinases and phosphatases such as calcineurin (Klee and Newton 1985). Which one is relevant during cell activation? Is there simply an increase in enzymatic cycling?

4. Phenothiazines, such as trifluoperazine, can inhibit calmodulin-activated processes in vitro. However, in the cell trifluoperazine may also inhibit protein kinase C (Takai et al. 1979; Nishizuka 1984), or act as a "local anaesthetic". Such drugs can therefore not be used as definitive evidence for a role for calmodulin in cell activation in vivo.

There is still much to be learnt about the molecular basis of intracellular Ca^{2+} as a cell activator.

The Pathology of Intracellular Calcium

Cause or Consequence?

Exogenous factors such as toxins, or viruses and bacteria responsible for infectious diseases, as well as endogenous factors such as cells and components of the immune system in diabetes, rheumatoid arthritis and multiple sclerosis, can cause increases in the permeability to, and the content of Ca^{2+} in cells. Disturbances in cell Ca^{2+} can also occur associated with genetic abnormalities such as sickle cell anaemia and cystic fibrosis (see Campbell 1983 for references). Malfunctioning of Ca^{2+}-dependent processes occurs in heart disease, muscle disease and endocrine disorders. Furthermore tissue calcification has been recognized for more than a century as a feature not only of disorders of whole body Ca^{2+} metabolism (metastatic), but also of severe tissue injury (dystrophic). Necrotic cells contain some 10–100 times their normal Ca^{2+} content (Majno 1964). The question therefore arises, are these disturbances in intracellular Ca^{2+} a cause or a consequence of cell injury? Secondary to this, could any of the rises in intracellular Ca^{2+} actually be beneficial to a cell under attack, enabling it to be activated, or to protect itself?

Five further pieces of evidence have been used to support the proposal that a rise in intracellular Ca^{2+} plays a part in cell injury:

1. Removal of external Ca^{2+} reduces or prevents the lethal effects of certain lymphotoxins, hepatotoxins, bacterial toxins and cytotoxic T cells.

2. Increases in organelle Ca^{2+} content can be correlated with the time course of cell damage. For example, anoxia, some genetic abnormalities, toxins and deficiency diseases can lead to increased mitochondrial and/or nuclear Ca^{2+} content.

3. Impairment of Ca^{2+} transport in mitochondria and endoplasmic reticulum can be induced by pathogens such as reactive oxygen metabolites (oxygen radicals).

4. Effects on the cell membrane such as shape change and vesiculation can be mimicked by Ca^{2+} ionophores.
5. Biochemical effects of Ca^{2+} in the pathological range (10–50 μM) have been identified in vitro. These include intracellular activation of neural proteases, phospholipases, nucleases and transglutaminases; chromatin condensation; and inhibition of enzymes such as those in intermediary metabolism or adenylate cyclase.

A major problem in establishing intracellular Ca^{2+} as a cause, as opposed to a consequence, of cell injury is the paucity of direct measurements of free Ca^{2+} in intact cells in relation to the time course of reversible and irreversible injury to the cell (Campbell et al. 1979a, b; Campbell and Luzio 1981; Cobbold and Bourne 1985; Morgan and Campbell 1985; Morgan et al. 1986b; Campbell 1987). A number of workers have attempted to use quin2 in blood cells, for example platelets and neutrophils (Lew et al. 1985b), isolated from patients. The aim was to compare the resting free Ca^{2+} in the patient's cells with that in cells from healthy volunteers. As yet, little of any value has come out of these studies.

Irreversible Cell Injury

Calcium could, in principle, be involved in irreversible injury to a cell in three ways:

1. The injury could involve impairment of a mechanism normally requiring intracellular Ca^{2+}.
2. Ca^{2+} could induce irreversible injury to a process in the cell required for its specialized function, e.g. contraction or secretion.
3. Ca^{2+} could direct a cell rapidly along a pathway to death, e.g. by initiating destruction of the cell membrane and by overloading the mitochondria with Ca^{2+}, thereby preventing ATP synthesis.

Two pathways leading to cell death have been defined morphologically and biochemically (Wyllie et al. 1980; Bowen and Lockshin 1981; Trump et al. 1981) – necrosis and apoptosis.

Necrosis is characterized by cell swelling, together with disruption of internal organelles and the plasma membrane. A drop in cellular ATP and increased membrane permeability to ions, leading to a rapid rise in intracellular free Ca^{2+} and cell swelling, is a common feature preceding the threshold for irreversible injury to the cell. Necrosis is provoked by agents such as toxins, ischaemia and anoxia, oxygen radicals and the membrane attack complex of complement. A particularly confusing example occurs in the myocyte. Ca^{2+} is, of course, necessary for excitation–contraction coupling in the heart. Removal of external Ca^{2+} causes loss of mechanical activity but not, initially, total electrical inactivation. However, during the period when no calcium is present, membrane damage occurs so that, on adding back Ca^{2+}, Ca^{2+} floods into the cell and kills it. This is the so-called "Ca^{2+} paradox" (see Parratt 1985 for references). A large increase in Ca^{2+} permeability also occurs in the hypoxic myocyte following reoxygenation, possibly caused by reactive oxygen metabolites (oxygen radicals). The resulting increase in intracellular Ca^{2+} could lead to mitochondrial overload, or to mechanical separation and disruption of cells. Though the precise role of Ca^{2+} in reoxygenation injury is still controversial it is important to clarify it if therapeutic regimes and cardioplegic solutions at operation are to be successful.

In contrast to necrosis, apoptosis is characterized by condensation of the cell, particularly the nucleus, and formation of surface protuberances, but with maintenance of organelle integrity and little acute loss of ATP or ion gradients. It can occur spontaneously, can be induced by target cells or can be provoked hormonally, for example by glucocorticoids acting on lymphocytes. This type of cell death involves activation of events within the cell, including protein synthesis. A role for intracellular Ca^{2+} release through inositol trisphosphate, or a role for protein kinase C, has yet to be established.

It is perhaps worth remembering that cell death, whilst damaging or even disastrous under some pathological conditions, may be beneficial or even essential in other circumstances. There are several natural killer mechanisms, both cellular and humoral, circulating in the blood, including phagocytes, T cells, NK cells, lymphotoxins, tumour necrosis factor and complement. Cell death may also occur as a result of internal programming. It plays an important role in embryogenesis, control of tissue size and differentiation, in the natural selection of cells and in removal of aged cells or cells infected with pathogens. In view of the number of potentially lethal agents circulating in the blood it is perhaps not surprising that our cells have evolved protection mechanisms to control the balance between reversible and irreversible injury to them.

Reversible Cell Injury and Protection

A rise in intracellular free Ca^{2+} during cell injury may not necessarily be disadvantageous. Gap junctions between cells are sealed by Ca^{2+} 1–10 μM (Loewenstein 1984), thereby protecting neighbouring cells. Furthermore, attacking agents or damaged membrane may be removed by endocytosis or budding. This mechanism appears to be particularly relevant in the protection of cells against attack by paramyxoviruses, the membrane attack complex of complement, bacterial toxins and analogous proteins in cytotoxic T lymphocytes called perforins (Campbell and Luzio 1981; Pasternak et al. 1985).

An initial insult to the cell, for example by complement, leads to a rapid increase in intracellular free Ca^{2+}, and formation of a pore through which intracellular constituents leak out. Several minutes after the initial attack, sometimes as long as 30–60 min, a threshold point is reached where the cell dies rapidly or explodes. This can be demonstrated by counting lysed cells in a fluorescence-activated cell sorter (Edwards et al. 1983). However, if the cell is able to protect itself, by removing the potentially lethal membrane complex, before the lethal threshold point is reached (Patel and Campbell 1987) then the cell has a chance to recover its ATP and ionic balance, as well as its normal morphology and function (Fig. 11.8).

Direct measurement of free Ca^{2+} in cells attacked by the membrane attack complex of complement has shown that the C5b-8 precursor forms within 2–5 min of complement activation, and that a free Ca^{2+} increase is detectable within 5 s of binding the terminal component C9 (Campbell et al. 1979a, b, 1981a, b; Wiedmer and Sims 1985). Some Ca^{2+} may also be released from

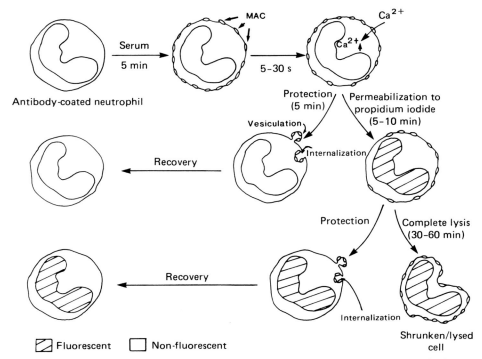

Fig. 11.8. Protection of nucleated cells against complement attack. Fluorescence = leakage to propidinium iodide which becomes fluorescent when it binds to DNA (Morgan et al. 1986b)

internal stores (Morgan and Campbell 1985) and Ca^{2+}-dependent events in the cell are activated, e.g. secretion or oxygen radical production (Hallett et al. 1981; Hallett and Campbell 1982). The membrane attack complexes, containing poly C9, patch and are removed by budding or endocytosis within a few minutes (Campbell and Morgan 1985; Morgan et al. 1986b). These cells recover, whereas cells which do not remove the complexes in time die. Adenosine, through cyclic AMP, inhibits the protection mechanism in neutrophils by slowing it down, thereby increasing the number of lysed cells (Roberts et al. 1984).

The balance between endocytosis and budding varies with cell type. The precise roles of intracellular Ca^{2+} and protein kinase C in the two processes have yet to be defined. The protection mechanism explains the long-known observation that nucleated cells are relatively resistant to complement attack (Boyle et al. 1976) and provides a mechanism for reversible cell injury in immune-based diseases such as diabetes, Graves' disease and multiple sclerosis where the role of complement was unclear because of the established dogma that "complement always kills cells". It does not.

There are thus four potential consequences of a rise in intracellular free Ca^{2+} following an insult to a cell:

1. Inappropriate activation of Ca^{2+}-dependent reactions or events, normally activated by physiological stimuli.

2. Injury to ATP, ionic balance, structure and function followed by recovery as a result of a protection mechanism.
3. Irreversible injury to the specialized function of a cell without cell death.
4. Necrotic cell death.

Pharmacology of Intracellular Ca^{2+}

A number of therapeutic agents work by interfering with the regulation of action of intracellular Ca^{2+} (Fig. 11.9). This can occur in any one of five ways (Campbell 1983):

1. Inhibition of binding of natural agonists to their receptor, e.g. α- and β-adrenergic blockers.
2. Binding of the drug to specific membrane receptors or ion channels normally activated by the physiological stimulus, e.g. the Ca^{2+} antagonists verapamil and nifedipine.
3. Binding of the drug to phospholipids or proteolipids thereby indirectly interfering with Ca^{2+} influx or internal Ca^{2+} release, e.g. anaesthetics.
4. A change in intracellular free Ca^{2+}, e.g. cardiac glycosides causing Na$^+$ increase, thereby increasing intracellular Ca^{2+} through Na$^+$/Ca^{2+} exchange, or methyl xanthines releasing Ca^{2+} from reticulum stores.

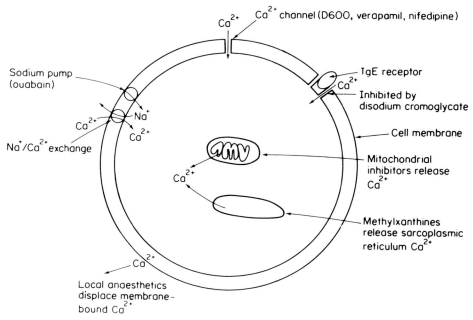

Fig. 11.9. The pharmacology of intracellular Ca^{2+} (Campbell 1983).

5. Inhibition of Ca^{2+} action, e.g. phenothiazines binding to calmodulin, though this action is not their only one and may not explain their therapeutic effects.

The establishment of the key role of intracellular Ca^{2+} in cellular processes which have been impaired or hyperactivated in disease, together with the role of Ca^{2+} in determining the balance between reversible and irreversible injury now provides a rational basis for designing new therapeutic agents capable of really attacking the major diseases of the twentieth century.

Perspectives

The aim of this chapter has been to establish the framework necessary for understanding how intracellular calcium can on the one hand provoke and regulate a diverse range of natural phenomena, and on the other play a key role in cell injury in disease. Two key features underpinning this framework have been identified, one conceptual, the other experimental. Firstly, in order to see the molecular perspective it is necessary to identify the threshold determining the point at which experience begins for an individual cell, be it activation of a physiological event, or the point of no return dividing reversible and irreversible cell injury. Secondly, the importance of carrying out measurement of, and manipulations on, free Ca^{2+} and Ca^{2+}-dependent processes in living cells, as opposed to the "grind and find" approach of the conventional biochemist, has been argued. One particularly ingenious emerging technique is the use of so-called "caged" compounds which release their active ingredient within the cell when exposed to a laser flash. Tsien has developed such compounds for releasing Ca^{2+} at discrete sites within the cell. Similarly Trentham has synthesized caged ATP and inositol trisphosphate. Recombinant DNA technology also offers exciting potential for the cell biologist. We have succeeded in incorporating obelin mRNA in human neutrophils, which then synthesize active Ca^{2+}-stimulatable photoprotein (Patel and Campbell, unpublished).

Much still needs to be learned about the molecular basis of the control and action of intracellular Ca^{2+}. In many cases of cell activation or injury it still remains to be established whether a rise in cytoplasmic Ca^{2+} is a cause or consequence of the cellular event. The role of intracellular Ca^{2+} in plants (Wyn-Jones and Lunt 1967; Trewavas 1985) and prokaryotes is still poorly characterized. A full understanding of the roles of intracellular Ca^{2+}–Ca^{2+}-binding protein versus cyclic nucleotide kinase (A and G), and versus diacylglycerol–protein kinase C is now required. I have called such interactions a "chemisymbiosis" (Campbell 1988). For example, Ca^{2+} can control the concentration of cyclic nucleotides in cells through effects on the respective cyclases and phosphodiesterases whereas cyclic nucleotides can regulate intracellular free Ca^{2+} through phosphorylation of proteins on internal organelles or the cell membrane. Furthermore, the intracellular regulators may coregulate the same enzyme, e.g. phosphorylase kinase–cyclic AMP and Ca^{2+}, protein kinase C–diacylglycerol and Ca^{2+}.

The rationalization of the functional and evolutionary significance of intracellular messengers requires a combination of the molecular and cellular approaches to living cells and whole organisms. It requires a realization that individual cells do different things at different times in response to different stimuli. What matters in the end is not the functioning or even the survival of an individual cell but the success of the whole tissue to the maintenance and reproduction of the whole organism. Cells do not necessarily behave anthropomorphically! Only when the necessary consummation of the two approaches to biology, the molecular and the cellular, has been achieved will their marriage lead to an embryo capable of growing into an adult with the wisdom to answer a question which has puzzled biologists and chemists for more than a century: "What is so special about calcium?"

Acknowledgements. I thank my research group for many years hard work, the Director and staff of the Marine Biological Association for their inspiring environment, and the late Dr. Leslie H. N. Cooper, FRS, whose wisdom I miss greatly. I also thank the MRC, SERC, ARC, MS Society and Royal Society for generous financial support, and Dr. Bob Dormer and Dr. Maurice Hallett for helpful comments on this manuscript and with them Drs. J. P. Luzio, A. Compston and B. P. Morgan for their stimulating collaboration.

References

Allen DG, Blinks JR (1978) Calcium transient in aequorin-injected frog cardiac muscle. Nature (London) 273:505–513

Ammann D (1985) Ca^{2+}-selective microelectrodes. Cell Calcium 6:39–55

Ammann D, Meier PC, Simon W (1979) Design and use of calcium-sensitive microelectrodes. In: Ashley CC, Campbell AK (eds) Detection and measurement of free Ca^{2+} in cells. Elsevier/North-Holland, Amsterdam, pp 117–129

Anderson JM, Cormier MJ (1978) Calcium-dependent regulator of NAD kinase in higher plants. Biochem Biophys Res Commun 84:595–602

Ashley CC, Campbell AK (1978) Effect of L-glutamate on aequorin light, tension and membrane response in barnacle muscle. Biochim Biophys Acta 512:429–435

Ashley CC, Campbell AK (eds) (1979) Detection and measurement of free Ca^{2+} in cells. Elsevier/North-Holland, Amsterdam

Ashley CC, Ridgway EB (1968) Simultaneous recording of membrane potential, calcium transient and tension in single muscle fibres. Nature (London) 219:1168–1169

Ashley CC, Ridgway EB (1970) On the relationship between membrane potential, calcium transient and tension in single barnacle muscle fibres. J Physiol (London) 209:105–130

Ashley CC, Timmerman MP (1986) Use of the new fluorescent Ca^{2+} indicator fura-2 to detect Ca^{2+} concentration changes in electrically stimulated *Balanus* muscle fibres. J Physiol (London) 367:64P

Ashley CC, Williamson RE (1981) Cytoplasmic free Ca^{2+} and streaming velocity in characean algae; measurements with micro injected aequorin. J Physiol (London) 319:103P

Ashley CC, Rink TJ, Tsien RY (1978) Changes in free Ca during muscle contraction, measured with an intracellular Ca-sensitive electrode. J Physiol (London) 280:27

Ashley CC, Potter JD, Strong P, Godber J, Walton A, Griffiths PJ (1985a) Kinetic investigations in single muscle fibres using luminescent and fluorescent Ca^{2+} probes. Cell Calcium 6:159–182

Bailey K (1942) Myosin and adenosine triphosphate. Biochem J 36:121–139

Baker PF, Blaustein MP (1968) Sodium-dependent uptake of calcium by crab nerve. Biochim Biophys Acta 150:167–170

Baker PF, Warner AE (1972) Intracellular calcium and cell cleavage in early embryos of *Xenopus*. J Cell Biol 53:579–581

Baker PF, Hodgkin AL, Ridgway EB (1971) Depolarization and calcium entry in squid giant axons. J Physiol (London) 218:708–785

Bennett HS, Porter KR (1953) Electron microscope study of sectioned breast muscle of domestic fowl. Am J Anat 93:61–105

Berliner K (1933) The action of calcium on the heart – a critical review. Am Heart J 8:548–562

Berridge MJ (1976) The interactions of cyclic nucleotides and calcium in the control of cellular activity. Advances in Cyclic Nucleotide and Protein Phosphorylation Research 6:1–96

Berridge MJ (1984) Inositol trisphosphate and diacylglycerol as second messengers. Biochem J 220:345–360

Berridge MJ, Irvine RF (1984) Inositol trisphosphate, a novel second messenger in cellular signal transduction. Nature (Lond) 312:315–321

Billah MM, Michell RH (1979) Phosphatidylinositol metabolism in rat hepatocytes stimulated by glycogenolytic hormones. Effects of angiotensin, vasopressin, adrenaline, ionophore A23187 and calcium-ion deprivation. Biochem J 182:661–668

Blaustein MP (1974) Interrelationships between sodium and calcium fluxes across cell membranes. Rev Physiol Biochem Pharmacol 70:33–82

Blinks JR, Prendergast F, Allen DG (1976) Photoproteins as biological calcium indicators. Pharmacol Rev 28:1–93

Blinks JR, Wier WG, Hess P, Prendergast FG (1982) Measurement of Ca^{2+} concentrations in living cells. Progress in Biophysics Molecular Biology 40:1–114

Borle AB, Snowdowne KW (1982) Measurement of intracellular free calcium in monkey kidney cells with aequorin. Science 217:252–254

Bowen ID, Lockshin RA (ed) (1981) Cell death in biology and pathology. Chapman and Hall, London and New York

Boyle MDP, Onanian SH, Borsos T (1976) Studies on the terminal stages of antibody-complement mediated killing of a tumour cell. II. Inhibition of transformation of T to dead cells by cyclic AMP. J Immunol 116:1276–1279

Brinley FJ, Tiffert T, Scarpa A (1979) Mitochondria and other calcium buffers of squid axon studied in situ. J Gen Physiol 72:101–127

Brostrom CO, Huang YC, Breckenridge BMcL, Wolff DJ (1975) Identification of calcium-binding protein as a calcium-dependent regulator of brain adenylate cyclase. Proc Natl Acad Sci USA 72:64–68

Brown HM, Pemberton JP, Owen JD (1976) A calcium-sensitive microelectrode suitable for intracellular measurement of calcium (II) activity. Anal Chim Acta 85:261–276

Brownlee CC, Wood JW (1986) A gradient of cytoplasmic free calcium in growing rhizoie cells of *Fucus serratus*. Nature 320:624–626

Campbell AK (1983) Intracellular calcium: its universal role as regulator. John Wiley, Chichester

Campbell AK (1986a) Living light: chemiluminescence in the research and clinical laboratory. TIBS 11:104–108

Campbell AK (1986b) Lewis Victor Heilbrunn: Pioneer of calcium as an intracellular regulator. Cell Calcium 6:287–296

Campbell AK (1987) Intracellular calcium: friend or foe? Clin Sci 72:1–10

Campbell AK (1988) Chemiluminescence: principles, biology and biomedical applications. Ellis Horwood/VCH, Chichester (in press)

Campbell AK, Dormer RL (1978) Inhibition of calcium of cyclic AMP formation in sealed pigeon erythrocyte "ghosts": a study using the photoprotein obelin. Biochem J 176:53–66

Campbell AK, Hales CN (1976) Some aspects of intracellular regulation. In: Beck F, Lloyd JB (eds) The cell in medical science, Vol 4. Academic Press, London and New York, pp 105–151

Campbell AK, Hallett MB (1983) Measurement of intracellular calcium ions and oxygen radicals in polymorphonuclear–leucocyte–erythrocyte "ghost" hybrids. J Physiol (London) 338:537–550

Campbell AK, Luzio JP (1981) Intracellular free calcium as a pathogen in cell damage initiated by the immune system. Experientia 37:1110–1112

Campbell AK, Morgan BP (1985) Monoclonal antibodies demonstrate protection of polymorpho-nuclear leucocytes against complement attack. Nature (London) 317:164–166

Campbell AK, Daw RA, Luzio JP (1979a) Rapid increase in intracellular Ca^{2+} induced by antibody plus complement. FEBS Letters 107:55–60

Campbell AK, Lea TJ, Ashley CC (1979b) Coelenterate photoproteins. In: Ashley CC, Campbell AK (eds) Detection and measurement of free Ca^{2+} in cells. Elsevier/North-Holland, Amsterdam, pp 13–72

Campbell AK, Daw RA, Hallett MB, Luzio JP (1981a) Direct measurement of the increase in intracellular free calcium ion concentration in response to the action of complement. Biochem J

194:551–560

Campbell AK, Hallett MB, Daw RA, Ryall MET, Hart RC, Herring PJ (1981b) Application of the photoprotein obelin to the measurement of free Ca^{2+} in cells. In: DeLuca M, McElroy WD (eds) Bioluminescence and chemiluminescence. Academic Press, New York, pp 601–607

Campbell AK, Dormer RL, Hallett MB (1985a) Coelenterate photoproteins as indicators of cytoplasmic free Ca^{2+} in small cells. Cell Calcium 61:69–82

Campbell AK, Hallett MB, Weeks I (1985b) Chemiluminescence as an analytical tool in cell biology and medicine. Methods of Biochemical Analysis 31:317–416

Capponi AM, Lew PD, Jarnot L, Vallotton MB (1985) Correlation between cytosolic free Ca^{2+} and aldosterone production in adrenal glomerulosa cells. J Biol Chem 259:8867–8869

Carafoli E, Crompton M (1976) Calcium ion and mitochondria. Symp Soc Exp Biol 30:89–115

Caswell AH (1979) Methods of measuring intracellular calcium. Int Rev Cytol 56:145–187

Chance B (1965) The energy-linked reaction of calcium with mitochondria. J Biol Chem 240:2729–2748

Cheung WY (1967) Cyclic 3′,5′-nucleotide phosphodiesterase: pronounced stimulation by snake venom. Biochem Biophys Res Commun 29:478–482

Cheung WY (1970) Cyclic 3′,5′-nucleotide phosphodiesterase. Demonstration of an activator. Biochem Biophys Res Commun 38:533–538

Cheung WY (1980) Calmodulin plays a pivotal role in cell regulation. Science 207:19–27

Cobbold PH, Bourne PK (1985) Aequorin measurements of free calcium in single heart cells. Nature (London) 312:444–446

Cobbold PH, Cutherbertson KSR, Goyns MH, Rice V (1983) Aequorin measurements of free calcium in single mammalian cells. J Cell Science 61:123–136

Cohen P, Burchell A, Foulkes JG, Cohen PTW (1978) Identification of the Ca^{2+}-dependent modulator protein as the fourth subunit of rabbit skeletal muscle phosphorylase kinase. FEBS Lett 92:287–293

Collins JH (1976) Structure and evolution of troponin C and related proteins. Soc Symp Exptl Biol 30:303–334

Davy H (1808) Electro-chemical researches, on the decomposition of the earths; with observations on the metals obtained from the alkaline earths, and on the amalgam from ammonium. Philos Soc Trans R Soc London 98:333–370

Dedman JR (1986a) The discovery of calmodulin – a historical perspective. J Cyclic Nucl Res (in press)

Dedman JR (1986b) Mediation of intracellular calcium: variances on a common theme. Cell Calcium 6:297–308

Denton RM, McCormack JG (1985) Ca^{2+} transport by mammalian mitochondria and its role in hormone action. Am J Physiol 249:E543

Denton RM, McCormack JG, Edyell NJ (1980) Role of calcium ions in the regulation of intramitochondrial metabolism. Biochem J 190:107–117

Dipolo R, Requena F, Brinley FJ Jr, Mullins LJ, Scarpa A (1976) Ionised calcium concentrations in squid axons. J Gen Physiol 67:433–467

Di Virgilio F, Lew DP, Pozzan T (1984) Protein kinase C activation of physiological processes in human neutrophils at vanishingly small cytosolic Ca^{2+} levels. Nature (London) 310:691–693

Dormer RL (1984) Direct demonstration of increases in cytosolic free Ca^{2+} during stimulation of pancreatic enzyme secretion. Bioscience Reports 3:233–240

Dormer RL, Hallett MB, Campbell AK (1985) Measurement of intracellular free Ca^{2+}. Control and manipulation of calcium movement. Raven Press, New York, pp 1–27

Douglas WW, Rubin RP (1961) The role of calcium in the secretory response of the adrenal medulla to acetylcholine. J Physiol (London) 159:40–57

Drabrowska R, Sherry JMF, Aromatorio D, Hartshorne DJ (1978) Modulator protein as a component of the myosin light chain kinase from chicken gizzard. Biochemistry 17:253–258

Ebashi S (1961) Calcium binding activity of vesicular relaxing factor. J Biochem 50:236–244

Ebashi S (1963) Third component participating in the superprecipitation of "natural actomyosin". Nature (London) 200:1010

Ebashi S (1980) Regulation of muscle contraction. Proc R Soc London Ser B 207:259–286

Ebashi S, Kodama A (1965) A new protein factor promoting aggregation of tropomyosin. J Biochem (Tokyo) 58:7–12

Ebashi S, Lipmann F (1962) Adenosine triphosphate-linked concentration of calcium ions in a particular fraction of rabbit muscle. J Cell Biol 14:389–400

Edwards S, Morgan BP, Hoy TG, Luzio JP, Campbell AK (1983) Complement-mediated lysis of pigeon erythrocyte ghosts analysed by flow cytometry. Biochem J 216:195–202

Eisen A, Reynolds GT (1985) Source and sinks for calcium released during fertilization of single sea urchin eggs. J Cell Biol 100:1522–1527

Eisen A, Kiehart DP, Wieland SJ, Reynolds GT (1984) Temporal sequence and spatial distribution of early events of fertilization in single sea urchin eggs. J Cell Biol 99:1647–1654

Fatt P, Katz B (1953) The electrical properties of crustacean muscle fibres. J Physiol (London) 120:171–204

Gilkey JC, Jaffe LF, Ridgway EB, Reynolds GT (1978) A free calcium wave traverses the activating egg of the medaka, *Oryzias latipes*. J Cell Biol 76:448–466

Gopinath RM, Vincenzi FF (1977) Phosphodiesterase protein activator mimics red blood cell cytoplasmic activator of $(Ca^{2+}-Mg^{2+})$ ATPase. Biochem Biophys Res Commun 77:1203–1209

Gorman ALF, Thomas MV (1978) Changes in intracellular concentration of free calcium ions in a pace-maker neurone, measured with metallochromic indicator dye arsenazo III. J Physiol (London) 275:357–376

Gray J (1922) The mechanism of ciliary movement. II. The effect of ions on the cell membrane. Proc R Soc London Ser B 93:122–131

Greengard P, Robison GA, Paoletti R, Nicosia S (1984) Cyclic nucleotides and protein phosphorylation. Advances in Cyclic Nucleotide and Protein Phosphorylation Res 17

Grynkiewicz G, Poenie M, Tsien RY (1985) A new generation of Ca^{2+} indicators with greatly improved fluoresence properties. J Biol Chem 260:3440–3450

Hagiwara S (1973) Ca spike. Adv Biophys 4:71–102

Hagiwara S, Naka KI, Chichibu S (1964) Membrane properties of barnacle muscle. Science 143:1446–1448

Hallam TJ, Rink TJ (1985) Responses to adenosine diphosphate in human platelets loaded with the fluorescent calcium indicator quin 2. J Physiol (London) 368:131–146

Hallam TJ, Sanchez A, Rink TJ (1984) Stimulus–response coupling in human platelets: changes evoked by platelet activating factor in cytoplasmic free calcium monitored with the fluorescent calcium indicator quin 2. Biochem J 218:819–827

Hallett MB, Campbell AK (1982) Measurement of changes in cytoplasmic free Ca^{2+} in fused cell hybrids. Nature (London) 295:155–158

Hallett MB, Campbell AK (1983a) Direct measurement of intracellular free Ca^{2+} in rat peritoneal macrophages: correlation with oxygen-radical production. Immunology 50:487–495

Hallett MB, Campbell AK (1983b) Two distinct mechanisms for stimulation of oxygen radical production by polymorphonuclear leucocytes. Biochem J 213:459–465

Hallett MB, Campbell AK (1985) Is intracellular Ca^{2+} the trigger for oxygen radical production by polymorphonuclear leucocytes? Cell Calcium 5:1–19

Hallett MB, Luzio JP, Campbell AK (1981) Stimulation of Ca^{2+}-dependent chemiluminescence in rat polymorphonuclear leucocytes by polystyrene beads and the non-lytic action of complement. Immunology 44:569–576

Harvey AM, MacIntosh FC (1940) Calcium and synaptic transmission in a sympathetic ganglion. J Physiol (London) 97:408–416

Hasslebach W, Makinose M (1963) Uber den Mechanismus des Calcium Transportes durch die Membranen des Sarkoplasmatischen Reticulums. Biochem Z 339:94–111

Heilbrunn LV (1927) Colloid chemistry of protoplasm. V. A preliminary study of the surface precipitation reaction of living cells. Arch f exper Zellforsch 4:246–263

Heilbrunn LV (1928) The colloid chemistry of protoplasm. Borntraeger, Berlin

Heilbrunn LV (1937 and 1943) An outline of general physiology. Saunders, Philadelphia, Pennsylvania

Heilbrunn LV (1940) The action of calcium on muscle protoplasm. Physiol Zool 13:88–94

Heilbrunn LV (1956) The dynamics of living protoplasm. Academic Press, London

Heilbrunn LV, Wiercinski FJ (1947) The action of various cations on muscle protoplasm. J Cell Comp Physiol 29:15–32

Heilbrunn LV, Wilbur KM (1937) Stimulation and nuclear breakdown in the Nereis egg. Biol Bull 73:557–564

Herbst C (1900) Uber das Anseinandergehen von Furchungs-und Gewebezellen in kalkfreiem Medium. Arch Entwicklungsmech 9:424–463

Hermann S (1932) Die Beeinflussbarkeit der Zustands Form des Calciums in Organismus durch Adrenalin. Naunyn-Schmiedebergs Arch Exp Pathol Pharmakol 167:82–84

Herzberg O, James MNG (1985) Structure of the calcium regulatory muscle protein troponin-C at 2.8 A^0 resolution. Nature (London) 313:653–659

Hodgkin AL, Keynes RD (1957) Movements of labelled calcium in squid giant axons. J Physiol

(London) 138:253–281

Hokin MR, Hokin LE (1953) Enzyme secretion and the incorporation of P^{32} into phospholipids of pancreas slices. J Biol Chem 203:967–977

Holz IV GG, Rane SG, Dunlop K (1986) GTP-binding proteins mediate transmitter inhibition of voltage-dependent calcium channels. Nature (London) 319:670–672

Houssay BA, Molinelli EA (1928) Excitabilité des fibres adrénalino-sécrétories du nerf grand splanchnique. Fréquences, seuil et optimum des stimulus. Role de L'ion calcium. C R Soc Biol 99:172–174

Inouye S, Noguchi M, Sakaki Y et al. (1985) Cloning and sequence analysis of cDNA for the luminescent protein aequorin. Proc Natl Acad Sci USA 82:3154–3158

Irvine RF, Moor RR (1986) Inositol 1,3,4,5 tetraplasphate opens Ca^{2+} channels in fertilised eggs. Biochem J 240: 917–920

Jarrett HW, Penniston JT (1977) Partial purification of the Ca^{2+}–Mg^{2+} ATPase activator from human erythrocytes: its similarity to the activator of 3':5'-cyclic nucleotide phosphodiesterase. Biochem Biophys Res Commun 77:1210–1216

Johnson PC, Ware JA, Clivedon OB, Smith M, Dvorak AM, Salzman EW (1985) Measurement of ionised calcium in blood platelets with the photoprotein aequorin. Comparison with quin 2. J Biol Chem 260:2069–2076

Kakiuchi S, Yamazaki R (1970) Calcium dependent phosphodiesterase. IV. Two enzymes with different properties from brain. Biochem Biophys Res Commun 41:1104–1110 (see also Proc Japan Acad 46:387–392)

Kamata T, Kinoshita H (1943) Disturbances initiated from naked surface of muscle protoplasm. Jpn J Zool 10:469–493

Keith CH, Ratan R, Maxfield FR, Bajer A, Shelanski ML (1985) Local cytoplasmic calcium gradients in living mitotic cells. Nature (London) 316:848–850

Klee CB, Newton DL (1985) Calmodulin: an overview in control and manipulation of calcium movement. Raven Press, New York, pp 131–146

Klee CB, Crouch TH, Richman PG (1980) Calmodulin. Annu Rev Biochem 49:489–515

Ketsinger RH (1976) Evolution and function of calcium-binding proteins. Int Rev Cytol 46:323–393

Kumagai H, Ebashi S, Takeda F (1955) Essential relaxing factor in muscle other than myokinase and creatine phosphokinase. Nature (London) 176:166

Lawaczek H (1928) Ueber das Verhalten des Kalziums unter Adrenalin. Dtsch Arch Klin Med 160:302–309

Lehninger AL (1962) Water uptake and extrusion by mitochondria in relation to oxidative phosphorylation. Physiol Rev 42:467–517

Lew DC, Andersson T, Hed T, Di Virgilio F, Pozzan T, Stendahl O (1985a) Ca^{2+}-dependent and Ca^{2+}-independent phagocytosis in human neutrophils. Nature (London) 315:509–511

Lew DC, Favre L, Waldvogel FA, Vallotton MB (1985b) Cytosolic free calcium and intracellular calcium stores in neutrophils from hypertensive subjects. Clinical Science 69:227–310

Locke FS (1894) Notiz uber den Einfluss, physiologischer Kochsalzlosung auf die Erregbarkeit von Muskel und Nerv. Zentralbl Physiol 8:166–167

Loewenstein O (1984) Junctional intracellular communication: the cell-to-cell membrane channel. Physiol Rev 61:829–913

Loewi O (1917) Uber den Zusammenhang zwischen Digitalis und Calcium-wirkung. Naunyn-Schmeidebergs. Arch Pharmacol 82:131–158

Loewi O (1918) Über den Zusammenhang zwischen Digitalis- und Kalziumwirkung. Arch Exp Pathol Pharmakol 82:131–158

Lux HD, Brown AM (1984) Single channel studies on inactivation of calcium currents. Science 225:432–434

MacLennan DH, Yip CC, Iles GH, Seeman P (1972) Isolation of sarcoplasmic reticulum proteins. Cold Spring Harbor Symp Quant Biol 37:469–477

Majno G (1964) Death of liver tissue. In: Rouiller C (ed) The liver, Vol. II. Academic Press, New York, pp 267–313

March BB (1951) A factor modifying muscle fibre syneresis. Nature (London) 167:1065–1066

Marston SB, Lehman W (1985) Caldesmon is a Ca^{2+}-regulatory component of native smooth-muscle thin filaments. Biochem J 231:517–522

Martonosi AN (1984) Mechanisms of Ca^{2+} release from sarcoplasmic reticulum. Physiol Rev 64:1240–1320

Mazia D (1937) The release of calcium in *Arbacia* eggs upon fertilisation. J Cell Comp Physiol 10:291–304

Means AR, Tash JS, Chafouleas JG (1982) Physiological implications of the presence, distribution, and regulation of calmodulin in eukaryotic cells. Physiol Rev 62:1–39

Meech RW (1979) Calcium-dependent potassium activation in nervous tissue. Annu Rev Biophys Bioeng 7:1–18

Michell RH (1975) Inositol phopholipids and cell surface receptor function. Biochim Biophys Acta 415:81–147

Mines GR (1911) On the replacement of calcium in certain neuro-muscular mechanisms by allied substances. J Physiol (London) 42:251–266

Moreau M, Guerrier P (1979) Free calcium changes associated with hormone action in oocytes. In: Ashley CC, Campbell AK (eds) Detection and measurement of free Ca^{2+} in cells. Elsevier, North-Holland, Amsterdam, pp 219–226

Morgan BP, Campbell AK (1985) The recovery of human polymorphonuclear leucocytes from sublytic complement attack is mediated by changes in intracellular free calcium. Biochem J 231:205–207

Morgan BP, Dankert JR, Esser AF (1986a) Recovery of human neutrophils from complement attack: removal of the membrane attack complex by endocytosis and exocytosis. J Immunol 138:246–251

Morgan BP, Luzio JP, Campbell AK (1986b) Intracellular Ca^{2+} and cell injury. The paradoxical role of Ca^{2+} in complement membrane attack. Cell Calcium 7:399–411

Morgan JP, Morgan KG (1982) Vascular smooth muscle: the first recorded calcium transient. Pflugers Arch 395:75–77

Morgan JP, Morgan KG (1984) Stimulus-specific patterns of intracellular calcium levels in smooth muscle of ferret portal vein. J Physiol 351:155–167

Morin J, Reynolds GT (1974) The cellular origin of bioluminescence in the colonial hydroid Obelia. Biol Bull 147:397–410

Needham DM (1942) The adenosine triphosphatase activity of myosin preparations. Biochem J 36:113–120

Nicchitta CV, Williamson JR (1984) A regulator of mitochondrial calcium cycling spermine. J Biol Chem 259:12978–12983

Nishizuka Y (1984) The role of protein kinase C in cell surface signal transduction and tumour promotion. Nature (London) 308:643–697

O'Doherty J, Stark RJ (1982) Stimulation of pancreatic acinar secretion: increases in cytosolic calcium and sodium. Am J Physiol 242:G513–G521

Oehme M, Kessler M, Simon W (1976) Neutral carrier Ca^{2+}-microelectrode. Chimia 30:204–206

Overton E (1902) Beiträge zur allgemeinen Muskel und Nerven-physiologie. II. Ueber die Unenbehr-lichkeit von Natrium (oder Lithium) Ionen fur den Contractions act des Muskels. Pflugers Arch Gesamte Physiol 92:346–386

Parratt JR (ed) (1985) Control and manipulation of calcium movement. Raven Press, New York

Pasternak CA, Bashford CL, Micklem KJ (1985) Ca^{2+} and the interaction of pore-formers with membranes. Proceedings of the International Symposium on Biomolecular Structural Interactions Supplement J Bioscience 8:2108–2189

Patel A, Campbell AK (1987) The membrane attack complex of complement induces permeability changes via thresholds in individual cells. Immunology 60:135–140

Poinie M, Alderton J, Tsien RY, Steinhardt RA (1985) Changes of free calcium levels with stages of the cell division cycle. Nature (London) 315:147–149

Pollack H (1928) Micrugical studies in cell physiology. VI. Calcium ions in living protoplasma. J Gen Physiol 11:539–545

Portzehl H, Caldwell PC, Ruegg JC (1964) The dependence of contraction and relaxation of muscle fibres from the crab Maia squinado on the internal concentration of free calcium ions. Biochim Biophys Acta 79:581–591

Potter JD, Gergeley J (1975) The calcium and magnesium binding sites of troponin and their role in the regulation of myofibrillar adenosine triphosphatase. J Biol Chem 250:4628–4633

Powers RE, Johnson PC, Houlihan MJ, Saluja AK, Steer ML (1985) Intracellular Ca^{2+} levels and amylase secretion in quin 2-loaded mouse pancreatic acini. Am J Physiol 248:C535–C541

Prasher D, McCann RO, Cormier MJ (1985) Cloning and expression of the cDNA coding for aequorin, a bioluminescent calcium binding protein. Biochem Biophys Res Commun 126:1259–1268

Reuter H (1983) Calcium channel modulation by neurotransmitters, enzymes and drugs. Nature (London) 301:569–574

Reynolds GT (1978) Application of photosensitive devices to bioluminescence studies. Photochem Photobiol 27:405–421

Ridgway EB, Ashley CC (1967) Calcium transients in single muscle fibres. Biochem Biophys Res Commun 29:224–234

Ringer S (1883) A further contribution regarding the influence of different constituents in the blood on the contractions of the ventricle. J Physiol (London) 4:26–43

Ringer S, Sainsbury H (1894) The action of potassium, sodium and calcium salts on *Tubifex rivulorum*. J Physiol (London) 16:1–9

Rink TJ, Hallam TJ (1984) What turns on platelets? Trends in Biochemical Sciences 9:215–219

Rink TJ, Smith SW, Tsien RY (1982) Cytoplasmic free Ca^{2+} in human platelets: Ca^{2+} thresholds and Ca-independent activation for shape change and secretion. FEBS Letters 148:21–26

Rink TJ, Sanchez A, Hallam TJ (1983) Diacyl glycerol and phorbol ester stimulate secretion without raising cytoplasmic free calcium in human platelets. Nature (London) 305:317–319

Roberts PA, Morgan BP, Campbell AK (1984) 2-chloroadenosine inhibits complement-induced reactive oxygen metabolite production and recovery of human polymorphonuclear leucocytes attacked by complement. Biochem Biophys Res Commun 126:692–697

Roberts PA, Newby AN, Hallett MB, Campbell AK (1985) Inhibition by adenosine of reactive oxygen metabolite production by human polymorphonuclear leucocytes. Biochem J 227:669–674

Rose B, Loewenstein WR (1975) Permeability of cell junction depends on local cytoplasmic calcium activity. Nature (London) 254:250–252

Rose B, Loewenstein WR (1976) Permeability of a cell junction and the local cytoplasmic free ionised calcium concentrations: a study with aequorin. J Membr Biol 28:87–119

Scarpa A (ed) (1985) Measurements of intracellular calcium. Cell Calcium 6

Scarpa A, Brinley FJ, Tiffert T, Dubyak GR (1978) Metallochromic indicators of ionized calcium. Ann NY Acad Sci 307:86–112

Schatzmann HJ (1966) ATP-dependent Ca^{2+} extrusion from human red cells. Experientia 22:364

Schatzmann HJ (1975) Active calcium transport and Ca^{2+}-activated ATPase in human red cells. Curr Top Membr Trans 6:125–168

Schulman H, Greengard P (1978) Ca^{2+}-dependent protein phosphorylation system in membranes from various tissues, and its activation by calcium-dependent regulator. Proc Natl Acad Sci USA 74:5432–5436

Schutt C (1985) Protein structure: hands on the calcium switch. Nature (London) 315:15

Severson DL, Denton RM, Bridges BJ, Randle PJ (1976) Exchangeable and total calcium pools in mitochondria or rat epididymal fat pads and isolated fat-cells. Role in the regulation of pyruvate dehydrogenase activity. Biochem J 154:209–223

Shimomura O, Johnson FH (1976) Regeneration of the photoprotein aequorin. Nature (Lond) 256:236–238

Smith GA, Hesketh TR, Metcalfe JC, Feeney J, Morris PG (1983) Intracellular calcium measurements by ^{19}F NMR of fluorine-labelled chelators. Proc Natl Acad Sci USA 80:7118–7182

Smoake JA, Song SY, Cheung WY (1974) Cyclic 3′5′ – nucleotide phosphodiesterase. Distribution and developmental changes of the enzyme and its protein activator in mammalian tissues and cells. Biochim Biophys Acta 341:402–411

Somlyo AP (1985) Excitation–contraction coupling and the ultrastructure of smooth muscle. Circulation Research 57:497–507

Somlyo AP (1986) The messenger across the gap. Nature (London) 316:298–299

Spat A, Fabiato A, Rubin RP (1986) Binding of inositol trisphosphate of a liver microsomal fraction. Biochem J 233:929–932

Stevens CF, Tsien RW, Yellen G (1982) Properties of single calcium channels in cardiac cell culture. Nature (London) 297:501–504

Streb H, Irvine RF, Berridge MJ, Schulz I (1983) Release of Ca^{2+} from a non-mitochondrial intracellular store in pancreatic acinar cells by inositol-1,4,5-trisphosphate. Nature (London) 306:67–69

Streb H, Heslop JP, Irvine RF, Schulz I, Berridge MJ (1984) Relationship between secretagogue-induced Ca^{2+} release and inositol polyphosphate production in permeabilized pancreatic acinar cells. J Biol Chem 260:7309–7315

Takai Y, Kishimoto A, Iwusa Y, Kawakara Y, Mori T, Nishizuka Y (1979) Calcium-dependent activation of a multi-functional protein kinase by phospholipids. J Biol Chem 254:3692–3695

Taylor DL, Reynolds GT, Allen RD (1975) Detection of free calcium ions in amoebae by aequorin bioluminescence. Biol Bull 149:448

Teo ST, Wang JH (1973) Mechanism of action of cyclic adenosine 3′:5′-monophosphate phosphodiesterase from bovine heart by calcium ions. J Biol Chem 248:5950–5955

Thomas MV (1982) Techniques in calcium research. Academic Press, London

Trewavas AJ (1985) Ca^{2+} in plants. Phytol Rev 31

Trump BF, Berezesky IK, Orsonio-Vargas AR (1981) Cell death and the disease process. The role of calcium. In: Bowen ID, Lockshin RA (eds) Cell death in biology and pathology. Chapman and Hall, London and New York, pp 209–242

Tsien RY (1980) New calcium indicators and buffers with selectivity against magnesium and protons: design, synthesis and properties of prototype structures. Biochemistry 19:2396–2404

Tsien RY (1981) A non-disruptive technique for loading calcium buffers and indicators into cells. Nature (London) 290:527–528

Tsien RY (1983) Intracellular measurements of ion activities. Ann Rev Biophys Bioeng 12:91–116

Tsien RY, Poinie M (1986) Fluorescence ratio imaging: a new window into intracellular ionic signalling. TIBS 11:450–455

Tsien RY, Pozzan T, Rink TJ (1982) T-cell mitogens cause early changes in cytoplasmic free Ca^{2+} and membrane potential in lymphocytes. Nature (London) 295:68–71

Tsien RY, Rink TJ, Peonie M (1985) Measurement of cytosolic free Ca^{2+} in individual small cells using fluorescent microscopy with dual excitation wavelengths. Cell Calcium 6:145–157

Verati E (1902) Ricerche sulla fine struttura della fibra muscolane striata. Arch Ital Biol 37:449–454

Waisman DM, Steven FC, Wang JH (1978) Purification and characterisation of a Ca^{2+} binding protein in *Lumbricus terrestris*. J Biol Chem 253:1106–1113

Watterson DM, Iverson DB, van Eldik LJ (1980) Spinach calmodulin: isolation, characterization and comparison with vertebrate calmodulin. Biochemistry 19:5762–5768

Weber A, Herz R, Reiss I (1964) The regulation of myofibrillar activity by calcium. Proc R Soc London Ser B 160:489–501

Weber A, Herz R, Reiss I (1966) Study of the kinetics of calcium transport by isolated fragmented sarcoplasmic reticulum. Biochem Z 345:329–369

Weeds A (1982) Actin-binding protein-regulators of cell architecture and motility. Nature (London) 296:811–816

Welsh MJ, Dedman JR, Brinkley BR, Means AR (1978) Calcium-dependent regulator protein: localization in mitotic apparatus of eukaryotic cells. Proc Natl Acad Sci USA 75:1867–1871

Wiedmer T, Sims PJ (1985) Effect of complement proteins C5b-9 on blood platelets. Evidence for reversible depolarization of membrane potential. J Biol Chem 260:8014–8019

Williams DA, Fogarty KE, Tsien RY, Fay FS (1986) Calcium gradients in single smooth muscle cells revealed by digital imaging microscopy using fura-2. Nature (London) 318:558–561

Williamson JR, Cooper RH, Joseph SK, Thomas AP (1985) Inositol trisphosphate and diacylglycerol as intracellular second messengers in liver. Am J Physiol 248:C203–216

Woods NM, Cutherbertson KSR, Cobbold PH (1986) Repetitive transient rises in cytoplasmic free calcium in hormone-stimulated hepatocytes. Nature (London) 319:600–602

Woodward AA (1949) The release of radioactive Ca^{45} from muscle during stimulation. Biol Bull (Woods Hole, Mass) 97:264

Wyllie AH, Kerr JFR, Currie AR (1980) Cell death: the significance of apoptosis. Int Rev Cytol 68:251–306

Wyn-Jones RG, Lunt OR (1967) The function of calcium in plants. Bot Rev 33:407–426

Chapter 12

Cellular Calcium: Muscle

S. Ebashi

Historical Survey of Calcium Research in Muscle

Nowadays, everyone knows that Ca^{2+} is the most important general intracellular regulatory factor. This fact was first deduced from studies on muscle, carried out by muscle scientists. Therefore, a historical review of calcium research in muscle will be of general relevance to all the biological sciences.

Dawn of Calcium Research

As is well known, the first person who recognized the role of Ca^{2+} in muscle contraction was Ringer (1883). It is now established that in heart muscle Ca^{2+} in the extracellular fluid enters into myoplasm as a slow inward current and participates in the activation of the contractile system. However, scientists at the end of the nineteenth century, including Ringer himself, merely interpreted their results as meaning that Ca^{2+} served to maintain the physiological state of the cell by antagonizing K^+. The need for Ca^{2+} in smooth muscle contraction was also explained on the basis of the ratio of Ca^{2+} to K^+ (Stiles 1903).

In 1940 Heilbrunn showed that an isolated frog muscle fibre, of which the ends had been cut, vigorously contracted when placed in isotonic $CaCl_2$. Inspired by this work, Kamada and Kinosita (1943) injected a minute amount of Ca^{2+} into the myoplasm of a single fibre through a micropipette and observed a reversible local contracture. Heilbrunn and Wiercinski (1947) also carried out a similar experiment and noticed the shortening of the fibre after a while.

Prior to this, Chambers and Hale (1932) had induced a vigorous contraction in a muscle fibre by applying calcium to once-frozen muscle fibre. This was certainly equivalent to the work of Heilbrunn. Keil and Sichel (1936) had also reported calcium injection experiments. In spite of these exciting findings, none of these earlier scientists suggested that Ca^{2+} might trigger physiological contraction. For the physiologists at that time it was perhaps impossible to imagine that such a common ion as Ca^{2+} could bring about such a phenomenon; the only agent conceivable to them would have been something related to electricity. (This reminds us of the fact that some physiologists were very reluctant to accept the chemical transmitter theory as an explanation for the process occurring at the neuromuscular junction of skeletal muscle.)

Heilbrunn, the first person to realize the true role of Ca^{2+} in muscle contraction, was a truly far-sighted scientist who deserves the name of the pioneer of calcium research.

Ca^{2+} on the Molecular Level

The experimental results obtained by Heilbrunn and his followers did not inspire muscle biochemists of the time to pursue further the study of the molecular mechanism of muscle contraction based on the calcium hypothesis. The work was overshadowed by the brilliant proposal of Szent-Györgyi (1942, 1947) and his school for an actin–myosin–ATP system for muscle contraction.

After Heilbrunn's work, it was nearly 20 years before the Ca^{2+} concept was revived on new grounds (for details, see Ebashi and Endo 1968), which can be summarized as follows:

1. Myofibrils or natural actomyosin (myosin B, a crude actomyosin system) require a minute amount of Ca^{2+} for their full activation (Weber 1959; Ebashi 1960). A few μM Ca^{2+} are enough to activate the contractile elements fully and even 0.2 μM Ca^{2+} definitely activates the actomyosin system from which contaminating calcium has been removed (Ebashi 1961a). The degree to which Ca^{2+} binds to myofibrils determines their contractility (Weber and Herz 1963).

2. The relaxing factor, identified with the fragmented sarcoplasmic reticulum, removes Ca^{2+} from the medium in an ATP-dependent reaction (Ebashi 1960; Ebashi and Lipmann 1962). This Ca^{2+}-depriving activity is strong enough to explain the mechanism of relaxation (Ebashi 1961a). It then became possible to explain the contraction–relaxation cycle by the simple scheme (Ebashi 1961b) now described in standard textbooks.

Thus the final explanation for muscle contraction/relaxation was very simple, but it was only arrived at after many twists and turns. It might be worthwhile looking back at the progress of the research.

First, the calcium concept could not have been explored before the development of apparatus made of hard glass or plastics from which no Ca^{2+} would come out. The Ca^{2+} concentration in the soft glass containers which were still in common use in biochemical laboratories in the 1950s easily exceeded the level required to activate the contractile system fully.

Secondly, it should be emphasized that the development of the calcium concept was made possible by the study on the relaxing factor (Marsh 1951), a study which was not considered to be in the main stream of muscle research. In 1955, the relaxing factor was identified by Kumagai et al. with the magnesium-activated granule ATPase discovered by Kielley and Meyerhof earlier (1948a, b). The microsomal nature of the ATPase eliminated the possibility that the relaxing factor would directly interact with the actomyosin system. Most biochemists then thought erroneously that a new substance of low molecular weight (a "soluble relaxing factor") was produced by ATPase, perhaps by ATP itself. It is very likely that this idea was suggested by the success of the work on cyclic AMP. Anyway, it was unfortunate that a number of biochemists spent much time pursuing this imaginary substance in vain.

Thirdly, the progress of calcium research was greatly facilitated by the discovery by Schwarzenbach (1955) of various kinds of chelating agents including EDTA and EGTA (GEDTA according to Schwarzenbach). Ethylenediaminetetraacetate (EDTA) was shown to mimic the natural relaxing factor by inducing the relaxation of glycerinated fibres (Bozler 1954; Watanabe 1955). Subsequently, the linear relationship between the calcium-removing capacities of various chelating agents and their relaxing actions on glycerinated fibres was revealed (Ebashi 1959, 1960).

Discovery of Troponin, a New Step in Calcium Research

The calcium hypothesis still had to overcome serious difficulties before being accepted by the majority of muscle biochemists. The main reason for this was that calcium regulation could certainly be seen in crude actomyosin systems, e.g. natural actomyosin, myofibrils and glycerinated fibres, but not in pure actomyosin reconstituted from separately extracted pure actin and myosin (Weber and Winicur 1961). The resistance to the calcium hypothesis by some muscle biochemists was due to the fact that the crude system could still be contaminated by sarcoplasmic reticulum which could produce "soluble" relaxing factor.

To overcome this opposition, the reason why pure actomyosin was not sensitive to Ca^{2+} had to be explained. This was possible following the discovery in the contractile system of a third factor, "native tropomyosin" (Ebashi 1963; Ebashi and Ebashi 1964), which was later separated into two components, classical tropomyosin (Bailey 1946) and a new globular protein, troponin, the first Ca^{2+} receptor protein to be found (Ebashi and Kodama 1965). Thus did the Ca^{2+} concept gain its firm foundation.

Troponin

Since the discovery of troponin (Ebashi and Kodama 1965), a tremendous amount of experimental research has been carried out on this protein (Ebashi and Endo 1968; Ebashi et al. 1969; Ebashi 1974b; Potter and Johnson 1982; Ohtsuki et al. 1986; the last article is a thorough review of troponin research).

There are many points of note, but only a few will be referred to in this chapter because of limited space.

Hybrid Troponins

Following the procedures used for separation of skeletal-muscle troponin, cardiac troponin was separated into its three components (Tsukui and Ebashi 1973). This enabled hybrid troponins to be made up (Ebashi 1974a, b).

The function of troponin is based on the three interactions of its three subunits: (a) a Ca^{2+}-dependent interaction of troponin I (inhibitory component) with troponin C (calcium-binding component); (b) a Ca^{2+}-independent interaction of troponin I with troponin C; (c) a Ca^{2+}-independent interaction of troponin I with troponin T (tropomyosin-binding component).

A puzzling but interesting interaction is that of troponin C with troponin T. In skeletal troponin, this interaction is fairly strong (Ebashi et al. 1972) and it was once believed that it played an essential role in physiological processes. An experiment using hybrid troponins has shown clearly, however, that the interaction is of no primary importance in the troponin mechanism (Table 12.1) (Ebashi 1974b). Interestingly, the hybrid troponin composed of skeletal troponin I, skeletal troponin C and cardiac troponin T shows a higher degree of regulation than do the original troponins but in this hybrid troponin the interaction is virtually nil.

Table 12.1. Actomyosin ATPase activity in relation to the troponin T–troponin C interaction and the troponin C–troponin I interaction

Origins of muscle from which troponin subunits were derived			Degrees of T–C inter-actions (A) (inhibiting)	Degrees of C–I inter-actions (B) (activating)	Maximum ATPase activities expected from A and B	Actual maximum ATPase activities
T	C	I				
c	c	c	+	+ +	+ + +	114
c	s	s	+	+ + +	+ + + +	153
s	s	c	+ + +	+	+	73
s	s	s	+ + +	+ + +	+ +	(100)

s, skeletal origin; c, cardiac origin; T, C and I, troponin T, C and I.
Expected maximum ATPase activities are estimated under the assumption that the ATPase activity is mainly dependent on the C–I interaction, but that the T–C interaction counteracts the C–I interaction repressing the maximum ATPase activity (i.e. the maximum ATPase activity should be something like B divided by A). As a result a good correlation between expected and actual ATPase activities was demonstrated. For the T–C and C–I interactions refer to Ebashi (1974) and for the ATPase activity, Yamamoto (1983).

However, there is circumstantial evidence that this interaction may play a subsidiary but useful role in skeletal muscle. At relatively high Ca^{2+} concentrations, say 10^{-4} M, troponin C binds to troponin I so strongly that it might overactivate the actin–myosin interaction. The affinity of troponin T for troponin C antagonizes this binding and, as a result, confines the contraction of skeletal muscle to a certain level. Thus the interaction of troponin T and troponin C may act as a self-defence mechanism preventing excess contraction in skeletal muscle, the muscle in which Ca^{2+} concentration can reach a high value (Ebashi 1984).

Conformation of Troponin with Special Reference to that of Troponin T

In contrast to tropomyosin, i.e. one of the most typical fibrous proteins, troponin was originally thought to be a globular protein because it can be dissolved in water to a concentration of more than 100 mg/ml without a

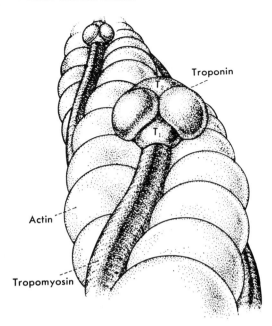

Fig. 12.1. The troponin–tropomyosin complex showing the arrangement of troponin subunits (the view from the Z-band side of the filament). It has not yet been decided which is troponin I and which troponin C, but it is more likely that the *left* one is troponin C. For the junctional point between two adjacent tropomyosin molecules, refer to Fig. 5 in Ebashi (1974a).

substantial increase in viscosity. The models for the troponin–tropomyosin–actin complex and the troponin–tropomyosin complex (Ebashi et al. 1969; Ohtsuki 1980) are based on this idea (Fig. 12.1).

However, Flicker et al. (1982) produced an electron microscopic profile of troponin, using the rotary shadowing technique, and concluded that it possessed an elongated conformation, the asymmetric part undoubtedly corresponding to troponin T.

Ohtsuki has re-examined this problem using the same technique (Ohtsuki et al. 1986; Ebashi and Ohtsuki 1983) and confirmed his previous view that troponin is in principle a globular protein; if aged preparations were studied in the same way, elongated forms could be seen (Fig. 12.2).

Troponin T has two binding sites for tropomyosin. The affinity of the N-terminal for tropomyosin is very strong, but that of the C-terminal is weak. Chymotryptic digestion splits troponin T into two fragments, named T_1 (N-terminal) and T_2 (C-terminal) (Fig. 12.3). Strangely enough, T_2 can almost fully perform the function of troponin T (if the weak binding site to tropomyosin is lost, the function disappears). In other words, the N-terminal part has no functional relevance in spite of its strong affinity for tropomyosin.

If troponin T is protruded from the main part of troponin as suggested by Flicker et al. (1982), its end can reach the junction of two adjacent tropomyosin molecules. This might suggest that troponin T serves a cooperative function with

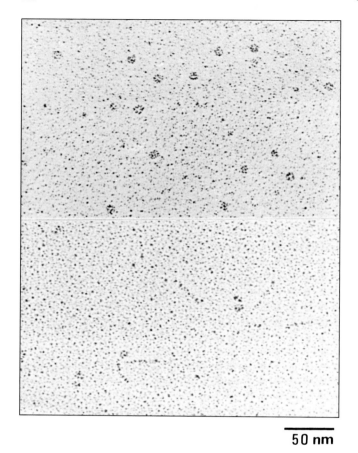

50 nm

Fig. 12.2. Electron micrograph of troponin using rotary shadowing technique. In a fresh preparation (*top*), globular forms are dominant, whereas elongated profiles are exclusively seen in an aged preparation (*bottom*). (Courtesy of I. Ohtsuki)

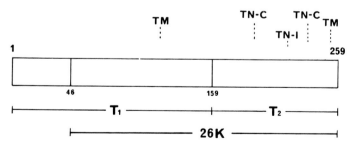

Fig. 12.3. Proteolytic fragments of troponin T. The portions having affinities for other proteins are indicated in the upper part of the corresponding fragments (treated with chymotrypsin). *TN–C*, *TN–I* and *TM*; troponin C, troponin I and tropomyosin respectively. The fragment with molecular weight 26 000 is formed during preservation by unidentified protease. No difference in biological activities could be found between the native troponin T and the fragment with molecular weight 26 000 (in the case of T_2, a slight difference could be detected) (Ohtsuki et al. 1984).

troponin. However, such speculation has lost ground in view of Ohtsuki's electron microscopic observations as well as proteolytic experiments.

Sr^{2+} Sensitivity

It is a remarkable fact that glycerinated skeletal and cardiac muscle fibres show a distinct difference in their sensitivity to Sr^{2+}, the latter being far more sensitive to it (Fig. 12.4). Such a difference can be demonstrated with natural actomyosin (myosin B), if superprecipitation is used as the index for in vitro contraction. (This property was successfully used to decide whether the troponin molecule could be fully responsible for calcium regulation; see Table 3 in Ebashi et al. 1968.) However, if actomyosin ATPase is used as the index of in vitro contraction, only a small difference (Berson 1974; Kohama in Ebashi et al. 1980; Yamamoto 1983) is seen. Furthermore, there is almost no difference in the Sr^{2+} concentration that binds half-maximally to troponin in cardiac and skeletal muscle (Kohama 1979; Kohama et al. 1986) (Table 12.2). No one doubts that glycerinated muscle is more akin to living muscle than is the ATPase activity. It is clear then that superprecipitation is a better index of in vitro contraction than is ATPase activity.

Myosin-Linked Regulation

Efforts to isolate troponin from molluscan striated muscle were unsuccessful. The final conclusion was that the calcium regulation site is to be found in the

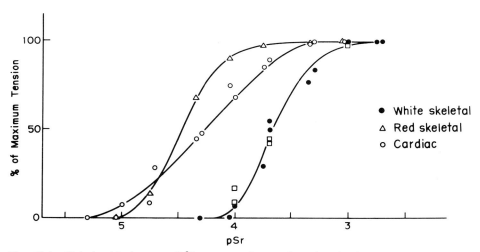

Fig. 12.4. Relationship between Sr^{2+} concentrations and tension development in glycerinated muscle fibres. In all cases, except for those represented by open squares, EGTA was present in the medium to eliminate the effect of Ca^{2+}. For other details, refer to the original paper (Kitazawa 1976). The upper two data in Table 12.2 were derived from this figure (Kitazawa 1976) but the Ca^{2+} concentrations have been corrected.

Table 12.2. Sr^{2+} sensitivities of various actomyosin systems

Actomyosin systems	Ratios of Sr^{2+} sensitivities of cardiac system to those of skeletal system[a]	References
Tension development		
Glycerinated fibres[b]	4.1	Kitazawa (1976)
	5.9[c]	Kitazawa (1976)
Superprecipitation		
Natural actomyosin	7.0	Ebashi et al. (1968)
	3.0	Kohama (unpublished see Ebashi et al. 1980)
Desensitized actomyosin with native tropomyosin	5.7[c]	Ebashi et al. (1968)
ATPase		
Desensitized actomyosin with native tropomyosin	1.6	Berson (1974)
Natural actomyosin	2.0	Kohama (unpublished see Ebashi et al. 1980)
Desensitized actomyosin with tropomyosin and reconstituted troponin	2.9	Yamamoto (1983)
Sr^{2+} binding to troponin		
Troponin alone	1.0	Kohama (1979)
Troponin bound to desensitized actomyosin	1.0	Kohama et al. (1986)

[a] Sr^{2+} concentrations to give half maximum activities were compared unless otherwise stated.
[b] Cardiac skinned fibres showed a slightly higher sensitivity than did cardiac glycerinated fibres, but comparative studies on skeletal skinned fibres under the same conditions were not made (Kohama, unpublished).
[c] Sr^{2+} concentrations to give 30% of maximum activities were compared.

myosin molecule itself (Kendrick-Jones et al. 1970). This has suggested that the calcium regulation mechanism differs from muscle to muscle, and it is now commonly believed that the diversity itself is a characteristic of calcium regulation.

Distribution of the Myosin-Linked System

Lehman and Szent-Györgyi (1975) performed elegant studies on the distribution of the troponin system and the myosin-linked system in the animal kingdom, using their unique method, the so-called "dilution" method. According to their results, the troponin system is found in the chordates or higher deuterostomia and in the horseshoe crab or higher protostomia. On the other hand, the myosin-linked system seems to exist in all the protostomia with a few exceptions. Their method is based on the assumption that the activation of actin–myosin–ATP interaction is based on a derepressive type of regulation (see the section on smooth muscle). However, the presence of an intermediate type of regulation (Endo and Obinata 1981), i.e. a type between derepression and activation (smooth muscle type), suggests that a final conclusion should await the actual demonstration of troponin or the myosin molecule with a calcium receptive site. Research in this field has so far shown that the prediction is right in principle. Indeed, the body-wall muscles of insects (K. Yagi, personal communication), *Ascaris* (Kimura and Obinata 1983; Kimura et al. 1987) and *C. elegans* (Nakae and Obinata 1985) certainly contain troponin. It is interesting that such a primitive animal as *Ascaris* is equipped with the troponin system; this may provide a clue to the phylogenetic position of this animal. On the other hand, efforts to obtain calcium-linked myosin from insects and crustaceans has so far been unsuccessful.

Troponin-like Protein in Molluscan Muscle

Lehman (1981) insists that all the muscles related to body movement in molluscs are equipped with both the troponin and myosin-linked systems. Yazawa (1985) has found that the calcium regulation of actomyosin reconstituted from separately isolated myosin and actin is not satisfactory and that the addition of a component with molecular weight 45 000, which has been isolated from molluscan thin filaments and is inhibitory in its nature like troponin I, confers the original Ca^{2+} sensitivity on reconstituted actomyosin. Furthermore, troponin itself has been isolated from molluscan muscle (Ojima and Nishita 1986a, b).

Thus the distribution of troponin, or troponin-like protein(s), is very widespread, and not confined to striated muscles of higher protostomia and deuterostomia, in partial agreement with Lehman's assertion (1981).

Ca^{2+} Regulation in Smooth Muscle

To discuss Ca^{2+} regulation in smooth muscle, it is necessary to refer to the characteristics of contractile proteins in this muscle, especially those of myosin. Smooth muscle myosin is quite different from the myosins of vertebrate striated muscles (cardiac and slow red muscle myosins are considerably different from fast white muscle myosin, but if compared with smooth muscle myosin, the difference is quite small).

1. Smooth muscle myosin is very soluble. It can be extracted from muscle mince even with 0.02 M KCl, if it contains ATP. This is necessary for dissociating myosin from actin, but at low ionic strength it does not prevent myosin molecules from forming filaments. The filaments of freshly prepared myosin do not aggregate with one another, staying in a soluble state, but after a while they start to form a conglomerate. The solubility then decreases rather rapidly and becomes less than that of skeletal muscle myosin, which is much less than that of fresh smooth muscle myosin but is preserved for a fairly long time. So far no way is known of maintaining the initial high solubility of pure smooth muscle myosin; the only way to maintain it is to keep the myosin in the state of actomyosin.

2. An intrinsic property of skeletal muscle myosin is to react with actin positively in the presence of ATP (contraction). This natural reaction of actin and myosin cannot be repressed (relaxation) without Ca^{2+}-free troponin and calcium is required for removing this repression (contraction) (Ebashi and Endo 1968; Ebashi et al. 1969). In sharp contrast with this, smooth muscle myosin does not react with actin in the presence of ATP (Ebashi et al. 1976). To activate it (contraction) some factor is necessary together with Ca^{2+} (Fig. 12.5). It is not clear how this property is related to that described in (1).

3. In contrast to skeletal muscle myosin, smooth muscle myosin cannot form filaments in vitro unless Mg^{2+} is present in fairly high concentrations of 2 mM or more. The smooth muscle myosin filament is easily dissociated by ATP to myosin monomers if the ionic strength is over 0.15; if it exceeds 0.25, the myosin filaments decompose into myosin monomers without ATP. Phosphorylation of the light chain elevates the upper range of ionic strength at which myosin stays in the filament form (Suzuki et al. 1978), but the magnesium requirement is the same. In vivo, where the free Mg^{2+} concentration is around 1 mM, the myosin filament is formed even in the resting state, i.e. without phosphorylation.

In addition to the above, smooth muscle tropomyosin is very different from the skeletal muscle tropomyosin which is required for activation of the actin–myosin interaction (Ebashi et al. 1976). Even smooth muscle actin is different from the skeletal one (Mikawa and Maruyama 1981).

The factor which activates the myosin–actin–ATP interaction is now widely believed to be the calmodulin-dependent myosin light chain kinase (Bremel et al. 1977; Hartshorne and Siemankowski 1981). Myosin in which the light chain of molecular weight 20 000 has been phosphorylated can react with actin in a similar way to skeletal muscle myosin. The interaction of myosin and actin is terminated when the phosphorylated light chain is dephosphorylated by a phosphatase. Thus the phosphorylation–dephosphorylation of the light chain appears to be the mechanism underlying the contraction–relaxation cycle.

However, one line of smooth muscle research has been presenting evidence opposing the above idea (Ebashi 1980; Ebashi et al. 1982). This asserts that an actin-linked factor, called "leiotonin", acts as the activator. Furthermore, there is a general consensus that after reaching peak tension the level of phosphoryla-

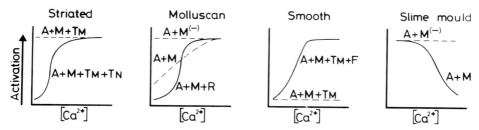

Fig. 12.5. Modes of calcium regulation in various muscles and slime mould. *Striated*, vertebrate skeletal and cardiac muscle. *Molluscan*, molluscan striated muscle. *Smooth*, vertebrate smooth muscle. *Slime mould*, both plasmodial and amoeba types. For "activating factor" and "regulatory factor" see text. *A*, actin; *M*, myosin; *TM*, tropomyosin; *TN*, troponin; *M⁻*, myosin without regulatory light chain; *F*, activating factor; *R*, regulatory factor.

tion in living fibres is reduced almost to the level that existed before stimulation while the tension is kept at the maximum level (e.g. Dillon et al. 1981). Thus the "light chain kinase" theory must be supported by more conclusive evidence.

In addition to the above, there is an opinion that caldesmon, a calmodulin-binding protein, is involved in the contractile processes (Sobue et al. 1982; Marston et al. 1985). Caldesmon in the absence of Ca^{2+}, i.e. caldesmon without calmodulin, binds to actin and inhibits the myosin–actin–ATP interaction. Ca^{2+} removes this inhibition by forming the calmodulin–caldesmon complex which cannot bind to actin. Thus the role of caldesmon is to inhibit the contraction and/or to facilitate the relaxation, but this idea assumes the presence of some other activating system. The evidence to support the physiological involvement of this factor is awaited.

Generally speaking, research on Ca^{2+} regulation in smooth muscle is still in a preliminary stage.

Actomyosin in Non-muscle Tissue

The most remarkable function performed by actomyosin in non-muscle tissues is control of cytoplasmic streaming, one of the characteristic features of some plant cells (another important one is amoeboid movement). There is now increasing evidence of the similarity between the contractile processes in muscle and cytoplasmic streaming; rabbit muscle myosin bound to beads streams along the actin rail of *Nitella* (Sheetz and Spudich 1983) and the sliding of rabbit muscle myosin along the rabbit actin filament is extremely rapid, comparable to cytoplasmic streaming, if the steric restraint of myosin molecules is released (Yanagida et al. 1985).

In this section, reference will be made mainly to the actin-myosin interaction in the slime mould, which exhibits marked cytoplasmic streaming when it is in the form of *Physarum polycephalum* (plasmodial form), but amoeboid movement if converted to *Physarum amoebae (amoeboid form)*.

Ca^{2+} in the Actin–Myosin–ATP Interaction of Slime Mould

As stated before, the mode of action of Ca^{2+} differs from muscle to muscle. Ca^{2+} acts as a derepressor of the actomyosin system of vertebrate striated muscles and a true activator of that of smooth muscle (Fig. 12.5). In any case, the final consequence is the activation of the contractile processes in muscle. Therefore, it was quite unexpected to find that the interaction of actin and myosin in plasmodial slime mould in the presence of ATP is inhibited by Ca^{2+} (Kohama et al. 1980; Kohama and Shimmen 1985). This is a very significant finding in the research on the actomyosin system. It is now realized that, although Ca^{2+} is a common regulator, its mode of action varies, no uniform principle being applicable in the systems controlled by Ca^{2+}.

Whether the effect of Ca^{2+} on cytoplasmic streaming is accelerating or repressing had been the subject of much discussion. Recently it has been widely agreed that Ca^{2+} exerts an inhibitory effect on streaming; using thread made of extracted cytoplasm, inhibition by Ca^{2+} of contractility could be observed (Yoshimoto et al. 1981; Yoshimoto and Kamiya 1984).

The Ca^{2+} Receptive Site

The Ca^{2+} receptive site of slime mould actomyosin is located in the myosin molecule (Kohama and Kohama 1984). However, the situation differs from that in molluscan myosin, where the calcium-binding site is not found in individual subunits of the myosin molecule, in spite of the fact that the whole myosin molecule binds calcium. The isolated light chain (molecular weight 14 000) of slime mould myosin shows an affinity for Ca^{2+} (Kessler et al. 1980), corresponding to the sensitivity of the whole actomyosin system to Ca^{2+}. On the whole it can be concluded that the light chain with molecular weight 14 000 is the calcium-receptive site of cytoplasmic streaming.

Another difference between molluscan muscle myosin and that in slime mould is that myosin in the latter is active only when it is in the phosphorylated state. However, the phosphorylation–dephosphorylation cycle is not involved in the physiological process (Kohama and Kendrick-Jones 1986).

Calcium Regulation in *Polycepharum amoebae*

If the plasmodial form is converted to the amoeboid form (this may be considered to be the conversion from a plant-like organism to an animal-like one) under some environmental conditions, the amoeboid slime mould forms its own heavy chain and light chain (molecular weight 18 000), distinguishable immunochemically from those of the plasmodial form, but the light chain with molecular weight 14 000 remains unchanged and its actomyosin system shows essentially the same response to Ca^{2+} as that of the plasmodial form (Kohama and Takano-Ohmuro 1984).

Excitation–Contraction Coupling

One of the important calcium-related events in muscle is the process by which the electrical signal at the surface membrane is transmitted to the cytoplasm so that Ca^{2+} becomes available for contractile elements. This process is called "excitation–contraction coupling". This is a complicated field so that it is difficult to describe the whole idea in a limited space and the reader is referred to review articles on the subject (e.g. Ebashi 1976; Endo 1985).

Skeletal Muscle

Skeletal muscle, particularly fast white muscle, has attained a high degree of differentiation. Reflecting this fact, the form of excitation–contraction coupling is simplified and limited to one route.

Almost all the Ca^{2+} of skeletal muscle utilized for contraction is dealt with by the sarcoplasmic reticulum. The ordinary surface membrane, sarcolemma, has virtually no direct role in carrying Ca^{2+} into sarcoplasm. In the resting state, calcium is almost exclusively stored in the terminal cistern of the sarcoplasmic reticulum.

The mechanism whereby the electrical signal at the T-system (T-tubule or transverse system) can elicit Ca^{2+} release from the cistern is the most fundamental and fascinating aspect of excitation–contraction coupling. Chandler has emphasized the role of feet protruded from the cistern to make a junction with the T-system and has proposed that the charge movement across the junction might play a crucial role (Chandler et al. 1976) but the evidence is circumstantial. Endo showed that depolarization (the interior of the cistern becomes more negative) can induce Ca^{2+} release (Endo et al. 1983; Endo 1985). However, the ionic composition inside the cistern is essentially the same as that of the sarcoplasm (Somlyo et al. 1977) so it is difficult to accept the role of depolarization.

Thus everything remains to be solved, although some truth may be hidden in the above observations.

Cardiac Muscle

In cardiac muscle, the sarcoplasmic reticulum and the T-system, which is much larger in diameter than that of skeletal muscle, do not form a typical triad but a complex composed of a poorly developed terminal cistern and the T-system, often called a "dyad". Below the surface membrane there is a similar structure, the subsurface cistern.

Ca^{2+} for the contractile system may be derived from (a) the outer medium: Ca^{2+} enters the myoplasm across the membrane as a slow inward current; (b) cisterns: the mechanism may be essentially the same as that in skeletal muscle. In the early days, it was thought that mitochondria might play a part, but it is now generally agreed that they do not play a role under physiological conditions (Kitazawa 1976; Chiesi et al. 1981).

Smooth Muscle

The sarcoplasmic reticulum is better developed in smooth muscle (Somlyo et al. 1971) than muscle scientists had earlier thought. The situation is therefore essentially the same as in heart muscle except that most of the action current is carried by Ca^{2+}. However, in the case of transmitter-induced contraction, say by catecholamine, inositol trisphosphate (IP_3) may play a role, affecting the sarcoplasmic reticulum to release Ca^{2+} (see Chap. 16).

Calcium-Induced Calcium Release

Endo's group and Podolsky's group have independently proposed (Endo et al. 1970; Ford and Podolsky 1970) that calcium release from the sarcoplasmic reticulum is facilitated by Ca^{2+} itself. This calcium-induced calcium release was once considered to be a strong candidate for the mechanism underlying excitation–contraction coupling. However, Endo himself has presented plenty of evidence that refutes this idea (Endo 1985). In skeletal muscle, this mechanism operates only in unusual states such as malignant hyperthermia (Takagi et al. 1976; Endo et al. 1983).

In cardiac muscle, there has been controversy as to whether or not this mechanism is physiological. Fabiato (1983) claims that this is a key to the release of Ca^{2+} from cardiac sarcoplasmic reticulum but Endo has not yet agreed with this idea (Endo 1984, 1985).

Endo and his colleagues have accepted that in smooth muscle, calcium-induced calcium release plays some role both in electrically and chemically induced contraction mainly on the basis of the inhibitory action of adenine and procaine on calcium-induced calcium release (Yagi et al. 1985). It is possible that this mechanism works with the IP_3 action in a cooperative manner, but further direct evidence is required.

Ca^{2+} Control of Metabolic Processes in Muscle

Various kinds of enzymes in many kinds of cells are subject to Ca^{2+} modulation and these matters are discussed in other chapters; references in this section will be confined to the metabolic activation associated with muscle contraction, or the "excitation–metabolism coupling".

It has been known since the nineteenth century that muscle contraction is associated with glycolysis which leads to lactic acid formation (Fletcher and Hopkins 1907) but earlier efforts to establish the link between muscle contraction and glycolysis were unsuccessful.

The advance of Ca^{2+} research allowed muscle scientists to reach the conclusion that Ca^{2+} in the same concentration range as that required for muscle contraction could activate phosphorylase b kinase and thus initiate glycolysis (Ozawa et al. 1967). It was a surprising finding at that time that even the cyclic AMP activation of this enzyme was dependent on Ca^{2+} (Ozawa and Ebashi 1967; Ozawa 1972), because most biochemists had believed that cyclic AMP was the most universal and fundamental regulator. This is now no longer surprising since the discovery that the enzyme contains calmodulin as its subunit (Cohen et al. 1978).

This Ca^{2+}-dependent activation of phosphorylase b kinase and existing knowledge about troponin encouraged Kakiuchi to pursue a possible role of Ca^{2+} in the enzyme actions of nervous tissue and he eventually discovered the Ca^{2+} activation of brain phosphodiesterase (Kakiuchi and Yamazaki 1970) and the "modulator protein", now called calmodulin, which is responsible for this activation (Kakiuchi et al. 1969, 1970). Cheung (1970) also found this modulator, calmodulin, as a result of his efforts to find the activator of phosphodiesterase.

The discovery of calmodulin has emancipated Ca^{2+} from the muscle field and it is now the common property of all biological scientists.

Calcium Antagonists

In cardiac and smooth muscle, Ca^{2+} utilized by the contractile system is, at least partly, derived from the outer medium and enters across the surface membrane. Therefore, if this Ca^{2+} entry is blocked, the contractility of the muscle is repressed. The chemicals which possess this kind of inhibitory action are now called "calcium antagonists" (calcium entry blockers or calcium channel blockers).

Historical Outline

As discussed earlier, the need for Ca^{2+} in the outer medium for the contractility of cardiac and smooth muscle was shown long ago but the role of Ca^{2+} was not properly understood. The first papers to interpret the role of Ca^{2+} in the modern sense that showed the need for Ca^{2+} in the action of a contraction-inducing drug appeared in 1960 (Yukisada 1960; Yukisada and Ebashi 1961). The Ca^{2+} requirement at the actomyosin level in cardiac and smooth muscle was shown later (Otsuka et al. 1964; Filo et al. 1965; Iwakura 1965).

Around that time, many drugs that are now classified as calcium antagonists, e.g. verapamil (Haas and Hartfelder 1962), were already used not only as pharmacological tools but also for clinical purposes; the mechanism of these drugs was explained in different ways at that time, say as β-adrenergic blockers or antihistaminics.

The papers which represent the earliest steps in research on calcium antagonists may be those by Godfraind and Kaba (1969a, b) working with cinnarizin and those by Fleckenstein et al. (1969) and Fleckenstein and Grün (1969) working with verapamil. Thereafter many papers appeared in quick succession and there now exist many review articles and monographs (e.g. Fleckenstein 1983; Godfraind 1985).

It may be noted that although the term "calcium antagonists" is now considered to mean drugs of an organic nature, we must not forget that manganese and related metals were the precursors of today's calcium antagonists (Fatt and Ginsborg 1958; Hagiwara and Nakajima 1965, 1966). Manganese played a crucial role in establishing the calcium-dependence of the action current in smooth muscle (Nonomura et al. 1966).

The Mechanism of Ca^{2+} Antagonists

At the moment there are three kinds of representative chemical entities among the specific calcium antagonists, i.e. verapamil derivatives (represented by verapamil), dihydropyridine derivatives (nifedipine) and benzodiazepine derivatives (diltiazem).

Generally speaking, the electrically induced Ca^{2+} channel (potential-dependent channel) is far more susceptible to calcium antagonists than is the chemically induced channel (receptor-operated channel).

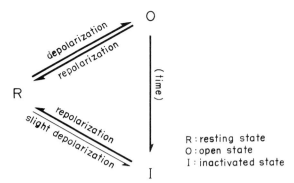

Fig. 12.6. The potential-dependent calcium channel. The process from the open state to the inactivated state is simply expressed by "time" in this figure, but it is actually a very complicated process, unlike that of the sodium channel. R, resting state; O, open state; I, inactivated state.

The potential-dependent channel has a property which is common to the sodium channel and to excitation–contraction coupling; it assumes three states as shown in Fig. 12.6: the resting state, the open state (activated state) and the inactivated state. Verapamil and diltiazem seem to bind to the open channel whereas nifedipine binds to the inactivated channel (Sanguinetti and Kass 1984). The pharmacological meaning of this is that the effect of the former two increases with repeated depolarization (use-dependent block) whereas the latter requires a weakly depolarized state for exhibiting its action, a state in which inactivation proceeds without substantial activation.

The binding of calcium antagonists to the calcium channel is itself an important subject of pharmacology but it also serves as a good guide for investigating the nature of the calcium channel. There is no space to refer to this interesting problem but one thing should be noted. In smooth muscle there is a good correlation between the drug concentrations which exhibit pharmacological action and those which bind to the receptor but a serious discrepancy has been shown in cardiac muscle with nifedipine-type drugs. This may well be explained by the difference in the mode of action of calcium antagonists mentioned above; smooth muscle is always in a somewhat depolarized state so that a significant part of the channel is inactivated, but the membrane potential of cardiac muscle is cyclically hyperpolarized so that the inactivated state of the channel is hardly observed under physiological conditions (Bean 1984).

There are a number of different kinds of calcium channel (even within a single cell) and the effects of calcium antagonists differ from channel to channel. It should be emphasized that even in a channel where no antagonist can be found, manganese and similar metals are effective in antagonizing calcium. Manganese is a truly competitive inhibitor, whereas calcium antagonists are modifiers of the channel.

Applications of Calcium Antagonists

The following are the areas where calcium antagonists are most frequently used: (a) various types of angina; (b) hypertension; (c) labile states of the conductive system of the heart (arrhythmia, fibrillation, etc.) and (d) vascular system-related disorders in the central nervous system (vertigo, migraine, etc.).

General agreement has been reached that the effectiveness of the drugs in (a) and (b) is due to their calcium antagonistic action. The favourable effect of some calcium antagonists in angina had already been established when the mechanism of their action was not known, but their effects in hypertension, comparable to those of existing drugs such as β-adrenergic blockades or diuretics, were rather unexpected.

Generally speaking, the problem of whether or not a chemical agent will be effective in hypertension is a matter of experience. Calcium antagonists naturally reduce vascular resistance but this effect by itself cannot clinically counteract hypertension. As is well known, vasodilation usually induces tachycardia as a reflex compensation through the autonomic nervous system but the calcium antagonists being now used as anti-hypertensive drugs, i.e. diltiazem and verapamil, hardly cause such an undesirable effect. The pharmacological basis for their favourable action in hypertension has not been clarified.

In the arrhythmias, the situation is more complicated. Since Ca^{2+} plays a crucial role in the pacemaker mechanism of the heart, it was not unexpected that Ca^{2+} antagonists would be effective in the arrhythmic state, but the detailed mechanism underlying their effect is not understood.

There is a serious controversy as to whether their effectiveness in (d), especially in vertigo, is truly due to calcium antagonism or some other action(s).

As a whole, the most promising target of calcium antagonists so far explored seems to be the cardiovascular system (a beneficial effect would be expected in Raynaud's disease and similar peripheral vascular diseases but this is not yet proven). It should be noted that visceral smooth muscles, used very often to assess calcium antagonistic action, are not the clinical target of this drug.

In addition to these disorders, protection of various tissues against anoxia and ischaemia, especially the heart and brain, have been mentioned as subjects to which the drug should be applied. In this area, calcium is to be considered as an agent toxic to living organisms, playing a key role in cascade reactions.

Since Ca^{2+} is the most common mediator of intracellular processes, both physiological and pathological, it is no wonder that the number of disorders to which calcium antagonists can be applied is increasing day by day; even in the field of oncology, some interesting effects of calcium antagonists have been reported. Thus calcium antagonist research will not only be beneficial in the pharmacological and clinical fields but it will also provide a good tool in research into the intracellular role of Ca^{2+} in individual cells. If it is to do this, precise criteria as to whether the chemical agent in question is truly a calcium antagonist or not must be agreed; otherwise unnecessary confusion will be introduced into the field of calcium research.

Addendum. A peptide, called ω-conotoxin, has been isolated from marine snails as a new type of calcium antagonist (Olivera et al. 1985). It is specific for neural tissues, having virtually no affinity for muscles. Since it is a fairly large peptide composed of 27 amino acids, there is little possibility of its being effective on the brain in situ but it will be a very useful tool for basic research on the calcium channel (Abe et al. 1986).

References

Abe T, Koyano K, Saisu H, Nishiuchi Y, Sakakibara S (1986) Binding of ω-conotoxin to receptor site associated with the voltage-sensitive calcium channel. Neurosci Lett 71:203–208

Bailey K (1946) Tropomyosin, a new asymmetric protein component of the muscle. Nature 158:368–369

Bean BP (1984) Nitrendipine block of cardiac calcium channels: high-affinity binding to the inactivated state. Proc Natl Acad Sci USA 81:6388–6392

Berson G (1974) Ca^{2+}, Sr^{2+} and Ba^{2+} sensitivity of tropomyosin–troponin complex from cardiac and fast skeletal muscle. In: Drabikowski W, Strzelecka-Golaszewska H, Carafoli E (eds) Calcium binding proteins. PWN-Polish Scientific, Warsaw, pp 197–201

Bozler E (1954) Relaxation in extracted muscle fibres. J Gen Physiol 38:149–159

Bremel RD, Sobieszek A, Small JV (1977) Regulation of actin–myosin interaction in vertebrate smooth muscle. In: Stephens NL (ed) The biochemistry of smooth muscle. University Park Press, Baltimore, pp 413–443

Chambers R, Hale HP (1932) The formation of ice in protoplasm. Proc R Soc Lond [Biol] 110:336–352

Chandler WK, Rakowski RF, Schneider MF (1976) Effects of glycerol treatment and maintained depolarization on charge movement in skeletal muscle. J Physiol 254:285–316

Cheung WY (1970) Cyclic 3′,5′-nucleotide phosphodiesterase: demonstration of an activator. Biochem Biophys Res Commun 38:533–538

Chiesi M, Ho MM, Inesi G, Somlyo AV, Somlyo AP (1981) Primary role of sarcoplasmic reticulum in phasic contractile activation of cardiac myocytes with shunted myolemma. J Cell Biol 91:728–742

Cohen P, Burchell A, Foulkes JG, Cohen PTW, Vanaman TC, Nairn AC (1978) Identification of the Ca^{2+}-dependent modulator protein as the fourth subunit of rabbit skeletal muscle phosphorylase kinase. FEBS Lett 92:287–293

Dillon PF, Aksoy MO, Driska SP, Murphy RA (1981) Myosin phosphorylation and the cross-bridge cycle in arterial smooth muscle. Science 211:495–497

Ebashi S (1959) The mechanism of relaxation in glycerinated muscle fibre. In: Natori R (ed) IVth symposium on physicochemistry of biomacromolecules, 1958. Nanko-do, Tokyo, pp 25–34

Ebashi S (1960) Calcium binding and relaxation in the actomyosin system. J Biochem 48:150–151

Ebashi S (1961a) Calcium binding activity of vesicular relaxing factor. J Biochem 50:236–244

Ebashi S (1961b) The role of "relaxing factor" in contraction–relaxation cycle of muscle. Prog Theor Phys 17:35–40

Ebashi S (1963) Third component participating in the superprecipitation of "natural actomyosin". Nature 200:1010

Ebashi S (1974a) Regulatory mechanism of muscle contraction with special reference to the Ca–troponin–tropomyosin system. In: Campbell PN, Dickens F (eds) Essays in biochemistry. Academic Press, London, pp 1–36

Ebashi S (1974b) Interactions of troponin subunits underlying regulation of muscle contraction by Ca ion. A study on hybrid troponins. In: Richter D (ed) Lipmann symposium. Energy, regulation and biosynthesis in molecular biology. Walter de Gruyter, Berlin, pp 165–178

Ebashi S (1976) Excitation–contraction coupling. Ann Rev Physiol 38:293–313

Ebashi S (1980) Regulation of muscle contraction. Proc R Soc Lond [Biol] 207:259–286 (Croonian Lecture, 1979)

Ebashi S (1984) Ca^{2+} and the contractile proteins. J Mol Cell Cardiol 16:129–136

Ebashi S, Ebashi F (1964) A new protein component participating in the superprecipitation of myosin B. J Biochem 55:604–613

Ebashi S, Endo M (1968) Calcium ion and muscle contraction. Prog in Biophys Mol Biol 18:123–183

Ebashi S, Kodama A (1965) A new protein factor promoting aggregation of tropomyosin. J Biochem 58:107–108

Ebashi S, Lipmann F (1962) Adenosine triphosphate-linked concentration of calcium ions in a particulate fraction of rabbit muscle. J Cell Biol 14:389–400

Ebashi S, Ohtsuki I (1983) Troponin, roles of its subunits. In: de Bernard B et al. (eds) Calcium-binding proteins. Elsevier, Amsterdam, pp 251–262

Ebashi S, Kodama A, Ebashi F (1968) Troponin. I. Preparation and physiological function. J Biochem 64:465–477

Ebashi S, Endo M, Ohtsuki I (1969) Control of muscle contraction. Q Rev Biophys 2:351–384

Ebashi S, Ohtsuki I, Mihashi K (1972) Regulatory proteins of muscle with special reference to troponin. Cold Spring Harbor Laboratory, pp 215–223 (Symposium on Quantitative Biology 37)

Ebashi S, Ohnishi S, Abe S, Maruyama, K (1974) A spin-label study on calcium-induced conformational changes of troponin components. J Biochem 75:211–213

Ebashi S, Nonomura Y, Toyo-oka T, Katayama E (1976) Regulation of muscle contraction by the calcium–troponin–tropomyosin system. In: Duncan CJ (ed) Calcium in biological systems. Cambridge University Press, pp 349–360 (Symposia of the Society for Experimental Biology, no. 30)

Ebashi S, Nonomura Y, Kohama K, Kitazawa T, Mikawa T (1980) Regulation of muscle contraction by Ca ion. In: Chapville F, Haenni H-L (eds) Chemical recognition in biology. Springer-Verlag, Berlin Heidelberg New York, pp 183–194

Ebashi S, Nonomura Y, Nakamura S, Nakasone H, Kohama K (1982) Regulatory mechanism in smooth muscle actin-linked regulation. Fed Proc 41:2863–2867

Endo M (1984) Control of Ca ion by the cardiac sarcoplasmic reticulum. In: Abe H et al. (eds) Regulation of cardiac function. Japanese Science Society Press, Tokyo, pp 33–40

Endo M (1985) Calcium release from sarcoplasmic reticulum. Curr Top Membr Trans 25:181–230

Endo M, Nakajima Y (1973) Release of calcium induced by "depolarization" of the sarcoplasmic reticulum membrane. Nature (New Biol) 246:216–218

Endo T, Obinata T (1981) Troponin and its components from ascidian smooth muscle. J Biochem 89:1599–1608

Endo M, Tanaka M, Ogawa Y (1970) Calcium-induced release of calcium from the sarcoplasmic reticulum of skinned skeletal muscle fibres. Nature 228:34–36

Endo M, Yagi S, Ishizuka T, Horiuti K, Koga Y, Amaha K (1983) Changes in the Ca-induced Ca release mechanism in the sarcoplasmic reticulum of the muscle from a patient with malignant hyperthermia. Biomed Res 4:83–92

Fabiato A (1983) Calcium-induced release of calcium from the cardiac sarcoplasmic reticulum. Am J Physiol 245:C1–14

Fatt P, Ginsborg BL (1958) The ionic requirements for the production of action potentials in crustacean muscle fibres. J Physiol (Lond) 142:516–543

Filo RS, Bohr DF, Ruegg JC (1965) Glycerinated skeletal and smooth muscle: calcium and magnesium dependence. Science 147:1581–1583

Fleckenstein A (1983) Calcium antagonism in heart and smooth muscle. Experimental facts and therapeutic prospects. Wiley, New York

Fleckenstein A, Grün G (1969) Reversible Blockierung der electromechanischen Koppelung-sprozesse in der Glättenmuskulatur des Rattenuterus mittels organischer Ca plus 2-antagonisten (iproveratril, D 600, prenylamin). Pfluegers Arch 307:26

Fleckenstein A, Tritthart H, Fleckenstein B, Herbst A, Grün G (1969) Eine neue Gruppe kompetitiver ziverwertiges Ca-antagonisten (iproveratril, D 600, prenylamin) mit starken Hemmeffekten auf die electromechanische Kappelung im Warmblüter-myokard. Pfluegers Arch 307:R25

Fletcher WM, Hopkins FG (1907) Lactic acid in amphibian muscle. J Physiol (Lond) 35:247–309

Flicker PF, Phillips GN Jr, Cohen C (1982) Troponin and its interactions with tropomyosin: an electron microscopy study. J Mol Biol 162:495–501

Ford LE, Podolsky RJ (1970) Regenerative calcium release within muscle cells. Science 167:58–59

Godfraind T (1985) Cellular and subcellular approaches to the mechanism of action of calcium antagonists. In: Rubin RP, Weiss GB, Putney JW Jr (eds) Calcium in biological systems. Plenum Press, New York, pp 411–421

Godfraind T, Kaba A (1969a) Inhibition by cinnarizine and chlorpromazine of the contraction induced by calcium and adrenaline in vascular smooth muscle. Br J Pharmacol 35:354

Godfraind T, Kaba A (1969b) Blockade or reversal of the contraction induced by calcium and adrenaline in depolarized arterial smooth muscle. Br J Pharmacol 36:549–560

Haas H, Hartfelder G (1962) α-Isopropyl-α-[(N-methyl-N-homoveratryl)-γ-aminopropyl]-3,4-dimethoxyphenylacetonitrile, eine Substanz mit coronargefasserweiternden Eigenschaften. Arzneimittelforsch 12:549–558

Hagiwara S, Nakajima S (1965) Tetrodotoxin and manganese ion: effects on action potential of the frog heart. Science 149:1245–1255

Hagiwara S, Nakajima S (1966) References in Na and Ca spikes as examined by application of tetrodotoxin, procaine, and manganese ions. J Gen Physiol 49:793–806

Hartshorne DJ, Siemankowski RF (1981) Regulation of smooth muscle actomyosin. Annu Rev Physiol 43:519–530

Heilbrunn LV (1940) The action of calcium on muscle protoplasm. Physiol Zool 13:88–94

Heilbrunn LV, Wiercinski FJ (1947) The action of various cations on muscle protoplasm. J Cell Physiol 29:15–32

Iwakura H (1965) New contractile protein of smooth muscle. 1. Native tropomyosin from smooth muscle. J Med Sci (Tokyo) 73:49–57

Kakiuchi S, Yamazaki R (1970) Stimulation of the activity of cyclic 3′,5′-nucleotide phosphodiesterase by calcium ion. Proc Japan Acad 46:387–392

Kakiuchi S, Yamazaki R, Nakajima H (1969) Studies on brain phosphodiesterase: 2. Bull Japan Neurochem Soc 8:17–20

Kakiuchi S, Yamazaki R, Nakajima H (1970) Properties of a heat stable phosphodiesterase activating factor isolated from brain extract. Proc Japan Acad 46:587–592

Kamada T, Kinosita H (1943) Disturbances initiated from naked surface of muscle protoplasm. Japan J Zool 10:469–493

Keil EM, Sichel FJM (1936) The injection of aqueous solutions, including acetylcholine, into the isolated muscle fiber. Biol Bull 71:402

Kendrick-Jones J, Lehman W, Szent-Györgyi AG (1970) Regulation in molluscan muscles. J Mol Biol 54:313–326

Kessler D, Eisenlohr LC, Lathwell MJ et al. (1980) *Physarum* myosin light chain binds calcium. Cell Motil 1:63–71

Kielley WW, Meyerhof O (1948a) A new magnesium-activated adenosine triphosphatase from muscle. J Biol Chem 174:387–388

Kielley WW, Meyerhof O (1948b) Studies on adenosine triphosphatase of muscle. 2. A new magnesium activated adenosine triphosphatase. J Biol Chem 176:591–601

Kimura K, Obinata T (1983) Troponin from nematode body wall muscle: purification and characterization. Zool Magazine 92(4):555

Kimura K, Tanaka T, Nakae H, Obinata T (1987) Troponin from nematode: purfication and characterization of troponin from ascaris body wall muscle. Comp Biochem Physiol (in press)

Kitazawa T (1976) Physiological significance of Ca uptake by mitochondria in heart in comparison with that by cardiac sarcoplasmic reticulum. J Biochem 80:1129–1147

Kohama K (1979) Divalent cation binding properties of slow skeletal muscle troponin in comparison with those of cardiac and skeletal muscle troponins. J Biochem 86:811–820

Kohama K, Kendrick-Jones J (1986) The inhibitory Ca^{2+}-regulation of the actin-activated Mg-ATPase activity of myosin from *Physarum polycephalum* plasmodia. J Biochem 99:1433–1446

Kohama K, Kohama T (1984) Myosin confers inhibitory Ca^{2+}-sensitivity on actin–myosin–ATP interaction of *Physarum polycephalum* plasmodia under physiological conditions. Proc Japan Acad 60:B435–439

Kohama K, Shimmen T (1985) Inhibitory Ca^{2+}-control of movement of beads coated with *Physarum* myosin along actin-cables in *Chara internodal cells*. Protoplasma 129:88–91

Kohama K, Takano-Ohmuro (1984) Stage specific myosins from amoeba and plasmodium of slime mold *Physarum polycephalum*. Proc Japan Acad 60:B431–434

Kohama K, Kobayashi K, Mitani S (1980) Effects of Ca ion and ADP on superprecipitation of myosin B from slime mold *Physarum polycephalum*. Proc Japan Acad 56:B591–596

Kohama K, Saida K, Hirata M, Kitaura T, Ebashi S (1986) Superprecipitation is a model for in vitro contraction superior to ATPase activity. Japan J Pharmacol 42:253–260

Kumagai H, Ebashi S, Takeda F (1955) Essential relaxing factor in muscle other than myokinase and creatine phosphokinase. Nature 176:166

Lehman W (1981) Thin-filament-linked regulation in molluscan muscles. Biochim Biophys Acta 668:349–356

Lehman W, Szent-Györgyi AG (1975) Regulation of muscular contraction, distribution of actin control and myosin control in the animal kingdom. J Gen Physiol 66:1–30

Marsh BB (1951) A factor modifying muscle fibre syneresis. Nature 167:1065–1066

Marston SB, Lehman W, Moody C, Smith CWJ (1985) Ca^{2+}-dependent regulation of smooth muscle thin filaments by caldesmon. Adv Prot Phosphatases 2:171–189

Mikawa T, Maruyama K (1981) Gizzard actin confers a magnesium sensitivity to skeletal muscle myosin at low ionic strengths. Proc Japan Acad 57:B23–28

Nakae H, Obinata T (1985) Expression of troponin during development of nematode. Proceedings of the 56th annual meeting of the zoological society of Japan, Zool Sci 2 (6):951

Nonomura Y, Hotta Y, Ohashi H (1966) Tetrodotoxin and manganese ions: effects on electrical activity and tension in taenia coli of guinea pig. Science 152:97–99

Ohtsuki I (1980) Functional organization of the troponin-tropomyosin system. In: Ebashi S, Maruyama K, Endo M (eds) Muscle contraction: its regulatory mechanisms. Japanese Science Society Press, Tokyo, pp 237–249

Ohtsuki I, Shiraishi F, Suenaga N, Miyata T, Tanokura M (1984) A 26K fragment of troponin T from rabbit skeletal muscle. J Biochem 95:1337–1342

Ohtsuki I, Maruyama K, Ebashi S (1986) Regulatory and cytoskeletal proteins of vertebrate skeletal muscle. Adv Protein Chem 38:1–67

Ojima T, Nishita K (1986a) Troponin from Akazara scallop striated adductor muscles. J Biol Chem 261:16749–16754

Ojima T, Nishita K (1986b) Isolation of troponins from striated and smooth adductor muscles of Akazara scallop. J Biochem 100:821–824

Olivera BM, Gray WR, Zeikus R et al. (1985) Peptide neurotoxins from fish-hunting cone snails. Science 230:1338–1343

Otsuka M, Ebashi F, Imai S (1964) Cardiac myosin B and calcium ions. J Biochem 55:192–194

Ozawa E (1972) Activation of muscular phosphorylase b kinase by a minute amount of Ca ion. J Biochem 71:321–331

Ozawa E, Ebashi S (1967) Requirement of Ca ion for the stimulating effect of cyclic 3',5'-AMP on muscle phosphorylase b kinase. J Biochem 62:285–286

Ozawa E, Hosoi K, Ebashi S (1967) Reversible stimulation of muscle phosphorylase b kinase by low concentration of calcium ions. J Biochem 61:531–533

Potter JD, Johnson JD (1982) Troponin. In: Chang WY (ed) Calcium and cell function. Academic Press, New York, pp 145–173

Ringer S (1883) A further contribution regarding the influence of the blood on the contraction of the heart. J Physiol (Lond) 4:29–42

Sanguinetti MC, Kass RS (1984) Voltage-dependent block of calcium channel current in the calf cardiac Purkinje fiber by dihydropyridine calcium channel antagonists. Circ Res 55:336–348

Schwarzenbach G (1955) Die komplexometrische Titration. Ferdinant Enke, Stuttgart

Sheetz JP, Spudich JA (1983) Movement of myosin-coated fluorescent beads on actin cables in vitro. Nature 303:31–35

Sobue K, Morimoto K, Inui M, Kanda K, Kakiuchi S (1982) Control of actin–myosin interaction of gizzard smooth muscle by calmodulin- and caldesmon linked flip-flop mechanism. Biomedical Research 3(2):188–196

Somlyo AP, Devine CE, Somlyo AV, North SR (1971) Sarcoplasmic reticulum and the temperature- dependent contraction of smooth muscle in calcium-free solutions. J Cell Biol 51:722–741

Somlyo AV, Shuman H, Somlyo AP (1977) Elemental distribution in striated muscle and the effects of hypertonicity. Electron probe analysis of cryo sections. J Cell Biol 74:828–857

Stiles PG (1903) On the rhythmic activity of the oesophagus and the influence upon it of various media. Am J Physiol 5:338–357

Suzuki H, Onishi H, Takahashi K, Watanabe S (1978) Structure and function of chicken gizzard myosin. J Biochem 84:1529–1542

Szent-Györgyi A (1942) Studies from Szeged: myosin and muscular contraction: 1. S Karger, Basel, pp 1–72

Szent-Györgyi A (1947) The chemistry of muscular contraction. Academic Press, New York

Takagi A, Sugita H, Toyokura Y, Endo M (1976) Malignant hyperpyrexia; effect of halothane on single skinned muscle fibers. Proc Japan Acad 52:603–606

Thorens S, Endo M (1975) Calcium-induced calcium release and "depolarization"-induced calcium release: their physiological significance. Proc Japan Acad 51:473–478

Tsukui R, Ebashi S (1973) Cardiac troponin. J Biochem 73:1119–1121

Watanabe S (1955) Relaxing effects of EDTA on glycerol-treated muscle fibers. Arch Biochem Biophys 54:559–562

Weber A (1959) On the role of calcium in the activity of adenosine 5'-triphosphate hydrolysis by actomyosin. J Biol Chem 234:2764–2769

Weber A, Herz R (1963) The binding of calcium to actomyosin systems in relation to their biological activity. J Biol Chem 238:599–605

Weber A, Winicur S (1961) The role of calcium in superprecipitation of actomyosin. J Biol Chem 236:3198–3202

Yagi S, Matsumura N, Endo M (1985) Effects of inhibitors of calcium-induced calcium release on receptor-mediated responses of smooth muscles and platelets. Proc Japan Acad [B] 61:399–402

Yamamoto K (1983) Sensitivity of actomyosin ATPase to calcium and strontium ion. Effect of hybrid troponins. J Biochem 93:1061–1069

Yanagida T, Arata T, Oosawa F (1985) Sliding distance of actin filament induced by a myosin cross-bridge during one ATP hydrolysis cycle. Nature 316:366–369

Yazawa Y (1985) Actin-linked regulation in scallop striated muscle. Proc Japan Acad 61:B497–500

Yoshimoto Y, Kamiya N (1984) ATP- and calcium-controlled contraction in a saponin model of *Physarum polycephalum*. Cell Struct Funct 9:135–141

Yoshimoto Y, Matsumura F, Kamiya N (1981) Simultaneous oscillations of Ca^{2+} efflux and tension generation in the permealized plasmodial strand of *Physarum*. Cell Motil 1:433–443

Yukisada N (1960) Roles of inorganic ions, especially Ca ion, in the mechanism of smooth muscle contraction. Folia Pharmacol Japan 56:936–945

Yukisada N, Ebashi F (1961) Role of calcium in drug action on smooth muscle. Japan J Pharmacol 11:46–53

Chapter 13

Cellular Calcium: Nervous System

M. P. Blaustein

Introduction

Calcium ions play numerous important roles in the function of nerve cells. Calcium's involvement in neurotransmitter release is one of its key roles, and was one of the first to be recognized. Early observations demonstrated that extracellular calcium was required for neurotransmitter release (Harvey and MacIntosh 1940). Subsequent studies established that transmitter release was evoked by calcium entry and a rise in $[Ca^{2+}]_i$, the intracellular free calcium concentration. This sequence of events accounted for the dependence of the reaction on external calcium (Katz 1969; Llinas et al. 1976; Drapeau and Blaustein 1983a). As we shall see, the modulation of synaptic transmission, and related fundamental aspects of transmitter release associated with memory and learning, are all profoundly influenced by calcium.

Calcium ions were also found to have a critical effect on the electrical excitability of neurons. Again, early observations concerned the effects of extracellular calcium concentrations, $[Ca^{2+}]_o$; elevation of $[Ca^{2+}]_o$ increases the firing threshold (i.e. reduces excitability) while reduction of $[Ca^{2+}]_o$ increases excitability. These effects could be explained, at least in part, by calcium's ability to screen surface charges and thereby alter the electrical potential profile between the bulk solution and the surface of the plasma membrane (Frankenhaeuser and Hodgkin 1957; Frankenhaeuser 1957). Subsequently, intracellular calcium was also found to modulate excitability: for example, by activating certain potassium conductances directly (Meech 1978) or by controlling calcium-activated protein kinases that, in turn, modulate ionic conductances (Rane and Dunlap 1986).

Intracellular calcium has a number of other effects as well. For example, it appears to control axoplasmic transport (Hammerschlag and Lavoie 1979; Chan et al. 1980). Intracellular calcium also regulates certain general cell functions such as cell development and repair (Meiri et al. 1981; Anglister et al. 1982), the control of the cytoskeletal structure (Schlaepfer and Zimmerman 1984) and mitochondrial enzyme activity (Denton and McCormack 1985). And finally, cell calcium overload may be associated with cell injury and death – either as cause or effect (Schlaepfer 1977; Trump and Berezesky 1985).

The critical importance of intracellular calcium for these many neuronal functions is readily apparent. Therefore, we will focus first on the mechanisms by which $[Ca^{2+}]_i$ is controlled in nerve cells.

The extracellular free calcium concentration, $[Ca^{2+}]_o$, is of the order of 1 mM in most vertebrate systems (Blaustein 1974), and is approximately 10 000 times greater than intracellular $[Ca^{2+}]_i$. As a consequence of this very large concentration gradient, and the fact that the cytoplasm of resting neurons is normally about 50–70 mV negative with respect to the extracellular fluid, calcium tends to leak into the cells and must be actively extruded. In this chapter we will consider the mechanisms of calcium entry into nerve cells, the disposition of intracellular calcium and the mechanisms involved in calcium extrusion and regulation of $[Ca^{2+}]_i$. We will then examine some of the physiological processes in neurons that are controlled by intracellular calcium.

Regulation of Intracellular Calcium in Nerve Cells

The Intracellular Free Calcium Concentration in Nerve Cells

The total calcium concentration in the blood plasma of a variety of mammals is of the order of 2 mM. About half of this calcium is bound to plasma proteins and about half is free: that is $[Ca^{2+}]_o \approx 1$ mM (Blaustein 1974). In contrast, $[Ca^{2+}]_i$ in "resting" mammalian nerve cells (and most other vertebrate and invertebrate nerve cells) is about 100–300 nM as determined by fluorescent dye methods (Connor 1986) and other optical and calcium electrode methods (Blinks et al. 1982; Connor et al. 1986). These $[Ca^{2+}]_i$ levels are comparable to those observed in a variety of other types of excitable and non-excitable cells (e.g. Blinks et al. 1982; Conway 1983). Indeed, the ability of cells to employ Ca^{2+} as a second messenger (see below) is, in part, dependent upon this low $[Ca^{2+}]_i$; the significance is that small changes in the absolute $[Ca^{2+}]_i$ ("signal" calcium or $\Delta[Ca^{2+}]_i$) can, nevertheless, represent a relatively large signal: background ratio ($\Delta[Ca^{2+}]_i/[Ca^{2+}]_i$) of the order of 5–10 or more (Blaustein 1985).

It is important to note that the free $[Ca^{2+}]_i$ does not account for all of the intracellular calcium. As discussed below, most of the intracellular calcium is bound to cytoplasmic proteins or is stored in intracellular organelles. Thus the free Ca^{2+} is only a very small fraction of the total intracellular calcium.

Calcium Entry into Nerve Cells

Nerve cells, like many other types of cells, contain a variety of gated, ion-selective channels in their plasma membranes, including some channels that are divalent cation-selective. These are, effectively, calcium-selective channels, since they are not very permeable to magnesium, the only other prevalent divalent cation in most vertebrate fluids. The best-studied divalent cation-selective channels are the voltage-gated calcium channels that open when the cells are depolarized (Katz and Miledi 1967, 1969; Llinas et al. 1976; Nachshen and Blaustein 1980; Nowycky et al. 1985; Tsien et al. 1987). The distribution of these channels may not be uniform over the cell surface. In some neurons, the density of these channels may be relatively high in the plasma membrane of axons and/or cell bodies (e.g. Hagiwara and Byerly 1981; Kostyuk 1981; Nowycky et al. 1985; Bolsover and Spector 1986; Ross et al. 1986). Thus, while some neurons have action potentials that are dependent primarily on an inward sodium current, other neurons have action potentials that depend primarily upon an inward calcium current, or upon both sodium and calcium currents (Hille 1984). Voltage-gated, calcium-selective channels are often found in abundance in the plasma membrane at presynaptic nerve terminals (Nachshen and Blaustein 1980); at least in some neurons, they tend to be clustered in this region (Katz and Miledi 1969; Ross et al. 1986). As we shall see, this has very important implications for (chemical) synaptic transmission.

In addition to the voltage-gated calcium channels, there are agonist-operated channels that are permeable to calcium ions as well as other cations. The best-studied of these are the channels associated with the (postsynaptic) nicotinic acetylcholine receptor. These channels, which are relatively selective for monovalent cations (sodium and potassium are nearly equally permeable: Takeuchi and Takeuchi 1960), also conduct calcium (e.g. Takeuchi and Takeuchi 1963; Miledi et al. 1980). The postsynaptic channels activated by glutamate also conduct calcium (Eusebi et al. 1985). These channels may be models for other agonist receptor-operated channels as well.

The potential importance of calcium entry through these postsynaptic channels is widely recognized. We will focus primarily on the presynaptic voltage-gated calcium channels, however, because these have been more extensively studied, and their physiological role (in transmitter release) is better understood.

Several types of voltage-gated calcium channels have been described, with different activation and inactivation kinetics, different divalent cation selectivities and different sensitivities to pharmacological agents (Bossu et al. 1985; Nowycky et al. 1985; Tsien et al. 1987). The calcium channels found at mammalian presynaptic nerve terminals appear to be predominantly of the "N-type" (according to the nomenclature of Nowycky and colleagues). Upon depolarization, they open rapidly, and then inactivate with a half-time of about 0.5–1 S (Nachshen and Blaustein 1980; Nachshen 1985). Neuronal calcium channels also appear to be blocked selectively by one of the ω-conotoxins from the venom of the cone snail, *Conus geographicus* (Reynolds et al. 1986).

The presynaptic calcium channels are permeable to strontium, barium and manganese, as well as to calcium, but not to magnesium (Nachshen and Blaustein 1982; Drapeau and Nachshen 1984). Also, in the presence of physiological concentrations of divalent cations in the extracellular fluid, they are relatively impermeable to sodium and potassium. The flux of ions through the channels is inversely related to the ion charge density, so that the maximum flux rates for the permeable divalent cations in the channels follows the sequence: barium>strontium>calcium>manganese (Nachshen and Blaustein 1982; Nachshen 1984; Drapeau and Nachshen 1984). Impermeable cations such as cadmium and nickel bind strongly to the cation-binding sites in the channels and block the channels, thereby preventing the passage of permeable ions (Nachshen 1984).

One of the striking characteristics of these neuronal calcium channels is that they are relatively insensitive to verapamil and to the dihydropyridine calcium channel blockers (Nachshen and Blaustein 1979; Suszkiw et al. 1986; Tsien et al. 1987). In contrast, most types of voltage-gated calcium channels found in mammalian cardiac and smooth muscle cells are sensitive to submicromolar concentrations of these agents (Fleckenstein 1985).

Calcium channels in some neurons (Pellmar 1981), like those in some other types of cells (Reuter 1983; Kameyama et al. 1986), can be regulated by cyclic AMP-dependent phosphorylation and by dephosphorylation. For example, phosphorylation of the channels may enhance their ability to open when the cells are depolarized (Flockerzi et al. 1986). Alternatively, phosphorylation by other protein kinases may prevent calcium channels from opening. This type of effect is illustrated by the inhibitory actions of a diacylglycerol analogue and a phorbol ester (both of which presumably activate a calcium-dependent protein kinase, protein kinase C) on chick sensory neurons (Rane and Dunlap 1986). Another example from the same laboratory is the gamma-aminobutyric acid- or norepinephrine-induced, cyclic GMP-mediated inhibition of voltage-gated calcium channels in chick dorsal root ganglion neurons (Holz et al. 1986). In some snail neurons, on the other hand, a cyclic GMP-dependent phosphorylation appears to enhance calcium channel activity (Paupardin-Tritsch et al. 1986).

These observations raise the possibility that cyclic AMP-dependent phosphorylations and cyclic GMP-dependent or protein kinase C-dependent phosphorylations may have opposite effects in the same cells, and perhaps even on the same channel molecules (Hartzell and Fischmeister 1986). Very recent evidence also indicates that some neuronal calcium channels may be modulated by the activation of receptors associated with GTP-binding proteins (Scott and Dolphin 1986); however, the possible relationship to cyclic AMP, which is regulated by GTP-binding proteins (Gilman 1986), has not yet been investigated. Furthermore, the specific physiological roles of these numerous types of modulation of calcium channels in the mammalian nervous system have not yet been explored.

When an action potential invades a nerve terminal, the depolarization transiently (for a few milliseconds) opens the voltage-gated calcium channels. There is a net influx of calcium through these calcium-selective channels because the calcium electrochemical gradient strongly favours the entry of calcium. This calcium entry causes an immediate rise in $[Ca^{2+}]_i$ which, in turn, triggers neurotransmitter release. Upon repolarization, the channels close rapidly, and transmitter release then declines with a half-time of a millisecond or less. This rapid decline in transmitter release is explained by a fall in $[Ca^{2+}]_i$ towards the resting level, with a half-time of <1 ms, which is a result of buffering and sequestration mechanisms that remove free calcium from the cytosol (see Fig. 13.1). The transmitter release process itself is not inactivated this rapidly; indeed, transmitter release can even be facilitated if the period of calcium entry is prolonged (Katz and Miledi 1968).

In the long run, calcium must be extruded from the neurons in order to maintain a steady calcium balance; however, as we shall discuss shortly, the calcium extrusion mechanisms cannot remove calcium from the cytosol fast enough to account for the very rapid decline in transmitter release rate that is

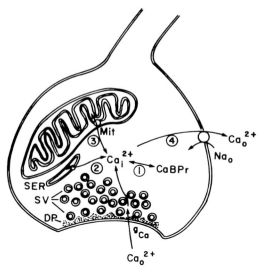

Fig. 13.1. Pathways for calcium movement and sites of calcium buffering in a presynaptic nerve terminal. Depolarization triggers an increase in the calcium conductance (g_{ca}) by opening voltage-gated calcium channels in the "active zone" of the plasmalemma (in the regions of synaptic contact where the synaptic vesicles fuse with the plasmalemma). Calcium enters the terminal, raising $[Ca^{2+}]_i$, and then rapidly diffuses to the high-affinity calcium-binding sites on cytosolic proteins (CaBPr; reaction 1). Intraterminal calcium sequestration systems in the smooth endoplasmic reticulum (SER; reaction 2) and (to a much lesser extent) in the mitochondria (Mit; reaction 3) take up calcium, further lowering $[Ca^{2+}]_i$. Ultimately, the calcium that entered during activity must be extruded from the terminal. The Na^+/Ca^{2+} exchange mechanism in the plasmalemma (reaction 4) appears to mediate most of this net extrusion (see text), although an ATP-driven calcium pump (not shown) may participate to a small extent. *SV*, synaptic vesicles; *DP*, dense projections in the region of the active zone. (Blaustein et al. 1980)

observed. Thus, other mechanisms must be responsible for the very rapid lowering of $[Ca^{2+}]_i$ that occurs during repolarization.

Intracellular Calcium Buffering in Nerve Cells

Several types of mechanisms apparently play a role in intracellular Ca^{2+} buffering (see Fig. 13.1): sequestration of calcium in smooth endoplasmic reticulum and mitochondria, and binding to cytoplasmic proteins and membrane structures.

Mitochondria

Under normal conditions, when $[Ca^{2+}]_i$ is within the dynamic physiological range (about 0.1–1.0 μM), mitochondria do not sequester much calcium (Hansford 1985). Although mitochondria can accumulate calcium, the affinity of the mitochondrial calcium transport system for calcium and the rate of calcium uptake are quite low in the presence of millimolar concentrations of magnesium (Vinogradov and Scarpa 1973; Scarpa 1976; Hutson et al. 1976); these are the magnesium concentrations normally present in the cytosol (e.g. DeWeer 1976; Alvarez-Leefmans et al. 1984). Thus, mitochondria will not normally take up much calcium from the cytosol unless the other buffering systems are saturated (Fig. 13.2). But even with a low affinity for calcium, they may take up small amounts when $[Ca^{2+}]_i$ rises during activity; indeed, this could play an important role in coupling mitochondrial enzyme activity to cellular activity (Denton and McCormack 1985; Hansford 1985; McCormack and Denton 1986). On the other hand, when the $[Ca^{2+}]_i$ exceeds about 5 μM (i.e. under conditions of calcium overload), mitochondria will begin to sequester substantial amounts of calcium (Blaustein and Rasgado-Flores 1981; Rasgado-Flores and Blaustein 1987b; and see Fig. 13.2); this will dissipate the energy that would normally be used for oxidative phosphorylation. This may occur under pathological conditions, and may, if not rapidly reversed, lead to cell death (Schlaepfer 1977; Trump and Berezesky 1985).

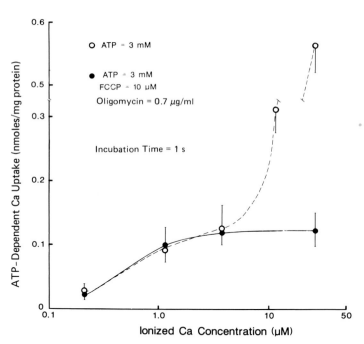

Fig. 13.2. Effect of mitochondrial blockers on the ATP-dependent calcium uptake into saponin-treated rat brain synaptosomes incubated at different free Ca^{2+} concentrations (as indicated on the abscissa). The saponin renders the plasma membrane leaky, but does not damage the cholesterol-poor mitochondria or smooth endoplasmic reticulum membranes. The leaky synaptosomes were incubated for 1 s with media containing ^{45}Ca and ATP, either without (○) or with (●) 10 μM FCCP (a mitochondrial uncoupler) and 0.7 μg/ml oligomycin. The ATP-dependent calcium uptake was obtained by subtracting the calcium uptake in the absence of ATP from the uptake measured in the presence of ATP. Note that below about 5 μM Ca^{2+}, the ATP-dependent uptake is unaffected by the mitochondrial inhibitors, whereas above this Ca^{2+} concentration the unpoisoned mitochondria take up a large amount of calcium. (Rasgado-Flores and Blaustein 1987)

Smooth Endoplasmic Reticulum

A second organelle that can sequester calcium is the smooth endoplasmic reticulum. The calcium-sequestering properties of this organelle have been most extensively studied in skeletal muscle (sarcoplasmic reticulum: Hasselbach 1977; Inesi 1985). Nevertheless, most other types of cells, including neurons, also contain an ATP-driven calcium sequestration system in the endoplasmic reticulum. Indeed, the calcium-uptake properties of this system in neurons appear to be very similar, if not identical, to those in muscle sarcoplasmic reticulum (Blaustein et al. 1978; Blaustein et al. 1980; Rasgado-Flores and Blaustein 1987a).

In muscles with a well-developed sarcoplasmic reticulum, cell activation may trigger the release of calcium from the sarcoplasmic reticulum, and this calcium may contribute to the activation of contraction (even in the absence of extracellular calcium). Although the mechanism of calcium release from the sarcoplasmic reticulum is not completely understood, in some muscles the release appears to be triggered by inositol triphosphate (Berridge and Irvine 1984; Somlyo et al. 1985; Hashimoto et al. 1986). At nerve terminals, however, depolarization does not normally trigger transmitter release in the absence of extracellular calcium (Harvey and MacIntosh 1940; Katz 1969); this implies that little, if any, calcium is released from the endoplasmic reticulum under these circumstances.

During a 1–2 ms action potential, calcium may be expected to enter mammalian brain nerve terminals at a rate of about 5 pmoles/mg protein/ms (Nachshen 1985). The maximum rate at which the endoplasmic reticulum in the terminals can sequester calcium is about 0.1–0.2 pmoles/mg protein/ms (Rasgado-Flores and Blaustein 1987a). Thus, the rate of sequestration is more than an order of magnitude too slow to account for the rate of calcium removal from the cytosol required to terminate transmitter release immediately following repolarization after an action potential. During a train of action potentials, however, the endoplasmic reticulum will slowly begin to accumulate calcium because calcium extrusion across the plasma membrane lags behind calcium entry (see below). This may be important for processes such as post-tetanic potentiation (PTP): as the endoplasmic reticulum begins to saturate with calcium, it will help to buffer the resting $[Ca^{2+}]_i$ to higher levels. Moreover, it is possible that subsequent action potentials may then be able to trigger the release of substantial amounts of calcium from these stores; this would, of course, significantly enhance the amount of transmitter release evoked by these action potentials (i.e. PTP). Clearly, these are points that need to be carefully explored.

Despite these questions about evoked calcium release from nerve terminal endoplasmic reticulum, evoked release of calcium from the endoplasmic reticulum in other portions of the neuron may be physiologically important. During the past several years, inositol triphosphate has been found to release calcium from the endoplasmic reticulum stores in a variety of cells (Streb et al. 1983; Berridge and Irvine 1984; Hashimoto et al. 1986). Inositol triphosphate also releases calcium from the endoplasmic reticulum of neuroblastoma cells (Chueh and Gill 1986). Recently, GTP has also been found to release calcium from the endoplasmic reticulum of these cells by a mechanism that involves GTP hydrolysis (Gill et al. 1986b); the GTP and inositol triphosphate act through different pathways (Chueh and Gill 1986; Gill et al. 1986b). The physiological role of these phenomena in neurons remains to be elucidated, but it is apparent that we now need to consider a number of newly discovered regulatory systems that participate in the control of $[Ca^{2+}]_i$.

Cytosolic Buffers

Calcium sequestration (in the endoplasmic reticulum and mitochondria) and calcium extrusion (see below) cannot remove calcium from the transmitter release sites sufficiently rapidly to account for the termination of transmitter release at the end of an action potential. Therefore, we must look elsewhere for the mechanisms responsible. The most likely candidates are neuronal cytoplasmic proteins that have a high affinity for calcium. Several such proteins have been identified, including calmodulin (Wood et al. 1980; DeLorenzo 1982), parvalbumin (Heizmann 1984) and vitamin D-dependent calcium-binding proteins (Pochet et al. 1985).

Calmodulin may be regarded as a "calcium-dependent second messenger"; when calcium is bound to calmodulin, the complex can, in turn, bind to, and activate, other "receptor" molecules. Parvalbumin may have a different function. It is present in relatively high concentrations in fast skeletal muscle cells, and in these cells it appears to serve as an intermediate repository for calcium: immediately after the activation of contraction, it pulls calcium off troponin C and binds the calcium until it is resequestered in the sarcoplasmic reticulum (Pechere et al. 1977; Somlyo et al. 1981; Heizmann 1984).

Some neurons also contain considerable amounts of parvalbumin (Heizmann 1984). Although there is no direct evidence for its role in calcium buffering in neurons, it is possible that this protein may function in neurons in a way comparable to its role in skeletal muscle: it may remove calcium from the transmitter release sites, and then release the bound calcium as it is sequestered (in the endoplasmic reticulum) or extruded across the plasma membrane. Some support for this view comes from the observation that parvalbumin is present in gamma-aminobutyric acid (GABA)-ergic neurons that fire at a high frequency, but not in those that fire at a low frequency (Celio 1986). The parvalbumin may be needed in the high-frequency neurons to buffer intracellular calcium in order to keep the calcium from activating calcium-dependent potassium conductances (see below) that may reduce excitability and prolong the refractory period in these neurons. The capacity of parvalbumin to bind calcium substantially exceeds that needed to bind the amount of calcium that enters during a single action potential. However, with a rapid train of action potentials, the parvalbumin will become more and more saturated with calcium with each succeeding action potential. Thus, this form of calcium buffering may play a very important role in the potentiation of the transmitter release that occurs when action potentials follow rapidly (within a few milliseconds) after one another (PTP).

Calcium Transport across the Neuronal Plasma Membrane

We have already noted that there is a very large electrochemical gradient which favours the entry of calcium across the nerve cell plasma membrane, and that there is a net gain of intracellular calcium during neuronal activity. Thus, to remain in long-term calcium balance, nerve cells must have calcium extrusion mechanisms which necessarily require the expenditure of metabolic energy. Like many other types of cells (Sheu and Blaustein 1986), neurons have two parallel, independent mechanisms in their plasma membranes for extruding calcium (Gill et al. 1981, 1986a): an ATP-driven calcium pump (analogous to the ATP-driven sodium pump), and an Na^+/Ca^{2+} exchange transport system (Fig. 13.3). The latter uses energy available from the sodium electrochemical gradient,

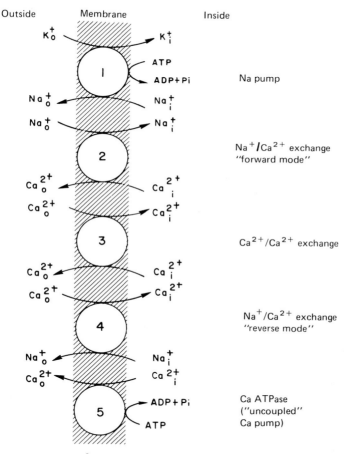

Fig. 13.3. Na^+/Ca^{2+} exchange (2 and 4) and ATP-driven calcium pump (5) modes of calcium transport across the plasma membrane. Both transport systems can mediate calcium extrusion (2 and 5, respectively). In addition, the Na^+/Ca^{2+} exchange system can operate in the "reverse mode" (also a normal mode of operation), bringing calcium into the cell in exchange for exiting sodium; the direction of net calcium movement depends upon the prevailing membrane potential and the sodium and calcium equilibrium potentials (see text). In the "reverse mode" of operation, the Na^+/Ca^{2+} exchanger moves calcium in parallel with the calcium movement through calcium-selective channels (not shown here; see Fig. 13.1). The Na^+/Ca^{2+} exchanger can also mediate Ca^{2+}/Ca^{2+} (tracer) exchange (3). The ATP-driven sodium pump (1) is shown at the *top* of the diagram. (Blaustein 1984)

generated by the ATP-driven sodium pump, to export calcium. This transport system is especially interesting because it can also promote calcium entry when the sodium concentration gradient is reduced or the membrane is depolarized (see below). In view of the parallel operation of these two transport systems, we need to determine the role that each of them plays in neuronal cell function.

Two general classes of ATP-driven calcium pump have been recognized: those located in the smooth endoplasmic reticulum that participate in intracellular calcium sequestration (see above), and those located in the plasma membrane, that help to extrude calcium from cells (Carafoli 1984; Sheu and Blaustein 1986). The latter transport systems are modulated by calmodulin, have stoichiometries of 1 calcium ion transported per ATP hydrolysed and are mediated by calcium-dependent ATPases with molecular weights of about 140 000 (Hincke and Demaille 1984). In contrast, the endoplasmic reticulum calcium transport systems are not modulated by calmodulin (but see Wulfroth and Peltzelt 1985); they have stoichiometries of two calcium ions transported per ATP hydrolysed and they are mediated by calcium-dependent ATPases with molecular weights of about 105 000.

In a number of types of cells, including some neurons, the plasma membrane ATP-driven calcium pump has a high affinity for calcium (i.e. a K_D of the order of 0.2–0.3 μM), but a low transport capacity (Gill et al. 1981; Carafoli 1984; Sanchez-Armass and Blaustein 1987; and see Fig. 13.4). On the other hand, the Na^+/Ca^{2+} exchange system appears to have a somewhat lower affinity for calcium (K_D of about 0.5–1 μM; Blaustein 1977), but a much larger capacity (or maximum velocity of transport) (Sanchez-Armass and Blaustein 1987; and see Fig. 13.4). Also, the Na^+/Ca^{2+} exchange system appears to be regulated by intracellular calcium: calcium concentrations within the dynamic physiological range (0.1–1.0 μM) are required to promote calcium entry ("reverse mode" exchange) as well as calcium exit ("forward mode" exchange) mediated by this transport system (DiPolo and Beauge 1986; Rasgado-Flores et al. 1986).

The importance of the sodium-dependent transport system for extruding calcium from mammalian neurons is illustrated by the data in Fig. 13.4. When nerve terminals are given a small load of calcium, equivalent to the net calcium entry during about 100–125 action potentials, calcium exit is almost entirely dependent on external sodium: the rate of tracer ^{45}Ca efflux is slowed by a factor of about ten when most of the external sodium is replaced by another monovalent cation (in this case, N-methyl glucamine or choline). In this experiment calcium was omitted from the external solution during the efflux period to minimize tracer ($^{40}Ca/^{45}Ca$) exchange. A number of monovalent cations were tested (including sodium, lithium, choline, N-methyl glucamine and tetramethylammonium): under these conditions, only sodium appeared to be able to activate calcium efflux. Moreover, sodium flux data from neuroblastoma cells (Wakabayashi and Goshima 1981) indicate directly that sodium is exchanged for the calcium.

Studies with the calcium-sensitive fluorochromes, quin2 and fura-2 (Nachshen et al. 1986), provide further evidence that the external sodium-dependent calcium efflux plays an important role in $[Ca^{2+}]_i$ regulation in presynaptic nerve terminals. After a period of calcium loading in low-sodium media, net efflux of calcium and restoration of $[Ca^{2+}]_i$ to its initial low value occur rapidly (within 1 min) when sodium is added back to the bathing medium. Neither the calcium uptake from low-sodium media, nor the sodium-dependent calcium

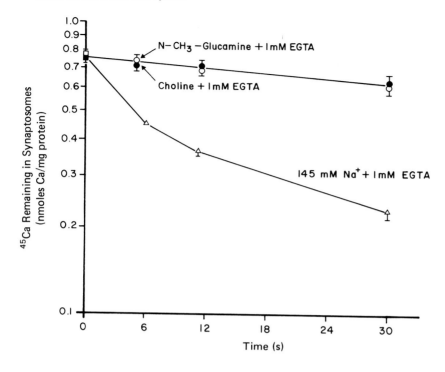

a

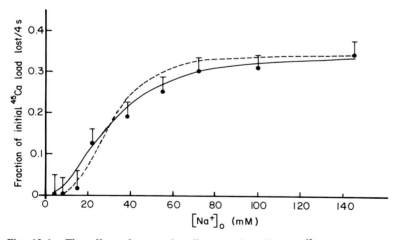

b

Fig. 13.4. The effect of external sodium on the efflux of ^{45}Ca from tracer-loaded rat brain synaptosomes in the absence of external sodium. **a** Time course of the calcium efflux into calcium-free media (with 1 mM EGTA) containing 145 mM sodium ($\triangle$), or only 4.5 mM sodium and either 140 mM choline ($\bullet$) or 140 mM N-methyl-glucamine ($\bigcirc$). **b** Relationship between the external sodium concentration (abscissa) and the rate of calcium efflux into calcium-free media. When external sodium was reduced below 145 mM it was replaced mole-for-mole by N-methyl-glucamine. The curves are drawn to fit the Hill equation with a Hill coefficient of 2 (*solid line*) or 3 (*broken line*). The very flat foot at low $[Na^+]_o$ fits the curve with the Hill coefficient of 3 best; this is consistent with a stoichiometry of 3 sodium : 1 calcium (Sanchez-Armass and Blaustein 1987)

extrusion, is affected by the mitochondrial uncoupler, FCCP; the implication is that the mitochondria do not normally contain large stores of calcium. However, FCCP, which should cause a marked reduction in intracellular ATP levels, does cause $[Ca^{2+}]_i$ to rise slightly; this may indicate that the ATP-driven calcium pump contributes to the control of resting $[Ca^{2+}]_i$ (see below).

The direction in which the exchanger moves (net) calcium across the plasma membrane is determined by the difference, ΔV, between the membrane potential, V_M, and the reversal potential for the Na^+/Ca^{2+} exchanger, $E_{Na/Ca}$ (Sheu and Blaustein 1986):

$$\Delta V = V_M - E_{Na/Ca} \tag{1}$$

When ΔV is negative, the exchanger will mediate net outward movement of calcium; the converse will be true when ΔV is positive. Thus, the direction of calcium transport is critically dependent upon the stoichiometry of the exchange, since:

$$(n-2)E_{Na/Ca} = nE_{Na} - 2E_{Ca} \tag{2}$$

where E_{Na} and E_{Ca} are, respectively, the equilibrium potentials for sodium and calcium, and n is the number of sodium ions exchanged for each calcium ion. Several observations indicate that n has a value of about 3 in neurons as well as

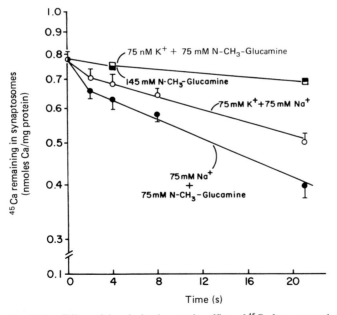

Fig. 13.5. Effect of depolarization on the efflux of ^{45}Ca from tracer-loaded rat brain synaptosomes. The calcium efflux was initiated by diluting the incubation media with a depolarizing (75 mM potassium) solution, without (□) or with (○) sodium, or with a low (5 mM) potassium solution without (■) or with (●) sodium. In all efflux solutions, the sum of the sodium + potassium + N-methyl-glucamine concentrations was 150 mM. All the efflux solutions were calcium-free. (Sanchez-Armass and Blaustein 1987)

in other types of cells. For example, there is evidence that external sodium-dependent calcium efflux is voltage-sensitive: it is inhibited by depolarization (Blaustein et al. 1974; Allen and Baker 1986; and see Fig. 13.5) and is stimulated by hyperpolarization (Mullins and Brinley 1975; Allen and Baker 1986). The stoichiometry has been measured directly with tracer flux methods in squid axons (Blaustein and Russell 1975) and barnacle muscle cells (Rasgado-Flores and Blaustein 1987b), and more indirectly in several other types of cells (Sheu and Blaustein 1986), and has been found to be very close to 3 Na^+ : 1 Ca^{2+} in all cases. Based on this exchange stoichiometry and the measured values for E_{Na}, E_{Ca} and V_M, ΔV should be slightly negative in resting neurons, so that calcium efflux will be favoured under these circumstances (see Fig. 13.6).

In addition to control by these thermodynamic factors, the Na^+/Ca^{2+} exchange is also influenced by kinetic factors based on the fractional occupancy of the carriers by transported ions as well as by activating (non-transported) ions (DiPolo and Beauge 1986; Kimura et al. 1986; Rasgado-Flores et al. 1986, 1987b). For example, studies in squid axons and barnacle muscle cells demonstrate that calcium entry mediated by the Na^+/Ca^{2+} exchanger is dependent on (non-transported) intracellular calcium concentrations in the dynamic physiological range (about 0.1–1.0 μM) (Rasgado-Flores and Blaustein 1987b). Thus, when $[Ca^{2+}]_i$ is low, as in the case of resting neurons (~0.1–0.3

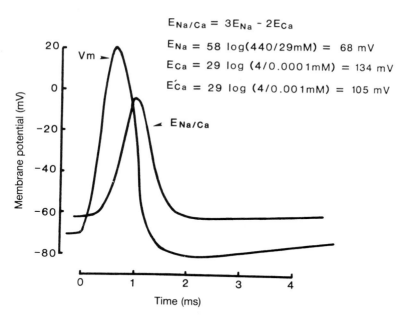

$$E_{Na/Ca} = 3E_{Na} - 2E_{Ca}$$

$$E_{Na} = 58 \log(440/29mM) = 68 \text{ mV}$$

$$E_{Ca} = 29 \log(4/0.0001mM) = 134 \text{ mV}$$

$$E'_{Ca} = 29 \log(4/0.001mM) = 105 \text{ mV}$$

Fig. 13.6. The membrane potential (V_m) and calculated Na^+/Ca^{2+} exchanger reversal potential ($E_{Na/Ca}$) for the squid giant synapse during an action potential. $E_{Na/Ca}$ was calculated on the assumption that the concentrations of the free ions at rest are: $[Na^+]_0 = 440$ mM, $[Na^+]_i = 29$ mM, $[Ca^{2+}]_0 = 4$ mM and $[Ca^{2+}]_i = 0.1$ μM. During the action potential, $[Ca^{2+}]_2$ is assumed to increase to 1.0 μM (i.e. E_{Ca} decreases from 134 mV to 105 mV). The exchanger mediates (net) calcium entry during the action potential (i.e. when V_m is more positive than $E_{Na/Ca}$). It mediates (net) calcium exit at the end of the action potential and during the hyperpolarization (i.e. when V_m is more negative than $E_{Na/Ca}$). See text for further details.

μM), the turnover of the exchanger will be very low because the internal calcium sites that participate in calcium extrusion ($K_{Ca} \sim 0.7$ μM; Blaustein 1977), as well as those that activate exchanger-mediated calcium entry ($K_{Ca} \sim 0.6$ μM; Rasgado-Flores et al. 1986) will be largely unoccupied. This means that, under resting conditions, even though the exchanger is poised to move calcium out of the cells (see Fig. 13.6), much of the net calcium extrusion may be mediated by the ATP-driven calcium pump. Nevertheless, the Na^+/Ca^{2+} exchange system will bias the distribution of calcium in the neurons: $[Ca^{2+}]_i$ and the amount of calcium on the buffers and in the endoplasmic reticulum will still be influenced by the sodium electrochemical gradient across the plasma membrane.

When the neurons are activated, a different situation prevails. Initially, when the membrane is depolarized, ΔV will become positive, and this driving force will now favour calcium entry via the Na^+/Ca^{2+} exchanger. Initially, however, the internal calcium activation sites will still be largely unoccupied. Once the voltage-gated calcium channels open, however, and calcium enters the cells, the exchanger will be activated and calcium will also enter by ("reverse mode") Na^+/Ca^{2+} exchange; the relative roles of the calcium channels and the Na^+/Ca^{2+} exchange system in mediating calcium entry is not yet clear. Now, with $[Ca^{2+}]_i$ still elevated, when the neurons repolarize, ΔV will become large and negative, and the exchanger will promote calcium extrusion until $[Ca^{2+}]_i$ declines and internal calcium-binding sites on the exchanger are no longer saturated. This anticipated relationship between V_M and $E_{Na/Ca}$ is shown diagrammatically in Fig. 13.6. The available data indicate that the Na^+/Ca^{2+} exchange system is turned on during depolarization, as a result of positive feedback (by increasing $[Ca^{2+}]_i$), and turned off as a result of negative feedback (by decreasing $[Ca^{2+}]_i$), when the cells repolarize.

The "Life Cycle" of Calcium at the Nerve Terminal

We can now put together all the available information on calcium entry, calcium buffering and calcium extrusion at nerve terminals, to obtain a comprehensive picture of how calcium is handled by the terminals during an action potential (see Fig. 13.1). As just discussed, during the rising phase of the action potential calcium will enter the terminals via voltage-gated calcium channels and, with a slight delay, via the Na^+/Ca^{2+} exchanger. The increase in $[Ca^{2+}]_i$ will trigger transmitter release. Then, as the membrane repolarizes, the calcium channels will close, and some calcium will exit through the Na^+/Ca^{2+} exchanger. However, much of the calcium that had just entered will diffuse away from the plasma membrane and will be buffered (perhaps by parvalbumin); these processes will markedly lower $[Ca^{2+}]_i$ in the region of the transmitter release sites with a half-time of about 1 ms. Then, during the next 10 ms or so, the calcium bound to the buffer proteins will be removed and sequestered in the endoplasmic reticulum. Finally, over the subsequent 0.1–1 s, the calcium extrusion mechanisms (Na^+/Ca^{2+} exchange and the ATP-driven calcium pump) will transport back to the extracellular fluid the remainder of the calcium that had entered and been sequestered.

During a rapid train of action potentials, intracellular $[Na^+]$ will tend to rise with time, due to a slight lag in the ability of the sodium pump to extrude the sodium that enters during the action potentials (Woodbury 1963). As a result of this reduction in the sodium concentration gradient, Na^+/Ca^{2+} exchange-mediated calcium entry will be enhanced, and calcium exit will be reduced, thereby causing the interspike $[Ca^{2+}]_i$ to rise slightly. In addition, a slight lag in the sequestration of calcium by the endoplasmic reticulum will cause the cytosolic buffers to begin to saturate. These factors will, in turn, augment neurotransmitter release – as manifested by facilitation and PTP (Charlton and Atwood 1977; Charlton et al. 1980; Atwood et al. 1983; Misler and Hurlbut 1983; Meiri et al. 1986).

Physiological Role of Intracellular Calcium in Nerve Cells

In the preceding sections of this chapter we examined the various mechanisms involved in the movements of calcium across the plasma membrane and in the regulation of $[Ca^{2+}]_i$. We will now turn our attention to some of the critical physiological actions of this intracellular calcium ("messenger calcium") in neurons.

Calcium and Neurotransmitter Release

A rise in $[Ca^{2+}]_i$ is the immediate trigger for secretion in most types of secretory cells where the secreted substances are stored in vesicles; this is usually the result of calcium entry from the extracellular fluid (Douglas 1974; Rubin 1982; Burgoyne 1984). This is also the case for neurotransmitter release: release is critically dependent upon extracellular calcium, and is normally initiated by the opening of calcium channels which permits a large increase in calcium entry and rise in $[Ca^{2+}]_i$ (Fig. 13.7; and Katz 1969; Llinas et al. 1981; Zucker and Lando 1986). The rise in $[Ca^{2+}]_i$ promotes fusion of synaptic vesicles with the plasma membrane, thereby inducing the exocytosis of vesicular stores of transmitter(s) and other molecules into the synaptic cleft (Katz 1969; Blaustein 1978; Drapeau and Blaustein 1983b). Nevertheless, the precise mechanism by which this calcium actually evokes release is uncertain. Data from various neuronal preparations indicate that the cooperative action of several calcium ions is required to activate the release of one packet (quantum) of transmitter, i.e. the amount contained in one synaptic vesicle (Dodge and Rahamimoff 1967; Augustine et al. 1985; Zucker and Fogelson 1986; but see Nachshen and Drapeau 1982).

The role of calcium does not appear to be limited to the neutralization of negative charges on the inner surface of the plasma membrane and outer surface of the vesicle membrane (Parsegian 1977). While such an effect would allow these two membranes to come together (as observed with high concentrations of magnesium or lanthanum), such apposition or fusion is not necessarily associated with exocytosis (Heuser et al. 1971; Heuser 1977).

Nerve terminals contain relatively high concentrations of contractile proteins and calmodulin, and several authors have suggested that these molecules may play a role in evoked transmitter release (e.g. Berl et al. 1973; Baker and Knight 1981; Burke and DeLorenzo 1982; DeLorenzo 1982, 1983; but see Wood et al.

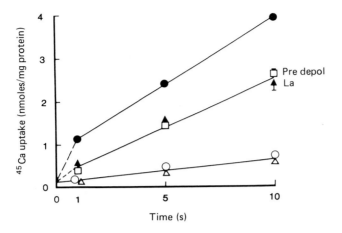

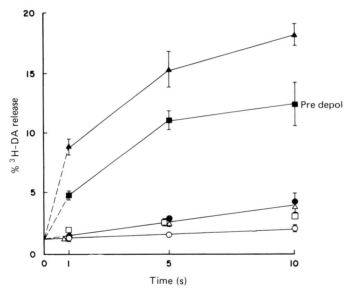

Fig. 13.7. Time course of ^{45}Ca uptake (A) and ^3H-dopamine (^3H-DA) release (B) in synaptosomes from rat brain corpus striatum. **a** Calcium uptake was measured in non-depolarizing (145 mM sodium + 5 mM potassium) media ($\bigcirc$, $\triangle$) or depolarizing (75 mM sodium + 75 mM potassium) media ($\bullet$, $\blacktriangle$, $\square$). In some instances the solutions contained 0.1 μM lanthanum ($\triangle$, $\blacktriangle$) to block calcium uptake through voltage-gated calcium channels. Some synaptosomes were predepolarized in calcium-free, 75 mM potassium media immediately before the ^{45}Ca uptake was initiated, to inactivate most of the calcium channels ($\square$). **b** Dopamine release (as a percentage of the ^3H-DA previously taken up by the synaptosomes) under conditions similar to those used for the calcium uptake experiment (in **a**). DA release was measured in 145 mM sodium + 5 mM potassium media ($\bigcirc$, $\bullet$) or in depolarizing 75 mM sodium + 75 mM potassium media ($\triangle$, $\blacktriangle$) that were either calcium-free ($\bigcirc$, $\triangle$) or contained 2 mM calcium ($\bullet$, $\blacktriangle$). Some synaptosomes were predepolarized in calcium-free, 75 mM sodium + 75 mM potassium media and then incubated (during the ^3H-DA efflux period) in calcium-free 75 mM potassium media ($\square$) or in 75 mM potassium media containing 2 mM calcium ($\blacksquare$). Note the parallelism, under comparable conditions, between the rates of calcium uptake (A) and DA release (B). (Drapeau and Blaustein 1983)

1980, regarding the distribution of calmodulin in neurons). However, the possible mechanisms by which these molecules may participate in the transmitter release process are not known.

Secretory mechanisms in adrenal chromaffin cells, which are homologous to neurons, are probably very similar to the mechanisms involved in neurotransmitter release. Thus, data on secretion from the adrenal chromaffin cells (Knight and Baker 1982, 1985; Burgoyne 1984) may be directly applicable to neurons. Available data from permeabilized chromaffin cells suggest that the calcium concentration that promotes half-maximal secretion is about 1 μM; magnesium inhibits calcium-dependent secretion with an apparent IC_{50} of about 1 mM (Baker and Knight 1981; Knight and Baker 1982). A phosphorylation step appears to be involved in secretion because removal of ATP causes a progressive and reversible inhibition of secretion, and non-hydrolysable analogues of ATP cannot substitute for the ATP (Baker and Knight 1981). Recently, GTP-binding proteins have also been implicated in calcium-dependent secretion from chromaffin cells (Knight and Baker 1985).

Attempts to elucidate the role of calmodulin in secretion, by employing calmodulin inhibitors (Baker and Knight 1981), have been inconclusive – in part because most of the agents employed, such as the phenothiazines, are not very selective. However, recent studies on insulin secretion have led to the suggestion that cyclic AMP and calmodulin may induce secretion by acting at different steps in the secretory process; the fusion of secretory granules with the plasma membrane may be a calcium/calmodulin-directed function (Steinberg et al. 1984). Whether or not this is applicable to neurotransmitter release as well, remains to be determined.

All of the aforementioned observations provide only indirect information about the mechanisms of secretion and neurotransmitter release. The precise molecules and mechanisms involved remain an enigma. However, possible clues to the mechanism may come from recent attempts to "reconstitute" the calcium-dependent exocytosis process with isolated secretory vesicles and plasma membrane from sea urchin eggs (Crabb and Jackson 1985). In this study, release of vesicular contents was evoked by calcium-dependent fusion of the secretory vesicles with patches (a "lawn") of plasma membrane. It may be possible to apply similar methods to the study of neurotransmitter release.

Calcium and the Control of Excitability in Neurons

In the Introduction, we noted that extracellular calcium influences membrane excitability. Under normal physiological conditions, however, extracellular divalent cation concentrations are usually well regulated within rather narrow limits. In contrast, as we have seen, $[Ca^{2+}]_i$ may vary substantially during cell activity, and this may markedly influence cell excitability.

Meech (1972) was the first to demonstrate clearly that intracellular calcium can activate a potassium conductance. Numerous investigators have expanded on these early observations and have shown that a rise in $[Ca^{2+}]_i$ increases membrane conductance and causes hyperpolarization in many types of vertebrate and invertebrate neurons (e.g. Meech 1978; Gorman and Hermann 1982; Higashi et al. 1984). Presynaptic nerve terminals in mammalian brain (Bartschat and Blaustein 1985) and in mammalian peripheral nerve (Mallart

1984) appear to be richly endowed with calcium-activated potassium channels. Indeed, data from the brain (Benishin et al. 1986; Lancaster and Adams 1986) and sympathetic ganglia (Pennefather et al. 1985) indicate that there may be several types of calcium-activated potassium channels.

Single channel studies have demonstrated that neuronal membranes contain calcium-activated potassium channels that are opened by $[Ca^{2+}]_i$ in the dynamic physiological range (about 0.1– 1 μM). In some cases, these channels appear to be blocked by phosphorylation that is promoted by the catalytic subunit of cyclic AMP-dependent protein kinase (Bartschat et al. 1986). In other instances, calcium-activated potassium conductance is enhanced by cyclic AMP-dependent protein phosphorylation that increases the probability of channel opening in the presence of calcium (de Peyer et al. 1982; Ewald et al. 1985).

The precise physiological roles of these forms of channel modulation are not completely understood. Nevertheless, it is apparent that, at least in some types of cells, membrane excitability can be controlled by "second messengers" such as cyclic nucleotides. In turn, the cyclic AMP may be controlled by receptor-regulated GTP-binding proteins, or "G-proteins" (Gilman 1986). Excitability can then be altered either by modulation of the calcium channels (see above) or by modulation of the calcium-activated potassium channels (and, perhaps, voltage-gated potassium channels: e.g. Breitwieser and Szabo 1985; Pfaffinger et al. 1985; Sakakibara et al. 1986). These covalent modifications of the ion channels (e.g. phosphorylation: see Levitan 1985; Nairn et al. 1985) would be expected to produce relatively long-lasting changes, and are thus a form of "memory". Indeed, this repertoire of ways to modify membrane excitability provides multiple opportunities for modulating and fine-tuning nervous system activity.

Intracellular calcium may also modulate other types of ion channels. For example, it inhibits the GABA-activated chloride conductance in sensory neurons from the bullfrog dorsal root ganglion. Calcium, when entering the cells through calcium-selective channels, appears to decrease the affinity of the membrane receptor for GABA (Inoue et al. 1986).

The relative contributions of the various types of conductances may differ greatly from cell type to cell type, and this will have a profound effect on the cell function. For example, in some cells, the calcium-activated potassium conductance may be quite long-lasting, especially after a burst of activity, and this may greatly suppress subsequent firing. Interesting examples of this kind of cell include the beating and bursting pacemaker cells in the abdominal ganglion of the marine mollusc, *Aplysia californica* (Gorman and Hermann 1982). The bursting pacemaker (cell R-15), whose cell cycle normally consists of a rapid series of action potentials followed by a long quiescent period, has a larger calcium conductance, and a larger calcium-activated potassium conductance than does the beating pacemaker (cell L-11). The interburst intervals in the bursting pacemaker can be accounted for by the large, long-lasting hyperpolar-ization, due to the increase in calcium-activated potassium conductance, that occurs after a burst of activity (with a relatively large influx of calcium). The implication is that $[Ca^{2+}]_i$ remains elevated for a substantial period of time following the burst. The low frequency, steady repetitive discharge that is normally seen in L-11 can be mimicked in R-15 simply by withdrawing extracellular calcium; this is consistent with the view that the large calcium influx and large calcium-activated potassium conductance in R-15 are responsible for the bursting activity.

In mammalian cells it may be possible to determine the role calcium plays in regulating activity by injecting EGTA or calcium intracellularly. For example, the inhibition of firing rate following a period of activation in spontaneously firing locus coeruleus neurons is associated with an afterhyperpolarization that is likely due to a calcium-dependent potassium conductance. The afterhyperpolarization is abolished and the spontaneous firing rate and reactivity to sensory stimulation are increased following injection of EGTA into these neurons (Aghajanian et al. 1983).

At many types of synapses, a period of high-frequency firing (tetanus) is often followed by a period of enhanced transmitter release in response to subsequent action potentials (i.e. PTP). This is usually attributed to the residual calcium in the cytoplasm after the burst of activity (see above). PTP is then usually followed by a period of post-tetanic depression. Although this depression has sometimes been attributed to an exhaustion of transmitter or inactivation of the release process, a likely alternative is that a large calcium-activated potassium conductance may cause an afterhyperpolarization and reduce membrane excitability (Kretz et al. 1982). The latter study, in fact, supports the "residual calcium" hypothesis, and suggests that calcium may be extruded only relatively slowly after a burst of neuronal activity (Bolsover and Spector 1986; Connor et al. 1986). Alternatively, however, we must also consider the possibility that a relatively brief elevation of $[Ca^{2+}]_i$ might trigger metabolic changes which could then activate potassium channels for a prolonged period, perhaps as a result of phosphorylation (Higashi et al. 1984; Miller and Kennedy 1986).

Calcium and Memory

Facilitation, post-tetanic potentiation and afterhyperpolarization are all examples of neuronal activity modulation that are a direct consequence of calcium retention in neurons (Bolsover and Spector 1986; Connor et al. 1986). We may consider these altered responses, that occur as a result of prior activation of the neurons, as manifestations of very brief "memory" – forms of "memory" that last many seconds to minutes. Somewhat longer term retention of information from prior neuronal activation (perhaps lasting minutes to hours), commonly referred to as "short-term memory", may result from covalent modification of ion channels – for example, channel phosphorylation, as mentioned above. In several instances long-term potentiation of transmitter release in the mammalian central nervous system has been attributed to increased calcium-dependent release of neurotransmitter (Feasey et al. 1986). It is possible that this may be the result of alterations in the calcium channels or calcium-activated potassium channels. In invertebrate systems, some conductance changes produced by protein kinase C activation have been found to mimic those induced by associative learning (Farley and Auerbach 1986). In the mammalian central nervous system, phorbol esters, which activate protein kinase C, block a calcium-dependent potassium conductance (Baraban et al. 1985) and potentiate synaptic transmission in much the same way as they affect long-term potentiation (Malenka et al. 1986).

There is also some evidence that long-term potentiation may involve postsynaptic mechanisms. Lynch and Baudry (1984) have suggested that a rise in $[Ca^{2+}]_i$ in the postsynaptic cells may activate a protease (calpain) that degrades a spectrin-like protein (fodrin) and thereby increase, irreversibly, the number of glutamate receptors in the plasma membrane.

Another mechanism by which transient elevation of $[Ca^{2+}]_i$ may lead to a relatively long-term increase in membrane channel phosphorylation involves the calcium/calmodulin-activated autophosphorylation of brain type II protein kinase (Miller and Kennedy 1986). With very brief elevation of $[Ca^{2+}]_i$ the protein kinase activity is greatly enhanced by interaction with calcium and calmodulin. With more prolonged elevation of $[Ca^{2+}]_i$, however, the calcium/calmodulin complex promotes autophosphorylation of the kinase and, in this state, the protein kinase activity becomes independent of calcium (and calmodulin). This calcium-independent state may be terminated by dephosphorylation or it may continue for the life of the molecule. Thus, this protein kinase system behaves like a molecular switch that can lead to a transiently activatable state or to a prolonged or permanently active state.

The forms of memory mentioned above do not involve protein synthesis, and are not disrupted by blockers of protein synthesis. In contrast, blockers of protein synthesis do interfere with the development of long-term memory that normally lasts for days, weeks or even years (Flexner et al. 1963; Davis and Squire 1984). While this implies that activation of genes and induction of messenger RNA and protein synthesis are probably involved in this type of memory, it does not rule out a role for calcium. Indeed calcium may be needed either because the induction of short-term memory may be a critical prerequisite for the induction of long-term memory and/or because calcium may be required to activate the genetic mechanisms involved in cell growth and differentiation that are associated with long-term memory. The evidence that short-term and long-term "memory" may involve the same synaptic connections (Goelet et al. 1986) is consistent with the view that different calcium-dependent mechanisms (see above), in the same neurons, may be responsible.

Recent observations in the pheochromocytoma PC12 cell line indicate that the proto-oncogene, c-*fos*, which is known to be involved in cell differentiation and growth (Alberts et al. 1983; Rosengurt 1986), is rapidly activated by stimulation of acetylcholine receptors, and by opening of calcium channels, or by growth factors that elevate $[Ca^{2+}]_i$ (Greenberg et al. 1986; Morgan and Curran 1986). This regulatory gene is normally activated by the same cytosolic messengers that are involved in short-term memory: calcium, cyclic AMP and diacylglycerol (Rosengurt 1986). Goelet and Kandel (1986; and see Goelet et al. 1986) have suggested that activation of some proto-oncogenes may be an early step in the sequence of events that leads to the establishment of long-term memory. While the precise sequence is not clear, there is evidence that long-term memory in both vertebrates (Greenough 1984) and invertebrates (Bailey and Chen 1983; Lnenicka et al. 1986) is associated with the growth of synaptic contacts.

Neuronal Growth and Development: The Influence of Calcium on Axoplasmic Transport and on Growth Cones

Neuronal sustenance requires the transport of proteins, and perhaps other materials, from the cell bodies to the periphery (Grafstein 1977). Calcium apparently plays a role in at least two steps in the transport process. The first step is the transfer of protein from the Golgi apparatus to the axonal transport system. This step is blocked when calcium is removed from the extracellular fluid (presumably because this lowers $[Ca^{2+}]_i$), or when cobalt is added to the bathing medium (Hammerschlag and Lavoie 1979; Lariviere and Lavoie 1982).

The second step, fast axonal transport of proteins down the axons to the periphery, is also inhibited by the removal of extracellular calcium (Chan et al. 1980), but this process is not inhibited by extracellular cobalt (Lariviere and Lavoie 1982). While little is known about the precise role of calcium, it is possible that calcium may activate these transport processes via an effect on the microtubules or the contractile proteins that are present in the cytoplasm (Lariviere and Lavoie 1982; Ochs and Jersild 1984).

In many neurons and other excitable cells, a dramatic change in membrane excitability takes place during development. In the early stages of development, inward current during action potentials is carried primarily by calcium. Then, as the cells mature, sodium channels appear and, in some instances, the calcium channels are lost (Spitzer 1979). This ontogenesis may, in part, reflect the fact that calcium channels evolved earlier than sodium channels in the course of philogenesis (Hille 1984). The presence of calcium channels during the early stages of neuronal development may be due to the fact that the growth of neurites is critically dependent upon calcium. Calcium-dependent inward currents and action potentials are a prominent feature of growth cones in developing and regenerating neurites (Strichartz et al. 1980; Meiri et al. 1981; Anglister et al. 1982). This is reflected in the elevated $[Ca^{2+}]_i$ levels observed in active growth cones (Bolsover and Spector 1986; Connor 1986). In addition to its role in axoplasmic transport, the intracellular calcium may participate in the extensions of the growth cones. This might occur either through an action of calcium on the contractile proteins present in the cytoplasm or by participation of calcium in the fusion of cytoplasmic vesicles with the plasma membrane, which increases the membrane surface area and introduces new proteins into the plasma membrane.

Calcium may also play a role in neuronal degeneration and cell death. When axons are damaged or transected, the Wallerian degeneration that occurs in the damaged regions is dependent upon calcium (Schlaepfer and Bunge 1973) and can be mimicked by treatment with the calcium ionophore, A-23187, which promotes calcium entry into the axoplasm (Schlaepfer 1977). The initial stage of the degenerative process appears to be massive calcium entry into the damaged region, which then may seal off and become much less permeable to calcium as the recovery process begins (Meiri et al. 1981). The initial very large increase in $[Ca^{2+}]_i$ will activate calcium-dependent neutral proteases, which then degrade various cytoskeletal and other axoplasmic proteins (Schlaepfer and Zimmerman 1984; Zimmerman and Schlaepfer 1984). Under more normal circumstances, the calcium-dependent neutral proteases may be activated (to a more limited extent) by endogenous calcium levels, and may participate in the normal remodelling and regulation of the cytoskeleton (Nixon et al. 1983).

Finally, as noted above, massive entry of calcium into the neurons, as a result of cell damage, may lead to cell death, possibly as a result of mitochondrial calcium overload (Trump and Berezesky 1985). It is possible that the decrease in ATP may, by reducing sodium pump activity, cause cells to accumulate sodium. This may, in turn, promote the net gain of calcium via Na^+/Ca^{2+} exchange

(Trump and Berezesky 1985; Sheu and Blaustein 1986); this would tend to augment the mitochondrial calcium overload and hasten cell death.

Summary and Conclusions

In this chapter we have reviewed the subject of calcium transport and regulation of $[Ca^{2+}]_i$ in neurons, and have examined some of the important "second messenger" roles of intracellular calcium in neuronal function. It is clear that intracellular calcium is essential for such activities as maintenance of the cytoskeleton and axonal transport, control of excitability, and the release of neurotransmitter substances. The cellular correlates of learning and memory are also critically dependent upon the messenger role of intracellular calcium.

Acknowledgements. The author is grateful to Drs. C. G. Benishin, K. A. Colby, B. K. Krueger, H. Rasgado-Flores and R. G. Sorensen for helpful comments on a preliminary version of this chapter. Some of the research described in this article was supported by a grant from the NIH (NS-16106).

References

Aghajanian GK, Vandermaelen CP, Andrade R (1983) Intracellular studies on the role of calcium in regulating the activity and reactivity of locus coeruleus neurons in vivo. Brain Res 273:237–243

Alberts B, Bray D, Lewis J, Raff M, Roberts K, Watson JD (1983) Molecular biology of the cell. Garland, New York, pp 621–628

Allen TJA, Baker PF (1986) Influence of membrane potential on calcium efflux from giant axons of *Loligo*. J Physiol (Lond) 378:77–96

Alvarez-Leefmans FJ, Gamino SM, Rink TJ (1984) Intracellular free magnesium in neurones of *Helix aspersa* measured with ion-selective micro-electrodes. J Physiol (Lond) 354:303–317

Anglister L, Farber IC, Shahar A, Grinvald A (1982) Localization of voltage-sensitive calcium channels along developing neurites: their possible role in regulating neurite elongation. Dev Biol 94:351–365

Atwood HL, Charlton MP, Thompson CS (1983) Neuromuscular transmission in crustaceans is enhanced by a sodium ionophore, monensin, and by prolonged stimulation. J Physiol (Lond) 335:179–195

Augustine GJ, Charlton MP, Smith SJ (1985) Calcium entry and transmitter release at voltage-clamped nerve terminals of squid. J Physiol (Lond) 367:163–181

Bailey CH, Chen M (1983) Morphological basis of long-term habituation and sensitization in *Aplysia*. Science 220:91–93

Baker PF, Knight DE (1981) Calcium control of exocytosis and endocytosis in bovine adrenal medullary cells. Philos Trans R Soc Lond [Biol] 296:83–103

Baraban JM, Snyder SH, Alger BE (1985) Protein kinase C regulates ionic conductance in hippocampal pyramidal neurons: electrophysiological effects of phorbol esters. Proc Natl Acad Sci USA 82:2538–2542

Bartschat DK, Blaustein MP (1985) Calcium-activated potassium channels in isolated presynaptic nerve terminals from rat brain. J Physiol (Lond) 361:441–457

Bartschat DK, French RJ, Nairn AC, Greengard P, Krueger BK (1986) Cyclic AMP-dependent protein kinase modulation of single calcium-activated potassium channels from rat brain in planar bilayers. Soc Neurosci Abstr 12:1198

Benishin CG, Krueger BK, Blaustein MP (1986) Low micromolar concentrations of phenothiazines and haloperidol selectively block Ca-activated K channels in rat brain synaptosomes. Soc Neurosci Abstr 12:1199

Berl S, Puszkin S, Nicklas WJ (1973) Actomyosin-like protein in brain. Science 179:441–446

Berridge MJ, Irvine RF (1984) Inositol triphosphate, a novel second messenger in cellular signal transduction. Nature 312:315–321

Blaustein MP (1974) The interrelationship between sodium and calcium fluxes across cell membranes. Rev Physiol Biochem Pharmacol 70:32–82

Blaustein MP (1977) Effects of internal and external cations and ATP on sodium–calcium exchange and calcium–calcium exchange in squid axons. Biophys J 20:79–111

Blaustein MP (1978) The role of calcium in catecholamine release from adrenergic nerve terminals. In: Paton DM (ed) The release of catecholamines from adrenergic neurons. Pergamon Press, Oxford, pp 39–58

Blaustein MP (1984) The energetics and kinetics of sodium–calcium exchange in barnacle muscles, squid axons and mammalian heart: the role of ATP. In: Blaustein MP, Lieberman M (eds) Electrogenic transport: fudamental principles and physiological implications. Raven Press, New York, pp 129–147

Blaustein MP (1985) Intracellular calcium as a second messenger. What's so special about calcium? In: Rubin RP, Weiss GB, Putney JW Jr (eds) Calcium in biological systems. Plenum Press, New York, pp 23–33

Blaustein MP, Rasgado-Flores H (1981) The control of cytoplasmic free calcium in presynaptic nerve terminals. In: Bronner F, Peterlik M (eds) Calcium and phosphate transport across biomembranes. Academic Press, New York, pp 53–58

Blaustein MP, Russell JM (1975) Sodium–calcium exchange and calcium–calcium exchange in internally dialyzed squid giant axons. J Membr Biol 22:285–312

Blaustein MP, Russell JM, De Weer P (1974) Calcium efflux from internally dialyzed squid axons: the influence of external and internal cations. J Supermolec Struct 2:558–581

Blaustein MP, Ratzlaff RW, Schweitzer ES (1978) Calcium buffering in presynaptic nerve terminals. 2. Kinetic properties of the non-mitochondrial Ca sequestration mechanism. J Gen Physiol 72:43–66

Blaustein MP, McGraw CF, Somlyo AV, Schweitzer ES (1980) How is the cytoplasmic calcium concentration controlled in nerve terminals? J Physiol (Paris) 76:459–470

Blinks JR, Wier WG, Hess P, Prendergast FG (1982) Measurement of Ca^{2+} concentrations in living cells. Prog Biophys Molec Biol 40:1–114

Bolsover SR, Spector I (1986) Measurements of calcium transients in the soma, neurite, and growth cone of single cultured neurons. J Neurosci 6:1934–1940

Bossu JL, Feltz A, Thomann JM (1985) Depolarization elicits two distinct calcium currents in vertebrate sensory neurones. Pfluegers Arch 403:360–368

Breitwieser GE, Szabo G (1985) Uncoupling of cardiac muscarinic receptors from ion channels by a guanine nucleotide analogue. Nature 317:538–540

Burgoyne RD (1984) Mechanisms of secretion from adrenal chromaffin cells. Biochim Biophys Acta 779:201–216

Burke BE, DeLorenzo RJ (1982) Ca^{2+}- and calmodulin-dependent phosphorylation of endogenous synaptic vesicle tubulin by a vesicle-bound calmodulin kinase system. J Neurochem 38:1205–1218

Carafoli E (1984) Plasma membrane Ca^{2+} transport, and Ca^{2+} handling by intracellular stores: an integrated picture with emphasis on regulation. In: Donowitz M, Sharp GWG (eds) Mechanisms of intestinal electrolyte transport and regulation by calcium. Alan R Liss, New York, pp 121–134

Celio MR (1986) Parvalbumin in most γ-aminobutyric acid-containing neurons of the rat cerebral cortex. Science 231:995–997

Chan SY, Ochs S, Worth RM (1980) The requirement for calcium and the effect of other ions on axoplasmic transport in mammalian nerve. J Physiol (Lond) 301:477–504

Charlton MP, Atwood HL (1977) Modulation of transmitter release by intracellular sodium in squid giant synapse. Brain Res 134:367–371

Charlton MP, Thompson CS, Atwood HL, Farnell B (1980) Synaptic transmission and intracellular sodium loading of nerve terminals. Neurosci Lett 16:193–196

Chueh S-H, Gill DL (1986) Inositol 1,4,5-triphosphate and guanine nucleotides activate calcium release from endoplasmic reticulum via distinct mechanisms. J Biol Chem 261:13883–13886

Connor J (1986) Digital imaging of free calcium changes and of spatial gradients in growing processes in single, mammalian central nervous system cells. Proc Natl Acad Sci USA 83:6179–6183

Connor JA, Kretz R, Shapiro E (1986) Calcium levels measured in a presynaptic neurone of *Aplysia* under conditions that modulate transmitter release. J Physiol (Lond) 375:625–642

Conway AK (1983) Intracellular calcium. Its universal role as regulator. John Wiley, Chichester

Crabb JH, Jackson RC (1985) In vitro reconstitution of exocytosis from plasma membrane and isolated secretory vesicles. J Cell Biol 101:2263–2273

Davis HP, Squire LR (1984) Protein synthesis and memory: a review. Psychol Bull 96:518–559

DeLorenzo RJ (1982) Calmodulin in neurotransmitter release and synaptic function. Fed Proc 41:2265–2272

DeLorenzo RJ (1983) Calcium–calmodulin systems in psychopharmacology and synaptic modulation. Psychopharmacol Bull 19:393–397

Denton RM, McCormack JG (1985) Physiological role of Ca^{2+} transport by mitochondria. Nature 315:635

de Peyer JE, Cachelin AB, Levitan IB, Reuter H (1982) Ca^{2+} activated K^+ conductance in internally perfused snail neurons is enhanced by protein phosphorylation. Proc Natl Acad Sci USA 79:4207–4211

De Weer P (1976) Axoplasmic free magnesium levels and magnesium extrusion from squid giant axons. J Gen Physiol 68:159–178

DiPolo R, Beauge L (1986) Reverse Na/Ca exchange requires internal Ca and/or ATP in squid axons. Biochim Biophys Acta 854:298–306

Dodge F, Rahamimoff R (1967) Co-operative action of calcium ions in transmitter release at the neuromuscular junction. J Physiol (Lond) 193:419–432

Douglas WW (1974) Involvement of calcium in exocytosis and the exocytosis vesiculation sequence. Biochem Soc Symp 39:1–28

Drapeau P, Blaustein MP (1983a) Initial release of ^{3}H-dopamine from rat striatal synaptosomes: correlation with Ca entry. J Neurosci 3:703–713

Drapeau P, Blaustein MP (1983b) Calcium and neurotransmitter release: what we know and don't know. In: Kalsner S (ed) Trends in autonomic pharmacology, vol 2. Urban and Schwarzenberg, Baltimore, pp 117–130

Drapeau P, Nachshen DA (1984) Manganese fluxes and manganese-dependent neurotransmitter release in presynaptic nerve endings from rat brain. J Physiol (Lond) 348:493–510

Eusebi F, Miledi R, Parker I, Stinnakre J (1985) Post-synaptic calcium influx at the giant synapse of the squid during activation by glutamate. J Physiol (Lond) 369:183–197

Ewald DA, Williams A, Levitan IB (1985) Modulation of single Ca^{2+}-dependent K-channel activity by protein phosphorylation. Nature 315:503–506

Farley J, Auerbach S (1986) Protein kinase C activation induces conductance changes in *Hemissenda* photoreceptors like those seen in associative learning. Nature 319:220–223

Feasey KJ, Lynch MA, Bliss TVB (1986) Long-term potentiation is associated with an increase in calcium-dependent, potassium-stimulated release of [^{14}C]glutamate from hippocampal slices: an ex vivo study in the rat. Brain Res 364:39–44

Fleckenstein A (1985) Calcium antagonism in heart and vascular smooth muscle. Med Res Rev 5:395–425

Flexner JB, Flexner LB, Stellar E (1963) Memory in mice as affected by intracerebral puromycin. Science 141:57–59

Flockerzi V, Oeken H-J, Hofmann F, Pelzer D, Cavalie A, Trautwein W (1986) Purified dihydropyridine-binding site from skeletal muscle t-tubules is a functional sodium channel. Nature 323:66–68

Frankenhaeuser B (1957) The effect of calcium on the myelinated nerve fibre. J Physiol (Lond) 137:245–260

Frankenhaeuser B, Hodgkin AL (1957) The action of calcium on the electrical properties of squid axons. J Physiol (Lond) 137:218–244

Gill DL, Grollman EF, Kohn LD (1981) Calcium transport mechanisms in membrane vesicles from guinea pig brain synaptosomes. J Biol Chem 256:184–192

Gill DL, Chueh S-H, Noel MW, Ueda T (1986a) Orientation of synaptic plasma membrane vesicles containing calcium pump and sodium–calcium exchange activities. Biochim Biophys Acta 856:165–173

Gill DL, Ueda T, Chueh S-H (1986b) Ca^{2+} release from endoplasmic reticulum is mediated by a guanine nucleotide regulatory mechanism. Nature 320:461–464

Gilman AG (1986) Receptor-regulated G proteins. Trends Neurosci 9:460–463

Goelet P, Kandel E (1986) Tracking the flow of learned information from membrane receptors to genome. Trends Neurosci 9:492–499

Goelet P, Castellucci V, Schacher S, Kandel ER (1986) The long and the short of long-term memory

– a molecular framework. Nature 322:419–422

Gorman ALF, Hermann A (1982) Quantitative differences in the currents of bursting and beating molluscan pace-maker neurones. J Physiol (Lond) 333:681–699

Gafstein B (1977) Axonal transport: the intracellular traffic of the neuron. In: Kandel E (ed) Handbook of physiology, section 1, The nervous system, vol 1, cellular biology of neurons, part 1. American Physiological Society, Bethesda, pp 691–717

Greenberg ME, Ziff EB, Greene LA (1986) Stimulation of neuronal acetylcholine receptors induces rapid gene transcription. Science 234:80–83

Greenough WT (1984) Possible structural substrates of phasic neuronal phenomena. In: Lynch G, McGaugh JL, Weinberg NM (eds) Neurobiology of learning and memory. Guildford Press, New York, pp 470–478

Hagiwara S, Byerly L (1981) Calcium channel. Annu Rev Neurosci 4:69–125

Hammerschlag R, Lavoie P-A (1979) Initiation of fast axonal transport: involvement of calcium during transfer of proteins from Golgi apparatus to the transport system. Neuroscience 4:1195–1201

Hansford RG (1985) Relation between mitochondrial calcium transport and control of energy metabolism. Rev Physiol Biochem Pharmacol 102:1–72

Hartzell HC, Fischmeister R (1986) Opposite effects of cyclic GMP and cyclic AMP on Ca^{2+} current in single heart cells. Nature 323:273–275

Harvey AM, MacIntosh FC (1940) Calcium and synaptic transmission in a sympathetic ganglion. J Physiol (Lond) 97:408–416

Hashimoto T, Hirata M, Itoh T, Kanmura Y, Kuriyama H (1986) Inositol 1,4,5-triphosphate activates pharmacomechanical coupling in smooth muscle of the rabbit mesenteric artery. J Physiol (Lond) 370:605–618

Hasselbach W (1977) The sarcoplasmic reticulum calcium pump – a most efficient ion translocating system. Biophys Struct Mech 3:43–54

Heizmann CW (1984) Parvalbumin, an intracellular calcium-binding protein; distribution, properties and possible roles in mammalian cells. Experientia 40:910–921

Heuser JE (1977) Synaptic vesicle exocytosis revealed in quick-frozen frog neuromuscular junctions treated with 4-aminopyridine and given a single electric shock. In: Cowan WM, Ferendelli JA (eds) Approaches to the cell biology of neurons. Society for Neuroscience, Bethesda, pp 215–239 (Society for Neuroscience Symposia vol. 2)

Heuser J, Katz B, Miledi R (1971) Structural and functional changes of frog neuromuscular junctions in high calcium solutions. Proc R Soc Lond [Biol] 178:407–415

Higashi H, Morita K, North RA (1984) Calcium-dependent after potentials in visceral afferent neurones of the rabbit. J Physiol (Lond) 355:479–492

Hille B (1984) Ionic channels of excitable cells. Sinauer, Boston

Hincke MT, Demaille JG (1984) Calmodulin regulation of the ATP-dependent calcium uptake by inverted vesicles prepared from rabbit synaptosomal plasma membranes. Biochim Biophys Acta 771:188–194

Holz GG, Rane SG, Dunlap K (1986) GTP-binding protein mediates transmitter inhibition of voltage-dependent calcium channels. Nature 319:670–672

Hutson SM, Pfeifer DR, Lardy HA (1976) Effect of cations and anions on the steady state kinetics of energy-dependent Ca^{2+} transport in rat liver mitochondria. J Biol Chem 251:5251–5258

Inesi G (1985) Mechanism of calcium transport. Annu Rev Physiol 47:573–601

Inoue M, Oomura Y, Yakushiji T, Akaike N (1986) Intracellular calcium ions decrease the affinity of the GABA receptor. Nature 324:156–158

Kameyama M, Hescheler J, Hofmann F, Trautwein W (1986) Modulation of Ca current during the phosphorylation cycle in the guinea pig heart. Pfluegers Arch 407:123–128

Katz B (1969) The release of neural transmitter substances. CC Thomas, Springfield, Illinois

Katz B, Miledi R (1967) A study of synaptic transmission in the absence of nerve impulses. J Physiol (Lond) 192:407–436

Katz B, Miledi R (1968) The role of calcium in neuromuscular facilitation. J Physiol (Lond) 195:481–492

Katz B, Miledi R (1969) Tetrodotoxin-resistant electrical activity in presynaptic terminals. J Physiol (Lond) 203:459–487

Kimura J, Noma A, Irisaya H (1986) Na–Ca exchange current in mammalian heart cells. Nature 319:596–598

Knight DE, Baker PF (1982) Calcium-dependence of catecholamine release from bovine adrenal medullary cells after exposure to intense electric fields. J Membr Biol 68:107–140

Knight DE, Baker PF (1985) Guanine nucleotides and Ca-dependent exocytosis. Studies on two adrenal cell preparations. FEBS Lett 189:345–349

Kostyuk PG (1981) Calcium channels in the neuronal membrane. Biochim Biophys Acta 650:128–150

Kretz R, Shapira E, Kandel ER (1982) Post-tetanic potentiation at an identified synapse in *Aplysia* is correlated with a Ca^{2+}-activated K^+ current in the presynaptic neuron: evidence for Ca^{2+} accumulation. Proc Natl Acad Sci USA 79:5430–5434

Lancaster B, Adams PR (1986) Calcium-dependent current generating the afterhyperpolarization of hippocampal neurons. J Neurophysiol 55:1268–1282

Lariviere L, Lavoie P-A (1982) Calcium requirement for fast axonal transport in frog motoneurons. J Neurochem 39:882–886

Levitan IB (1985) Phosphorylation of ion channels. J Membr Biol 87:177–190

Llinas R, Steinberg IZ, Walton K (1976) Presynaptic calcium currents and their relation to synaptic transmission: voltage clamp study in squid giant synapse and theoretical model for the calcium gate. Proc Natl Acad Sci USA 73:2918–2922

Llinas R, Steinberg IZ, Walton K (1981) Relationship between presynaptic calcium current and postsynaptic potential in squid giant synapse. Biophys J 33:323–352

Lnenicka GA, Atwood HL, Marin L (1986) Morphologic transformation of synaptic terminals of a phasic motoneuron by long-term tonic stimulation. J Neurosci 6:2252–2258

Lynch G, Baudry M (1984) The biochemistry of memory: a new and specific hypothesis. Science 224:1057–1063

Malenka RC, Madison DV, Nicoll RA (1986) Potentiation of synaptic transmission in the hippocampus by phorbol esters. Nature 321:175–177

Mallart A (1984) Calcium-activated potassium current in presynaptic terminals. Biomed Res 5:287–290

McCormack JG, Denton RM (1986) Ca^{2+} as a second messenger within mitochondria. Trends Biochem Sci 11:258–262

Meech RW (1972) Intracellular calcium injection causes increased potassium conductance in *Aplysia* nerve cells. Comp Biochem Physiol [A] 42:493–499

Meech RW (1978) Calcium-dependent potassium activation in nervous tissues. Annu Rev Biophys Bioeng 7:1–18

Meiri H, Parnas I, Spira M (1981) Membrane conductance and action potential of a regenerating axonal tip. Science 211:709–711

Meiri H, Zellingher J, Rahamimoff R (1986) A possible involvement of the Na–Ca exchanger in regulation of transmitter release at the frog neuromuscular junction. In: Rahamimoff R, Katz B (eds) Calcium, neuronal function and transmitter release. Martinus Nijhoff, Amsterdam, pp 239–254

Miledi R, Parker I, Schalow G (1980) Transmitter induced calcium entry across post-synaptic membrane at frog end-plates measured using Arsenazo III. J Physiol (Lond) 300:197–212

Miller SG, Kennedy MB (1986) Regulation of brain type II Ca^{2+}/calmodulin-dependent protein kinase by autophosphorylation: a Ca^{2+}-triggered molecular switch. Cell 44:861–870

Misler S, Hurlbut WP (1983) Post-tetanic potentiation of acetylcholine release at the frog neuromuscular junction develops after stimulation in Ca^{2+}-free solutions. Proc Natl Acad Sci USA 80:315–319

Morgan JI, Curran T (1986) Role of ion flux in the control of c-*fos* expression. Nature 322:552–555

Mullins LJ, Brinley FJ, Jr (1975) Sensitivity of calcium efflux from squid axons to changes in membrane potential. J Gen Physiol 65:135–152

Nachshen DA (1984) Selectivity of the Ca binding site in synaptosome Ca channels. J Gen Physiol 83:941–967

Nachshen DA (1985) The early time course of potassium-stimulated calcium uptake in presynaptic nerve terminals from rat brains. J Physiol (Lond) 361:251–268

Nachshen DA, Blaustein MP (1979) The effects of some organic "calcium antagonists" on calcium influx in presynaptic nerve terminals. Molec Pharmacol 16:579–586

Nachshen DA, Blaustein MP (1980) Some properties of potassium-stimulated calcium influx in presynaptic nerve endings. J Gen Physiol 76:709–728

Nachshen DA, Blaustein MP (1982) The influx of calcium, strontium and barium in presynaptic nerve endings. J Gen Physiol 79:1065–1087

Nachshen DA, Drapeau P (1982) A buffering model for calcium-dependent neurotransmitter release. Biophys J 38:205–208

Nachshen DA, Sanchez-Armass S, Weinstein AM (1986) The regulation of cytosolic calcium in rat brain synaptosomes by sodium-dependent calcium efflux. J Physiol (Lond) 381:17–28

Nairn AC, Hemmings HC Jr, Greengard P (1985) Protein kinases in the brain. Annu Rev Biochem 54:931–976

Nixon RA, Brown BA, Marotta CA (1983) Limited proteolytic modification of a neurofilament protein involves a proteinase activated by endogenous levels of calcium. Brain Res 275:381–388

Nowycky MC, Fox AP, Tsien RW (1985) Three types of neuronal calcium channel with different calcium agonist sensitivity. Nature 316:440–443

Ochs S, Jersild RA Jr (1984) Calcium localization in nerve fibers in relation to axoplasmic transport. Neurochem Res 9:823–835

Parsegian VA (1977) Considerations in determining the mode of influence of calcium on vesicle membrane interaction. In: Cowan WM, Ferendelli JA (eds) Approaches to the cell biology of neurons. Society for Neuroscience, Bethesda, pp 161–171 (Society for Neuroscience Symposia, vol 2)

Paupardin-Tritsch D, Hammond C, Gerschenfeld HM, Nairn AC, Greengard P (1986) cGMP-dependent protein kinase enhances Ca^{2+} current and potentiates the serotonin-induced Ca^{2+} current increase in snail neurones. Nature 323:812–814

Pechere J-F, Derancourt J, Haiech J (1977) The participation of parvalbumins in the activation-relaxation cycle of vertebrate fast skeletal muscle. FEBS Lett 75:111–114

Pellmar TC (1981) Ionic mechanism of a voltage dependent current elicited by cyclic AMP. Cell Molec Neurobiol 1:87–97

Pennefather P, Lancaster B, Adams PR, Nicoll RA (1985) Two distinct Ca-dependent K currents in bullfrog sympathetic ganglion cells. Proc Natl Acad Sci USA 82:3040–3044

Pfaffinger PJ, Martin JM, Hunter DD, Nathanson NM, Hille B (1985) GTP-binding proteins couple cardiac muscarinic receptors to a K channel. Nature 317:536–538

Pochet R, Parmentier M, Lawson DEM, Pasteels JL (1985) Rat brain synthesizes two "vitamin D-dependent" calcium-binding proteins. Brain Res 345:251–256

Rane SG, Dunlap K (1986) Kinase C activator 1,2-oleylacetylglycerol attenuates voltage-dependent calcium current in sensory neurons. Proc Natl Acad Sci USA 83:184–188

Rasgado-Flores H, Blaustein MP (1987a) Na/Ca exchange in barnacle muscle cells has a stoichiometry of 3 Na : 1 Ca^{2+}. Am J Physiol 252 (Cell Physiol 21):C499–C504

Rasgado-Flores H, Blaustein MP (1987b) ATP-dependent regulation of cytoplasmic free Ca^{2+} in nerve terminals. Am J Physiol 252 (Cell Physiol 21):C588–C594

Rasgado-Flores H, Santiago EM, Blaustein MP (1986) Calcium influx and sodium efflux mediated by the Na/Ca exchanger in giant barnacle muscle cells are promoted by intracellular Ca^{2+}. Biophys J 49:546a

Reuter H (1983) Calcium channel modulation by neurotransmitters, enzymes and drugs. Nature 301:569–574

Reynolds IJ, Wagner JA, Snyder SH, Thayer Sa, Olivera BM, Miller RJ (1986) Brain voltage-sensitive calcium channel subtypes differentiated by ω-conotoxin fraction GVIA. Proc Natl Acad Sci USA 83:8804–8807

Rosengurt E (1986) Early signals in the mitogenic response. Science 234:161–166

Ross WN, Stockbridge LL, Stockbridge NL (1986) Regional properties of calcium entry in barnacle neurons determined with arsenazo III and a photodiode array. J Neurosci 6:1148–1159

Rubin RR (1982) Calcium and cellular secretion. Plenum Press, New York

Sakakibara M, Alkon DL, DeLorenzo R, Goldenring JR, Neary JT, Heldman E (1986) Modulation of calcium-mediated inactivation of ionic currents by a Ca^{2+}/calmodulin-dependent protein kinase II. Biophys J 50:319–327

Sanchez-Armass S, Blaustein MP (1987) Role of sodium/calcium exchange in the regulation of intracellular Ca^{2+} in nerve terminals. Am J Physiol (Cell Physiol) 21:C595–C603

Scarpa A (1976) Kinetic and thermodynamic aspects of mitochondrial calcium transport. In: Packer L, Gomez-Puyou A (eds) Mitochondria. Bioenergetics, biogenesis and membrane structure. Academic Press, New York, pp 31–45

Schlaepfer WW (1977) Structural alterations of peripheral nerve induced by the calcium ionophore A23187. Brain Res 136:1–9

Schlaepfer WW, Bunge RP (1973) Effects of calcium ion concentration on the degeneration of amputated axons in tissue culture. J Cell Biol 59:456–470

Schlaepfer WW, Zimmerman U-JP (1984) Calcium-activated protease and the regulation of the axonal cytoskeleton. In: Elam JS, Cancalon P (eds) Axonal transport in neuronal growth and regulation. Plenum Press, New York, pp 261–273

Scott RH, Dolphin AC (1986) Regulation of calcium currents by a GTP analogue: potentiation of (−)-baclophen-mediated inhibition. Neurosci Lett 69:59–64

Sheu S-S, Blaustein MP (1986) Sodium/calcium exchange and the regulation of cell calcium and contractility in cardiac muscle, with a note about vascular smooth muscle. In: Fozzard HA, Haber E, Jennings RB, Katz AM, Morgan HE (eds) The heart and cardiovascular system. Raven

Press, New York, pp 509–535

Somlyo AV, Gonzales-Serratos H, Shuman H, McClellan G, Somlyo AP (1981) Calcium release and ionic changes in the sarcoplasmic reticulum of tetanized muscle: an electron probe study. J Cell Biol 90:577–594

Somlyo AV, Bond M, Somlyo AP, Scarpa A (1985) Inositol triphosphate-induced calcium release and contraction in vascular smooth muscle. Proc Natl Acad Sci USA 82:5231–5235

Spitzer NC (1979) Ion channels in development. Annu Rev Neurosci 2:363–397

Steinberg JP, Leitner JW, Draznin B, Sussman KE (1984) Calmodulin and cyclic AMP. Possible different sites of action of these two regulatory agents in exocytotic hormone release. Diabetes 33:339–344

Streb H, Irvine RF, Berridge MJ, Schulz I (1983) Release of Ca^{2+} from a nonmitochondrial intracellular store in pancreatic acinar cells by inositol-1,4,5-triphospate. Nature 306:67–69

Strichartz G, Small R, Nicholson C, Pfenninger KH, Llinas R (1980) Ionic mechanisms for impulse propagation in growing nonmyelinated axon: saxitoxin binding and electrophysiology. Soc Neurosci Abstr 6:660

Suszkiw JB, O'Leary ME, Murawsky MM, Wang T (1986) Presynaptic calcium channels in rat cortical synaptosomes: fast-kinetics of phasic calcium influx, channel inactivation and relationship to nitrendipine receptors. J Neurosci 6:1349–1357

Takeuchi A, Takeuchi N (1960) On the permeability of end-plate membrane during the action of the transmitter. J Physiol (Lond) 154:52–67

Takeuchi A, Takeuchi N (1963) Effects of calcium on the conductance change of the end-plate during the action of the transmitter. J Physiol (Lond) 167:141–155

Trump BF, Berezesky IK (1985) Cellular ion regulation and disease: a hypothesis. Curr Top Membr Transp 25:279–319

Tsien RW, Hess P, McClescky EW, Rosenberg RL (1987) Calcium channels: mechanisms of selectivity, permeation and block. Annu Rev Biophys Biophys Chem 16:265–290

Vinogradov A, Scarpa A (1973) The initial velocities of calcium uptake by rat liver mitochondria. J Biol Chem 248:5527–5531

Wakabayashi S, Goshima K (1981) Kinetic studies on sodium-dependent calcium uptake by myocardial cells and neuroblastoma cells in culture. Biochim Biophys Acta 642:158–172

Wood JG, Wallace RW, Cheung WY (1980) Immunocytochemical studies of the localization of calmodulin and CaM-BP[80] in brain. In: Cheung WY (ed) Calcium and cell function, vol 1. Academic Press, New York, pp 291–303

Woodbury W (1963) Interrelationships between ion transport mechanism and excitatory events. Fed Proc 22:31–35

Wulfroth P, Peltzelt C (1985) The so-called anticalmodulins, fluphenazine, calmidazolium, and compound 48/80 inhibit the Ca^{2+}-transport system of the endoplasmic reticulum. Cell Calcium 6:295–310

Zimmerman U-JP, Schlaepfer WW (1984) Calcium-activated neutral protease (CANP) in brain and other tissues. Prog Neurobiol 23:63–78

Zucker RS, Fogelson AL (1986) Relationship between transmitter release and presynaptic calcium influx when calcium enters through discrete channels. Proc Natl Acad Sci USA 83:3032–3036

Zucker RS, Lando L (1986) Mechanism of transmitter release: voltage hypothesis and calcium hypothesis. Science 231:574–579

Chapter 14

Cellular Calcium: Secretion of Hormones

W. J. Malaisse

Introduction

During the late 1960s, being interested in the regulation of insulin release from isolated islets of Langerhans prepared from the pancreatic gland, I became convinced, like a few other scientists, that calcium plays a critical role in the process of insulin secretion by the pancreatic β-cell (Grodsky and Bennett 1966; Milner and Hales 1967). I thought, therefore, that it would be interesting to study the effect of D-glucose and other secretagogues upon the handling of radioactive ^{45}Ca by the isolated islets (Malaisse-Lagae et al. 1969). Hence, I asked the director of the department, a leading clinician not too far from retirement, for permission to order some ^{45}Ca and to introduce this isotope in the laboratory. The answer came, in essence, as follows. "Are you not confused? Calcium may play a role in the regulation of hormonal release by the parathyroid gland. But it is glucose and not calcium which stimulates insulin secretion."

Today, it would be impossible within the framework of a rather short account to review all the information available on the handling of calcium by isolated islets and the relevance of changes in calcium fluxes to the process of insulin release evoked by D-glucose or other secretagogues in the islet cells (Malaisse et al. 1978a; Wollheim and Sharp 1981; Hellman and Gylfe 1986).

This dramatic progress in knowledge over less than two decades is the topic of the present review. Being still interested in the field of insulin secretion, I intend to use the pancreatic β-cell as the model to discuss the role of calcium in hormonal secretion. It should be realized, however, that the fundamental concepts reviewed in this account also apply, *mutatis mutandis*, to a great variety of secretory processes in other endocrine cells.

In dealing with the role of cellular calcium in the regulation of hormonal secretion, we will consider three major aspects. Firstly, we will review the evidence for the fact that the concentration of ionized calcium in the cytoplasm plays a key role in triggering hormonal release. Secondly, we will consider those factors which control the concentration of cytosolic ionized calcium. Lastly, we will discuss which systems in the secreting cell act as targets for cytosolic Ca^{2+}.

Evidence that Calcium Participates in Secretion

The first indication that cellular calcium may play a role in the process of glucose-stimulated insulin release was found in studies of the influence of extracellular cationic concentrations on the rate of insulin secretion evoked by D-glucose in the pancreatic β-cell (Grodsky and Bennett 1966; Milner and Hales 1967). Further evidence of such a role came from the observation that, under suitable experimental conditions, Ca^{2+} itself may provoke insulin release (Devis et al. 1975, 1977). A final crucial piece of information came from the measurement of cytosolic Ca^{2+} in intact normal islet cells stimulated by D-glucose (Deleers et al. 1985). These three major steps in unveiling the participation of Ca^{2+} in insulin release will be reviewed in the first part of this chapter.

Effect of Extracellular Ca^{2+}

Glucose-induced insulin release does not occur in the absence of extracellular Ca^{2+}. As a rule, the secretory response to other secretagogues is also impaired at low extracellular Ca^{2+} concentrations. The relative extent of inhibition of insulin release depends on a number of environmental factors such as the nature of the secretagogue(s) used to stimulate the β-cell, the time course of changes in extracellular Ca^{2+} relative to the moment at which the secretagogue is administered, the precise concentration of Ca^{2+} and other ions in the extracellular fluid and the period during which the secretory activity of the β-cell is monitored. For instance, at a given extracellular Ca^{2+} concentration, the secretory response to D-glucose is less severely affected when an activator of adenylate cyclase (e.g. forskolin) or inhibitor of phosphodiesterase (e.g. theophylline) is used together with D-glucose to stimulate insulin release than when the hexose is used alone for the same purpose (Brisson et al. 1972; Malaisse et al. 1984b).

The inhibition of insulin release attributable to the absence of extracellular Ca^{2+} is more marked, especially during the early peak-shaped phase of the secretory response, when the islets were already deprived of Ca^{2+} for some time prior to administration of the hexose rather than simultaneously deprived of extracellular Ca^{2+} and exposed to D-glucose (Wollheim et al. 1978). Even in the nominal absence of extracellular Ca^{2+}, a sizeable secretory response to D-glucose is observed if Mg^{2+} is also omitted from the extracellular medium (Somers et al. 1979b). As a rule, however, such is not the case if all extracellular Ca^{2+} is chelated, e.g. by EGTA or EDTA. Conversely, even in the presence of Ca^{2+}, increasing concentrations of Mg^{2+} cause a dose-related inhibition of insulin release. These and other findings indicate that extracellular Mg^{2+} and Ca^{2+} compete for a common transport system (Malaisse et al. 1976).

Under suitable conditions, insulin release can be stimulated in the absence of Ca^{2+}, suggesting that the presence of the latter cation in the extracellular fluid may not itself be the essential prerequisite for stimulation of hormonal release. For instance, in the absence of Ca^{2+} and even in the presence of EGTA, Ba^{2+} dramatically stimulates insulin release (Somers et al. 1976a). It could be argued, however, that Ba^{2+} merely acts as a substitute for Ca^{2+}. Therefore, further information on the role of Ca^{2+} in glucose-induced insulin release was sought for in experiments designed to establish whether a facilitated entry of Ca^{2+} into the pancreatic β-cell may be sufficient to provoke insulin release, even in the absence of D-glucose.

Stimulation of Secretion by Ca^{2+}

One of the first attempts to stimulate secretion by Ca^{2+} involved the incubation of islets in an alkaline medium (pH 7.9 instead of pH 7.4), deprived of Mg^{2+} (instead of containing 1 mM Mg^{2+}) and enriched with K^+ (20 mM instead of 5 mM). Under these conditions, stimulation of both insulin release and ^{45}Ca net uptake was observed even when the islets were deprived of extracellular glucose (Malaisse et al. 1971). It was later shown that Ca^{2+} (2 to 10 mM) provokes a short-lived release of insulin in perfused rat pancreases first exposed to EGTA in glucose-free media (Devis et al. 1975). The secretory response to Ca^{2+} was abolished by the organic calcium antagonist verapamil. A number of further observations have confirmed that insulin release can be stimulated, even in the absence of exogenous nutrient, provided that a sufficient amount of Ca^{2+} is allowed to enter the islet cells. For the sake of brevity, I will here mention only some of the various procedures used to facilitate the entry of Ca^{2+} into the cell and indeed found to cause or enhance insulin release:

1. Prior exposure of the cells to media deprived of Ca^{2+} (with or without EGTA) and then exposure to high concentrations of extracellular Ca^{2+} (Devis et al. 1975, 1977).
2. Depolarization of the plasma membrane by an increase in the extracellular concentration of K^+ (e.g. from 5 to 20 mM or more) (Herchuelz et al. 1980c).
3. Exposure of the islet cells to calcium ionophores (e.g. A23187 or X537A) (Wollheim et al. 1975; Somers et al. 1976b).
4. Permeabilization of the plasma membrane by prior treatment with either digitonin or electric discharge followed by exposure of the cells to low Ca^{2+} concentrations close to physiological cytosolic values (Pace et al. 1980; Yaseen et al. 1982).
5. Exposure of islet cells to organic calcium agonists such as BAY K8644 or CGP 28392 (Malaisse-Lagae et al. 1984; Malaisse and Mathias 1985).
6. Manipulation aimed at increasing intracellular Na^+ and hence mobilizing Ca^{2+} from intracellular pools, e.g. by exposing the islets to veratridine or K^+-deprived media (Lowe et al. 1976; Herchuelz and Malaisse 1980a, b).

Measurement of Cytosolic Ca^{2+} Concentration

The development of the fluorescent Ca^{2+} indicator quin2 has made it possible to measure changes in cytosolic free ionized calcium in suspensions or small aggregates of isolated islet cells. D-glucose was indeed found to increase cytosolic Ca^{2+} activity in normal islet cells (Rorsman et al. 1984; Deleers et al.

1985). The basal cytosolic free Ca^{2+} concentration is close to 100–150 nM and D-glucose raises the concentration to about 170–240 nM. It should be emphasized that these values, and especially the glucose-induced increment in cytosolic Ca^{2+} activity, are derived from the fluorescent signal and assume an even distribution of Ca^{2+} in the cellular space occupied by quin2. The latter assumption is open to question since it is now believed that, especially under non-steady-state conditions, the cytosolic concentration of Ca^{2+} may vary within a given cell. In fact, an onion leaf sheet model of stratification of the cytosolic compartment in terms of its Ca^{2+} concentration had to be introduced into the mathematical modelling of stimulus–secretion coupling in the pancreatic β-cell to allow for simulation of the dynamics of insulin release in response to a square-waved increase in D-glucose concentration (Scholler et al. 1983).

Further progress in our understanding of the role of Ca^{2+} in insulin release is to be expected following more extensive measurements of cytosolic Ca^{2+} activity. To cite one example, under suitable experimental conditions, namely when the glucose-stimulated entry of Ca^{2+} into islet cells is prevented, it is found that the sugar can also lower the cytoplasmic Ca^{2+} activity, probably as a result of Ca^{2+} sequestration in intracellular organelles (Rorsman et al. 1984).

Is Stimulation of Secretion Invariably Attributable to Changes in Cytosolic Ca^{2+} Activity?

The knowledge so far reviewed could give the impression that changes in cytosolic Ca^{2+} activity represent the sole and obligatory determinant of insulin release. This does not appear to be true. Through recent determinations of cytosolic Ca^{2+} concentration in cells exposed to various secretagogues, the concept has evolved that, under selected conditions, dramatic changes in insulin release may occur in the absence of sizeable alterations in cytosolic Ca^{2+} activity. Two illustrations of this will be mentioned.

A possible interaction between cyclic AMP and Ca^{2+} in the process of insulin secretion has been suggested (Malaisse and Malaisse-Lagae 1984). First, it was proposed that cyclic AMP facilitated the entry of Ca^{2+} into the β-cell via gated voltage-dependent Ca^{2+} channels (Henquin and Meissner 1983; Henquin et al. 1983). Second, cyclic AMP was thought not to affect the net balance between Ca^{2+} entry into and exit from the β-cell but, instead, to cause an intracellular redistribution of the cation favouring the cytosolic accumulation of Ca^{2+} at the expense of its storage by intracellular organelles (Brisson et al. 1972). Third, it was proposed that calcium–calmodulin modulates the generation rate of cyclic AMP via activation of adenylate cyclase (Valverde et al. 1979). Last, it was suggested that cyclic AMP may fail to affect the fluxes of Ca^{2+} into the β-cell but may increase the sensitivity of the insulin-releasing effector system to cytosolic Ca^{2+} (Malaisse and Malaisse-Lagae 1984).

The first of these proposals is difficult to reconcile with the knowledge that an increase in cyclic AMP availability exerts little if any effect upon the net uptake of ^{45}Ca by the islet cells (Brisson et al. 1972; Malaisse et al. 1984b). The second concept was invoked to account for the fact that an increase in cyclic AMP net production tends to restore the secretory response to D-glucose in islets incubated either at low extracellular Ca^{2+} concentration or in the absence of

Ca^{2+}. The latter finding is also compatible with the idea that cyclic AMP increases the responsiveness to Ca^{2+} of the insulin-releasing apparatus. The possible involvement of calcium–calmodulin in cyclic AMP generation is considered in greater detail elsewhere in this chapter (see p. 378).

The existence of multiple modes of interrelationship between cyclic AMP and Ca^{2+} should not be ruled out and was indeed introduced in a recent mathematical model of stimulus–secretion coupling in the β-cell (Malaisse et al. 1982a). Nevertheless, since recent findings suggest that cyclic AMP fails to exert any major effect upon cytosolic Ca^{2+} activity in islet cells (Rorsman and Abrahamsson 1985), the postulated increased reactivity of the effector system to cytosolic Ca^{2+} may well represent, from the quantitative standpoint, the major modality of action of the cyclic nucleotide in the β-cell. It would also account for the fact that agents causing cyclic AMP accumulation exert little or no effect upon insulin release when the β-cell is not yet stimulated or primed to stimulation by exposure to other suitable secretagogues.

A second situation in which a change in cytosolic Ca^{2+} activity may not represent the major determinant of insulin release involves the secretory response of the β-cell to the tumour-promoting phorbol ester 12-O-tetradeca-noylphorbol 13-acetate (TPA). This phorbol ester is a potent activator of the Ca^{2+}-responsive enzyme protein kinase C. At high concentrations, TPA may even cause activation of protein kinase C in the absence of Ca^{2+} (Hubinont and Malaisse 1985). It is thus conceivable that the secretory response to TPA occurs without any real change in cytosolic Ca^{2+} activity. TPA does not reduce ^{86}Rb outflow from prelabelled islets (Malaisse et al. 1980a), suggesting that it does not reproduce the inhibitory action of nutrient secretagogues on K^+ permeability; the latter inhibitory action is assumed to account for depolariza-tion of the plasma membrane and subsequent gating of voltage-dependent Ca^{2+} channels. Although this observation does not rule out the possibility that TPA affects calcium handling by islet cells, stimulation of insulin release by another activator of protein kinase C, namely 1-oleyl-2-acetylglycerol, was observed in islets deprived of extracellular Ca^{2+} (Malaisse et al. 1985a).

In conclusion, although the cytosolic accumulation of Ca^{2+} may well represent the major determinant of insulin release evoked by nutrient secretagogues, it does not necessarily represent the sole modality for stimulation of hormone secretion.

Calcium Fluxes in Endocrine Cells

The second major topic to be dealt with in considering the participation of cellular calcium in hormone secretion is the regulation of Ca^{2+} fluxes in endocrine cells. The essential question is which mechanism(s) increase(s) cytosolic Ca^{2+} activity? There are at least three mechanisms which could theoretically account for such a cytosolic accumulation of Ca^{2+}: an increased inflow of Ca^{2+} into the cell, a decreased outflow of Ca^{2+} from the cell or an intracellular redistribution of the cation between the ionized cytosolic pool and either unionized or organelle-stored pools.

Regulation of Ca^{2+} Inflow

An essential component in the nutrient-induced increase in cytosolic Ca^{2+} concentration in the pancreatic β-cell is an increased inflow of Ca^{2+} through gated Ca^{2+} channels (Henquin and Meissner 1984; Findlay and Dunne 1985). The gating of these voltage-dependent channels is itself attributable to depolarization of the plasma membrane which is the result mainly of a decrease in K^+ conductance leading to a decrease in the efflux of K^+ from the islet cells into the extracellular fluid through appropriate K^+ channels. It should be underlined that the two ionic movements under consideration, namely the efflux of K^+ and the influx of Ca^{2+}, are both mediated by channels along the prevailing transmembrane electrochemical gradient. Neither of these movements consumes ATP and their efficiency depends mainly on the open or closed states of the relevant channels.

According to the model just defined, a crucial question in islet cytophysiology concerns the mechanism by which exposure of the β-cell to a high concentration of glucose eventually leads to a decrease in K^+ conductance, i.e. the closing of K^+ channels. Current concepts on such a mechanism have been summarized by Malaisse et al. (1984a).

The identification of D-glucose and other nutrient secretagogues as stimuli for insulin release appears to be closely connected to their capacity to augment oxidative fluxes in the islet cells. In this respect, the β-cell should be looked upon as a fuel-sensor organ (Malaisse et al. 1979a). The coupling of metabolic to ionic events remains a matter of debate. Possible second messengers generated at an increased rate, as a result of the enhanced oxidation of nutrients, and susceptible to control the efflux of K^+, include protons (H^+), reducing equivalents (such as NADH, NADPH or reduced glutathione) and high-energy phosphate intermediates (e.g. ATP). Recent patch-clamp recordings from β-cells showed a K^+ channel controlled by ATP (Ashcroft et al. 1984; Cook and Hales 1984; Rorsman and Trube 1985). It has been proposed, therefore, that the glucose-induced closing of K^+ channels may be the consequence of an increase in cytoplasmic ATP concentration. Much work, including the measurement of cytosolic ATP concentration, remains to be done in order to assess the validity of the latter proposal.

An alternative proposal, made a few years ago, postulates that the nutrient-induced decrease in K^+ conductance could be due to closing of Ca^{2+}-sensitive K^+ channels (Atwater et al. 1980). This concept implies that the initial effect of D-glucose would be to lower cytosolic Ca^{2+} activity as a result, for instance, of increased sequestration of the ion by intracellular organelles. However, there is no evidence that, in normal islet cells and at normal extracellular Ca^{2+} concentrations, glucose provokes an initial fall, preceding the well-documented and sustained rise in cytosolic Ca^{2+} activity.

Whatever its precise determinant, both radioisotopic and bioelectrical methods are being used to assess the stimulation of Ca^{2+} inflow evoked, in the islet cells, by nutrient secretagogues.

The uptake of ^{45}Ca by islets can be measured during short periods of exposure to the isotope. A more dynamic but indirect procedure involves the measurement of ^{45}Ca outflow from prelabelled and perifused islets. When the data collected from islets perifused in the presence of extracellular unlabelled Ca^{2+} are compared with those obtained in islets perifused in the absence of extracellular Ca^{2+}, the difference in the pattern of ^{45}Ca efflux, in response to stimulation by D-glucose or other nutrient secretagogues, consists mainly of a marked stimulation of ^{45}Ca outflow in the presence of extracellular Ca^{2+}. This nutrient-induced increase in ^{45}Ca efflux displays a biphasic pattern, namely an early peak followed by a secondary reascension, virtually superimposable on the biphasic pattern of insulin release (Herchuelz and Malaisse 1980a, b; Herchuelz et al. 1980b).

Under suitable experimental conditions, e.g. at low temperature, the cationic response is maintained, whereas the secretory response is abolished (Atwater et al. 1984). This indicates that the increase in ^{45}Ca efflux is not merely a consequence of the stimulation of insulin release. Instead, it is believed that the increase in ^{45}Ca efflux is secondary to stimulation of ^{40}Ca influx into the islet cells, leading to a process of exchange between influent ^{40}Ca and effluent ^{45}Ca.

Although this radioisotopic procedure represents a valuable tool for assessing the influence of nutrients and other secretagogues upon Ca^{2+} inflow into the islet cells, the interpretation of the experimental data may require a rather sophisticated model. Indeed, an exchange between influent ^{40}Ca and effluent ^{45}Ca could only be simulated by postulating the existence of a process of Ca^{2+}-stimulated Ca^{2+} release from intracellular organelles of the vacuolar system (Scholler et al. 1985).

The second current approach to the study of the inflow of Ca^{2+} into islet cells involves recording the bioelectrical activity of impaled β-cells (Atwater et al. 1980; Meissner and Preissler 1980). This bioelectrical technique offers several advantages. It permits the identification of rapid events, specifically located at the plasma membrane, in individual β-cells. The bioelectrical activity of the β-cells is characterized by bursts of spikes separated by silent, polarized periods at intermediate glucose concentrations, or by continuous spiking at higher hexose concentrations. Each burst of spikes is preceded first by a slow and then by a rapid phase of depolarization. The ascending segment of each spike, and possibly the rapid phase of depolarization at the onset of each burst, are thought to coincide with the facilitated entry of Ca^{2+} into the cell (Malaisse et al. 1982b). Hence, the time course for electrical events may reveal changes in Ca^{2+} influx.

A last point about the regulation of Ca^{2+} inflow into the islet cells needs to be made. The knowledge in this field has been presented so far as if there existed only one modality for Ca^{2+} influx, namely through gated voltage-sensitive Ca^{2+} channels. This is probably an oversimplification. Indeed, the inhibitory effect of verapamil, an organic Ca^{2+} antagonist, upon glucose-stimulated ^{45}Ca efflux from prelabelled islets suggests that the hexose facilitates the entry of Ca^{2+} into the β-cell through distinct classes of Ca^{2+} channels (Lebrun et al. 1982). It is also premature to rule out the possibility that native ionophores, acting as Ca^{2+} carriers, may participate in the transport of Ca^{2+} from the extracellular to the intracellular milieu (Deleers and Malaisse 1982). It should also be borne in mind that ^{45}Ca is taken up by islet cells even when they are deprived of exogenous nutrient and, hence, in the absence of any bioelectrical activity.

Regulation of Ca^{2+} Outflow

The outflow of Ca^{2+} from the islet cells occurs against a steep electrochemical gradient. It is thought to be mediated mainly by two distinct systems.

The first system involves the operation of a process of Na$^+$/Ca^{2+} countertransport, in which the influx of Na$^+$ into the cell along its electrochemical gradient is coupled to the uphill exit of Ca^{2+} from the cell. Such a countertransport process might be mediated by native ionophores. The countertransport process can indeed by simulated by ionophores in artificial membrane models (Malaisse and Couturier 1978). Moreover, it is possible to extract from islet membranes a lipophilic material displaying ionophoretic capacity and this material is indeed able to mediate Na$^+$/Ca^{2+} exchange (Anjaneyulu et al. 1980).

The second modality for the outflow of Ca^{2+} from the cell involves the operation of a Ca^{2+}-dependent ATPase. This enzyme was characterized in islet cell subcellular fractions and, according to some authors, is susceptible to activation by the calcium–calmodulin complex (Pershadsingh et al. 1980).

A third and quantitatively minor mechanism for the extrusion of Ca^{2+} from the cell could involve the release of calcium sequestered in secretory granules at the time and site of exocytosis (Malaisse et al. 1973).

To my knowledge, the sole approach so far used to study the influence of environmental factors upon the outflow of calcium from islet cells involves the study of ^{45}Ca efflux from prelabelled islets (Malaisse et al. 1973). In perifused prelabelled islets, D-glucose provokes a rapid, sustained, rapidly reversible and dose-related inhibition of ^{45}Ca efflux. This inhibitory action is most evident when the islets are perifused in the absence of extracellular Ca^{2+}. Indeed, in the presence of extracellular Ca^{2+}, the early fall in ^{45}Ca outflow is soon masked by a superimposed increase in ^{45}Ca outflow, as already discussed in this chapter. The two effects of D-glucose, namely its inhibitory action and its stimulatory action on ^{45}Ca outflow, can be distinguished from one another not solely by their different time courses and different dependencies on extracellular Ca^{2+}, but also by their vastly different dose–response relationships at increasing glucose concentrations and their respective sensitivity to suitable pharmacological manipulation (Herchuelz and Malaisse 1980a, b; Herchuelz et al. 1980b).

Basically, two distinct explanations have been offered to account for the inhibition by D-glucose of ^{45}Ca outflow from prelabelled islets. According to the first hypothesis, the decrease in ^{45}Ca outflow reflects a primary inhibition of Ca^{2+} efflux as mediated by the Na$^+$/Ca^{2+} countertransport process. This interpretation has been both defended and rebuked in the light of the effects upon ^{45}Ca efflux of changes in extracellular Na$^+$ concentration (Herchuelz et al. 1980a; Hellman and Gylfe 1984). According to the second hypothesis, glucose-induced decrease in ^{45}Ca efflux represents a secondary phenomenon resulting from the stimulated sequestration of ^{45}Ca into intracellular organelles (Hellman 1985). The possible consequence of such a hypothetical phenomenon has already been mentioned in this chapter. The two hypotheses are not necessarily exclusive of one another. As a matter of fact, they were both recently included in a mathematical model of Ca^{2+} handling by islet cells (Malaisse et al. 1985b).

Little is known about the factors responsible for the nutrient-induced decrease in Ca^{2+} outflow from islet cells. The inhibition of the process of Na^+/Ca^{2+} countertransport could be the result of an increased generation of H^+, resulting from the accelerated oxidation of nutrients. Thus, H^+ might compete with Ca^{2+} for a common ionophore, a phenomenon which can be simulated by native ionophoretic material extracted from the islets (Anjaneyulu et al. 1980). It should be underlined, however, that the changes in cytosolic pH caused by D-glucose and other nutrients remain to be unambiguously established.

The facilitated sequestration of Ca^{2+} by intracellular organelles could be viewed as the consequence of an increase in ATP availability in nutrient-stimulated islet cells. However, the total ATP content of islet cells is little affected by D-glucose, at least in the range of hexose concentrations in which insulin release is regulated by the sugar (Malaisse et al. 1979b). Hence, the question must be raised whether cytosolic and total ATP display a changing relationship at increasing concentrations of D-glucose, and whether and how an increase in ATP generation could be coupled to an increase in ATP utilization in the absence of any change in ATP concentration (Malaisse et al. 1984a).

Regulation of Ca^{2+} Intracellular Distribution

When the net uptake of ^{45}Ca is measured at close-to-isotopic equilibrium, and the result expressed relative to the intracellular H_2O space, the apparent cellular concentration of calcium falls in the millimolar range, namely four orders of magnitude higher than the cytosolic concentration of ionized calcium. The sequestration of Ca^{2+} by cellular organelles may not be sufficient to account fully for such a difference (Malaisse et al. 1973). Hence, a significant amount of calcium may be located in the cytosolic domain, but bound to cytoplasmic proteins or other cytosolic molecules. The regulation of cytosolic Ca^{2+} activity by environmental factors is not currently attributed to alteration in this binding phenomenon. Instead, emphasis is given to those factors which may affect the subcellular repartition of calcium between the cytosolic pool and intracellular organelles of the vacuolar system.

In the latter context, several potential regulatory mechanisms must be mentioned. However, it should first be realized that a number of distinct organelles may participate in "buffering" the cytosolic Ca^{2+} concentration. They include mitochondria, the endoplasmic reticulum and secretory granules. In addition, a plasma membrane-associated calcium pool may also play a role in the control of cytosolic Ca^{2+} activity (Täljedal 1980). In each of these types of organelles, a distinct series of regulatory factors is probably operative. Therefore, the precise quantitative estimation of Ca^{2+} inflow into and outflow from each organelle of a given cell, in both the resting and stimulated state, represents a difficult and complex task. Nevertheless, simplified models for Ca^{2+} handling have been designed in which all organelles involved in the handling of Ca^{2+} are treated as a single, or integrated, vacuolar system (Malaisse et al. 1985b).

Mitochondria isolated from transplantable insulinomas can maintain an ambient free Ca^{2+} concentration of 0.3 μM in the absence of Mg^{2+} and 0.9 μM in the presence of Mg^{2+} (1.0 mM). The steady-state extramitochondrial Ca^{2+} concentration is increased when either the pH is lowered from 7.0 to 6.9 or Na^+ added, the latter effect reaching half-maximal magnitude at a Na^+

concentration close to 4 mM and being apparently attributable to stimulation of Ca^{2+} efflux from the mitochondria (Prentki et al. 1983). Rat insulinoma microsomes can maintain a lower ambient Ca^{2+} concentration (0.2 μM) through the operation of a magnesium–ATP-dependent Ca^{2+}-transporting activity. Such is not the case with secretory granules, which nevertheless display a high calcium content (0.1 $\mu mol/mg$ protein) which is released by the Ca^{2+} ionophore A23187 (Prentki et al. 1984c). Coordinated regulation of free Ca^{2+} by distinct isolated organelles from the rat insulinoma has been investigated in combined fractions enriched respectively with mitochondria, microsomes and secretory granules (Prentki et al. 1984a). These studies support the observations made in intact islet cells and suggest that the intracellular distribution of Ca^{2+} may be affected by ATP availability (and also by ADP, which increases ambient Ca^{2+} concentration in isolated microsomes), cytosolic Na^+ concentration and pH.

Another factor which may favour the mobilization of Ca^{2+} from non-mitochondrial organelles is inositol 1,4,5-triphosphate (Prentki et al. 1984b). This metabolite is generated from phosphatidylinositol 4,5-bisphosphate, an inositol-containing phospholipid located, in part, in the plasma membrane. Certain secretagogues, especially cholinergic agents acting at the level of muscarinic receptors, activate the enzyme phospholipase C which catalyses the hydrolysis of phosphoinositides into diacylglycerol and inositol-phosphate (Best and Malaisse 1984). Nutrient secretagogues also provoke the hydrolysis of polyphosphoinositides (Best and Malaisse 1984). The mechanism leading to activation of phospholipase C in nutrient-stimulated islet cells remains to be elucidated. It is conceivable that the nutrient-induced increase in cytosolic Ca^{2+} activity plays a modulatory role in phospholipase C activity (Mathias et al. 1985). If so, the liberation of inositol 1,4,5-triphosphate and the resulting mobilization of Ca^{2+} from non-mitochondrial organelles would provide a device for amplification of the changes in cytosolic Ca^{2+} concentration that occur in the course of nutrient-stimulated insulin release.

A final mechanism for mobilization of Ca^{2+} from intracellular organelles may involve a process of Ca^{2+}-stimulated Ca^{2+} release, which was recently included in a mathematical model of stimulus–secretion coupling in the β-cell (Scholler et al. 1985). It should be noted, however, that this process has not yet been demonstrated in isolated organelles, both mitochondria and microsomes taking up Ca^{2+} when exposed to an increase in ambient Ca^{2+} concentration. It is conceivable, therefore, that the postulated process of Ca^{2+}-stimulated Ca^{2+} release is mediated by some cellular second messenger.

Targets for Cytosolic Ca^{2+}

I have so far reviewed the evidence indicating that an increase in cytosolic Ca^{2+} activity plays a critical role in triggering insulin release from the pancreatic β-cell and analysed the mechanisms possibly involved in the cytosolic accumulation of Ca^{2+}. In the last part of this chapter, emphasis will be given to the systems which act in the β-cell as a target for cytosolic Ca^{2+}. In other words, attention will now be paid to the regulatory role of Ca^{2+} in hormone secretion rather than on the regulation of Ca^{2+} fluxes in endocrine cells.

Microtubular–Microfilamentous System

Proposals have been made about the way in which an increase in cytosolic Ca^{2+} activity may cause the release of insulin from the pancreatic β-cell.

It was first postulated that the cytoplasmic accumulation of Ca^{2+} could cause the collapse of a potential energy barrier between the inner face of the plasma membrane and the outer face of secretory granules, both of which carry electrostatic negative charges (Dean 1974).

An alternative proposal is that the increase in cytosolic Ca^{2+} concentration causes the contraction of a network of microfilaments located beneath the plasma membrane and described as a cell web (Malaisse et al. 1972; Orci et al. 1972). Such a cell web could act as a sphincter and control the access of secretory granules to exocytic sites on the plasma membrane. It was in fact shown, in monolayer cultures of endocrine pancreatic cells examined by time-lapse cinematography, that cellular motile events, including the movements of secretory granules, are under the control of a microtubular–microfilamentous system. Thus, secretory granules undergo saltatory movements along oriented microtubular pathways and stimulation of insulin release coincides with an increase in the frequency of both these saltatory movements and contractile events located at the cell boundary (Somers et al. 1979a; Lacy et al. 1975).

The participation of the β-cell microtubular–microfilamentous system in the control of those motile events involved in both the conversion of proinsulin to insulin and the release of insulin into the extracellular space, is also supported by an array of ultrastructural, biochemical and physiological findings (Lacy and Malaisse 1973; Malaisse et al. 1975a, b; Malaisse and Orci 1979) including the effect of drugs acting upon the function of either microtubules (e.g. colchicine or vincristine) or microfilaments (cytochalasin B). Moreover an abnormality of the β-cell microtubular system was encountered in spiny mice (*Acomys cahirinus*) and was proposed to account, in part at least, for the perturbed dynamics of insulin release in these diabetes-prone animals (Malaisse-Lagae et al. 1975). In this mechanical model, the microtubular–microfilamentous system may provide both the structural framework for the functional segregation and oriented translocation of β-granules and the motive force for their intracellular migration and eventual release by exocytosis.

An increase in cytosolic Ca^{2+} activity may, in addition to triggering motile events, exert other regulatory influences in pancreatic endocrine cells. With this dual action in mind, the following information refers to selected systems susceptible to changes in Ca^{2+} handling by the pancreatic β-cell.

Calmodulin

The calcium-dependent regulatory protein calmodulin is present in rat pancreatic islets in an amount close to 0.1 pmol/islet or a concentration close to 50 μM (Valverde and Malaisse 1984). The specific binding of ^{125}I-calmodulin to a subcellular particulate fraction derived from rat islets represents a calcium-dependent process. Calmodulin-binding proteins were also identified in islet cells. The activity of several enzymes in islet homogenates or subcellular fractions is affected by calmodulin in a Ca^{2+}-dependent manner. It was first shown that calcium–calmodulin activates adenylate cyclase (Valverde et al. 1979). In the presence of Ca^{2+} (0.2 mM), half-maximal activation is observed at a calmodulin concentration close to 0.1 μM. In the presence of calmodulin, the threshold concentration for activation by Ca^{2+} is close to 10^{-7} M and a half-maximal response occurs at a concentration of Ca^{2+} close to 10^{-5} M. These findings suggest that physiological variations in cytosolic Ca^{2+} activity may regulate the activity of adenylate cyclase via calcium–calmodulin. This may have two important implications.

First, the activation of adenylate cyclase by calcium–calmodulin would provide a device for amplification of the secretory response, since cyclic AMP itself enhances insulin release evoked by various secretagogues. Second, the activation of adenylate cyclase could account for the fact that several secretagogues, which, like D-glucose, do not exert any direct effect upon enzyme activity in the subcellular fraction, increase cyclic AMP production by intact islets, the latter effect being then the consequence of an increase in cytosolic Ca^{2+} activity (Malaisse and Malaisse-Lagae 1984). The latter interpretation is supported by the finding that the nutrient-induced increment in cyclic AMP production by intact islets is suppressed in the absence of extracellular Ca^{2+} (Valverde et al. 1983). Incidentally, calcium–calmodulin may also cause a limited activation of cyclic AMP phosphodiesterase, so that the overall effect of the regulatory protein on cyclic AMP metabolism could be viewed as an increase in the turnover rate of the cyclic nucleotide.

Calcium–calmodulin may also modulate the activity of other enzymes in the pancreatic β-cell. For instance, under suitable experimental conditions, calmodulin may increase the activity of a calcium-activated ATPase located at the plasma membrane and mediating the extrusion of Ca^{2+} from the cell into the extracellular fluid, against the prevailing electrochemical gradient (Valverde and Malaisse 1984). This could represent a way of restoring the resting cytosolic Ca^{2+} activity in response to a prior stimulation of the cell with coinciding inflow of Ca^{2+} through gated calcium channels. In other words, calcium–calmodulin could participate in cyclic or rhythmic shifts between the stimulated (high cytosolic Ca^{2+} activity) and resting (low cytosolic Ca^{2+} activity) functional states of the islet cells.

Islet cell homogenates also display calcium–calmodulin-responsive protein kinase activity. Endogenous proteins which could act as a substrate for the latter kinase may include either the alpha and beta subunits of tubulin or myosin light chains (Colca et al. 1983; Kowluru and MacDonald 1984). In the latter case, it could be that, when the calcium-binding sites of calmodulin are saturated, myosin light chain kinase would catalyse the phosphorylation of myosin which, in turn, should permit myosin to interact with actin to allow contraction of microfilaments in the β-cell.

Transglutaminase

The enzyme transglutaminase catalyses the cross-linking between endo-γ-glutamyl and endo-ε-lysyl residues of protein. This Ca^{2+}-responsive enzyme is present in islet homogenates (Gomis et al. 1983; Bungay et al. 1984). Its activity is modulated by the Ca^{2+} concentration of the assay medium, with a threshold Ca^{2+} concentration for activation close to 10^{-7} M and a half-maximal activation at a Ca^{2+} concentration close to 60 μM. Calmodulin fails to affect transglutaminase activity in islet homogenates. It was proposed that transglutaminase participates in the control of mechanical events in the islet cells. The following findings support the latter proposal.

D-glucose increases the activity of transglutaminase in intact cells. This was documented by an increased incorporation of $[2,5-^4H]$histamine in trichloroacetic acid-precipitable material in islets incubated in the presence, as distinct from absence, of D-glucose. The stimulatory effect of D-glucose upon $(2,5-^3H)$histamine incorporation was observed in the presence but not the absence of extracellular Ca^{2+}, suggesting a key role for Ca^{2+} in the activation of the enzyme. The activity of transglutaminase may also be affected by changes in redox state (Gomis et al. 1986a) and cyclic AMP availability (Gomis et al. 1986b). The presence of endogenous substrates for transglutaminase was also documented in islet homogenates.

Several inhibitors of transglutaminase, including alkylamines, bacitracin, monodansylcadaverine, N-p-tosylglycine and hypoglycaemic sulphonylureas (Gomis et al. 1984a, b; Lebrun et al. 1984; Sener et al. 1984, 1985b), were examined for their effect upon islet function. The most convincing results were obtained with glycine methylester, control experiments being performed with sarcosine methylester to assess the specificity for the primary amino group functionality (Sener et al. 1985a). Glycine methylester, but not sarcosine methylester, was found to cause a dose-related inhibition of transglutaminase activity in islet homogenates, to decrease $[^{14}C]$methylamine incorporation in endogenous proteins of intact islets and to provoke a rapid and sustained inhibition of insulin release evoked by D-glucose or other secretagogues.

These findings support the view that, in the pancreatic β-cell, transglutaminase participates in the machinery controlling the access of secretory granules to the exocytic sites. Moreover, glycine methylester, but not sarcosine methylester, delayed the conversion of proinsulin to insulin in islets pulse-labelled with L-[4-^3H]phenylalanine (Alarcon et al. 1985). The incorporation of L-[4-^3H]phenylalanine into the islet peptides, the ratio of hormonal to total tritiated peptides and the insulin content of the islets failed to be affected by either of these methylesters. Thus, transglutaminase may also participate in the control of motile events involved in the transfer of proinsulin from its site of synthesis to its site of conversion.

A Further Regulatory Role of Ca^{2+}

Targets affected by cytosolic Ca^{2+} and/or calcium–calmodulin include Ca^{2+}-responsive K^+ channels; two distinct Ca^{2+}-ATPases, located respectively in the plasma membrane and endoplasmic reticulum; Ca^{2+}- and/or calcium–calmodulin-responsive kinases; transglutaminase; and adenylate cyclase. The responses of these enzymes to Ca^{2+} itself and to calcium–calmodulin need not be identical. For instance, in the absence of calmodulin, increasing Ca^{2+} concentrations inhibit rather than activate adenylate cyclase (Valverde et al. 1979). Most of these enzymes have already been mentioned and their possible participation in the process of stimulus–secretion coupling was discussed in preceding sections of this chapter. The present section will be restricted, therefore, to the consideration of a further possible regulatory role of Ca^{2+} not yet considered.

Several Ca^{2+} movements in the islet cells are apparently mediated by systems which do not directly consume ATP. Such is the case for the inflow of Ca^{2+} into the cell through gated Ca^{2+} channels, and for the extrusion of Ca^{2+} from the cell by a process of Na^+/Ca^{2+} exchange. To the extent that the latter exchange, as well as other Ca^{2+} movements, may be mediated by the intervention of native ionophores, such ionophoretic processes may also occur without a direct and concomitant consumption of ATP. Yet other Ca^{2+} movements are mediated by Ca^{2+}-sensitive ATPases.

The precise quantitation of these ATP-consuming Ca^{2+} movements in intact islet cells represents a difficult task. It has nevertheless been proposed that a significant fraction of the ATP turnover rate in islet cells may coincide with the activity of Ca^{2+}-sensitive ATPases (Owen et al. 1983). Hence, the movements of Ca^{2+} in intact cells would account for a far-from-negligible fraction of the total O_2 uptake. This is an important consideration, especially in a cell supposed to act as a fuel-sensor organ and mainly regulated in its functional activity by the availability of circulating nutrients. Indeed, it suggests that a primary alteration in Ca^{2+} fluxes may result in a feedback control of metabolic events in the islet cells. This view is supported by the finding that the oxidation rates of such exogenous nutrients as D-[U-^{14}C]glucose or L-[U-^{14}C]glutamine are indeed decreased in islets deprived of extracellular Ca^{2+} (Malaisse et al. 1978b; Sener et al. 1982). A dual interplay between metabolic and cationic events would thus be operative in the islet cells. On one hand, the catabolism of nutrients would regulate the fluxes of cations. On the other hand, the latter fluxes would exert a feedback control upon metabolic events. It is conceivable that such an interplay somehow participates in the rhythmogenesis of bioelectrical and secretory phenomena in the pancreatic β-cell (Malaisse et al. 1980b).

Concluding Remarks

The present account of the role of cellular calcium in the secretion of hormones is far from exhaustive. I have purposefully neglected, for the sake of simplicity, studies performed in systems other than the insulin-releasing pancreatic β-cell. Even the latter subject has been treated in an integrated rather than an analytic manner. I hope, nevertheless, that sufficient information has been provided to indicate that the study of the role of cellular calcium in the secretion of hormones represents a vast and complex field, wide open for further investigations.

The conclusion seems obvious. The grand old man was right. It is glucose that stimulates insulin release. Yet it would be unwise to ignore the role of calcium in this and other secretory processes.

Acknowledgements. The experimental work from this laboratory reviewed here was supported by grants from the Belgian Foundation for Scientific Medical Research. We thank C. Demesmaeker for secretarial help.

References

Alarcon C, Valverde I, Malaisse WJ (1985) Transglutaminase and cellular motile events: retardation of proinsulin conversion by glycine methylester. Biosci Rep 5:581–587

Anjaneyulu K, Anjaneyulu R, Malaisse WJ (1980) The stimulus–secretion coupling of glucose-induced insulin release. XLIII. Na–Ca countertransport mediated by pancreatic islet native ionophores. J Inorg Biochem 13:178–188

Ashcroft FM, Harrison DE, Ashcroft SJH (1984) Glucose induces closure of single potassium channels in isolated rat pancreatic β-cells. Nature 312:446–448

Atwater I, Dawson CM, Scott A, Eddlestone G, Rojas E (1980) The nature of the oscillatory behaviour in electrical activity from pancreatic β-cell. Horm Metab Res [Suppl] 10:100–107

Atwater I, Ferrer R, Goncalves A et al. (1984) Cooling dissociates insulin release from electrical activity and cationic fluxes in pancreatic islets. J Physiol (Lond) 348:615–627

Best L, Malaisse WJ (1984) Nutrient and hormone-neurotransmitter stimuli induce hydrolysis of polyphosphoinositides in rat pancreatic islets. Endocrinology 115:1814–1820

Brisson GR, Malaisse-Lagae F, Malaisse WJ (1972) The stimulus–secretion coupling of glucose-induced insulin release. VII. A proposed site of action for adenosine-3′,5′-cyclic monophosphate. J Clin Invest 51:232–241

Bungay PT, Potter JM, Griffin M (1984) The inhibition of glucose-stimulated insulin secretion by primary amines. A role for transglutaminase in the secretory mechanism. Biochem J 219:819–827

Colca JR, Brooks CL, Landt M, McDaniel ML (1983) Correlation of Ca^{2+}- and calmodulin-dependent protein kinase activity with secretion of insulin from islets of Langerhans. Biochem J 212:819–827

Cook DL, Hales N (1984) Intracellular ATP directly blocks K^+ channels in pancreatic B-cells. Nature 311:271–273

Dean PM (1974) Surface electrostatic-charge measurements on islet and zymogen granules: effect of calcium ions. Diabetologia 10:427–430

Deleers M, Mahy M, Malaisse WJ (1982) Calcium ionophoresis by pancreatic islet extracts in model membranes. Int Biochem 4:47–57

Deleers M, Mahy M, Malaisse WJ (1985) Glucose increases cytosolic Ca^{2+} activity in pancreatic islet cells. Biochem Int 10:97–103

Devis G, Somers G, Malaisse WJ (1975) Stimulation of insulin release by calcium. Biochem Biophys Res Commun 67:525–529

Devis G, Somers G, Malaisse WJ (1977) Dynamics of calcium-induced insulin release. Diabetologia 13:531–536

Findlay I, Dunne MJ (1985) Voltage-activated Ca^{2+} currents in insulin secreting cells. FEBS Lett 189:281–285

Gomis R, Sener A, Malaisse-Lagae F, Malaisse WJ (1983) Transglutaminase activity in pancreatic islets. Biochim Biophys Acta 760:384–388

Gomis R, Deleers M, Malaisse-Lagae F et al. (1984a) Metabolic and secretory effects of methylamine in pancreatic islets. Cell Biochem Funct 2:161–166

Gomis R, Mathias PCF, Lebrun P et al. (1984b) Inhibition of transglutaminase by hypoglycaemic sulphonylureas in pancreatic islets and its possible relevance to insulin release. Res Commun Chem Pathol Pharmacol 46:331–349

Gomis R, Arbos MA, Sener A, Malaisse WJ (1986a) Glucose-induced activation of transglutaminase in pancreatic islets. Diab Res 3:115–117

Gomis R, Arbos MA, Malaisse WJ (1986b) Activation of transglutaminase by dibutyryl-cyclic AMP and theophylline in rat pancreatic islets. IRCS Med Sci 14:134–135

Grodsky GM, Bennett LL (1966) Cation requirement for insulin secretion in the isolated perfused pancreas. Diabetes 15:910–913

Hellman B (1985) β-Cell cytoplasmic Ca^{2+} balance as a determinant for glucose-stimulated insulin release. Diabetologia 28:494–501

Hellman B, Gylfe E (1984) Glucose inhibits ^{45}Ca efflux from pancreatic β-cells also in the absence of Na^+–Ca^{2+} counter-transport. Biochim Biophys Acta 770:136–141

Hellman B, Gylfe E (1986) Calcium and the control of insulin secretion. In: Chiung W (ed) Calcium and cell biology, vol 6, Academic Press, New York, Chapter 8

Henquin JC, Meissner HP (1983) Dibutyryl cyclic AMP triggers Ca^{2+} influx and Ca^{2+}-dependent electrical activity in pancreatic B-cell. Biochem Biophys Res Commun 112:614–620

Henquin JC, Meissner HP (1984) Significance of ionic fluxes and changes in membrane potential for stimulus–secretion coupling in pancreatic B-cells. Experientia 40:1043–1052

Henquin JC, Schmeer W, Meissner HP (1983) Forskolin, an activator of adenylate cyclase, increases Ca^{2+}-dependent electrical activity induced by glucose in mouse pancreatic B-cells. Endocrinology 112:2218–2220

Herchuelz A, Malaisse WJ (1980a) Regulation of calcium fluxes in pancreatic islets: two calcium movements' dissociated response to glucose. Am J Physiol 238:E87–95

Herchuelz A, Malaisse WJ (1980b) Regulation of calcium fluxes in rat pancreatic islets: dissimilar effects of glucose and of sodium ion accumulation. J Physiol (Lond) 302:263–280

Herchuelz A, Sener A, Malaisse WJ (1980a) Regulation of calcium fluxes in rat pancreatic islets. Calcium extrusion by sodium–calcium counter transport. J Membr Biol 57:1–12

Herchuelz A, Couturier E, Malaisse WJ (1980b) Regulation of calcium fluxes in pancreatic islets: glucose-induced calcium–calcium exchange. Am J Physiol 238:E96–103

Herchuelz A, Thonnart N, Sener A, Malaisse WJ (1980c) Regulation of calcium fluxes in pancreatic islets: the role of membrane depolarization. Endocrinology 107:491–497

Hubinont CJ, Malaisse WJ (1985) Protein kinase C activity in pancreatic islets: effects of Ca^{2+}, calmodulin and retinoic acid. Biochem Int 10:577–584

Kowluru A, MacDonald MJ (1984) Protein phosphorylation in pancreatic islets: evidence for separate Ca^{2+} and cAMP-enhanced phosphorylation of two 57000 Mr proteins. Biochem Biophys Res Commun 118:797–804

Lacy PE, Malaisse WJ (1973) Microtubules and beta cell secretion. Recent Prog Horm Res 29:199–222

Lacy PE, Finke EH, Codilla RC (1975) Cinematographic studies of β granule movement in monolayer culture of islet cells. Lab Invest 33:570–576

Lebrun P, Malaisse WJ, Herchuelz A (1982) Evidence for two distinct modalities of Ca^{2+} influx into the pancreatic B-cell. Am J Physiol 242:E59–66

Lebrun P, Gomis R, Deleers M et al. (1984) Methylamines and islet function: cationic aspects. J Endocrinol Invest 7:347–355

Lowe DA, Richardson BP, Taylor P, Donatsch P (1976) Increasing intracellular sodium triggers calcium release from bound pools. Nature 260:337–338

Malaisse WJ, Couturier E (1978) An ionophoretic model for Na–Ca counter transport. Nature 275:664–665

Malaisse WJ, Malaisse Lagae F (1984) The role of cyclic AMP in insulin release. Experientia 40:1068–1075

Malaisse WJ, Mathias PCF (1985) Stimulation of insulin release by an organic calcium agonist. Diabetologia 28:153–156

Malaisse WJ, Orci L (1979) The role of the cytoskeleton in pancreatic B-cell function. In: Gabbiani G (ed) Methods of achievements in experimental pathology, vol 9. Karger, Basel, pp 112–136

Malaisse WJ, Brisson GR, Malaisse-Lagae F (1971) Effet insulinotrope du calcium. Ann Endocrinol 32:621–622

Malaisse WJ, Hager DL, Orci L (1972) The stimulus–secretion coupling of glucose-induced insulin release. IX. The participation of the β-cell web. Diabetes 21:594–604

Malaisse WJ, Brisson GR, Baird LE (1973) The stimulus–secretion coupling of glucose-induced insulin release. X. Effect of glucose on ^{45}Ca efflux from perifused islets. Am J Physiol 224:389–394

Malaisse WJ, Leclercq-Meyer V, Van Obberghen E et al. (1975a) The role of the microtubular–microfilamentous system in insulin and glucagon release by the endocrine pancreas. In: Borgers M, De Branbander M (eds) Microtubules and microtubule inhibitors. North-Holland, Amsterdam, pp 143–152

Malaisse WJ, Malaisse-Lagae F, Van Obberghen E et al. (1975b) Role of microtubules in the phasic pattern of insulin release. Ann NY Acad Sci 253:630–652

Malaisse WJ, Devis G, Herchuelz A, Sener A, Somers G (1976) Calcium antagonists and islet function. VIII. The effect of magnesium. Diabete Métab 2:1–4

Malaisse WJ, Herchuelz A, Devis G et al. (1978a) Regulation of calcium fluxes and their regulatory roles in pancreatic islets. Ann NY Acad Sci 307:562–582

Malaisse WJ, Hutton JC, Sener A et al. (1978b) Calcium-antagonists and islet function. VII. Effect of calcium deprivation. J Membr Biol 38:193–208

Malaisse WJ, Sener A, Herchuelz A, Hutton JC (1979a) Insulin release: the fuel hypothesis. Metabolism 28:373–386

Malaisse WJ, Hutton JC, Kawazu S, Herchuelz A, Valverde I, Sener A (1979b) The stimulus–secretion coupling of glucose-induced insulin release. XXXV. The links between metabolic and cationic events. Diabetologia 16:331–341

Malaisse WJ, Sener A, Herchuelz A et al. (1980a) Insulinotropic effect of the tumor promotor 12-O-tetradecanoylphorbol-13-acetate in rat pancreatic islets. Cancer Res 40:3827–3831

Malaisse WJ, Sener A, Herchuelz A et al. (1980b) The interplay between metabolic and cationic events in islet cells: coupling factors and feedback mechanisms. Horm Metab Res [Suppl] 10:61–66

Malaisse WJ, Valverde I, Owen A, Verhulst D, Cantraine F (1982a) Mathematical modelling of cyclic AMP–Ca^{2+} interactions in pancreatic islets. Diabetes 31:170–177

Malaisse WJ, Lebrun P, Herchuelz A (1982b) Ionic determinants of bioelectrical spiking activity in the pancreatic B-cell. Pfluegers Arch 395:201–203

Malaisse WJ, Malaisse-Lagae F, Sener A (1984a) Coupling factors in nutrient-induced insulin release. Experientia 40:1035–1043

Malaisse WJ, Garcia-Morales P, Dufrane SP, Sener A, Valverde I (1984b) Forskolin-induced activation of adenylate cyclase, cyclic adenosine monophosphate production and insulin release in rat pancreatic islets. Endocrinology 115:2015–2020

Malaisse WJ, Dunlop ME, Mathias PCF, Malaisse-Lagae F, Sener A (1985a) Stimulation of protein kinase C and insulin release by 1-oleoyl-2-acetyl-glycerol. Eur J Biochem 149:23–27

Malaisse WJ, Scholler Y, De Maertelaer V (1985b) Mathematical modelling of stimulus–secretion coupling in the pancreatic B-cell. IV. Dissociated time course for the glucose-induced changes in distinct Ca^{2+} movements. Diabetes Res 2:195–199

Malaisse-Lagae F, Mahy M, Malaisse WJ (1969) Effect of epinephrine upon ^{45}Ca uptake by isolated islets of Langerhans. Horm Metab Res 1:319–320

Malaisse-Lagae F, Ravazzola M, Amherdt M et al. (1975) An apparent abnormality of the B-cell microtubular system in spiny mice (*Acomys cahirinus*). Diabetologia 11:71–76

Malaisse-Lagae F, Mathias PCF, Malaisse WJ (1984) Gating and blocking of calcium channels by dihydropyridines in the pancreatic B-cell. Biochem Biophys Res Commun 123:1062–1068

Mathias PCF, Best L, Malaisse WJ (1985) Stimulation by glucose and carbamylcholine of phospholipase C in pancreatic islets. Cell Biochem Funct 3:173–177

Meissner HP, Preissler M (1980) Ionic mechanisms of the glucose-induced membrane potential changes in B-cells. Horm Metab Res [Suppl] 10:91–99

Milner RDG, Hales CN (1967) The role of calcium and magnesium in insulin secretion from rabbit pancreas studied in vitro. Diabetologia 3:47–49

Orci L, Gabbay KH, Malaisse WJ (1972) Pancreatic β-cell web: its possible role in insulin secretion. Science 175:1128–1130

Owen A, Sener A, Malaisse WJ (1983) Stimulus–secretion coupling of glucose-induced insulin release. LI. Divalent cations and ATPase activity in pancreatic islets. Enzyme 29:2–14

Pace CS, Tarvin JT, Neighbors AS, Pirkel JA, Greider MH (1980) Use of a high voltage technique to determine the molecular requirements for exocytosis in islet cells. Diabetes 29:911–918

Pershadsingh HA, McDaniel ML, Landt M, Bry GC, Lacy PE, McDonald JM (1980) Ca^{2+}-activated ATPase and ATP-dependent calmodulin-stimulated Ca^{2+} transport in islet cell plasma membrane. Nature 288:492–495

Prentki M, Janjic D, Wollheim CB (1983) The regulation of extramitochondrial steady state free Ca^{2+} concentration by rat insulinoma mitochondria. J Biol Chem 258:7597–7602

Prentki M, Janjic D, Wollheim CB (1984a) Coordinated regulation of free Ca^{2+} by isolated organelles from a rat insulinoma. J Biol Chem 259:14054–14058

Prentki M, Biden TJ, Janjic D, Irvine RF, Berridge MJ, Wollheim CB (1984b) Rapid mobilization of Ca^{2+} from rat insulinoma microsomes by inositol-1,4,5-trisphosphate. Nature 309:562–564

Prentki M, Janjic D, Biden TJ, Blondel B, Wollheim CB (1984c) Regulation of Ca^{2+} transport by isolated organelles of a rat insulinoma. J Biol Chem 259:10118–10123

Rorsman P, Abrahamsson H (1985) Cyclic AMP potentiates glucose-induced insulin release from mouse pancreatic islets without increasing cytosolic free Ca^{2+}. Acta Physiol Scand 125:639–647

Rorsman P, Trube G (1985) Glucose dependent K^+-channels in pancreatic β-cells are regulated by intracellular ATP. Pfluegers Arch 405:305–309

Rorsman P, Abrahamsson H, Gylfe E, Hellman B (1984) Dual effects of glucose on the cytosolic Ca^{2+} activity of mouse pancreatic β-cells. FEBS Lett 170:196–200

Scholler Y, De Maertelaer V, Malaisse WJ (1983) Mathematical modelling of stimulus–secretion coupling in the pancreatic B-cell. I. Dynamics of insulin release. Acta Diabetol Lat 20:329–340

Scholler Y, De Maertelaer V, Malaisse WJ (1985) Mathematical modelling of stimulus–secretion coupling in the pancreatic B-cell. II. Calcium-stimulated calcium release. Comput Programs Biomed 19:119–126

Sener A, Malaisse-Lagae F, Malaisse WJ (1982) The stimulus–secretion coupling of glucose-induced insulin release. LII. Environmental influences on L-glutamine oxidation in pancreatic islets. Biochem J 202:309–316

Sener A, Gomis R, Lebrun P, Herchuelz A, Malaisse-Lagae F, Malaisse WJ (1984) Methylamine and islet function: possible relationship to Ca^{2+}-sensitive transglutaminase. Mol Cell Endocrinol 36:175–180

Sener A, Dunlop ME, Gomis R, Mathias PCF, Malaisse-Lagae F, Malaisse WJ (1985a) Role of transglutaminase in insulin release. Study with glycine and sarcosine methylesters. Endocrinology 117:237–242

Sener A, Gomis R, Billaudel B, Malaisse WJ (1985b) Facilitation of insulin release by N-p-tosylglycine. Biochem Pharmacol 34:2495–2499

Somers G, Devis G, Van Obberghen E, Malaisse WJ (1976a) Calcium-antagonists and islet function. VI. Effect of barium. Pfluegers Arch 365:21–28

Somers G, Devis G, Malaisse WJ (1976b) Analogy between native and exogenous ionophores in the pancreatic B-cell. FEBS Lett 66:20–22

Somers G, Blondel B, Orci L, Malaisse WJ (1979a) Motile events in pancreatic endocrine cells. Endocrinology 104:255–264

Somers G, Devis G, Malaisse WJ (1979b) Calcium-antagonists and islet function. IX. Is extracellular calcium required for insulin release? Acta Diabetol Lat 16:9–18

Täljedal I-B (1980) Fluorescent probing of calcium ions in islet cells. Horm Metab Res [Suppl] 10:130–137

Valverde I, Malaisse WJ (1984) Calmodulin and pancreatic B-cell function. Experientia 40:1061–1068

Valverde I, Vandermeers A, Anjaneyulu R, Malaisse WJ (1979) Calmodulin activation of adenylate cyclase in pancreatic islets. Science 206:225–227

Valverde I, Garcia-Morales P, Ghiglione M, Malaisse WJ (1983) The stimulus–secretion coupling of glucose-induced insulin release. LIII. Calcium-dependency of cyclic AMP response to nutrient secretagogues. Horm Metab Res 15:62–68

Wollheim CB, Sharp GWG (1981) Regulation of insulin release by calcium. Physiol Rev 61:914–973

Wollheim CB, Blondel B, Trueheart PA, Renold AE, Sharp GWG (1975) Calcium induced insulin release in monolayer culture of the endocrine pancreas. Studies with ionophore A23187. J Biol Chem 250:1354–1360

Wollheim CB, Kikuchi M, Renold AE, Sharp GWG (1978) The roles of intracellular and extracellular Ca^{2+} in glucose-stimulated biphasic insulin release by rat islets. J Clin Invest 62:451–458

Yaseen MA, Pedley KC, Howell SL (1982) Regulation of insulin secretion from islets of Langerhans rendered permeable by electric discharge. Biochem J 206:81–87

Chapter 15

Cellular Calcium: Action of Hormones

R. H. Wasserman

Introduction

An appreciation of the indispensability of the calcium ion in physiological systems began with the experiments of Sydney Ringer, who, in the 1880s, noted that calcium was required to maintain the contractility of the frog heart, essential to the development of fertilized eggs and tadpoles and important in cell adhesion (Campbell 1983). Subsequent studies by Locke, Loeb, Loewi and others in the early 1900s demonstrated that calcium was required for the action of hormones such as adrenaline. That calcium was necessary for the survival of cells in culture and for the maintenance of colonies of cells in tissues and organs became known in later years. More recently, it has become widely recognized that the actions of various hormones and other agonists involve changes in intracellular calcium levels.

Multicellular organisms require communication between tissues and organs to coordinate growth and development and to respond to internal and external stimuli and stresses. The channels of communication at the organismal level are primarily the nervous and circulatory systems. The messages are growth factors, neurotransmitters and hormones. Hormones have traditionally been defined as chemical substances synthesized at one site and transferred through the circulatory system to affect a function at another site. Implicit in this definition is that their secretion is "feedback" controlled, so the availability of these stimulatory (or inhibitory) substances is closely regulated and tuned to the requirements of the organism.

Other ways of communicating are by paracrine and autocrine signalling

(Darnell et al. 1986). Paracrines are substances secreted from cells that affect the behaviour of near-neighbour target cells, and include neurotransmitters and neurohormones. For example, upon depolarization of a presynaptic neuron, the neurotransmitter acetylcholine is released from the terminal, diffuses across a synaptic cleft and affects the functioning of an ion channel on the postsynaptic membrane.

In autocrine signalling, the stimulatory substances bind to receptors on the same cell that released them. For example, growth factors elaborated by tumour cells result in unregulated growth of the tumour cell itself (Darnell et al. 1986).

Tissue cells have other ways of communicating. Adjacent cells can signal one another through gap junctions which form aqueous channels between cells. The average diameter of the aqueous channels allows passage only of molecules with a molecular weight of approximately 1000 or less. Small peptide hormones and the cyclic nucleotides can readily pass from one cell to another through these channels. Electrical coupling between cells also occurs through these junctional complexes. Calcium exerts a profound effect on the diameter of gap junctions and therefore on the size of the molecules capable of being transferred from cell to cell (Rose and Lowenstein 1975). The diameter of the aqueous channel decreases and eventually closes as the intracellular free calcium concentration increases.

The association of a hormone, neurotransmitter or other agonist with its specific cell surface receptor or receptors initiates a series of events which yields a biological response. The concept of a "second messenger" comes into play when intermediary substances produced by the agonist–receptor interaction promote a biological response. The first well-documented and well-accepted "second messengers" were the cyclic nucleotides, cyclic adenosine $3',5'$ monophosphate (cAMP) and cyclic guanosine $3',5'$-monophosphate (cGMP).

Considerable attention has now been given to the role played by calcium as a second messenger in the physiological action of hormones, neurotransmitters and other agonists and in the pharmacological action of specific drugs. The source of calcium for this second-messenger function might be extracellular or from intracellular storage sites. The stimulus somehow increases the permeability of the outer plasma membrane to calcium or causes the release of calcium from intracellular stores, or both. The result is a transient increase in the free calcium concentration in the cytosol. The next step is the binding of calcium to proteins, such as calmodulin and troponin C. The activated calcium–protein complex then directly stimulates a physiological or biochemical event.

Many agonists operate through the calcium–second messenger system, for example, acetylcholine through the muscarinic receptor, norepinephrine and epinephrine through the α-1-receptor, dopamine through the D_2 receptor, histamine via the H1 receptor, and thromboxane, thrombin and various growth factors. Also acting through the calcium system are angiotensin II which stimulates the synthesis of aldosterone by adrenal tissue, glucose which activates the release of insulin from β-islet cells of the pancreas and insulin which controls gluconeogenesis by the liver (see reviews by Exton 1984; Nishizuka et al. 1984; Rasmussen 1986; Williamson et al. 1986). Agonists that act through the cAMP system (Table 15.1) include norepinephrine and epinephrine via the β-receptor, dopamine via the D_1 receptor, vasopressin at the V_2 receptor, and adrenocorticotropic hormone (ACTH) and luteinizing hormone (LH). Other agonists, as noted in Table 15.1, inhibit cAMP formation.

Table 15.1. Hormones and neurotransmitters that stimulate or inhibit adenylate
cyclase activity (adapted from Taylor and
Merritt 1986)

Stimulatory through G_s protein
   Epinephrine (beta)
   Follicle-stimulating hormone (FSH)
   Glucagon
   Adrenocorticotropic hormone (ACTH)
   Dopamine (D_1 receptor)
   Luteinizing hormone (LH)
   Vasopressin (V_2 receptor)

Inhibitory through G_i protein
   Epinephrine (α-2-receptor)
   Acetylcholine (muscarinic receptor)
   Opiates
   Dopamine (D_2 receptor)
   Angiotensin II

Control of Intracellular Calcium

The concentration of calcium in the *extracellular* fluid is tightly controlled at about 1.25 mM, primarily by the action of the calcium-regulating hormones: parathyroid hormone, calcitonin and the vitamin D series. This aspect of calcium metabolism is reviewed in more detail in Chapters 2 and 3.

The concentration of *intracellular* free calcium in resting, unstimulated cells is also constrained at concentrations of about 10^{-7} M (Rasmussen 1986) in the face of an extracellular unbound calcium concentration of about 10^{-3} M and an intracellular negative electropotential of perhaps 30–90 mV. The driving force for the entry of the positively charged calcium ion into the cell therefore is enormous – the 10 000-fold concentration gradient of calcium plus the negative intracellular electrical potential. The total energy expenditure required by the cell to maintain a resting concentration of 10^{-7} M depends on the *rate* of entry of calcium into the cell as well as the extremely large electropotential difference against which intracellular calcium must be extruded.

Figure 15.1 presents a summary of mechanisms that are involved in the regulation of calcium by a typical cell. The calcium ion can enter the cell by three general mechanisms, shown on the left aspect of the figure. Calcium can enter the cell (a) by simple or facilitated diffusion, the latter implying the involvement of a membrane translocator that facilitates transmembrane movement; (b) through a calcium channel activated by the binding of an agonist to its receptor; or (c) via a voltage-dependent calcium channel activated by depolarization of the cell.

The mechanisms of extrusion of the calcium ion are represented on the right of Fig. 15.1. Shown are: (a) a calcium pump, depending upon the hydrolysis of ATP to provide the energy to drive calcium uphill against the considerable energy gradient; and (b) a Na^+/Ca^{2+} exchange mechanism, with the energy for extruding calcium derived from the potential energy associated with the sodium gradient maintained by the sodium pump.

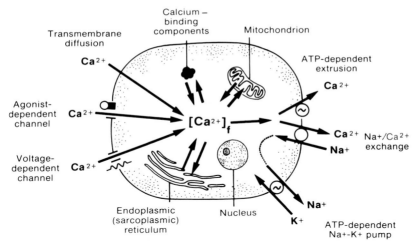

Fig. 15.1. Cellular control of cytosolic free calcium concentrations. The cytosolic free calcium concentration $[Ca^{2+}]_f$, is a function of four mechanisms: the rate of influx by diffusion across the plasma membrane; the passage through agonist-dependent and voltage-dependent channels; the rate of efflux by ATP-dependent extrusion (the calcium pump) and Na^+/Ca^{2+} exchange; and the intracellular sequestration by endoplasmic reticulum, mitochondria or binding components.

Within the cell proper, calcium can be sequestered by the endoplasmic reticulum (or sarcoplasmic reticulum in muscle cells), by mitochondria or by binding to high-affinity molecules, such as proteins and acidic carbohydrates. Calcium within these intracellular stores is available for release under the appropriate conditions.

Modulation of Intracellular Free Ca^{2+} Concentrations

The enhancement of cytosolic free calcium concentrations in a hormone-responsive tissue could theoretically be due to several different mechanisms. Referring again to Fig. 15.1 they are (a) an increase in Ca^{2+} entry by activation and opening of the receptor-operated calcium channels or voltage-dependent channels; (b) release of Ca^{2+} from intracellular membrane-bound stores; (c) inhibition of the calcium extrusion systems, the calcium pump and the Na^+/Ca^{2+} exchange mechanism; and (d) release of Ca^{2+} from intracellular binding sites. Each of these mechanisms has been proposed to operate in different tissues in response to various Ca^{2+}-mobilizing agonists, and this chapter includes some of the evidence that links receptor occupancy by calcium-mobilizing agonists to membrane-associated events and the subsequent elevation of intracellular free calcium levels.

The return of Ca^{2+} concentrations to resting levels after stimulation is brought about in essence by the reversal of these events, i.e. by the release of hormone from its receptor, the destruction of intracellular second messengers, the active extrusion of Ca^{2+} from the cell and the sequestration of Ca^{2+} by intracellular organelles and binders.

Calmodulin and Other Calcium-Binding Proteins

Certain hormones and other agonists influence target cells by binding to so-called calcium-mobilizing receptors. After they occupy these receptors, the concentration of intracellular free calcium increases and thereby signals a biological response. A rise in intracellular calcium does not usually trigger a response itself but mediates a response through calcium receptors or calcium-binding proteins. The most significant of these proteins in terms of diversity of its effects is calmodulin (Cheung 1970, 1980; Means and Dedman 1980; Means et al. 1982). Calmodulin is present in most or all tissues of vertebrates, invertebrates and plants and in some unicellular organisms. Upon binding calcium, calmodulin undergoes a conformational change, and is then capable of interacting with, and stimulating the activity of, various enzymes and other macromolecular processes (Table 15.2).

Table 15.2. Some calcium–calmodulin-dependent enzymes and functions (adapted from Cheung 1980; Manalan and Klee 1984; Means et al. 1982)

Enzymes	Function
Adenylate cyclase	Neurotransmitter release
Phosphodiesterase	Membrane phosphorylation
Phospholipase A_2	Microtubule assembly
Ca^{2+} ATPase	
Myosin light chain kinase	
Phosphorylase kinase	
Guanylate cyclase	
Ca^{2+}-dependent protein kinase	
NAD kinase	

Calmodulin, with a molecular weight of about 16 700, is an acidic, heat-stable protein that binds four calcium ions with an association constant of about 10^6 M^{-1}. This association constant is in the appropriate range for calmodulin to serve as a modulator when a stimulus results in an increase in the intracellular level of cytosolic free Ca^{2+} from the resting level of about 0.1 μM up to 0.5 μM or higher.

The amino acid sequence of calmodulin has been determined by conventional protein-sequencing techniques and by sequence analysis of calmodulin cDNAs (Watterson et al. 1980; Sasagawa et al. 1982; Putkey et al. 1983). Kretsinger (1980) deduced from the sequences of calmodulin and other high-affinity calcium-binding proteins the putative calcium-binding sites. Crystallographic studies (Kretsinger 1980) and theoretical analysis of the sequence for β-turns by the method of Chou and Fasman (1979) gave supportive evidence that the Ca^{2+}-binding region lies in a "pocket" formed by the fore and aft α-helical regions, which Kretsinger designated the "E–F hand" configuration.

The conformational change in calmodulin upon binding Ca^{2+} is revealed by different techniques, one of which is measuring the change in electrophoretic mobility in sodium dodecyl sulphate-polyacrylamide gels (Burgess et al. 1980) and another is studying calcium effects using nuclear magnetic resonance (Krebs and Carafoli 1982; Ikura et al. 1983). Further, the anti-psychotic drugs, the phenothiazines, avidly bind to calmodulin in a Ca^{2+}-dependent manner,

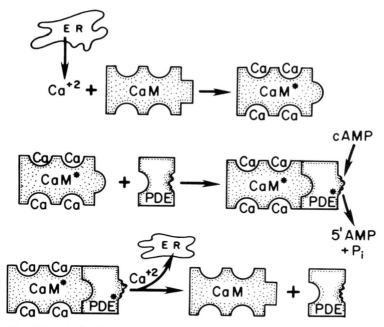

Fig. 15.2. Activation of phosphodiesterase by calcium and calmodulin. Calcium, released from an intracellular compartment, raises intracellular free calcium concentration from about 10^{-7} M to about 10^{-6} M. The calcium ions bind to calcium-binding sites on calmodulin (*CaM*), resulting in a conformational change in the protein. This alteration in calmodulin structure exposes a region of the protein capable of binding a phosphodiesterase (*PDE*). The association of CaM and PDE activates the enzyme, stimulating the degradation of cAMP to 5' AMP and phosphate (P_i). A decrease of intracellular free Ca^{2+} by sequestration and extrusion reverses the process. *ER*, endoplasmic reticulum. (* signifies activation.)

providing additional evidence of a Ca^{2+}-induced change in conformation (Weiss et al. 1980; Head et al. 1982).

The Ca^{2+}-induced conformational change in calmodulin is a prerequisite to its binding with, and activating of, target enzymes (Fig. 15.2). Ca^{2+} binds to the protein in an ordered step-wise fashion and induces sequential conformational changes in calmodulin (Klee and Haiech 1980; Demaille 1982). The number of bound calcium ions required for activation of a given enzyme might differ, but

Table 15.3. Some high-affinity intracellular calcium-binding proteins (other than calmodulin) (adapted from Manalan and Klee 1984)

Protein	Proposed function
Parvalbumin	Calcium buffer
Regulatory myosin light chain	Regulation of contraction
Troponin C	Regulation of contraction
Calbindin-D (vitamin D-induced calcium-binding protein)	Epithelial calcium transport; calcium buffer
Calcineurin B	Protein dephosphorylation
S-100	Regulation of cytoskeleton
Oncomodulin	Unknown; present in tumour cells

for calmodulin-stimulated phosphodiesterase, activation occurs when the protein contains three or four calcium ions (Manalan and Klee 1984).

Other proteins with high-affinity calcium-binding sites are known, but they have more specific tissue localizations and are not known to be as universally distributed as calmodulin. Some calcium-binding proteins and their putative functions are listed in Table 15.3. The exact functions of some of them are unknown, but the direct involvement of troponin C in the contraction of muscle has been appreciated for some time (Kretsinger 1980).

The Phosphoinositide Cycle

The association of a calcium-mobilizing agonist with its receptor activates a phosphodiesterase, phospholipase C (Table 15.4). This enzyme preferentially hydrolyses the inositol-containing phospholipids. The observations that first suggested a relationship between hormonal effects and phosphatidylinositol turnover were those of Hokin and Hokin (1953, 1954, 1955). When the neurotransmitter acetylcholine (or carbamylcholine) stimulated enzyme secretion by the pancreas, it was accompanied by a tenfold increase in the incorporation of ^{32}P into phospholipids. Studies undertaken after methods of identifying phospholipids became available showed that the most rapidly and extensively labelled phospholipids upon agonist stimulation were phosphatidylinositol (PI) and phosphatidic acid (PA) (Hokin 1985). The hormone-stimulated labelling of phosphatidylinositol also occurs in the brain, sympathetic ganglia, pineal gland, parotid gland and smooth muscle (Abdel-Latif 1986). Subsequent studies demonstrated that PI actually decreases with hormone stimulation (Hokin-Neaverson 1974), and the earlier observed increase in the ^{32}P labelling of PI and PA reflected an increase in their turnover.

Phosphoinositides present in membranes include PI and phosphorylated derivatives of PI. These derivatives include two polyphosphoinositides, phosphatidylinositol-4-phosphate (PIP) and phosphatidyl-4,5-bisphosphate (PIP$_2$),

Table 15.4. Hormones, neurotransmitters and growth factors that stimulate phospholipase C (adapted from Anderson and Salomon 1985)

Agent	Cell type
Acetylcholine	Exocrine tissues, smooth muscle, neural tissues
α-1-adrenergic agonists	Exocrine tissues, neural tissues, liver, adipose tissues
Bradykinin	Porcine aortic endothelial cells
Angiotensin II	Liver
Vasopressin	Rat mammary tumour epithelial cells, aorta, liver
TRH	GH$_3$ pituitary cells
TSH	Thyroid
ACTH	Adrenal cortex
Insulin	Adipose tissue
Thrombin, PAF	Platelets
EGF	A431 Human epidermoid carcinoma cells, murine 3T3 cells
PDGF	Murine 3T3 cells
NGF	Sympathetic ganglia

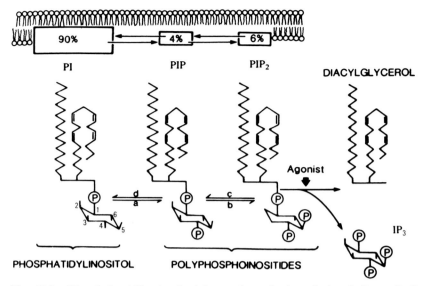

Fig. 15.3. Phosphoinositides involved in agonist activation of phospholipase C. Depicted are schematic representations of phosphatidylinositol (*PI*), phosphatidylinositol-4-phosphate (*PIP*), phosphatidylinositol-4,5-diphosphate (*PIP$_2$*), myo-inositol-1,4,5-trisphosphate (*IP$_3$*) and diacylglycerol. At the *top* of the figure are the relative percentage distributions of the phosphoinositides. (Modified from Berridge 1985 and reproduced by permission of author and publisher)

which are present as minor components of cellular membranes (Fig. 15.3). It is these polyphosphoinositides that are the preferred substrates of the phospholipase C enzyme.

The formation of the polyphosphoinositides is a consequence of phosphorylation of PI by ATP in the presence of specific kinases at the plasma membrane to

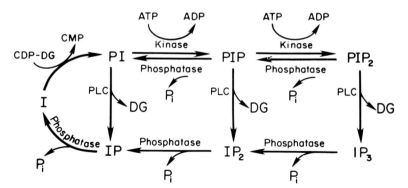

Fig. 15.4. Formation and degradation of the phosphoinositides. *PI*, phosphatidylinositol; *PIP*, phosphatidylinositol-4-phosphate; *PIP$_2$*, phosphatidylinositol-4,5-bisphosphate; *PLC*, phospholipase C; *DG*, diacylglycerol; *IP$_3$*, myo-inositol-1,4,5-trisphosphate; *IP$_2$*, myo-inositol-4,5-bisphosphate; *IP*, myo-inositol-1-phosphate; *P$_i$*, inorganic phosphate; *CDP-DG*, cytidyl diphosphate diacylglycerol; *CMP*, cytidyl monophosphate. (Modified from Abdel-Latif 1986 and reproduced by permission of the American Society for Pharmacology and Experimental Therapeutics)

form PIP and subsequently to form PIP$_2$ from PIP (Fig. 15.4). These reactions are reversible through the hydrolytic activity of specific phosphatases to form PIP from PIP$_2$ and PI from PIP. Phosphoinositol is also hydrolysable by phospholipase A$_2$ to form phosphatidic acid (PA) and free fatty acid, usually arachidonic acid, from carbon-2 of the glycerol backbone of PI. Arachidonic acid is the precursor for the biosynthesis of a number of stimulatory eicosanoids, the prostaglandins, thromboxanes and prostacyclins.

G Proteins Linking Receptors to Enzyme Activation

Hormone binding to receptor is presumed to elicit a conformational change in the receptor, and this rearrangement of the receptor molecule constitutes the way that a signal is transmitted from the external surface to internal effector molecules. In the adenylate cyclase system, the transmembrane hormonally induced signal modifies the behaviour of a G protein complex that lies subjacent to the receptor (Taylor and Merritt 1986). The G proteins are so named because their activity is dependent on guanine nucleotide binding. Two general types of G proteins in the adenylate cyclase system have been described. One form (G$_s$) is involved in stimulating adenylate cyclase, and the other (G$_i$) is inhibitory. The G proteins are comprised of three subunits, designated alpha, beta and gamma (Fig. 15.5). The beta–gamma subunits of the G$_s$ and G$_i$ proteins appear to be similar. The G proteins are activated when the GDP bound to the alpha subunit in the "resting" state is replaced by GTP, resulting in a disassociation of the alpha subunit from the beta–gamma complex. The alpha$_s$ subunit–GTP complex

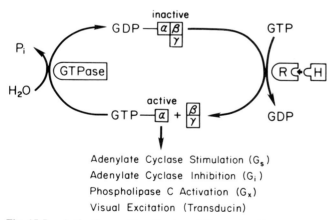

Adenylate Cyclase Stimulation (G$_s$)
Adenylate Cyclase Inhibition (G$_i$)
Phospholipase C Activation (G$_x$)
Visual Excitation (Transducin)

Fig. 15.5. Activation and inactivation of G proteins. Receptor (*R*) binding of hormone (*H*) causes the displacement of guanosine diphosphate (*GDP*) by guanosine triphosphate (*GTP*) from the alpha subunit of the G protein. The GTP–alpha subunit dissociates from the beta and gamma subunits and can stimulate or inhibit biological reactions. The innate guanosine triphosphatase (*GTPase*) associated with the alpha subunit hydrolyses GTP to form GDP, inactivating the alpha subunit, which reassociates with the beta and gamma subunits, reforming the inactive G protein trimer. (Adapted from Stryer and Bourne 1986 and reproduced by permission of the author and the Annual Review of Cell Biology)

is apparently the direct stimulator of adenylate cyclase in the synthesis of cAMP from its precursor, adenosine triphosphate (ATP). An intrinsic GTPase hydrolyses GTP to GDP, whereupon the alpha subunit reassociates with the beta–gamma subunit, reconstituting the inactive holo-G_s trimer.

A similar series of events pertains to the activation of the G_i regulatory protein, in which $alpha_i$ disassociates from the other subunits upon binding GTP. However, the mechanism by which $alpha_i$–GTP inhibits adenylate cyclase has not been exactly defined (review by Stryer and Bourne 1986). One proposal suggests that the inhibition is due to the increased concentration of the beta–gamma complex of the G protein formed after $alpha_i$ disassociates. The free beta–gamma complex might regulate adenylate cyclase activity by binding, and thereby decreasing the availability of the $alpha_s$ subunit. There is other evidence that the alpha subunit of G_i directly inhibits the activation of adenylate cyclase and that the beta–gamma subunit might also directly inhibit cAMP formation (Stryer and Bourne 1986).

Cholera toxin exerts its effect on intestinal fluid secretion by modifying the operation of the G_s protein associated with adenylate cyclase activation. The toxin, upon binding to a cell surface receptor, catalyses, through one of its subunits, the ADP-ribosylation of the $alpha_s$ subunit of G_s. This reaction inhibits GTPase activity of the $alpha_s$ subunit, thereby maintaining this subunit in its activated state. The "excess" cAMP produced in the presence of toxin is considered responsible for altering electrolyte and water transport by the intestine and causing the life-threatening diarrhoea of cholera (Fishman 1980).

The involvement of a G protein in the activation of phospholipase C by Ca^{2+}-mobilizing agonists was established by examining the effect of non-hydrolysable analogues of GTP on permeabilized cells and on isolated membrane preparations from various cell types. As reviewed by Abdel-Latif (1986), these analogues duplicated the physiological effect of Ca^{2+}-dependent agonists, releasing serotonin from permeabilized platelets and stimulating PIP_2 hydrolysis by membrane preparations from hepatocytes, smooth muscle cells, neutrophils and other isolated membranes. In certain cases, the presence of GTP in the membrane preparation was required for hormone activation of phospholipase C. In other systems, hormone and GTP acted synergistically.

Additional evidence for a G protein involvement in phospholipase C activation was derived from the use of pertussis toxin, which inhibits adenylate cyclase activity through interaction with G_i. In some cells pertussis toxin inhibits the action of Ca^{2+}-mobilizing agonists. A case in point is the response of mast cells. Prior treatment of mast cells with this toxin inhibits histamine release (Nakamura and Ui 1985; Abdel-Latif 1986; Stryer and Bourne 1986). Pertussis toxin also promotes the ADP-ribosylation of a protein of the same molecular size as the alpha subunit of G_i.

In other cells, pertussis toxin does not inhibit the effect of Ca^{2+}-mobilizing hormones on PIP_2 hydrolysis, and more recent evidence from some systems suggests that the G protein linking Ca^{2+}-mobilizing receptor to phospholipase C is distinct from G_i, the adenylate cyclase-inhibitor protein. For example, Merritt et al. (1986) studied the effect of caerulein (acting at the cholecystokinin receptor) and carbachol (acting at the muscarinic cholinergic receptor) on the formation of one product of phospholipase C activity, inositol-1,4,5-trisphos-phate (IP_3), in electrically permeabilized rat pancreatic acinar cells. In the presence of either agonist and GTP or non-hydrolysable analogues of GTP, specifically GDP(S) [guanosine 5'-(beta-thio) diphosphate] and p(NH)ppG [guanosine 5'-(beta, gamma-imido)triphosphate], a dose-dependent increase in IP_3 formation was seen. In the absence of guanosine nucleotides, there was little or no effect, which implicated a G protein in the transduction process. Significantly, neither cholera nor pertussis toxin inhibited these agonist-stimulated reactions, implicating a G protein distinct from the G_s and G_i proteins.

The GTPase activity associated with GH_3 rat anterior pituitary-cell membranes was assessed by Wojcikiewicz et al. (1986). The effect of thryrotropin-releasing hormone (TRH), which increases GTPase activity, was not inhibited by cholera or pertussis toxin, suggesting that TRH, known to stimulate phospholipase C, operates through a G protein different from G_s and G_i.

Although there are still gaps in our knowledge of the various G proteins, studies on some systems clearly indicate that the G protein linked to phospholipase C is different from G proteins linked to adenylate cyclase (Joseph 1985; Michell and Kirk 1986). The complexity of the family of G proteins was revealed by recent studies on their properties and molecular biology, as summarized by Bourne (1986).

Phospholipase C Activation and Phosphoinositide Second Messengers

Receptor occupancy by a calcium-mobilizing agonist, operating via a G protein, activates phospholipase C. Phospholipase C, as suggested by a number of studies, catalyses the hydrolysis of each of the three inositol lipids, although the formation of IP_2 from PIP and IP_3 from PIP_2 precedes the fall of PI (Fig. 15.3). The difference in the rates of hydrolysis of the phosphoinositides could be due to the ease of access of the specific inositol lipids to the active site of hormone-activated phospholipase C. The difference could also be related to the differences in the Ca^{2+} requirements of phospholipase C for the hydrolysis of PI and the polyphosphoinositides PIP and PIP_2. For the hydrolysis of PI, the K_m of the enzyme is about 1 μM, whereas the hydrolysis of PIP_2 is met by resting levels of Ca^{2+} of around 0.1 μM (Williamson 1986). The Ca^{2+} for the stimulation of PI hydrolysis by phospholipase C might be derived from the release of Ca^{2+} from intracellular stores by inositol-1,3,5-trisphosphate (IP_3).

The immediate substrate for hormone-dependent phospholipase C activity appears to be PIP_2 which, upon hydrolysis, yields two substances that serve as second messengers, diacylglycerol and IP_3 (Takai et al. 1984; Nishizuka et al. 1984; Nishizuka 1986). The IP_3 generated by hormone-activated phospholipid C effects the release of Ca^{2+} from an intracellular store, raising intracellular Ca^{2+} levels. Diacylglycerol is the endogenous activator of protein kinase C. These and subsequent reactions are shown schematically in Fig. 15.6. The Ca^{2+} released by IP_3 elevates cytosolic free Ca^{2+} to concentrations appropriate to the activation of a calmodulin-sensitive protein kinase and other Ca^{2+}-dependent reactions (Berridge 1984a; Berridge and Irvine 1984). The phosphorylation of enzymes or other proteins by the catalytic activity of these kinases yields covalently modulated proteins that result in activation or inactivation.

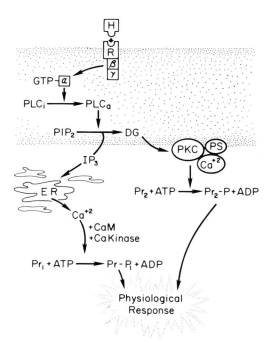

Fig. 15.6. Formation of second messengers by Ca^{2+}-mobilizing agonists. Hormone (*H*) or other agonists binds to receptor (*R*), which activates the G protein by inducing GTP binding to the alpha subunit. The GTP–alpha subunit stimulates phospholipase C activity which, in turn, hydrolyses PIP_2 to the two second messengers, inositol-1,4,5-trisphosphate (*IP₃*) and diacylglycerol (*DG*). IP_3 interacts with an intracellular Ca^{2+} store (endoplasmic reticulum), releasing Ca^{2+} and elevating the intracellular free Ca^{2+} concentration. Through calmodulin (*CaM*), the Ca^{2+}-dependent protein kinase catalyses the phosphorylation of appropriate substrates, the latter being responsible for a biological response. Diacylglycerol, in the presence of phosphatidylserine and Ca^{2+}, activates another kinase, protein kinase C (*PKC*), which catalyses the phosphorylation of proteins and also leads to a biological response. The second messengers, IP_3 and DG, may act synergistically or independently.

The production of aldosterone by bovine adrenal glomerulosa cells in response to angiotensin II provides us with a useful example of the interplay between the two second messengers generated by phospholipase C activation (Kojima et al. 1984; Rasmussen et al. 1984; Rasmussen 1986). As depicted in Fig. 15.7, the addition of angiotensin II to adrenal cells increased the rate of production of aldosterone monotonically after a delay of about 9 min, followed by a plateau over the period of observation. The addition of a calcium ionophore to the preparation yielded a transient rise in aldosterone production which peaked at about 20 min. The phorbol ester yielded a different pattern of response. With this activator of protein kinase C, there was no immediate effect, but after about 20 min, a very slow increase in aldosterone production occurred. Together, the calcium ionophore and the phorbol ester produced a response amazingly similar to the effect of angiotensin II itself, illustrating synergism between the Ca^{2+} and protein kinase C pathways and, in addition, the temporal integration of a physiological response.

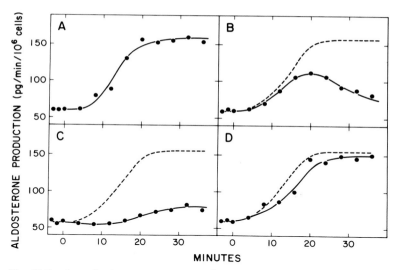

Fig. 15.7. Interplay between cytosolic Ca^{2+} and protein kinase C activation in the production of aldosterone by adrenal glomerulosa cells. *Panel A* depicts the effect of the hormone angiotensin II on aldosterone production, *panel B* the effect of the calcium ionophore A23187, *panel C* the effect of the phorbol ester TPA and *panel D* the response to calcium ionophore and phorbol ester together. Note that the early effect of the ionophore on aldosterone production was not sustained. The phorbol ester effected a response after a significant delay. A23187 and TPA essentially duplicated the effect of hormone, illustrating temporal and integrative control of physiological process. (From Rasmussen et al. 1984 and reproduced by permission of author and Raven Press)

Hormone-Stimulated Ca^{2+} Fluxes

Michell (1975) initially suggested a relationship between the hydrolysis of PI and the generation of a calcium signal. The calcium signal would take the form of an increased concentration of cytosolic free Ca^{2+}, and the source of the Ca^{2+} could be the external bathing medium or intracellular sources or both. It was proposed that a product of PI metabolism increased the permeability of the plasma membrane to calcium, that a "calcium gate" was opened. Thus the elevated intracellular Ca^{2+} due to hormone stimulation was from the rapid influx of Ca^{2+}. Subsequent studies have modified this notion.

Ca^{2+} from Intracellular Stores

Studies in which IP_3 was added to permeabilized cells provided substantial evidence that the polyphosphoinositide was responsible for the release of Ca^{2+} from internal stores (Streb et al. 1983; Thomas et al. 1984; Suematsu et al. 1984). The intracellular organelle now considered to be the source of Ca^{2+} is the endoplasmic reticulum (ER) or a component thereof (Streb et al. 1983; Berridge and Irvine 1984). Fractionation studies demonstrated that fractions containing ER released Ca^{2+} upon addition of IP_3 (Streb et al. 1984). In a more recent report, Ueda et al. (1986) demonstrated that IP_3 stimulated the release of Ca^{2+} from a microsomal preparation from neuronal cells, but the plasma membrane fraction did not respond in the same fashion. The mitochondrion was eliminated as the IP_3-sensitive organelle since inhibitors of mitochondrial calcium uptake do not prevent the subsequent release of Ca^{2+} from intracellular stores by IP_3 (e.g. Streb et al. 1983).

The location of the IP_3-releasable Ca^{2+} pool is not precisely known, but evidence summarized by Putney (1986) led to the model that follows (Fig. 15.8). In order to account for various observations, Putney proposed that a component of the endoplasmic reticulum that is affected by IP_3 is located close to the plasma membrane. In this way extracellular Ca^{2+} could directly enter this IP_3-responsive pool. The controlling element in this model is Ca^{2+} itself. In the basal state, sufficient Ca^{2+} is present in the intermembrane space between the plasma membrane and the membrane of the specialized compartment of the endoplasmic reticulum to maintain Ca^{2+} channels in the plasma membrane in the closed state. After Ca^{2+} is released from the Ca^{2+} pool by IP_3, the

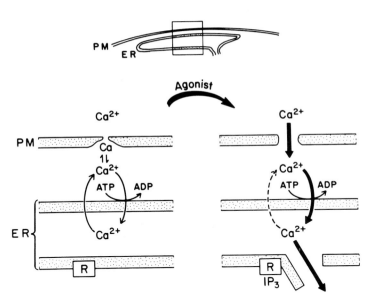

Fig. 15.8. A mechanism by which Ca^{2+} release from endoplasmic reticulum could activate Ca^{2+} entry. Presumably this occurs in regions where plasma membrane (*PM*) and endoplasmic reticulum (*ER*) are in physical apposition (*top*). In the region between PM and ER in the resting condition (*left*), the balance of leak and ATP-dependent uptake of Ca^{2+} by the ER maintains cytosolic free Ca^{2+} in the 100–200 nM range, under which condition Ca^{2+} is bound to sites, restricting Ca^{2+} entry. Following the addition of an agonist (*right*) IP_3 binds to its receptor (*R*), and Ca^{2+} is released to the cytosol. With Ca^{2+} in the ER decreased, net Ca^{2+} transport from the region between ER and PM occurs. Ca^{2+} in this region is decreased, leading to the dissociation of Ca^{2+} from the sites restricting Ca^{2+} entry. Ca^{2+} channels (or carriers) are activated. Because of the net driving force moving Ca^{2+} from the extracellular space to the cytosol, net Ca^{2+} influx will occur as long as IP_3 is present to maintain a low level of Ca^{2+} in the ER. (From Putney 1986 and reproduced with permission of authors and publisher)

ATP-dependent uptake of calcium by the endoplasmic reticular membrane replenishes the Ca^{2+} store and also depletes the Ca^{2+} in the intermembrane space. Lowering the Ca^{2+} concentration in this intermembrane region opens Ca^{2+} gates, allowing external Ca^{2+} to enter the hormone-responsive intracellular storage pool. This model is particularly appealing because some Ca^{2+}-mobilizing hormones are known to increase Ca^{2+} fluxes from both the endoplasmic reticulum and from the external milieu. Part of the evidence to support the model came from Poggioli et al. (1985) who reported that lowering intracellular Ca^{2+} in rat hepatocytes promotes Ca^{2+} entry. Putney's scheme does require evidence of a close physical association between the two membranes, and such an association was observed in smooth muscle and other cell types (Devine et al. 1971).

The mechanism by which IP_3 promotes the release of Ca^{2+} from the endoplasmic reticular compartment is not known. There is evidence that there is an IP_3 receptor on intracellular membranes of bovine adrenal cortex cells, hepatocytes and neutrophils. The receptor displays saturation and, for bovine adrenal cortex microsomes, high-affinity and low-capacity binding (Abdel-Latif 1986). Smith et al. (1985) provide information that IP_3 in cultured vascular smooth muscle cells activates a Ca^{2+} channel. Similarly, the activation of a Ca^{2+} channel in the rough ER membranes from hepatocytes by IP_3 was reported by Muallem et al. (1985). All this suggests that IP_3, by its interaction with a receptor on the ER membrane, opens a calcium channel through which intravesicular calcium is released.

Using photoaffinity-labelled phosphoinositides, Hirata et al. (1985) showed that IP_3, but not IP_2, binds to intracellular membranes. This observation indicates a certain degree of specificity of the IP_3 receptor and suggests a mechanism for the inactivation of the endoplasmic reticular calcium channel. Hydrolysis of membrane-bound IP_3 by a phosphoinositide phosphatase to IP_2 would result in disassociation of the IP_2 from the receptor and, hence, channel closure.

Ca^{2+} Influx

The hormone-dependent formation of IP_3 from PIP_2 hydrolysis causes a rapid and transient increase in cytosolic free Ca^{2+} following the release of Ca^{2+} from intracellular stores. This Ca^{2+} can be quickly sequestered by reaccumulation by ER, as well as the extrusion by the ATP-dependent calcium pump and the Na^+/Ca^{2+} exchanger located in the plasma membrane (Fig. 15.1). In order to sustain hormone-mediated, calcium-dependent physiological responses, other mechanisms come into play to maintain elevated Ca^{2+} levels.

The requirement of an increased calcium influx to maintain intracellular Ca^{2+} levels at an "excitatory" state was documented by Exton (1984) and Williamson et al. (1986). In a study reported by Williamson et al. (1986) intact rat hepatocytes were stimulated by vasopressin and changes in cytosolic free Ca^{2+} were monitored by the quin2 method. When the cells were incubated in Ca^{2+}-deficient buffer mixture (due to the presence of EGTA), there was an initial rapid elevation of intracellular free Ca^{2+}, but the Ca^{2+} concentration returned to baseline in about 4 min (Fig. 15.9). When the cells were incubated in a buffer solution containing unbound Ca^{2+}, there was a similar rapid increase in cytosolic Ca^{2+} which remained elevated over the entire period of observation. This, plus the additional evidence that Ca^{2+} influx from extracellular fluid is greater after vasopressin stimulation than in its absence, substantiates an effect of hormone on Ca^{2+} entry across the plasma membrane. Supporting this concept are the observations of Mauger et al. (1984), who showed that various Ca^{2+}-mobilizing hormones (vasopressin, noradrenaline and angiotensin II) increase Ca^{2+} influx from extracellular sources within 30 s of adding hormone to isolated rat hepatocytes.

It has been known for some time that epinephrine binding to the α-1-adrenergic receptors (Gill 1985) increases influx of Ca^{2+} and stimulates contraction of smooth muscle cells. Plasma membranes of smooth muscle cells contain voltage-dependent Ca^{2+} channels, but studies with Ca^{2+}-channel

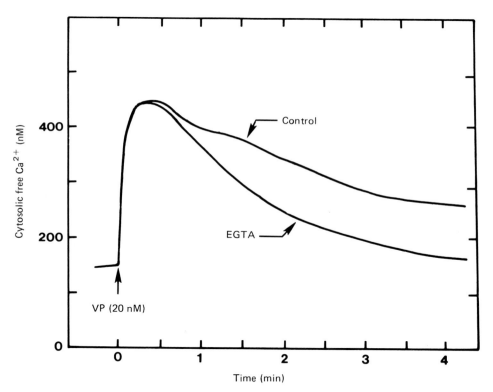

Fig. 15.9. Effect of a calcium-mobilizing hormone (vasopressin) on release of intracellular Ca^{2+} and increase in the influx of Ca^{2+} in hepatocytes. The hepatocytes were incubated in the presence of a buffer solution containing 1.3 mM Ca^{2+}. In the treated cells, EGTA at a concentration of 1.4 mM was added 30 s before vasopressin to complex most of the buffer Ca^{2+}. In the presence of EGTA, vasopressin caused a rapid increase in intracellular Ca^{2+} which fell to about base-line values at 4 min, indicating the mobilization of intracellular Ca^{2+}. In the absence of EGTA, the early, rapid increase in intracellular Ca^{2+} was similar, but Ca^{2+} levels remained elevated over the period of observation, indicative of the influx of extracellular Ca^{2+} in addition to Ca^{2+} from intracellular stores. (Redrawn from Williamson et al. 1986 and reproduced with permission of the author and Plenum Press)

blockers suggest that the α-1-receptor activates a receptor-operated channel distinct from the voltage-operated one (Hurwitz 1986; Hughes et al. 1986). Occupancy of the α-1-receptor by an agonist also stimulates contraction in the absence of available external Ca^{2+}, indicating the additional release of intracellular Ca^{2+} (Smith et al. 1985).

These and other studies (Gill 1985; Williamson et al. 1985; Abdel-Latif 1986) clearly substantiate an effect of hormones on the permeation of Ca^{2+} across the plasma membrane from extracellular fluid. The contribution of extracellular Ca^{2+} to the cytosolic free Ca^{2+} pool might be small relative to the contribution from intracellular stores, but it is sufficient to prolong the hormone-stimulated rise in intracellular Ca^{2+}.

The mechanism by which the influx of Ca^{2+} is hormonally controlled is not known. Phosphatidic acid (PA) has been proposed as a possible endogenous calcium ionophore because of its lipophilic properties and ability to bind calcium. But this proposal is controversial (Holmes and Yoss 1983). As mentioned previously, IP_3 does not itself appear to stimulate Ca^{2+} fluxes across the plasma membrane as it does across endoplasmic reticular membranes. A reasonable proposal is that certain calcium-mobilizing receptors are closely associated with plasma-membrane calcium channels and, when occupied by their specific agonists, cause channel opening by conformational modulation of the channel proteins. Kinase-catalysed phosphorylation of the channel proteins is another proposed mechanism of channel activation (Hurwitz 1986).

Inhibition of the Ca^{2+} Pump

Another effect of hormone action on liver cells that helps maintain higher concentrations of cytosolic free Ca^{2+} is an apparent inhibition of the activity of the plasma-membrane calcium extrusion pump (Lin et al. 1983; Williamson et al. 1985, 1986; Abdel-Latif 1986). The results of an experiment demonstrating this effect are depicted in Fig. 15.10. Isolated hepatocytes incubated in a low Ca^{2+} (about 10 μM) buffer were stimulated with vasopressin, and changes in cytosolic Ca^{2+} were monitored with quin2. The rate of efflux was determined with another Ca^{2+}-sensitive indicator, arsenazo III, present in the extracellular buffer. Vasopressin elicited a rapid increase in intracellular Ca^{2+} levels, followed by a slower decrease. After a delay of 10 s, the extrusion of Ca^{2+} from the cell increased, peaking at about 60 s. This delay has been attributed to the transient inhibition of the plasma membrane Ca^{2+} pump. The decrease in cytosolic Ca^{2+} after initial stimulation can be attributed to the efflux of Ca^{2+} from the cell, reaccumulation by endoplasmic reticulum and possibly uptake by mitochondria. The mechanism responsible for this inhibitory effect might be related to a decrease in the PIP_2 content of the plasma membrane due to hormone action, since PIP_2 is a possible activator of the calcium pump.

In any event, the rapid rise of cytosolic free Ca^{2+} and its sustained elevation after hormone treatment of target cells might be due to: (a) IP_3 formation and the release of Ca^{2+} from intracellular stores; (b) the increased influx of Ca^{2+} across the plasma membrane; and (c) a decrease in the efflux of cytosolic Ca^{2+} via inhibition of the ATP-dependent plasma membrane Ca^{2+} pump.

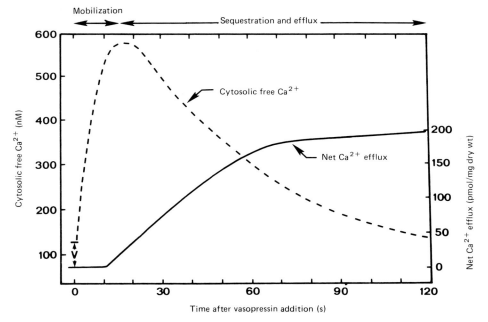

Fig. 15.10. Effect of a calcium-mobilizing hormone (vasopressin) on the release of intracellular Ca^{2+} and a transient delay of Ca^{2+} efflux in hepatocytes. Hepatocytes were incubated in a buffer mixture containing a low concentration of Ca^{2+} (about 10 μM) and, at zero time, vasopressin (10 μM) was added. The rapid and transient elevation of cytosolic Ca^{2+} was primarily derived from intracellular stores. The delay and slow rise in Ca^{2+} efflux, measured with external arsenazo III, is considered indicative of a transient inhibition of the plasma calcium pump. (Redrawn from Williamson et al. 1986 and reproduced with permission of the author and Plenum Press)

Diacylglycerol and Protein Kinase C

Diacylglycerol, the other product of hormone-stimulated PIP_2 hydrolysis, is an endogenous activator of protein kinase C (PKC) (see Fig. 15.6). The activity of PKC is dependent on Ca^{2+}, which is further enhanced by the presence of phospholipids, especially phosphatidylserine (Nishizuka et al. 1984). Protein kinase C is a single polypeptide chain that, by SDS-PAGE, has a molecular weight of about 77 000, has an isoelectric point of about pH 5.6, an optimal pH in the 7.5–8.0 range and requires Mg^{2+}. The molecule has two functional domains, a hydrophobic one that binds PKC to membranes and a hydrophilic domain that has the catalytic site. Presumably the three enzyme modulators, Ca^{2+}, phosphatidylserine and diacylglycerol change the conformation of the enzyme, exposing the active catalytic site. Activation can also be accomplished by the limited proteolysis by Ca^{2+}-dependent neutral proteases, the calpains.

A significant effect of diacylglycerol on PKC is to increase its binding affinity for Ca^{2+}, so the Ca^{2+} requirement of the enzyme can be met by resting levels of the divalent cation. In its absence, much higher concentrations of Ca^{2+} are required for PKC activation.

Protein kinase C is a ubiquitous enzyme, found in all tissues and organs of mammals examined thus far (Nishizuka et al. 1984; Nishizuka 1986). Relatively large concentrations occur in platelets, brain tissue and vas deferens muscle; smaller ones in liver, adipocytes and skeletal muscle.

Some of the potential roles of protein kinase C in physiological responses are summarized in Table 15.5. The protein kinase C pathway in the activation of biological responses, shown in Fig. 15.6, is distinct from, but often synergistic with, activation through the IP_3–calcium-mobilizing pathway.

Table 15.5. Potential roles of protein kinase C in cellular responses (adapted from Kikkawa and Nishizuka 1986)

Tissues and cells	Responses
Platelets	Serotonin release
Neutrophils	Superoxide generation
Mast cells	Histamine release
Adrenal medulla	Catecholamine secretion
Adrenal cortex	Aldosterone secretion
Pancreatic islets	Insulin release
Pituitary cells	Growth hormone release
Leydig cells	Steroidogenesis
Parotid gland	Protein secretion
Neurons	Neurotransmitter release
Smooth muscle	Contraction
Adipocytes	Glucose transport
Hepatocytes	Glycogenolysis

The stimulation of a cellular response by protein kinase C involves the transfer of phosphate from ATP to a receptor protein. Phosphorylation of an enzyme or another type of protein leads to its activation (and sometimes inhibition); the amino acids phosphorylated by the kinase are serine and threonine residues. A number of possible phosphate acceptor proteins have been identified, and Table 15.6 emphasizes the wide range of physiological and biochemical reactions possibly stimulated or activated by protein kinase C. However, the identification of many of these receptor proteins has come mainly from in vitro studies, and the definitive physiological significance of their activation by protein kinase C requires additional exploration (Kikkawa and Nishizuka 1986). Deactivation of the phosphorylated receptor proteins to the "resting" state is accomplished by dephosphorylation of the protein, a process catalysed by specific protein phosphatases.

One fascinating finding is that tumour-promoting agents, the phorbol esters (e.g. PMA, phorbol-12-myristate-13-acetate and TPA, 12-O-tetradecanoylphorbol-13-acetate), are potent activators of protein kinase C. Serving as a substitute for diacylglycerol (Kikkawa and Nishizuka 1986), phorbol esters substantially increase the affinity of the kinase for Ca^{2+}. These non-hydrolysable tumour agents have been used extensively to investigate the activity and substrates of protein kinase C. However, Kikkawa and Nishizuka (1986) stress that caution is required in interpreting data derived from studies with the phorbol esters. The concentration of diacylglycerol is only transiently increased in response to hormones whereas the non-hydrolysable phorbol esters persist and may distort cell responses. At high concentrations, targets other than protein kinase C may

Table 15.6. Possible acceptor proteins of protein kinase C (adapted from Kikkawa and Nishizuka 1986)

Agonist-receptor proteins
  Epidermal growth factor receptor
  Insulin receptor
  Nicotinic acetylcholine receptor
  β-adrenergic receptor

Membrane proteins
  Ca^{2+} transport ATPase
  Sodium–potassium-ATPase
  Na^+ channel protein
  Glucose transporter
  GTP-binding protein

Contractile and cytoskeletal proteins
  Myosin light chain
  Troponin T and I
  Vinculin
  Microtubule-associated proteins

Enzymes
  Glycogen phosphorylase kinase
  Glycogen synthase
  Myosin light chain kinase

be affected by phorbol esters, and lipophilic phorbol esters could also act by perturbing membrane structure.

Physiological and Biochemical Responses to Ca^{2+}-Mobilizing Agonists in Selected Systems

Table 15.4 lists a wide variety of systems that involve phospholipase C activation. In this section, four systems influenced by Ca^{2+} and affected by Ca^{2+}-mobilizing hormones are briefly discussed: platelet activation; glycogen metabolism by hepatocytes; the epithelial transport of monovalent ions; and cellular growth and proliferation.

Platelets

These blood cells, involved in the blood-clotting mechanism, have frequently been used as a model system in which to study the physiological and biochemical basis of hormone regulation. Activation of platelets by thrombin, collagen, adenosine diphosphate (ADP), platelet-activating factor (PAF) and other substances leads to a "release reaction". Components of the release reaction include aggregation of the platelets, shape change and the secretion of various substances contained in intracellular granules and vesicles. Among these elaborated substances are serotonin (5-hydroxytryptamine), ATP and acid hydrolases; their release has been used experimentally to monitor platelet activation. Other agonists such as adenosine, prostaglandin E_1 and prostaglandin I_2 inhibit platelet activation and appear to modulate platelet function by stimulating adenylate cyclase activity and cAMP formation.

Activation of platelets is a Ca^{2+}-dependent process and the general features of Ca^{2+} metabolism by platelets were reviewed by Rubin (1982). These and other aspects of platelet function and calcium can be found in the papers of Turner and Kuo (1985), Kikkawa et al. (1985), Daniel (1985) and Rink et al. (1985).

The stimulation of hydrolysis of the phosphoinositides in response to thrombin and other platelet activators has been well documented. Diacylglycerol and IP_3 are formed from the degradation of PIP_2 as in other systems (Nishizuka et al. 1984; Majerus et al. 1985; Abdel-Latif 1986), and IP_3 was shown in human platelets to release Ca^{2+} from intracellular stores (Authi and Crawford 1985).

Phospholipase C in thrombin-activated platelets hydrolyses PI and PIP_2 so that more equivalents of diacylglycerol are generated than of IP_3; equimolar amounts of diacylglycerol and IP_3 are expected from PIP_2 hydrolysis alone (Majerus et al. 1985). Since the hydrolysis of PI by phospholipase C depends on intracellular concentrations of Ca^{2+} in the micromolar range ($K_m = 1$–2 μM), this suggests that the initial activation by thrombin results in the hydrolysis of PIP_2; the resultant IP_3 releases Ca^{2+} from internal stores in sufficient amounts to stimulate phospholipase C hydrolysis of PI.

The diacylglycerol produced by phosphoinositide breakdown activates PKC, as well as constituting a source of arachidonic acid substrate for the synthesis of thromboxane A_2 (by cyclooxygenase and thromboxane synthetase) and other eicosanoids. Thromboxane A_2 is another substance, like IP_3, that releases Ca^{2+} from internal stores in platelets (Rink et al. 1985). Arachidonic acid can also arise from the activity of phospholipase A_2, another Ca^{2+}-dependent enzyme that removes fatty acids from carbon-2 of phospholipids. In platelets, phospholipase A_2 is also activated by thrombin (Rittenhouse-Simmons et al. 1977).

Phospholipase C activation of platelets by thrombin is followed by the phosphorylation of at least two proteins, one of molecular weight 40 000 and the other of molecular weight 20 000 (Nishizuka et al. 1984). The protein with molecular weight 40 000 was shown to be a substrate for PKC. The agonist-dependent phosphorylation of the protein with molecular weight 20 000, identified as myosin light chain, is a consequence of the activity of the substrate-specific myosin light chain kinase activated by Ca^{2+} and calmodulin. The duality of agonist effects was delineated by studies done under in vitro conditions. The addition of a permeating diacylglycerol to platelets resulted in the preferential phosphorylation of the protein of molecular weight 40 000, whereas the addition of a calcium ionophore resulted in the preferential phosphorylation of myosin light chain. Other experiments with platelets further demonstrated, and most impressively, that the maximal release of serotonin, an indicator of platelet activation, is related to the phosphorylation of both the 40 000 and 20 000 molecular weight proteins (Fig. 15.11). Although a cause-and-effect relationship between the phosphorylation of these proteins and serotonin release cannot be established absolutely, the evidence for an interdependency is seemingly apparent.

Activation of platelets by collagen is a significant part of the blood-clotting mechanism when collagen is exposed as a result of tissue injury. Collagen

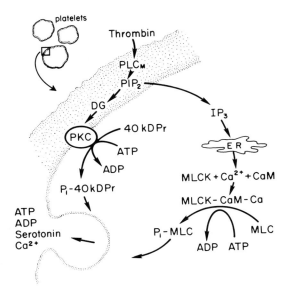

Fig. 15.11. Protein phosphorylation associated with the thrombin-induced "release reaction" of platelets. Shown are the stimulated phosphorylation of a protein with molecular weight 40 000 by protein kinase C (*PKC*) and a protein with molecular weight 20 000 (myosin light chain) by myosin light chain kinase (*MLCK*) activated by diacylglycerol and Ca^{2+} (with calmodulin), respectively. Phosphorylation of these proteins is correlated with serotonin release. (Based on the report of Nishuzuka et al. 1984)

apparently activates platelets in a different fashion to thrombin, according to Rink et al. (1985). Binding of collagen to its receptor on the platelet surface does not appear to elevate intracellular Ca^{2+} levels by the same mechanism as thrombin. The release of Ca^{2+} from intracellular stores by collagen is blocked by aspirin, which is an inhibitor of cyclooxygenase. Thromboxane A_2, produced from arachidonic acid, is the proposed messenger responsible for the Ca^{2+} rise. The source of diacylglycerol for the activation of PKC after collagen binding to its receptors does not seem to be PIP_2 hydrolysis by phospholipase C but possibly the hydrolysis of phospholipids by another phospholipase. If PIP_2 had been hydrolysed, the IP_3 messenger itself would presumably release Ca^{2+} from intracellular stores by a non-aspirin-inhibitable process.

The inhibitory effects of adenosine and prostaglandins on platelet function are mediated by elevated cAMP levels and are considered to be due to the cAMP-dependent activation of a calcium pump, thereby reducing intracellular Ca^{2+} levels (Daniel 1985). A cAMP-dependent kinase also phosphorylates myosin light chain kinase, which reduces its capacity to be activated by Ca^{2+} and calmodulin. In addition, there are reports that cAMP in some manner inhibits phospholipase C activity (Billah et al. 1979).

A disease in Bassett hounds called hereditary canine thrombocytopenia bears on the inhibitory effect of cAMP in platelets. Adding thrombin to platelets from these dogs causes a delayed release reaction (Bordeaux et al. 1986). The defect could be due to a deficiency of receptor, abnormal transduction after receptor occupancy or a defect in the response to second messengers. A clue to the defect in this complex system is a higher than normal level of cAMP in these platelets in response to the inhibitory agonist prostaglandin I_2, or to forskolin, a substance that directly activates adenylate cyclase. A series of experiments pointed to an abnormally low activity of the enzyme that hydrolyses and inactivates cAMP, the cAMP phosphodiesterase. Insufficient destruction would tend to maintain cAMP at high levels and thereby depress the usual response to thrombin.

Glycogen Metabolism in Liver and Muscle

An important source of energy for various tissues is derived from the metabolism of glucose by the Embden–Meyerhof and other pathways in the formation of ATP. Glucose can be absorbed directly from the diet, synthesized from precursors or derived from the breakdown of glycogen stored in muscle, liver and other tissues. Glycogen degradation is stimulated by agonists that act through the cyclic nucleotides and the phosphoinositide second-messenger (diacylglycerol and IP_3) systems to activate kinases involved in glycogen breakdown. The kinases are phosphorylase kinase and phosphorylase, the former activating the latter by phosphorylation (Fig. 15.12). The synthesis of glycogen from glucose is catalysed by glycogen synthase, an enzyme that is more active in the dephosphorylated state than when phosphorylated. The simultaneous stimulation of the kinases involved in glycogenolysis and inhibition of the glycogen synthesis serve to provide a higher concentration of glucose for energy production. Dephosphorylation by appropriate protein phosphatases inactivates phosphorylase kinase and phosphorylase and activates glycogen synthase. This system constitutes another example of agonist stimulation of the phospholipase C–Ca^{2+}-mobilizing pathway and interaction with cyclic nucleotides.

In the degradation of glycogen, the first enzyme activated in the chain of events is phosphorylase kinase, a multimeric molecule composed of four dissimilar subunits with a stoichiometry of (alpha, beta, gamma, delta)$_4$ (Chan and Graves 1984). The gamma subunit has amino acid sequences homologous with the catalytic subunit of cAMP-dependent protein kinase, suggesting that it is responsible for the kinase activity of the molecule. Further, detailed studies by Kee and Graves (1986) on purified gamma subunits from different sources substantiated gamma's role as the catalytic subunit, with its dependence on calmodulin for enhanced activity.

The delta subunit of phosphorylase kinase from skeletal muscle is almost identical in its primary structure to bovine brain calmodulin (Grand et al. 1981). Unlike the case in other calmodulin-dependent enzymes, the delta subunit of phosphorylase kinase cannot be readily disassociated from the other subunits by the complexation of Ca^{2+} with chelators such as EGTA. The delta calmodulin-like subunit also binds to the gamma subunit in a manner that makes it inaccessible to binding by trifluoperazine, whereas this anti-psychotic drug readily inhibits most other calmodulin-stimulated enzymes. This tight interaction provides the molecule with a built-in Ca^{2+} sensor (Chan and Graves 1984), which allows for a faster response of the molecule to changes in intracellular Ca^{2+} levels than if, like in other calmodulin-dependent enzymes, there is a sequence of steps required for calmodulin activation, i.e.:

$$Ca^{2+} + CaM \rightleftharpoons Ca–CaM + E \rightleftharpoons Ca–CaM–E^*$$

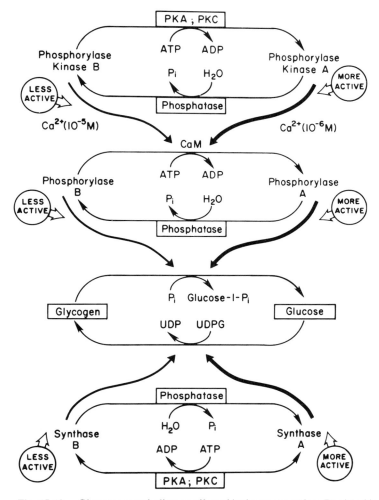

Fig. 15.12. Glycogen metabolism as affected by hormone action. Depicted is a simplified scheme of the activation by phosphorylation of phosphorylase kinase B by cAMP-dependent protein kinase A (*PKA*) and/or diacylglycerol–Ca^{2+}–phosphatidylserine protein kinase C (*PKC*) to form phosphorylase kinase A. Phosphorylase catalyses the degradation of glycogen to glucose-1-phosphate, and glucose is subsequently utilized for energy production. The formation of glycogen from glucose is under the control of glycogen synthase, which is more active in the dephosphorylated form. Dephosphorylation of the enzymes is catalysed by specific phosphatases.

Phosphorylase kinase undergoes reversible covalent modification through a phosphorylation–dephosphorylation cycle (Chan and Graves 1984). The alpha and beta subunits of the multimeric molecule contain the phosphorylation sites. Phosphorylation of these subunits is catalysed by a cAMP-dependent kinase (Cohen 1983), although phosphorylation can also be catalysed by other kinases, including the Ca^{2+}–diacylglycerol-dependent protein kinase C (Kishimoto et al. 1977).

Phosphorylation of phosphorylase kinase increases the Ca^{2+} sensitivity of the enzyme (Chan and Graves 1984). The non-activated and dephosphorylated enzyme requires 10^{-5} M Ca^{2+} for half-maximal activation. When activated, phosphorylase kinase can be stimulated by a tenth of the concentration of Ca^{2+} (10^{-6} M). Exogenous calmodulin further activates the kinase in the presence of Ca^{2+}, and the sites of calmodulin interaction are considered to be the alpha and beta subunits.

The physiological relevance of Ca^{2+} activation of phosphorylase kinase in muscle is suggested by the following. Upon depolarization, an early event in the contraction–relaxation cycle is the release of Ca^{2+} from the sarcoplasmic reticulum, transiently increasing cytosolic free Ca^{2+}. Binding to troponin C, Ca^{2+} initiates the contraction of myofibrils, an ATP-requiring reaction. The elevated Ca^{2+} at the same time activates phosphorylase kinase, and thus phosphorylation and activation of phosphorylase. The resultant breakdown of glycogen provides glucose, a substrate for ATP production. ATP is then available for myofibril contraction and subsequently for relaxation due to the accumulation of cytosolic Ca^{2+} by the ATP-dependent calcium pump of the sarcoplasmic reticulum.

Ca^{2+} Electrolyte Transport and Hormones

Epithelial tissues such as the intestine and kidney are sites of transcellular movement of Na^+, Cl^-, K^+, Ca^{2+} and other ions. The net absorption of NaCl follows this route: transference across the brush-border membrane, diffusion through the cytosolic compartment, followed by the ATP-dependent uphill transport of Na^+ across the basolateral membrane (Fig. 15.13).

Consideration has been given to the overall control of NaCl transport by epithelia. The need for some type of control derives from the notion that the transport capacity of the exit mechanism on the basolateral membrane should be the limiting step in overall transport and that the rate of entry of Na^+ should not exceed this capacity. Otherwise, NaCl might flood unchecked into the epithelial cell, causing abnormal and potentially disastrous swelling. The concentration of cytosolic free Ca^{2+} is considered to be the link between the Na^+ exit and entry processes. This concept is derived from several different experimental protocols, using transporting epithelia from various species. Raising intracellular Ca^{2+} concentrations with calcium ionophores in toad bladders and proximal renal tubules of rabbits decreased the transcellular movement of Na^+ (Windhager and Taylor 1983). Quinidine, a compound that raises cytosolic free Ca^{2+} by releasing it from internal stores, was also observed to depress epithelial transport of Na^+.

It has been suggested that the mechanism by which cytosolic Ca^{2+} effects a decrease in transcellular Na^+ is the depression of the Na^+ permeability of the apical brush-border membrane by Ca^{2+} (Fig. 15.13) (Chase and Al-Awqai 1983; Fan et al. 1983; Fan and Powell 1983). The connection between an inappropriately high concentration of Na^+ in the epithelial cell and increased Ca^{2+} might involve the Na^+/Ca^{2+} exchange system. Under normal physiological conditions, this system extrudes Ca^{2+} in exchange for Na^+ that enters the cell down its electrochemical gradient. When the Na^+ concentration within the cell increases, either by excess influx or by decrease in the transport

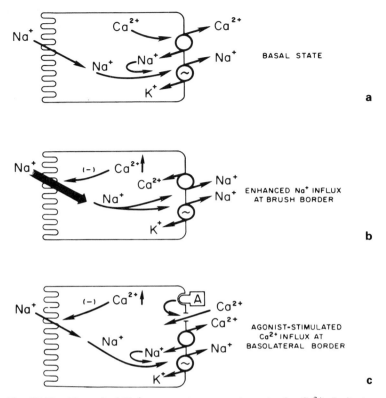

Fig. 15.13. Control of Na$^+$ transport across enterocytes by Ca^{2+}. In its transcellular transport, Na$^+$ enters the cell across the brush border, moves through the intracellular compartment and is then extruded across the basolateral membrane by the Na$^+$/K$^+$ pump. **a** When minimal Na$^+$ is being transported, the Ca^{2+} concentration is maintained at basal levels through the functioning of the Na$^+$/Ca^{2+} exchanger and the ATP-dependent Ca^{2+} pump (not shown). **b** An increase in Na$^+$ entrance elevates intracellular Na$^+$ concentration, resulting in the less efficient extrusion of Ca^{2+} by the Na$^+$/Ca^{2+} exchange and a rise of intracellular Ca^{2+} of sufficient magnitude to depress the influx of Ca^{2+} across the brush border. **c** In response to certain agonists, the influx of Ca^{2+} across the basolateral membrane (and possibly through calcium channels) is increased, resulting in the elevation of cytosolic Ca^{2+} and inhibition of Na$^+$ entry at the brush border.

activity of the sodium pump, the Na$^+$/Ca^{2+} exchanger operates less efficiently, and intracellular Ca^{2+} concentration increases. The higher Ca^{2+} concentration might therefore inhibit or at least decrease the rate of Na$^+$ entry across the luminal border, bringing the rate of Na$^+$ entry down to a level comparable to its rate of exit (Taylor and Windhager 1979; cf. also the review by Donowitz 1983).

Neurohumoral substances, which include serotonin, carbachol, substance P and neurotensin, inhibit NaCl absorption and increase the rate of entry of Ca^{2+} across the basolateral membrane of the intestine (Donowitz and Welsh 1986), providing evidence for the agonist control of sodium transport by Ca^{2+}.

Several proteins associated with the brush-border membrane have been shown to be phosphorylated under the catalytic action of both Ca^{2+}–calmodulin- and cyclic nucleotide-dependent kinases (Donowitz et al. 1984). Whether one or more of these phosphorylated proteins modulates Ca^{2+} entry across the brush-border membrane has yet to be determined.

Phorbol esters and other activators of protein kinase C were shown to decrease electrolyte transport by a cell line (A6) from *Xenopus* kidney, apparently by an inhibition of amiloride-sensitive sodium channels on the brush-border membrane through a phosphorylation reaction (Yanase and Handler 1986).

Ca^{2+} in Cell Growth and Proliferation

The significance of calcium in cell proliferation and growth was long suspected, and more recent studies confirm the essential role of this mineral ion. Second messengers besides Ca^{2+}, such as cAMP and diacylglycerol, have signalling functions in DNA synthesis and cell replication.

DNA Synthesis

Cells in culture progress through at least four identifiable stages: (a) the G_1 phase, in which synthesis of various substances (e.g. calmodulin and cAMP) and other reactions (e.g. membrane protein phosphorylation and increase in endoplasmic reticular Ca^{2+} occur); (b) the S phase, in which DNA synthesis and replication of the chromosomes take place; (c) the G_2 phase, which precedes (d) the M or mitotic phase.

An early demonstration of the Ca^{2+} requirement in cell replication stemmed from studies on the effect of hypocalcaemia on the formation of new liver cells after partial hepatectomy of rats (Whitfield et al. 1982). The lower circulating Ca^{2+} levels after parathyroidectomy inhibited cell division and the initiation of DNA synthesis. Other events preceding DNA synthesis, such as cAMP synthesis, membrane protein phosphorylation and the increase in soluble calmodulin, were unaffected. This inhibitory effect of lowered extracellular Ca^{2+} on DNA synthesis could be reversed by the injection of Ca^{2+}, if the Ca^{2+} pulse was given during the DNA-synthetic (S) phase of the cell cycle.

In studies with Chinese hamster ovary cells (CHO-K_1) in culture, it was shown that the concentration of calmodulin within these cells increases relatively early in the cell cycle and just prior to the DNA-synthetic phase (Means et al. 1982; Means 1982). This suggests that the Ca^{2+}-receptor protein, calmodulin, plays an important part in the stimulation of DNA synthesis at the G_1/S boundary. To reaffirm the significance of calmodulin in this process, adding an inhibitor of calmodulin, W-13, to growing CHO-K_1 cells was shown to reduce cell numbers by 50% after a 24-h incubation period. Using synchronized growing cells, the addition of W-13 resulted in a halt in the progression of the cells from the G_1 to the S phase, as monitored by the degree of incorporation of labelled precursors into DNA. The rise in intracellular Ca^{2+} concentration and increased synthesis of calmodulin within the G_1 phase of the cell replication cycle provide evidence of a functional role for the Ca^{2+}–calmodulin complex in DNA synthesis in the S phase.

Mitosis

During mitosis, the chromosomes of the cell divide to form pairs of chromatids, each composed of one of the daughter chromosomes. The movement of the chromosomes involves microtubules, some of which extend from a pole to the chromatids at the kinetochore and others that reach from the pole to the equator of the cell. Both types of microtubule act to move the chromatids to opposite poles of the cell. Calcium and calmodulin are thought to have important roles in this process (Ratan and Shelanski 1986). Physiological concentrations of Ca^{2+} in the presence of calmodulin depolymerize microtubules and inhibit their assembly (Marcum et al. 1978; Nishida et al. 1979). Calmodulin was shown, using immunohistochemical techniques, to be associated only with microtubules connecting the pole of the dividing cell with the kinetochore of the chromatid, and not with those running from pole to equator (Ratan and Shelanski 1986). That is, only the microtubules that "shorten" or disassemble as the chromatids move towards the pole contain bound calmodulin. The calcium for activating calmodulin is apparently derived from ramifying endoplasmic reticulum adjacent to the spindle poles and the mitotic apparatus, and the release of calcium may be mediated by IP_3.

The proposal that calmodulin, with its bound calcium, might be directly responsible for microtubule disassembly has been modified to account for the possible involvement of other factors, such as the microtubule-associated proteins (MAPs) (Manalan and Klee 1984). These MAPs have a functional role in the assembly–disassembly of the microtubules, and different MAPs are associated with microtubules from different sources. It has been suggested that calmodulin exerts its inhibitory effect on tubular polymerization by interacting with two MAPs, the *tau* factor and MAP2 (Yamamoto et al. 1983; Sobue et al. 1985). It is also proposed that one primary effect of Ca^{2+} and calmodulin on microtubule assembly and disassembly is due to the phosphorylation of specific proteins catalysed by a Ca^{2+}–calmodulin-stimulated protein kinase.

Oncogenes and Growth Factors

The capacity of cells to respond to growth-promoting substances (mitogenic factors) is dependent upon the products of specific genes, the proto-oncogenes. Defects in these genes lead to the formation of oncogenes, the products of which lead to uncontrolled cell growth, or cancer (Berridge 1984b; Macara 1985). The involvement of phosphoinositides and Ca^{2+} in cell growth has been documented, with cAMP also playing a significant role in cell proliferation (Sekar and Hokin 1986).

A considerable advancement in our understanding of the molecular basis of tumour formation arose from the study of certain viruses that transform cells from a resting state to a proliferative one (Berridge 1984b; Macara 1985). These transforming viruses contain oncogenes (v-*onc*) that control the growth of the host cells. After infection, the viral oncogenes might be incorporated into the genome of an otherwise normal cell, and the products from the oncogenes might very well be expressed, which could result in the uncontrolled production of some component that regulates replication. Other means have been suggested by which viral infections can cause tumour production, such as the insertion of viral genetic components into a chromosome at a point that, under normal circumstances, would control the expression of proto-oncogenes (Berridge 1984b).

The products of retrovirus oncogenes have, in certain instances, been identified, and it is of considerable interest that some of them are related to components of the phosphoinositide pathway (Berridge and Irvine 1984; Macara 1985; Sekar and Hokin 1986). The *sis* oncogene, associated with Simian sarcoma retrovirus, is coded for the synthesis of a protein almost identical to platelet-derived growth factor (PDGF); PDGF activates the phosphoinositide cycle. The *src* oncogene of Rous avian sarcoma and the *ros* oncogene of avian sarcoma URII appear to function as phosphatidylinositol kinases, increasing available amounts of PIP and PIP_2, the substrates of phospholipase C. Another viral oncogene, *v.erb B* from the avian erythroblastosis retrovirus, codes for a product that is similar to a segment of the EGF receptor; receptor occupancy by EGF also stimulates phosphoinositide metabolism. Still another oncogene, the *ras* oncogene from Harvey and Kirsten murine sarcoma, codes for a protein, p 21, of molecular weight 1000, that has high GTPase activity and is related, and possibly identical, to the G protein involved in the activation of phosphoinositide metabolism (Wakelam et al. 1986).

Thus cell transformation and cell growth induced by some viral or cellular oncogenes involve components of the phosphoinositide cycle. The products of specific oncogenes can apparently serve as a growth factor, a growth factor receptor, a G protein that couples receptor occupancy with phospholipase C activation, and kinases that phosphorylate phosphoinositides to produce increased amounts of PIP and PIP_2. Cell growth and cell division accompanying tumour formation could then result in the uncontrolled formation of the second messengers, diacylglycerol and IP_3.

The exact functions of PKC activation and the Ca^{2+} released by IP_3 in stimulating DNA synthesis and cell replication are not known, but it has been proposed that one effect is to stimulate a Na^+/H^+ exchange mechanism, resulting in an increase in the intracellular pH, i.e. alkalinization. Alkalinization of the cytosol, characteristic of the action of growth factors, is considered to be an important signal in stimulating cell growth. Mitogens, such as PDGF, vasopressin and thrombin, cause alkalinization of the cytosol of fibroblasts and induce DNA synthesis. Amiloride, an inhibitor of the Na^+/H^+ exchanger, inhibits this effect (Vicentini and Villereal 1986). Further, submaximal levels of the Ca^{2+} ionophore and a phorbol ester (PMA) synergistically stimulate Na^+ influx, suggesting that Ca^{2+}–calmodulin and PKC can, in some fashion, activate the Na^+/H^+ exchanger that leads to alkalinization and cell proliferation (Fig. 15.14).

Oncomodulin

Oncomodulin is a unique calcium-binding protein present in some tumour cells; it is homologous with calmodulin (MacManus 1979). Oncomodulin has been detected in hepatomas from different species and in some cell lines that have spontaneously transformed or were transformed by certain viruses. Not all tumours, however, produce this protein, and it has not been detected in normal tissues. Oncomodulin differs from calmodulin by virtue of its lower molecular

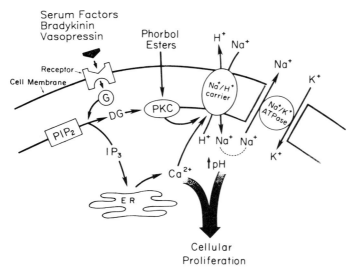

Fig. 15.14. Membrane and intracellular events affected by growth factors. Depicted schematically is the stimulation of the hydrolysis of phosphatidyl-4,5-diphosphate (*PIP₂*) to diacylglycerol (*DG*) and inositol-1,4,5-trisphosphate (*IP₃*) by agonists upon activating their receptor and the associated G protein. DG activates protein kinase C (*PKC*) which, through protein phosphorylation, promotes the extrusion of H⁺ by the Na⁺/H⁺ exchange, resulting in cytosolic alkalinization, one stimulus for cell proliferation. IP₃ stimulates the release of Ca²⁺ from an intracellular compartment and the elevated cytosolic Ca²⁺ constitutes another stimulus for cell proliferation, as well as affecting Na⁺/H⁺ exchange. (Based on Vicentini and Villereal 1986; and redrawn with permission of the authors and Pergamon Press)

weight ($\sim$12 000 vs $\sim$17 000) and its two high-affinity calcium-binding sites versus calmodulin's four. It bears a considerable resemblance to β-parvalbumin in its amino acid sequence and other properties. The function of oncomodulin in carcinogenesis is not precisely known, but the addition of oncomodulin to calcium-deprived rat-liver cells stimulates DNA synthesis, as determined by the incorporation of labelled thymidine. Calmodulin was ten times less active in this assay (MacManus and Whitfield 1983).

Epilogue

The activation of phospholipase C and the formation of the two second messengers, diacylglycerol and inositol-1,3,4-trisphosphate (IP₃), are central to the biological effect of specific hormones and other agonists. The resultant elevation of intracellular free calcium by IP₃ and activation of protein kinase C by diacylglycerol represent bifurcating pathways that, in some tissues, are required for the integration of physiological responses. Both paths are required for aldosterone production by adrenal glomerulosa cells (Kojima et al. 1984), the "release reaction" of platelets (Nishizuka et al. 1984) and glucose stimulation of insulin secretion by pancreatic β-cells (Zawalich et al. 1984).

Hormonal activation of adenylate cyclase constitutes another path of cell regulation that can elicit responses similar to Ca^{2+} and protein kinase C or that can modulate the effect of the Ca^{2+}-mobilizing hormones. Elevation of cAMP in platelets inhibits the thrombin-induced release reaction. In this case, cAMP apparently exerts two effects, a decrease in cytosolic Ca^{2+} by stimulating the reaccumulation of Ca^{2+} by intracellular stores and an inhibition of phosphoinositide hydrolysis (Rasmussen 1986). Cyclic AMP, on the other hand, acts synergistically with Ca^{2+} in the activation of glycogen phosphorylase by phosphorylase kinase. The cAMP-dependent protein kinase phosphorylation of phosphorylase kinase considerably decreases the concentration of Ca^{2+} required for activation. The relationship between Ca^{2+} and cAMP is further accentuated by the well-known enhancement of cAMP synthesis by a Ca^{2+}–calmodulin-stimulated adenylate cyclase and the stimulation of cAMP hydrolysis by a phosphodiesterase. Thus, the Ca^{2+}, diacylglycerol and cAMP pathways are intimately involved in the control of cellular responses of some tissues and cells. Although not detailed here, the products of arachidonic acid metabolism, arachidonic acid itself and cGMP, are other messengers that exert profound effects on physiological and biochemical functions.

Again, the modulation of Ca^{2+} concentrations by hormone action and the stimulation of Ca^{2+}-dependent events are central to the biological effects elicited by the Ca^{2+}-mobilizing hormones and other agonists. We look forward to significant developments in our understanding of the bases of the effect of hormones and other agonists on biochemical and physiological systems. New knowledge and the findings of the future will prove valuable in deciphering in the minutest detail the fundamental processes of living organisms, in addition to having great potential for application to disease.

References

Abdel-Latif AA (1986) Calcium-mobilizing receptors, polyphosphoinositides and the generation of second messengers. Pharmacol Rev 38:227–272

Anderson WB, Salomon DS (1985) Calcium, phospholipid dependent protein kinase C as a cellular receptor for phorbol ester tumor promotors: possible role in modulating cell growth and tumor promotion. In: Kuo JF (ed) Phospholipids and cellular regulation, vol 2. CRC Press, Boca Raton, Florida, pp 127–170

Authi KS, Crawford N (1985) Inositol 1,4,5-trisphosphate-induced release of sequestered Ca^{2+} from highly purified human platelet intracellular membranes. Biochem J 230:247–253

Berridge MJ (1984a) Inositol trisphosphate and diacylglycerol as second messengers. Biochem J 220:345–360

Berridge MJ (1984b) Oncogenes, inositol lipids and cellular proliferation. Biotechnology 2:541–546

Berridge MJ (1985) Calcium-mobilizing receptors: membrane phosphoinositides and signal transduction. In: Rubin RP, Weiss GB, Putney JW Jr (eds) Calcium in biological systems. Plenum Press, New York, pp 37–44

Berridge MJ, Irvine RF (1984) Inositol trisphosphate, a novel second messenger in cellular signal transduction. Nature 312:315–321

Billah MM, Lapetina EG, Cuatrecasas P (1979) Phosphatidylinositol-specific phospholipase-C of platelets: association with 1,2-diacylglycerol-kinase and inhibition by cyclic AMP. Biochem Biophys Res Comm 90:92–98

Bourdeaux MK, Dodds J, Slauson DO, Catalfamo JL (1986) Impaired cAMP metabolism associated with abnormal function of thrombopathic canine platelets. Biochem Biophys Res Comm 140:595–601

Bourne HR (1986) GTP-binding proteins: one molecular machine can transduce diverse signals. Nature 321:814–816

Burgess WH, Jemiolo DK, Kretsinger RH (1980) Interaction of calcium and calmodulin in the presence of sodium dodecyl sulfate. Biochim Biophys Acta 623:257–270

Campbell AK (1983) Intracellular calcium, its universal role as regulator. John Wiley, New York

Chan K-FJ, Graves DJ (1984) Molecular properties of phosphorylase kinase. In: Cheung WY (ed) Calcium and cell function, vol 5. Academic Press, New York, pp 1–31

Chase HS Jr, Al-Awqai Q (1983) Calcium reduces the sodium permeability of luminal membrane vesicles from toad bladder. Studies using a fast-reaction apparatus. J Gen Physiol 81:643–665

Cheung WY (1970) Cyclic 3',5'-nucleotide phosphodiesterase: demonstration of an activator. Biochem Biophys Res Comm 38:533–538

Cheung WY (1980) Calmodulin plays a pivotal role in cellular regulation. Science 207:19–27

Chou PY, Fasman GD (1979) Prediction of beta-turns. Biophys J 26:367–373

Cohen P (1983) The role of protein phosphorylation in neural and hormonal control of cellular activity. Nature 296:613–620

Daniel JL (1985) Protein phosphorylation and calcium as mediator in human platelets. In: Rubin RP, Weiss GB, Putney JW Jr (eds) Calcium in biological systems. Plenum Press, New York, pp 165–171

Darnell J, Lodish H, Baltimore D (1986) Molecular cell biology. Scientific American Books, New York

Demaille JG (1982) Calmodulin and calcium-binding protein: evolutionary diversification of structure and function. In: Cheung WY (ed) Calcium and cell function, vol 2. Academic Press, New York, pp 111–144

Devine CE, Somlyo AV, Somlyo AP (1971) Sarcoplasmic reticulum and excitation–contraction coupling in mammalian smooth muscle. J Cell Biol 52:690–718 (Cited by JW Putney Jr, et al. (1986) Fed Proc 45:2634–2638)

Donowitz M (1983) Ca^{2+} in the control of active intestinal Na and Cl transport: involvement in neurohumoral action. Am J Physiol 245:G165–177

Donowitz M, Welsh MJ (1986) Ca^{2+} and cyclic AMP in regulation of intestinal Na, K and Cl transport. Annu Rev Physiol 48:135–150

Donowitz M, Cohen ME, Gudewich R, Taylor L, Sharp GWG (1984) Ca^{2+}-calmodulin, cyclic AMP- and cyclic GMP-induced phosphorylation of proteins in purified microvillus membranes of rabbit ileum. Biochem J 219:573–581

Exton JH (1984) Mechanisms involved in the actions of calcium dependent hormone in liver. In: Ebashi S, Endo M, Imahori K, Kakiuchi S, Hishizuka Y (eds) Calcium regulation in biological systems. Academic Press, Tokyo, pp 141–156

Fan C-C, Powell DW (1983) Calcium/calmodulin inhibition of coupled NaCl transport in membrane vesicles from rabbit ileal brush border. Proc Natl Acad Sci USA 80:5248–5252

Fan C-C, Faust RG, Powell DW (1983) Coupled sodium-chloride transport by rabbit ileal brush border membrane vesicles. Am J Physiol 244:G375–385

Fishman PH (1980) Mechanism of action of cholera toxin: events on the cell surface. In: Fordtran JS, Schulz SG (eds) Secretory diarrhoea. American Physiological Society, Bethesda, pp 86–106

Gill DL (1985) Receptors coupled to calcium mobilization. In: Cooper DMF, Seamon KB (eds) Advances in cyclic nucleotide and protein phosphorylation research, vol 19. Raven Press, New York, pp 307–324

Grand RJ, Shenolikar S, Cohen P (1981) The amino acid sequence of the delta subunit (calmodulin) of rabbit skeletal muscle phosphorylase kinase. Eur J Biochem 113:359–367

Head JF, Masure HR, Kaminer B (1982) Identification and purification of a phenothiazine binding fragment from bovine brain calmodulin. FEBS Lett 137:71–74

Hirata M, Sasaguri T, Hamachi T, Hashimoto T, Kukita M, Koga T (1985) Irreversible inhibition of Ca^{2+} release in saponin-treated macrophages by a photoaffinity derivative of inositol-1,4,5-trisphosphate. Nature 317:723–725

Hokin LE (1985) Receptors and phosphoinositide-generated second messengers. Annu Rev Biochem 54:205–235

Hokin LE, Hokin MR (1955) Effects of acetylcholine on the turnover of phosphoryl units in individual phospholipids of pancreas slices and brain cortex slices. Biochim Biophys Acta 18:102–110

Hokin MR, Hokin LE (1953) Enzyme secretion and the incorporation of ^{32}P into phospholipids of pancreatic slices. J Biol Chem 203:967–977

Hokin MR, Hokin LE (1954) Effects of acetylcholine and phospholipids in the pancreas. J Biol Chem 209:549–559

Hokin-Neaverson M (1974) Acetylcholine causes a net decrease in phosphatidylinositol and a net increase in phosphatidic acid in mouse pancreas. Biochem Biophys Res Comm 58:763–768

Holmes RP, Yoss NL (1983) Failure of phosphatidic acid to translocate Ca^{2+} across phosphatidylcholine membranes. Nature 305:637–638

Hughes BP, Milton SE, Barritt GJ, Auld AM (1986) Studies with verapamil and nifedipine provide evidence for the presence in the liver cell plasma membrane of two types of Ca^{2+} inflow transporter which are dissimilar to potential-operated Ca^{2+} channels. Biochem Pharmacol 35:3045–3052

Hurwitz L (1986) Pharmacology of calcium channels and smooth muscle. Annu Rev Pharmacol Toxicol 26:225–258

Ikuri M, Hiraoki T, Hikichi K, Mikuni T, Yazawa M, Yagi K (1983) Nuclear magnetic resonance studies of calmodulin: calcium-induced conformational change. Biochemistry 22:2573–2579

Joseph SK (1985) Receptor-stimulated phosphoinositide metabolism: a role for GTP-binding protein. Trends Biochem Sci 10:297–298

Kee SM, Graves DJ (1986) Isolation and properties of the active gamma subunit of phosphorylase kinase. J Biol Chem 261:4732–4737

Kikkawa U, Kaibuchi K, Takai Y, Nishizuka Y (1985) Phospholipid turnover in signal transduction: protein kinase C and calcium ion as two synergistic mediators. In: Kuo JF (ed) Phospholipids and cellular regulation, vol 2. CRC Press, Boca Raton, Florida, pp 111–126

Kikkawa Y, Nishizuka Y (1986) The role of protein kinase C in transmembrane signalling. Annu Rev Cell Biol 2:149–178

Kishimoto A, Takai Y, Nishizuka Y (1977) Activation of glycogenphosphorylase kinase by a calcium-activated, cyclic nucleotide-independent protein kinase system. J Biol Chem 252:7449–7452

Klee CB, Haiech J (1980) Concerted role of calmodulin and calcineurin in calcium regulation. Ann NY Acad Sci 356:43–54

Kojima I, Kojima K, Kreutter D, Rasmussen H (1984) The temporal integration of the aldosterone secretory response to angiotensin occurs via two intracellular pathways. J Biol Chem 259:14448–14457

Krebs J, Carofoli E (1982) Influence of Ca^{2+} and trifluoperazine on the structure of calmodulin: a 'H-nuclear magnetic resonance study. Eur J Biochem 124:619–627

Kretsinger RH (1980) Structure and evolution of calcium-modulated proteins. CRC Crit Rev Biochem 8:119–174

Lin S-H, Wallace MA, Fain JN (1983) Regulation of Ca^{2+}–Mg^{2+}-ATPase activity in hepatocyte plasma of membranes by vasopressin and phenylephrine. Endocrinology 113:2268–2275

Macara IG (1985) Oncogenes, ions and phospholipids. Am J Physiol 248:C3–11

MacManus JP (1979) Occurrence of a low-molecular-weight calcium-binding protein in neoplastic liver. Cancer Res 39:3000–3005

MacManus JP, Whitfield JF (1983) Oncomodulin: a calcium-binding protein from hepatoma. In: Cheung WY (ed) Calcium and cell function. Academic Press, New York, chapter 11

Majerus PW, Wilson DB, Connally TM, Boss TE, Neufeld EJ (1985) Phosphoinositide turnover provides a look at stimulus–response coupling. Trends Biochem Sci 10:168–171

Manalan AS, Klee CB (1984) Calmodulin. In: Greengard P, Robison GA (eds) Advances in cyclic nucleotides and protein phosphorylation, vol 18. Raven Press, New York, pp 227–278

Marcum JM, Dedman JR, Brinkley BR, Means AR (1978) Control of microtubule assembly–disassembly by calcium-dependent regular protein. Proc Natl Acad Sci USA 75:3771–3775

Mauger J-P, Poggioli J, Guesdon F, Claret M (1984) Noradrenaline, vasopressin and angiotensin increase Ca^{2+} influx by opening a common pool of Ca^{2+} channels in isolated rat liver cells. Biochem J 221:121–127

Means AR (1982) Calmodulin: an intracellular calcium receptor involved in regulation of cell proliferation. In: Corradino RA (ed) Functional regulation at the cellular and molecular levels. Elsevier/North-Holland, New York, pp 47–68

Means AR, Dedman J (1980) Calmodulin: an intracellular receptor. Nature 285:73–77

Means AR, Lagace L, Guerriero V Jr, Chafouleas JG (1982) Calmodulin as a mediator of hormone action and cell regulation. J Cell Biochem 20:317–330

Merritt JE, Taylor CW, Rubin RP, Putney JW Jr (1986) Evidence suggesting that a novel guanine nucleotide regulatory protein couples receptors to phospholipase C in exocrine pancreas. Biochem J 236:337–343

Michell B, Kirk C (1986) G-protein control of inositol phosphate hydrolysis. Nature 323:112–113

Michell RH (1975) Inositol phospholipids and cell surface receptor function. Biochim Biophys Acta 415:81–147

Muallem S, Schoeffield M, Pandol S, Sachs G (1985) Inositol trisphosphate modification of ion transport in rough endoplasmic reticulum. Proc Natl Acad Sci USA 82:4433–4437

Nakamura T, Ui M (1985) Simultaneous inhibitions of inositol phospholipid breakdown, arachidonic release, and histamine secretion in mast cells by islet-activating protein, pertussis toxin. J Biol Chem 260:3584–3593

Nishida E, Kumagai H, Ohtsuki I, Sakai H (1979) The interactions between calcium-dependent regulator protein of cyclic nucleotide phosphodiesterase and microtubule proteins. I. Effect of calcium-dependent regular protein on the calcium sensitivity of microtubule assembly. J Biochem (Tokyo) 85:1257–1266

Nishizuka Y (1986) Studies and perspectives of protein kinase C. Science 233:305–312

Nishizuka Y, Takai Y, Kishumoto A, Kikkawa U, Kaibuchi K (1984) Phospholipid turnover in hormone action. Recent Prog Hormone Res 40:301–345

Poggioli J, Mauger J-P, Guesdon F, Claret M (1985) A regulatory calcium-binding site for calcium channel on isolated rat hepatocytes. J Biol Chem 260:3289–3294

Putkey JA, Ts'iu KF, Tanaka T et al. (1983) Chicken calmodulin genes: a species comparison of cDNA sequences and isolation of a genomic clone. J Biol Chem 258:11864–11870

Putney JW Jr (1986) A model for receptor regulated calcium entry. Cell Calcium 7:1–12

Rasmussen H (1986) The calcium messenger system. N Engl J Med 314:1094–1101, 1164–1170

Rasmussen H, Kojima I, Kojima K, Zawalich W, Apfeldorf W (1984) Calcium as intracellular messenger: sensitivity modulation, C-kinase pathway and sustained cellular response. In: Greengard P, Robison GA (eds) Advances in cyclic nucleotide and protein phosphorylation research, vol 18. Raven Press, New York, pp 159–193

Ratan RR, Shelanski ML (1986) Calcium and the regulation of mitotic events. Trends Biochem Sci 11:456–459

Rink TJ, Tsien RY, Sanchez A, Hallam TJ (1985) Calcium and diacylglycerol: separable and interacting intracellular activators in human platelets. In: Rubin RP, Weiss GB, Putney JW Jr (eds) Calcium in biological systems. Plenum Press, New York, pp 153–164

Rittenhouse-Simmons S, Russell FA, Deykin D (1977) Mobilization of arachidonic acid in human platelets: kinetics and Ca^{2+} dependency. Biochim Biophys Acta 488:370–380

Rose B, Lowenstein WR (1975) Permeability of cell junction depends on local cytoplasmic calcium activity. Nature 254:250–252

Rubin RP (1982) Calcium and cellular secretion. Plenum Press, New York, pp 93–98

Sasagawa T, Ericsson LH, Walsh KA, Schreiber WE, Fischer EH, Titani K (1982) Complete amino acid sequence of human brain calmodulin. Biochemistry 21:2565–2569

Sekar MC, Hokin LE (1986) The role of phosphoinositides in signal transduction. J Membr Biol 89:193–210

Smith JB, Smith L, Higgins BL (1985) Temperature and nucleotide dependence of calcium release by myo-inositol 1,4,5-trisphosphate in cultured vascular smooth muscle cells. J Biol Chem 260:14413–14416

Sobue K, Tanaka T, Ashino N, Kakiuchi S (1985) Ca^{2+} and calmodulin regulate microtubule-associated protein–actin filament interaction in a flip-flop switch. Biochim Biophys Acta 845:366–372

Streb H, Irvine RF, Berridge MJ, Schulz I (1983) Release of Ca^{2+} from a non-mitochondrial intracellular store in pancreatic acinar cells by inositol-1,4,5-trisphosphate. Nature 306:67–69

Streb H, Bayerdorffer E, Hoase W, Irvine RF, Schulz I (1984) Effects of inositol-1,4,5-trisphosphate on isolated subcellular fractions of rat pancreas. J Membr Biol 81:241–253

Stryer L, Bourne HR (1986) G Proteins: a family of signal transducers. Annu Rev Cell Biol 2:391–419

Suematsu E, Hirata M, Hashimoto T, Kuriyama H (1984) Inositol 1,4,5-trisphosphate releases Ca^{2+} from intracellular store sites in skinned single cells of porcine coronary artery. Biochem Biophys Res Comm 120:481–485

Takai Y, Kikkawa U, Kaibuchi K, Nishizuka Y (1984) Membrane phospholipid metabolism and signal transduction for protein phosphorylation. In: Greengard P, Robison GA (eds) Advances in cyclic nucleotide and protein phosphorylation research, vol 18. Raven Press, New York, pp 119–158

Taylor A, Windhager EE (1979) Possible role of cytosolic calcium and Na–Ca exchange in regulation of transepithelial sodium transport. Am J Physiol 236:F505–512

Taylor CW, Merritt JE (1986) Receptor coupling to polyphosphoinositide turnover: a parallel with the adenylate cyclase system. Trends Pharmacol Res 7:238–242

Thomas AP, Alexander J, Williamson JR (1984) Relationship between inositol polyphosphate production and the increase of cytosolic free Ca^{2+} induced by vasopressin in isolated hepatocytes. J Biol Chem 259:5574–5584

Turner RS, Kuo JF (1985) Phospholipid-sensitive Ca^{2+}-dependent protein kinase (protein kinase C): the enzyme, substrates and regulation. In: Kuo JF (ed) Phospholipids and cellular regulation, vol 2. CRC Press Inc, Boca Raton, Florida, pp 75–110

Ueda T, Chueh SH, Noel MW, Gill DL (1986) Influence of inositol-1,4,5-trisphosphate and guanine nucleotides on intracellular calcium release within the N1E–115 neuronal cell line. J Biol Chem 261:3184–3192

Vicentini LM, Villereal ML (1986) Inositol phosphates turnover, cytosolic Ca^{++} and pH: putative signals for the control of cell growth. Life Sci 38:2269–2276

Wakelam MJO, Davies SA, Houslay MD, McKay I, Marshall CJ, Hall A (1986) Normal p21 (*N-ras*) couples bombesin and other growth factor receptors to inositol phosphate production. Nature 323:173–176

Watterson DM, Sharief F, Vanaman TC (1980) The complete amino acid sequence of the Ca^{2+}-dependent modulator protein (calmodulin) of bovine brain. J Biol Chem 255:962–975

Weiss B, Prozialeck W, Cimino M, Barnette MS, Wallace TL (1980) Pharmacological regulation of calmodulin. Ann NY Acad Sci 356:319–345

Whitfield JF, MacManus JP, Boynton AL, Durkin J, Jones A (1982) Futures of calcium, calcium-binding proteins, cyclic AMP and protein kinases in the quest for an understanding of cell proliferation and cancer. In: Corradino RA (ed) Functional regulation at the cellular and molecular levels. Elsevier/North-Holland, New York, pp 59–87

Williamson JR (1986) Role of inositol lipid breakdown in the generation of intracellular signals. Hypertension 8:II140–156

Williamson JR, Cooper RH, Joseph SK, Thomas AP (1985) Inositol trisphosphate and diacylglycerol as intracellular second messengers in liver. Am J Physiol 248:C203–216

Williamson JR, Joseph SK, Coll KE, Thomas AP, Verhowen A, Prentki M (1986) Hormone-induced inositol lipid breakdown and calcium-mediated cellular responses in liver. In: Poste G, Crooke ST (eds) New insights into cell and membrane transport process. Plenum Press, New York, pp 217–247

Windhager EE, Taylor A (1983) Regulatory role of intracellular calcium ions in epithelial Na transport. Annu Rev Physiol 45:519–532

Wojcikiewicz RJ, Kent PA, Fain JA (1986) Evidence that thyrotropin-releasing hormone-induced increases in GTPase activity and phosphoinositide metabolism in GH_3 cells are mediated by a guanine nucleotide-binding protein other than G_s or G_i. Biochem Biophys Res Comm 138:1383–1389

Yamamoto H, Fukunaga K, Tanaka E, Miyamoto E (1983) Ca^{2+}- and calmodulin-dependent phosphorylation of microtubule-associated protein-2 and tau factor, and inhibition of microtubule assembly. Neurochemistry 41:1119–1125

Yanase M, Handler JS (1986) Activators of protein kinase C inhibit sodium transport in A6 epithelia. Am J Physiol 250:C517–522

Zawalich W, Zawalich K, Rasmussen H (1984) Insulin secretion: combined tolbutamide, forskolin and TPA mimic action of glucose. Cell Calcium 5:551–558

Chapter 16

Cellular Calcium: Cell Growth and Differentiation

T. Fujita and Y. Nakao

Introduction

Among the various constituents of the living organism, calcium occupies a unique position in the control of cell function, including cell growth and proliferation, as seen in the essential role of calcium in the life cycle of a cell from fertilization, parthenogenesis and mitosis to meiosis. The great concentration gradient of calcium between various compartments, as well as the dynamic equilibrium between rapid exchanges, such as influx and efflux, are no doubt important, in addition to the intrinsic properties of the calcium ion itself. Such dynamic exchanges of calcium across the cell are the basis for excitation and signal transduction. These cellular events are under the control of nutritional factors in the host organism, such as calcium intake. The intracellular factors, such as calmodulin and other calcium-binding proteins, cyclic nucleotides, phospholipids and protein kinase C, also control cell growth and proliferation through interactions with calcium, in conjunction with growth factors, cytokines and oncogenes.

It is the purpose of this chapter to delineate this enormous range of the cytokinetic actions of calcium in the light of a unified concept of the dynamic distribution of calcium from the ultramicroscopic cellular level to the macroscopic level of the whole organism.

Extracellular Calcium and Cellular Calcium Requirement

The importance of calcium in the growth of prokaryotic and eukaryotic cells has been known for a long time (Campbell 1983). Heilbrunn (1937) was the first to show the importance of extracellular calcium in the breakdown of the germinal vesicles of marine annelids by inhibiting it through the addition of sodium citrate. Egg fertilization, a calcium-requiring process, is also associated with a release of calcium ions (Mazia 1937). Calcium apparently also plays a decisive role in parthenogenesis (cell division without sperm) since this is also inhibited in the absence of calcium (Tyler 1941).

Calcium apparently stimulates the transition from the telophase of the first meiosis to the prophase of the second meiosis. A23187 is capable of initiating meiosis in oocytes (Schuetz 1972). A role for calcium in stimulating Go–S transition, activation to the G_1 and G_2 phase (Whitfield et al. 1966; Whitfield and Youdale 1966; Whitfield 1982; Izant 1983; Hepler 1985) and mitosis has been suggested. Calcium is said to be necessary to maintain the activity of DNA polymerase for DNA synthesis (Zwierchowski et al. 1984). Mitosis itself is rather insensitive to extracellular calcium deprivation. Figure 16.1 summarizes the actions of calcium in a typical cell cycle.

At the level of the whole organism Rixon et al. (1955) and Rixon and Whitfield (1961, 1963) called attention to the radioprotective action of parathyroid hormone-induced hypercalcaemia in rats. The most life-threatening effects of irradiation, bone marrow depression and intestinal epithelial necrosis, were effectively antagonized by the cell-proliferating effect of calcium mobilized from bone by parathyroid extract. Parenteral calcium was as effective as parathyroid extract (Whitfield and Rixon 1962; Sanfelippo et al. 1965; Perris and Whitfield 1967; Perris et al. 1967, 1968). While the mechanism of action of calcium in stimulating DNA synthesis is not fully understood, calcium might share the role of natural condensing agents (polyamines and basic proteins), initiating cell division by promoting coiling and aggregation of chromosomes, possibly by combining with the phosphate group of the DNA molecule.

There is an important difference in the extracellular calcium requirement of normal and neoplastic cells (Swierenga et al. 1976a; Boynton et al. 1977a). While normal cells require calcium in the extracellular environment for proliferation and mitosis, neoplastic cells have apparently lost such dependency on calcium (Boynton and Whitfield 1976). Loss of calcium dependency was found in hepatic neoplastic cells (Swierenga et al. 1978), NRK cells transformed

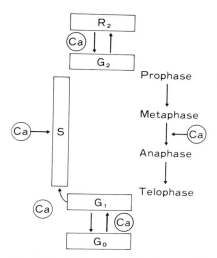

Fig. 16.1.   Actions of calcium in a typical cell cycle.

by avian sarcoma virus B77 and cells infected by Rous sarcoma virus (Boynton et al. 1977b; Boynton and Whitfield 1978). The requirement of calcium appears to be an important factor in cell proliferation in view of the action of TPA (12-O-tetradecanoyl-phorbol-13-acetate) and EGF in reducing calcium requirement (McKeehan and McKeehan 1979; Lechner and Kaighn 1979; Fisher and Weinstein 1981).

In cells which have lost calcium dependence, the total calcium content seems to be unchanged, but a calcium redistribution within the cell remains a possibility (Tupper and Zorgniotti 1977; Hazelton and Tupper 1979; Eilam and Szydel 1981). Normal mouse cells showing growth arrest at the G_1 phase in a low calcium medium (Paul and Ristow 1979) are freed of the block by a tumour-promoting phorbol ester (Boynton et al. 1975). Extracellular calcium deprivation leads to a rearrangement of microtubules and a perinuclear accumulation of cytoskeleton and vimentin filaments in normal rat liver T51B epithelial cells but not in Morris No. 7795 hepatoma cells (Swierenga et al. 1984), providing a morphological basis for the change in calcium dependence.

Calcium mediates the actions of acetylcholine, bradykinin, cortisol, concanavalin A, detergents, growth hormone, lysine vasopressin, polyamines and TPA in promoting cell proliferation (Perris and Whitfield 1969; MacManus and Whitfield 1969b; Whitfield et al. 1973; McManus et al. 1975), whereas all these compounds are inactive in calcium-free medium.

Intracellular Calcium and Calcium Influx

A transient increase in intracellular free calcium occurs at the time of fertilization of fish or sea urchin eggs. Intracellular injection of calcium stimulates mammalian oocytes to divide (Fulton and Whittingham 1978), indicating that cell growth and differentiation are actually controlled by intracellular calcium or calcium influx rather than by extracellular calcium. Peaks of intracellular calcium were noted during pronuclear migration, nuclear envelope breakdown, metaphase/anaphase transition and cleavage (Poenie et al. 1985).

According to the hypothesis of Lansing (1947), the age-related accumulation of calcium within the cell stops growth and the low intracellular calcium in tumour cells (Coman 1944; Braunschweig et al. 1946; Dunham et al. 1946; deLong et al. 1950) prevents this growth-inhibiting action causing unlimited growth. Anghileri and Miller (1971), however, reported an increased uptake of calcium by a growing tumour. Seltzer et al. (1970) found an increased calcium content in human breast carcinoma, and Hickie and Galant (1967) reported a higher calcium content in the Morris hepatoma of the rat than in normal liver tissue. High intracellular calcium levels and increased calcium influx were also found in HeLa cells (Dasdia et al. 1979). Necrotic tissue may take up more calcium because of the formation of chelating agents during the process of dystrophic calcification.

One of the most dramatic developments in immunology in recent years was the discovery of a rise in intracellular free calcium in proliferating lymphocytes stimulated with lectin or antigen (Whitney and Sutherland 1972a, b; Lichtman et al. 1983). Such calcium influx across the cell membrane and the consequent rise of intracellular free calcium, representing an important signal for the initiation of DNA synthesis, is accompanied by a change in membrane potential (Gelfand et al. 1984). The magnitude of the increase in free cytosolic calcium corresponds to the proliferative response and lectin-induced calcium flux accompanying membrane depolarization. Epidermal growth factor also raises the intracellular calcium of human fibroblasts (Mix et al. 1984; Moolenaar et al. 1984).

Along with the calcium influx and rise of intracellular free calcium, there is a rise of intracellular pH brought about by growth factors and protein kinase C or TPA through the Na^+/H^+ exchange system (Hesketh et al. 1985). Although the rise of intracellular pH induced by protein kinase C may occur without any measurable rise in intracellular free calcium, the release of calcium from the endoplasmic reticulum may be a response to phospholipid breakdown producing inositol triphosphate associated with protein kinase C activation (Grinstein et al. 1985).

An increase of intracellular calcium has also been proposed as a trigger for cell differentiation (Brown 1984). Excessive calcium entry into the cell induced by 25-dehydrocholesterol, an inhibitor of cholesterol biosynthesis, depressed DNA synthesis by p815 mastocytoma cells, through a change of plasma membrane cholesterol concentration (Boissonneault and Heiniger 1984). Arslan et al. (1985) measured intracellular free calcium in Ehrlich and Yoshida carcinoma cells by the quin2 method, obtaining subnormal values. They pointed out, however, that these low values might be due to artifactual quenching of quin2 fluorescence by heavy metals. After correction for this factor, cytosolic free calcium homeostasis in these cells was indistinguishable from that of normal cells. Cheung and McCarty (1985) introduced calcium-containing hydroxyapatite crystals into cultured human fibroblasts to stimulate mitogenesis, and ascribed the effect to their intracellular dissolution raising intracellular calcium, which was blocked by chloroquine or ammonium chloride.

Calmodulin and Other Calcium-Binding Proteins

A large proportion of intracellular calcium is bound to protein. Calmodulin, a ubiquitous calcium-binding protein in eukaryotic cells, plays an important part in cell division. Calmodulin was shown to be localized in the mitotic spindle (Means and Dedman 1980), as was rhodamine isothiocyanate-labelled calmodulin (Zavortink et al. 1983). At prometaphase and metaphase, calmodulin is located mainly near the poles and in the space between chromosomes and poles in early anaphase. No specific localization of calmodulin has been noted in the interphase cells. Rapidly growing tumour cells contain more calmodulin than their slowly growing non-tumourous counterparts (MacManus et al. 1981; Wei and Hickie 1981; Wei et al. 1982). Calmodulin has also been reported to stimulate DNA synthesis (Boynton et al. 1980) and to stimulate lymphocyte proliferation (Gorbackhevaskaya et al. 1983).

Transformation of mammalian tissue culture cells by oncogenic viruses caused a twofold increase of calmodulin (Chafouleas et al. 1981; Veigl et al. 1982) through increased synthesis rather than decreased degradation, while the tubulin levels were unchanged. Calmodulin has been reported to mediate the calcium-dependent assembly–disassembly of microtubules, so that an increase of calmodulin in transformed cells could destabilize the microtubules with a consequent decrease in their number.

Other calcium-binding proteins are also found in tumours (MacManus 1979, 1980, 1981). A specific calcium-binding protein, oncomodulin, has been found in many of the chemically induced rat hepatomas and in rat kidney cells virally transformed by avian sarcoma virus, but not in uninfected rat kidney cells (Durkin et al. 1983). Such high levels of intracellular calcium-binding proteins are thought to be involved in the uncontrolled growth of tumour cells. Pfyffer et al. (1984), applying immunohistochemical methods in conjunction with a non-specific anti-parvalbumin anti-serum on three human cancer cell lines, LICR (Loud)-HN1, HN2 and -NH6, with various degrees of motility, demonstrated high levels of calcium-binding protein in cells with translocative motility. Unique calcium-binding proteins, absent from non-neoplastic cells of the same tissue origin, were also found. One of them showed electrophoretic motility similar to parvalbumin with homology in the amino acid sequence and a molecular weight of 12 000. Spermatocytes also contain a very high concentration of calmodulin, localized in the head region (Jones et al. 1980).

The calmodulin concentration, in fact, appears to be directly related to the proliferative response. The mitogenic response of T lymphocytes stimulated with lectin is accompanied by an increase in calmodulin (Yates and Vanaman 1980), and phenothiazines which antagonize calmodulin inhibit the mitogenic response of lymphocytes. In rapidly growing Morris hepatoma cells, a tenfold increase in the ratio of soluble to particulate-bound calmodulin has been reported, suggesting a shift of calmodulin to a predominantly cytoplasmic localization in rapidly growing cells.

Since phosphodiesterase activation is one of the major actions of calmodulin, the increased calmodulin level in rapidly growing tumour cells is usually associated with an augmented cyclic AMP (cAMP)-degrading activity by phosphodiesterase with consequent decrease of intracellular cAMP content. Cyclic GMP (cGMP) degradation, on the other hand, is usually decreased, raising cGMP levels (Vensshi et al. 1980; Criss and Kakiuchi 1982). Human leukaemia cells also have a high calmodulin content (Takemoto and Jilka 1983). However, Moon et al. (1983) questioned the functional significance of calmodulin in the growth of human cancer cells, based on their failure to find a difference in calmodulin content between human renal carcinoma cell lines and foetal kidney cell lines.

Disturbance of cellular calcium homeostasis with calcium antagonists and calmodulin antagonists increases the susceptibility of tumour cells to cytotoxic chemotherapeutic agents. In the light of the important role of intracellular calcium and calmodulin in cell proliferation, this might be related to antagonism of the uncontrolled cell growth induced by disturbed calcium homeostasis (Tsuruo et al. 1982; Ito and Hidaka 1983; Gonopathi et al. 1984; Kikuchi et al. 1984; Lazo et al. 1984; Yanovich and Preston 1985; Yalowich and Ross 1985).

The mechanism of the action of calcium–calmodulin on mitosis has not been fully elucidated. Froscio et al. (1981) demonstrated a block by a calmodulin antagonist of early responses of cultured mammalian cells to the tumour-promotor TPA. In the presence of 40 μM trifluoperazine (TFP), mouse epidermal cells were insensitive to the TPA inhibition of epidermal growth factor binding. Trifluoperazine also markedly inhibited the basal rate of ^{3}H-choline incorporation into HeLa cell phospholipids. The association of calmodulin with microtubules does not appear to require calcium, since fluorescein-labelled anti-calmodulin anti-serum stains microtubules after lysis and incubation with 10 mM EGTA. Deery et al. (1984) demonstrated the association of calmodulin with both the cytoplasmic microtubule complex and the centrosomes. By interacting directly with the microtubules, calmodulin apparently influences microtubular assembly and increases the calcium sensitivity of both mitotic and cytoplasmic microtubules. Owada et al. (1984) observed a reduced content of caldesmon 77, a calcium-binding protein with molecular weight of 77 000 associated with actin filaments in Rous sarcoma virus-transformed cells.

Calcium and Cyclic Nucleotides

Adenosine-3',5'-cyclic monophosphate (cAMP) controls mammalian cell growth. Growth-inducing factors decrease cAMP and intracellular cAMP is also reduced in virally transformed cells (Burk 1968; Friedman 1976; Ralph 1983). The growth-promoting action of serum is reversed by cAMP, and agents which increase cellular cAMP reduce proliferation, suggesting an inverse relationship between intracellular cAMP and cell proliferation (MacManus and Whitfield 1969a; Heidrik and Ryan 1970). Johnston et al. (1971) treated transformed fibroblasts with dibutyryl cAMP, causing a change from a polygonal shape to the original elongated shape. High intracellular cAMP usually stops the cycle of the G_1 phase. According to Friedman (1976), intracellular cAMP rises as the cell enters the G_0 quiescent state and falls as the cell enters the G_1 growing state. Agents used to raise cellular cAMP may also change the intracellular concentrations of calcium and other substances affecting cell growth. Cells made quiescent by serum restriction usually contain abundant cAMP (Kram et al. 1973; Moens et al. 1975).

Cyclic AMP, on the other hand, stimulates cell proliferation. Very low concentrations of cAMP (10^{-8}–10^{-6} M) stimulate the lymphoblasts occupying 20% of the lymphoid cells in the thymus (MacManus and Whitfield 1971). When thymocyte proliferation is stimulated by an extracellular calcium concentration over 1 mM, intracellular cAMP concentration rises. Imidazole decreases cAMP by activating phosphodiesterase and inhibits the mitogenic action of calcium and caffeine which increases cAMP. Cholera toxin and prostaglandin E_1 also increase intracellular cAMP and act synergistically with various mitogens. Cyclic AMP stimulates DNA synthesis in rat thymocytes and hepatocytes. Adenylate cyclase stimulators, such as epinephrine and prostaglandin E_1 and A_1, stimulate blocked thymic lymphoblasts in a calcium-free medium to initiate DNA synthesis (Franks et al. 1971; MacManus et al. 1971; Whitfield et al. 1972a, b).

The inhibition by cAMP or dibutyryl cAMP of growth initiation by serum or by mitogens is apparently mediated by inhibition of calcium influx at the critical stage of the G_1 phase. Cyclic AMP also reduces calcium availability by stimulating extrusion of calcium from the cell, and this action is overcome by the calcium ionophore A23187. While cyclic AMP generally decreases intracellular calcium and antagonizes its mitogenic action, cAMP-dependent phosphorylation may be required for actual DNA synthesis after the cell passes the G_1/S boundary. In fact, it is difficult to distinguish between the actions of calcium and cAMP because their actions are frequently intermingled or "synerchic" (Berridge 1976; Rasmussen 1981).

Cyclic GMP has been thought to antagonize cAMP in a "yin-yang" fashion (Goldberg et al. 1973). Cyclic GMP levels fall markedly as cells enter quiescence (Rudland et al. 1974). Serum and growth factors cause a dramatic rise of cellular cGMP (Rudland et al. 1974). In the meiotic maturation of the amphibian oocyte, progesterose increases calcium influx and the release of calcium from intracellular reservoirs, and this is followed by a rise of cGMP. An increase in the cGMP/ cAMP ratio is thought to be important in cell division and proliferation (Kostellow and Morril 1980).

In density-dependent inhibition, cGMP levels fall but cAMP levels do not. Calcium appears to be involved in the action of both these nucleotides (Berridge 1975b). Very low concentrations of cAMP or cGMP can replace calcium to induce mitotic stimulation, DNA synthesis and cell proliferation in human and rat bone marrow cells (Rixon et al. 1970; Byron 1972; Tisman and Herbert 1973). Calcium causes a sudden rise of cGMP in thymic lymphocytes after pretreatment with concanavalin A (ConA) (Whitfield et al. 1974). A direct role of cGMP in stimulating DNA synthesis, however, is unlikely, in view of the inverse relationship between cGMP and proliferation of lymphoblasts exposed to acetylcholine.

Involvement of cAMP in liver regeneration has also been suggested (Short et al. 1975). Exogenous cAMP stimulates DNA synthesis in cultures of proliferative active hepatocytes (Medoff and Parker 1971; Armato et al. 1974a, b). The calcium-mediated action of cAMP may be connected with its capacity to release calcium from mitochondria (Borle 1974).

Adenylate cyclase activity and cAMP levels are higher in hepatoma cells than in normal cells (Chayroth et al. 1972; Thomas et al. 1973; Boyd et al. 1974). Intraperitoneal injection of cAMP in mice facilitates the appearance of papillomas and squamous cell carcinomas of the skin induced by 7,12-dimethylbenzanthracene (Curtis et al. 1974). It is possible that cAMP and calcium act synergistically in cell proliferation, but the action of cGMP still remains uncertain. Calcium inhibits the cAMP-mediated stimulation of thymic lymphocytes by prostaglandin E_1 (Whitfield et al. 1972a). Cyclic AMP may thus play a decisive role at a later stage of prereplicative development. The later prereplicative cAMP surge observed in BALB/3T3 cells occurs in various normal and neoplastic cells, and elimination of such surges reduces DNA synthesis (Boynton et al. 1978).

According to Berridge (1975a, b), calcium rather than either cAMP or cGMP is the central regulator of all the phenomena in cell division. Cyclic AMP may increase intracellular calcium by stimulating the release of calcium stored within the cell or prolong the influx of extracellular calcium, but it may have a completely opposite effect under other circumstances. Inconsistencies in the action of cyclic nucleotides may therefore be explained by the variable response of intracellular calcium to cyclic nucleotides. Calcium, in turn, may stimulate the formation of cGMP.

Phospholipid and Protein Phosphorylation

Cell membranes do not function adequately unless the inositol phospholipids maintain normal turnover. Michell (1975) pointed out the causal role of inositol phospholipid breakdown in the rise of cytosolic free calcium, mainly through the action of inositol-1,4,5-triphosphate mobilizing calcium from the endoplasmic reticulum, which also plays a role in the increase of intracellular free calcium in sea urchin eggs on fertilization (Berridge and Irvine 1984).

The ubiquitous enzyme activated by calcium and diacylglycerol, probably acting as a receptor for tumour-promoting phorbol ester (TPA), has linked the wide gap between phospholipid, calcium and cell proliferation (Castagna et al. 1982; Kikkawa et al. 1983; Niedel et al. 1983; Nishizuka 1984; Truneh et al. 1985). The effect of TPA on lymphocyte proliferation occurs in the absence of extracellular calcium, or a detectable change in intracellular free calcium in response to mitogen, probably because the calcium requirement of protein kinase C is extremely small (Gelfand et al. 1985). Instead of a straightforward rise of intracellular calcium, TPA-induced oscillations of cytosolic calcium may be important in the initiation of proliferation (Cuthbertson and Cobbold 1985).

The mechanism of viral transformation of a cell is also related to protein kinase C and phorbol esters (Macara et al. 1984). Oncogene kinases may increase phosphatidylinositol turnover which leads to increased production of diacylglycerol activating protein kinase C. A feedback regulation may also occur in the signal transduction system involving the phosphatidylinositol breakdown pathway. In cultured human 1321 N_1 astrocytoma cells, muscarinic receptor stimulation leads to phosphoinositide hydrolysis and subsequent mobilization of intracellular calcium, which is in turn blocked by TPA (Orellana et al. 1985).

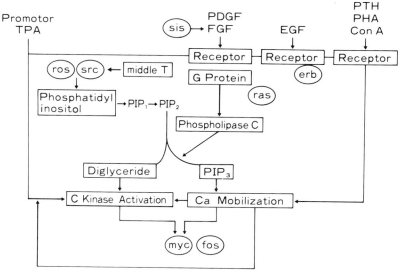

Fig. 16.2.   The effect of a rise of intracellular calcium on the potency of phorbol ester in protein kinase C activation.

The phorbol ester–protein kinase C system also stimulates differentiation. Very low concentrations of phorbol ester can induce differentiation of some leukaemia cells but cause proliferation of melanoma cells. Protein kinase C may mediate the TPA-induced differentiation of the human leukaemia cell line HL-60 (Nakai et al. 1984), whereas the methylation inhibitor, 3-deazaadenosine, inhibits differentiation at a later stage (Feuerstein and Cooper 1984).

The G_0–G_1 and G_1–S transitions of $T_{51}B$ rat liver epithelial cells in serum-stimulated confluent culture, requiring a high concentration of extracellular calcium, are accompanied by an increase of protein kinase C. A release of protein kinase C from the restraint of the normal controlling frame may be the key to neoplastic transformation (Boynton et al. 1985). In fact, it is important not to forget the complex and multifactorial action of protein kinase C which further confuses the already complex mechanism of cell proliferation.

Intracellular calcium may modify the phorbol ester-induced alterations in cell morphology as shown by Smith et al. (1983) in murine macrophages. Protein kinase C together with calcium controls the affinity of the EGF receptor (Fearn and King 1985) and activates lymphocytes increasing mIg-A expression (Monvae et al. 1984). A rise of intracellular calcium may increase the potency of phorbol ester in protein kinase C activation (May et al. 1985) and lead to recruitment of protein kinase C to the plasma membrane (Fig. 16.2).

Calcium and Immune Function

One of the most dramatic events in the immunological defence mechanism of the living organism is the functional activation and subsequent clonal proliferation of quiescent lymphocytes in response to lectin or antigen stimulation. The role of calcium as the messenger in lymphocyte mitogenesis has been well documented (Freedman 1979; Lichtman et al. 1983). Lymphocyte mitogenesis is usually accompanied by ^{3}H-thymidine incorporation 48–72 hours after the contact with mitogens. Lectin binding by lymphocytes requires calcium (Lindahl-Kiessling 1976). Calcium is also required during the first 20 hours of lectin stimulation (Whitney and Sutherland 1972b).

Exposure of lymphocytes to PHA or ConA, on the other hand, increases calcium uptake (Allwood et al. 1971). Although net cell calcium does not necessarily increase, calcium exchange always increases on mitogenic stimulation, including antigen-specific stimulation (Nisbet-Brown et al. 1985), probably through an increased membrane permeability for calcium. The calcium ionophore A23187 can also stimulate lymphocytes to mitogenesis, by increasing the total and radiolabelled cell calcium four to five times. The increased calcium uptake by T lymphocytes induced by ConA is modulated by cyclic nucleotides (Freedman and Raff 1975); cAMP inhibits and cGMP enhances it. Phorbol esters also activate lymphocytes in a calcium-dependent fashion (Mastro and Smith 1983). Interleukin-2 (IL-2) stimulation is also accompanied by calcium influx (Gearing et al. 1985) as well as by a rise in intracellular pH (Mills et al. 1985). The response to IL-2 is blocked by the calcium antagonists nifedipine and verapamil as well as by calmodulin antagonists.

Calmodulin is thus apparently involved in lymphocyte mitogenesis. Transmembrane signalling by the T lymphocyte antigen receptor complex is linked to a membrane-potential-sensitive calcium influx. Using murine Lyt-2$^+$ cytotoxic or human T$_4$ helper T cell clones, Albert et al. (1985) showed that the activation of protein kinase C can replace the antigen-receptor triggering events leading to interleukin secretion and expression of IL-2 receptors. Calcium ionophores and TPA synergistically stimulate antigen-dependent lymphocyte activation. The action of A23187 on lymphocytes appears to be interleukin-independent, because the stimulation by A23187 takes place in a monocyte-depleted population without addition of IL-1 (Koretzky et al. 1983).

The antigen receptor on the surface of B lymphocytes, the surface immunoglobulin, undergoes a series of movements in response to specific antigen stimulation, including a redistribution called capping which is subsequently cleared by endocytosis (Braun et al. 1979). Since such a process requires a contractile response probably involving the cytoskeleton, a relationship with calcium has been suggested. Calcium ionophore is capable of dispersing the cap of surface immunoglobulin, possibly through the contraction of microfilaments. Since capping may occur in the absence of extracellular calcium, mobilization of calcium from intracellular sources probably takes place (Schreiner and Unanue 1976).

Calmodulin, representing 0.58% of the total cellular protein, is mainly found in the region of the Golgi apparatus and diffusely over the cytoplasm, but not in the nuclei of T and B lymphocytes (Bachivaroff et al. 1984). Lymphocytes cultured in the presence of trifluoperazine (TFP), a calmodulin antagonist, failed to undergo blastic transformation in response to antigens ConA and PHA. Calcium-binding proteins with molecular weights of 68 000, 33 000 and 28 000 have also been found on lymphocyte plasma membranes. The largest of them may be related to calelectrin, identified in the synaptic membrane of electric cells (Davies and Crumpton 1985). Among various ion channels in the lymphocyte plasma membrane, a calcium channel is required for immunoglobulin synthesis by B cells (Chandy et al. 1985).

Calcium is also involved in cytotoxicity. Excessive calcium influx induced by A23187 is cytotoxic to lymphocytes. The presence of extracellular calcium is essential to this phenomenon. Exposure of natural killer cells to TFP prevents the effective phase of killing. The well-known lymphotoxic action of glucocorticoids is also mediated by calcium (Kaiser and Edelman 1977). Removal of calcium from the medium attenuates the effect of triamcinolone acetate on lymphocytes. DNA cleavage by endonuclease activation induced by the entrance of calcium into the nucleus is probably responsible for this lymphotoxic action of corticosteroids (Cohen and Duke 1984). The lysis of the tumour cells by lymphocytes is also apparently dependent on calcium (Gately and Martz 1979). This action of calcium on cell lysis is specific and cannot be replicated by strontium and other alkaline earth metals.

Oncogenes and Growth Factors

Cell growth depends on numerous growth factors which act in three different ways: on distant cells via the blood stream in the classical endocrine fashion; on neighbouring cells in direct contact in the paracrine fashion; and on the cells from which they were secreted in the autocrine fashion (Sporn and Todaro 1980). Oncogenes found in tumour cells are apparently responsible for the synthesis of growth factors themselves, growth factor receptors and some of the enzymes, including protein kinases, for postreceptor events. Among the oncogenes recently shown to be involved in cell proliferation, c-*sis* encodes the platelet-derived growth factor and c-*erb*-B encodes nothing but the epidermal growth factor (EGF) receptor.

The differentiation of cultured human keratinocytes is regulated by calcium through EGF (O'Keefe and Payne 1983). Changes of morphology and decreased differentiation of human keratinocytes in low calcium medium is mediated by the increase of EGF receptor numbers without changes in affinity; the binding of ConA and somatostatin is also enhanced.

The endocytosis of EGF is also controlled by cytosolic calcium (Tupper and Bodine 1983). The ability of serum or foetal bovine serum protein to support fibroblast cell growth depends on the presence of extracellular calcium. The well-known phenomenon of growth arrest in human fibroblasts (W138) in response to lowering of extracellular calcium, and the disappearance of such calcium dependency in SV-40-transformed cells, may be explained by the difference in calcium requirement for receptor-mediated endocytosis between these cells. Both internalization and degradation of EGF receptor complex are reduced by calcium deprivation in intact but not in virus-transformed cells. Calmodulin antagonists inhibit EGF internalization and degradation both in normal and transformed W138 cells, with a marked inhibition of EGF binding to cell surface receptors.

Since phorbol esters induce phenotypic changes similar to those caused on transformation by virus, including activation of tyrosine-specific protein kinases, these tumour promotors also inhibit EGF interaction with the cells. Phorbol esters inhibit EGF binding by intact cells, but not the binding to membrane preparations, suggesting the importance of cytosolic factors in the control of EGF-binding to the receptor. Preventing calcium entry with EGTA, or increasing intracellular calcium with A23187, controls EGF affinity to receptors. Activation of protein kinase C appears to modulate EGF receptor affinity (Fearn and King 1985) probably by controlling protein phosphorylation (Cochet et al. 1984).

A rise of ecto-calcium-ATPase activity is a useful indicator of the neoplastic transformation of the cell surface, and the increased expression of the ecto-calcium-ATPase depends on EGF, cholera toxin and hydrocortisone in a clonal cell line from primary cultures of a human hepatoma xenograft (Knowles et al. 1985). EGF reduces the requirement for serum and calcium in human epithelial cell cultures, but neither serum nor calcium change EGF requirement (Lechner and Kaighn 1979; McKeehan and McKeehan 1979; Lechner 1984).

To explain the action of platelet-derived growth factor (PDGF) on 3T3 cells, Frantz (1985) reported a decrease in cell ^{45}Ca content and increased prelabelled ^{45}Ca efflux, suggesting a release of Ca^{2+} from intracellular stores; this is similar to the action of A23187, but unlike that of EGF, insulin and TPA. Platelet-derived growth factor also increases cytosolic free calcium. Because TPA initiates 3T3 cell growth without affecting calcium stores, protein kinase C activation may represent the final common pathway in the stimulation of cell proliferation. While PDGF does not stimulate DNA synthesis in human neonatal skin fibroblasts plated sparsely in MCD8, PDGF becomes mitogenic in low calcium medium or in the presence of the calmodulin inhibitor, W-7, or in confluent fibroblasts under physiological pH, suggesting the importance of extracellular calcium and cell density in the action of growth factors. Both EGF and PDGF act additively on this system (Betsholz and Westermark 1984).

Calcium may also release other factors. Cultured human alveolar macrophages produce various factors stimulating fibroblast proliferation, prostaglandin E_2 and collagenase production. Calcium ionophore A23187 and leukotriene B_4 stimulate the release of these growth factors (Polla et al. 1985).

The *scr* and *ros* oncogene products facilitate the phosphorylation of phosphatidyl inositol to phosphatidyl ionositon 4,5-phosphate (Macara et al. 1984; Sugimoto et al. 1984). The *ras* oncogene product might be related to G_s protein-connecting receptor and adenylate cyclase. Table 16.1 summarizes various factors influencing cell proliferation.

Table 16.1.   Factors involved in cell proliferation and differentiation

Proliferation	Anti-proliferation differentiation
Calcium	
Cyclic GMP	Cyclic AMP
Alkalinity	Acidity
Calcium ionophores	Calcium antagonists
Calmodulin	Calmodulin antagonists
Protein kinase C	Protein kinase C-antagonists
Phorbol esters	
Interleukins	
Interferons	
Parathyroid hormone	Calcitonin, $1,25(OH)_2$ vitamin D_3

This simplified classification is not intended to be universal; it is only applicable to certain kinds of cells at certain stages of development.

Bone Cell Differentiation

Bone is formed by osteoblasts of mesenchymal origin (like fibroblasts) (Owen 1978, 1980), and resorbed by osteoclasts of haematopoietic stem cell origin (like monocytes or macrophages) in a process of remodelling (Seiff et al. 1983; Chambers and Morton 1984). For some unknown reason, haematopoiesis takes place in the bone marrow in land-based animals. Sea-based animals like fish, and the foetus in the amniotic fluid of land-based animals, depend on extramedullary haematopoiesis. This might be explained by a special need for calcium in haematopoiesis, a process which requires continuous cell proliferation and differentiation. Since seawater contains abundant calcium, sea-based animals are never deficient in calcium, and haematopoiesis can operate in many different sites. In land-based animals, on the other hand, bone marrow in close proximity to the bone, containing 99% of the whole calcium store of the body, is the only place with a sufficient calcium supply.

The absence of parathyroids in fish and their presence in land-based animals might indicate the need for this endocrine gland to compensate for the calcium-deficient environment and to provide the bone marrow cavity (Fujita 1985, 1986). The high level of parathyroid hormone in the bone marrow blood of osteoporotic patients compared to the normal level in peripheral blood (Atkinson et al. 1984) might also indicate the intimate functional relationship between parathyroid hormone and bone marrow cavity.

Osteoclast-activating factor (OAF), described for the first time by Horton et al. (1972, 1974), is released from activated T lymphocytes after their interaction with macrophages (Luben et al. 1974; Mundy et al. 1977; Yoneda and Mundy 1979; Horowitz et al. 1984). Osteoclast-activating factor is also secreted from myeloma cells as well as from normal peripheral blood mononuclear cells and probably plays an important role in the control of bone resorption in normal and pathological states (Mundy and Salmon 1974; Trummel et al. 1975; Coccia 1984; Raisz and Kream 1984). Osteoclast-activating factor is probably a group of cytokines acting on the bone – one of them was recently identified as interleukin 1.

Calcium-Regulating Hormones

The calcium-regulating hormones, parathyroid hormone (PTH), calcitonin (CT) and $1,25(OH)_2$ vitamin D_3 control the homeostasis of calcium within the cell as well as in the whole living organism. Through calcium, these hormones influence growth and proliferation of not only bone and kidney cells and their immediate target organs, but also many other kinds of cells throughout the organism.

$1,25(OH)_2$ Vitamin D_3

Following the reports of Abe et al. (1981) and Miyaura et al. (1981) on the action of $1,25(OH)_2$ vitamin D_3 [$1,25(OH)_2D_3$] in inducing differentiation and inhibition of proliferation of mouse leukaemia (M1) and human leukaemia (HL-60) cell lines, many studies were conducted on the cytokinetic action of this hormone and these are summarized in Table 16.2.

As to the clinical implication of the cytostatic action of $1,25(OH)_2D_3$, the possible use of this hormone in treating human neoplasms has been explored but so far with little success (Colston et al. 1981; Frampton et al. 1983; Koeffler 1983; Cunningham et al. 1985). Fujita et al. (1984) found an increased $T4^+/T8^+$ ratio in the lymphocyte subset of the peripheral blood of patients with osteoporosis and this is also the case in many patients with autoimmune diseases. $1,25(OH)_2D_3$ administration caused a fall of this ratio to normal along with clinical effects. Psoriasis possibly involving abnormal proliferation of immune cells responded favourably to $1\alpha(OH)$ vitamin D_3 administration (Morimoto et al. 1986).

Calcitonin

Calcitonin is produced mainly by the parafollicular C-cells of the thyroid in mammals and the ultimobranchial body in submammalian vertebrates. A

Table 16.2. Cytokinetic action of 1,25(OH)$_2$ vitamin D$_3$

Author	Year	Findings
Abe et al.	1981	1,25(OH)$_2$D differentiates M1
Miyaura et al.	1981, 1982	1,25(OH)$_2$D differentiates HL-60
Tanaka et al.	1982	Receptor-mediated action of 1,25(OH)$_2$D
Sato et al.	1982, 1984	1,25(OH)$_2$D inhibition of Lewis lung carcinoma and rat hepatoma K231
Kuribayashi et al.	1983	Variant HL-60 clone resistant to 1,25(OH)$_2$D
Reitsma et al.	1983	C-myc suppression by 1,25(OH)$_2$D in HL-60
Murao et al.	1983	Comparison of 1,25(OH)D and TPA action of HL-60
Homma et al.	1983	Prolonged survival of M1 bearing mice by 1,25(OH)$_2$D
Wood et al.	1983	1,25(OH)$_2$D inhibition of mouse skin carcinogenesis
Mangelsdorf et al.	1984	HL-60 differentiation into macrophage by 1,25(OH)$_2$D
Matsui et al.	1984	Phenotypic analysis of HL-60 differentiation by 1,25(OH)$_2$D
Amento et al.	1984	
Rigby et al.	1984	1,25(OH)$_2$D effect on U-937
Mazzetti et al.	1984	
Suda et al.	1984	1,25(OH)$_2$D inhibition of differentiation of Friend erythroleukaemia cells
Suda et al.	1985	1,25(OH)$_2$D inhibits mouse skin carcinoma induction
Ohta et al.	1985	Stimulation of human peripheral monocyte proliferation by 1,25(OH)$_2$D
Matsui et al.	1985a	Failure of calmodulin antagonist inhibition of 1,25(OH)$_2$D induced HL-60 differentiation
Matsui et al.	1985b	Protein kinase C inhibitor H7 suppresses phenotypic changes of HL-60 induced by 1,25(OH)$_2$D
Matsui et al.	1985c	IFN-γ and 1,25(OH)$_2$D cooperatively inhibits c-myc expression of HL-60
Koizumi et al.	1985	1,25(OH)$_2$D inhibits corticosteroid-resistant thymic lymphocytes

smaller amount of calcitonin is also produced from the cells of the so-called APUD system, scattered all over the body in the lungs, pancreas, thymus and intestine. Various neoplasms, especially those originating from this cell group, produce calcitonin. The receptor for calcitonin is also found in various neoplasms. The human breast cancer cell line (T47D) has a specific, high-affinity calcitonin receptor and the growth of this tumour is inhibited by calcitonin (Iwasaki et al. 1983). Calcitonin and 1,25(OH)$_2$D$_3$ have an additive effect. Cells with calcitonin receptors frequently secrete calcitonin themselves, suggesting an autocrine effect (Zajac et al. 1984). Pertussis toxin, which inactivates the inhibitory regulatory component nickel, is capable of inhibiting adenylate cyclase responses to prostaglandin E$_2$ and to calcitonin in human breast cancer cells (Michelangeli et al. 1984).

In patients with liver cirrhosis, a rise in plasma calcitonin may indicate the development of hepatoma (Conte et al. 1984). Ectopically produced calcitonin is thus utilized as a cancer marker. When pulmonary endocrine cell hyperplasia is induced by diethylnitrosamine in a hamster, serum calcitonin again serves as the indicator for the hyperplastic change (Linnoila et al. 1984). Calcitonin inhibits metastasis of D_s sarcoma cells (Anghileri et al. 1980) and epinephrine–cAMP-induced proliferation of thymocytes (Whitfield et al. 1971). Calcitonin also sensitizes the hepatocyte to the action of calcium in the regenerating liver (Rixon et al. 1979). The calcitonin deficiency that results from thyroidectomy in normocalcaemic rats with functional parathyroid transplants reduces DNA synthesis by regenerating hepatocytes. Injection of calcitonin restores DNA synthesis by hepatocytes. Serum calcium restoration by parathyroid hormone and $1,25(OH)_2D_3$ is also effective in stimulating DNA synthesis.

Parathyroid Hormone (PTH)

The last of the three calcium-regulating hormones, parathyroid hormone, has been known for the longest time, but was the latest to appear in evolution. The participation of PTH in tumour proliferation was clearly demonstrated by Kaplan et al. (1971) who compared 118 patients with primary hyperparathyroidism due to parathyroid adenoma with 117 control patients of the same age with thyroid carcinoma. All were autopsy cases. A gastrointestinal tumour was found in 30% of patients with parathyroid adenoma but only in 12% of patients with thyroid carcinoma. Carcinoma of the breast was found in 29% of the former and 7% of the latter. Such a high incidence of malignancy in primary hyperparathyroidism would indicate a tumour-promoting effect of excess PTH. Hypercalcaemia is the most consistent finding in primary hyperparathyroidism and parathyroid hormone stimulates calcium influx, at least in its target cells. This would stimulate cell proliferation and eventually malignant neoplasia. Ziv et al. (1985) found pancytosis with 20 g/dl haemoglobin, 59% haematocrit, $25\,000/mm^3$ WBC and $450\,000/mm^3$ platelets in a patient with primary hyperparathyroidism. After removal of the parathyroid adenoma, all the haematological findings returned to normal, suggesting PTH plays a role in stimulating blood cell proliferation.

Parathyroid hormone also appears to be important in lymphocyte proliferation and cell-mediated immunity. The increase of lymphocyte numbers promoted by non-specific mitogenic lymphokines is markedly impaired in hypocalcaemic parathyroidectomized rats (Perris et al. 1984). The parathyroid glands are essential for the development of the hypercalcaemic episode which follows an antigenic challenge causing cell proliferation in primary lymphoid tissues.

Parathyroid hormone stimulates mitotic activity in HeLa cells in culture (Borle and Neuman 1965), possibly through increasing calcium influx (Borle 1968). Parathyroidectomy 24 hours before the injection of sheep red cells into rats inhibits the immune response, including splenic DNA synthesis and the subsequent increase in the proportion of plaque-forming cells, but parathyroidectomy after antigen injection is without effect (Swierenga et al. 1976b). Parathyroid hormone thus seems to control the proliferative phase of the immune response, although a possible role of extracellular calcium concentration cannot be excluded.

Like liver regeneration after partial hepatectomy, renal compensatory growth after unilateral nephrectomy is enhanced by PTH. Calcium restriction, causing secondary hyperparathyroidism, also stimulates compensatory renal hypertrophy, so that PTH itself or $1,25(OH)_2D_3$ is probably involved in the stimulation of compensatory renal hypertrophy. The participation of calcium itself is rather unlikely because hypocalcaemia also stimulates this process. Calcitonin levels decrease on calcium restriction, but the role of calcitonin does not appear important in controlling renal cell proliferation (Jobin et al. 1984). The mitotic stimulation of bone marrow cells by PTH is an important mechanism in the recovery from acute haemorrhage (Byron 1972).

Calcium Deficiency and Neoplasia

Nutritional calcium deficiency causes secondary hyperparathyroidism, probably as the cumulative result of incipient and transient hypocalcaemia. Since the only calcium supply received by land-based animals is obtained by mouth, they appear to be relatively more deficient in calcium than fish living in seawater, which take as much calcium as they need through their gills. The absence of parathyroids in seawater-based fish and their appearance in land-based animals might indicate that the parathyroid glands compensate for calcium deficiency (Roth and Schiller 1976).

In order to maintain a normal serum calcium level, parathyroid hormone mobilizes calcium from bone, so that the loss of bone calcium with consequent calcium deposition in soft tissues represents a characteristic change of ageing. Excessive PTH stimulates calcium influx which leads to an increase in intracellular calcium, blunting the sharp difference in calcium concentration between the extracellular and intracellular compartments. Such blunting would decrease the excitability of nerve, endocrine and immune cells, probably contributing to the general decrease of cell reactivity and cell function seen in ageing. The loss of immune function in ageing, especially the loss of the immune surveillance mechanism, would increase the risk of cancer. Increased calcium influx induced by PTH might also stimulate cell proliferation and neoplastic changes. Calcium deficiency might thus explain at least part of the increased incidence of neoplasia with ageing.

The incidence of gastric cancer, which is the most common malignant neoplasm among the Japanese who have a notoriously low calcium intake, is exceptionally low in those who drink 380 ml milk daily compared to those who drink no milk at all [71 versus 153 per 100 000 population (Hirayama 1977)]. This would suggest that high calcium intake offers some protection against the development of gastric cancer. In chronic renal failure with deficiency of $1,25(OH)_2D_3$, disturbance of intestinal calcium absorption, calcium deficiency and secondary hyperparathyroidism, an increased incidence of neoplasm has been repeatedly reported, along with a disturbance of cell-mediated immunity (Matas et al. 1975; Jacobs et al. 1979).

Low calcium and vitamin D intake, as well as insufficient exposure to sunlight, increases the prevalence of colorectal cancer (Garland et al. 1985). The incidence of colon cancer in Finland, Norway, Sweden and Denmark is inversely related to milk consumption (Garland and Garland 1980). Intraluminal calcium ions could reduce the carcinogenic effects of fatty acids and free bile acids in the colon by converting them to soaps. The possibility that calcium deficiency and secondary hyperparathyroidism facilitate the development of cancer should also be considered (Newmark et al. 1985). Supplementation with 1.25 g/day calcium changes the pattern of cell proliferation in the epithelial cells lining the colonic crypts in subjects at high risk for familial colon cancer, reducing the risk of cancer (Lipkin and Newmark 1985).

Summary

Calcium is essential for many cell functions, including proliferation. To draw a comprehensive picture of the role of calcium in cell growth and replication, proliferation and neoplasia, a distinction should be made between the cellular and whole organism levels. The great difference between the concentrations of calcium in the skeletal, extracellular and intracellular compartments is well known. There is an almost 10 000-fold step-down in calcium concentration between each of these compartments, making the distribution of calcium unique in living organisms.

The relationship between cell calcium and body calcium may provide a key to the understanding of the control of cell growth and proliferation within the living organism. Because of this unique distribution, calcium serves as a messenger between different cells and compartments in the organism. In contrast, the calcium concentration in the extracellular fluid is one of the most strictly maintained biological constants, essential for the function of many vital organs. Since each cell is bathed constantly in the extracellular calcium, the cell environment should be studied first when trying to understand the control of cell growth. Parathyroid hormone, calcitonin and $1,25(OH)_2D_3$ are calcium-regulating hormones which are essential for the maintenance of the constant calcium concentration in the extracellular fluid. At the same time they play an important role in controlling cellular calcium homeostasis and proliferation through the regulation of calcium fluxes and the synthesis of calcium-binding proteins. Calmodulin and other calcium-binding proteins, as well as phospholipid degradation products, are also important in calcium transport and the maintenance of intracellular calcium homeostasis.

Studies on protein kinase C and phorbol esters have recently brought many loose ends together in the field of cell growth and proliferation, contributing a great deal to a unified concept of mitogenesis and proliferation under the influence of calcium, growth factors and oncogenes. Calcium plays a central role in cell proliferation and in the concerted action of all these factors. Further progress in our understanding of cellular calcium metabolism and oncogenesis could be relevant to the prevention and treatment of cancer.

References

Abe E, Miyaura C, Sakagami H et al. (1981) Differentiation of mouse myelid leukemia cells induced by 1α25(OH)₂ vitamin D₃. Proc Natl Acad Sci USA 78:4990–4994

Albert F, Hua C, Truneh A et al. (1985) Distinction between antigen receptor and IL-2 receptor triggering events in the activation of alloreactive T cell clones with calcium ionophore and phorbol ester. J Immunol 134:3649–3655

Allwood G, Asherson GL, Gaveey MJ (1971) The early uptake of radioactive calcium by human lymphocytes treated with phytohemagglutinin. Immunology 21:509–516

Amento EP, Bhalla AK, Kurnick JT et al. (1984) $1\alpha25$-dihydroxyvitamin D_3 induces maturation of the human monocyte cell line U937, and, in association with a factor from human T lymphocytes, augments production of monokine, mononuclear cell factor. J Clin Invest 73:731–739

Anghileri LJ, Miller ES (1971) Calcium metabolism in tumors: its relationship with chromium complex accumulation. I. Uptake of calcium and phosphorus by experimental tumors. Oncology 25:119–136

Anghileri LJ, Delbrück HG, Cressent M et al. (1980) Effect of calcitonin on metastasis formation by experimental tumors. Arch Geschwulstforsch 50:105–110

Armato U, Andreis PG, Belloni AS et al. (1974a) Long term stimulatory effects of adenosine 3′,5′-cyclic monophosphate on nuclear and cytoplasmic protein synthesis of rat hepatocyte in primary culture. Acta Anat 88:456–480

Armato U, Andreis PG, Draghi E et al. (1974b) cAMP and cGMP enhance hepatocyte DNA synthesis in vitro. In Vitro 9:357–361

Arslan PD, Virgilio F, Beltrame M (1985) Cytosolic Ca^{2+} homeostasis in Ehrlich and Yoshida carcinoma. J Biol Chem 260:2719–2727

Atkinson MJ, Schettler T, Bodenstein H et al. (1984) Osteoporosis. A bone turnover defect resulting from an elevated parathyroid hormone concentration within the bone narrow cavity. Klin Wochenschr 62:129–132

Bachivaroff RJ, Miller F, Rappaport FT (1984) The role of calmodulin in the regulation of human lymphocyte activation. Cell Immunol 85:135–153

Berridge MJ (1975a) Control of cell division: a unifying hypothesis. J Cyclic Nucleotide Res 1:305–320

Berridge MJ (1975b) The interaction of cyclic nucleotides and calcium in the control of cellular activity. Adv Cyclic Nucleotide Res 6:1–98

Berridge MJ (1976) Calcium cyclic nucleotide and cell division. In: Duncan CJ (ed) Calcium in biological systems. Cambridge University Press, pp 219–231 (Society for Experimental Biology Symposium XXX)

Berridge MJ, Irvine RF (1984) Inositol triphosphate, a novel second messenger in cellular signal transduction. Nature 312:315–320

Betsholz C, Westermark B (1984) Growth factor-induced proliferation of human fibroblasts in serum-free culture depends on cell density and extracellular calcium concentration. J Cell Physiol 118:203–210

Boissonneault GA, Heiniger HJ (1984) 25-hydroxycholesterol-induced elevation in ^{45}Ca uptake. Correlation with depressed DNA synthesis. J Cell Physiol 120:151–156

Borle AB (1968) calcium metabolism in HeLa cells and the effect of parathyroid hormone. J Cell Biol 36:562–582

Borle AB (1974) Cyclic AMP stimulation of calcium efflux from kidney, liver and heart mitochondria. J Membr Biol 17:221–236

Borle AB, Neuman WF (1965) Effects of parathyroid hormone on HeLa cell cultures. J Cell Biol 24:316–323

Boyd H, Louis CJ, Martin TJ (1974) Activity and hormone responsiveness of adenyl cyclase during induction of tumors in rat liver with 3′,5′-methyl-4-dimethyl-aminoazo-benzene. Cancer Res 34:1720–1725

Boynton AL, Whitfield JF (1976) Different calcium requirement for proliferation of conditionally and unconditionally tumorigenic mouse cells. Proc Natl Acad Sci USA 73:1651–1654

Boynton AL, Whitfield JF (1978) Calcium requirement for the proliferation of cells infected with a temperature-sensitive mutant of Rous sarcoma virus. Cancer Res 38:1237–1240

Boynton AL, Whitfield JF, Isaacs RJ (1975) Calcium-dependent stimulation of Balb/c 3T3 mouse cell DNA synthesis by a tumor-promoting phorbol ester (PMA). J Cell Physiol 87:25–32

Boynton AL, Whitfield JF, Isaacs RJ et al. (1977a) The control of human W1-38 cell proliferation by extra-cellular calcium and its elimination by SV-40 virus-induced proliferative transformation. J Cell Physiol 72:241–247

Boynton AL, Whitfield JF, Isaacs RJ et al. (1977b) Different extracellular calcium requirement for proliferation of non-neoplastic, preneoplastic and neoplastic mouse cells. Cancer Res 37:2657–2661

Boynton AL, Whitfield JF, Isaacs RJ et al. (1978) An examination of the roles of cyclic nucleotides in the initiation of cell proliferation. Life Sci 22:703–710

Boynton AL, Whitfield JF, MacManus JP (1980) Calmodulin stimulates DNA synthesis by rat liver cells. Biochem Biophys Res Comm 95:745–749

Boynton AL, Kleine LP, Whitfield JF et al. (1985) Involvement of the Ca^{2+} phospholipid-dependent protein kinase in the G_1 transit of T51B rat liver epithelial cells. Exp Cell Res 160:197–205

Braun J, Shaafi RI, Unanue EK (1979) Crosslinking by ligands to surface immunoglobulin triggers mobilization of intracellular $^{45}Ca^{2+}$ in B lymphocytes. J Cell Biol 82:755–766

Braunschweig A, Dunham LJ, Nichols S (1946) Potassium and calcium content of gastric carcinoma. Cancer Res 6:230–232

Brown WF (1984) The basic connection. Nature 312:312–313

Burk BR (1968) Reduced adenyl cyclase activity in a polyoma virus-induced cell line. Nature 219:1272–1275

Byron JW (1972) Evidence for a β-adrenergic receptor initiating DNA synthesis in hemopoietic stem cells. Exp Cell Res 71:228–232

Campbell AK (1983) Intracellular calcium. Its universal role as regulator. John Wiley, New York, pp 362–392

Castagna M, Takai Y, Kaibuchi K et al. (1982) Direct activation on calcium-activated, phospholipid dependent protein kinase by tumor-promoting phorbol esters. J Biol Chem 257:7847–7851

Chafouleas IG, Pardue RL, Brinkley BR et al. (1981) Regulation of intracellular levels of calmodulin and tubulin in normal and transformed cells. Proc Natl Acad Sci USA 78:996–1000

Chambers TJ, Morton MA (1984) Failure of cells of the mononuclear phagocyte series to resorb bone. Calcif Tissue Int 36:556–558

Chandy KG, Decoursey TE, Cahalan MD et al. (1985) Ion channels in lymphocytes. J Clin Immunol 5:1–4

Chayroth R, Epstein S, Field JB (1972) Increased cyclic AMP levels in malignant hepatic nodules of ethionine-treated rats. Biochem Biophys Res Comm 49:1663–1670

Cheung HS, McCarty DJ (1985) Mitogenesis induced by calcium containing crystals. Role of intracellular dissolution. Exp Cell Res 157:63–70

Coccia CF (1984) Cells that resorb bone. N Engl J Med 310:456–457

Cochet C, Gill GN, Meisenhelder J et al. (1984) C-kinase phosphorylates the EGF receptor and reduces its EGF-stimulated tyrosine protein kinase activity. J Biol Chem 259:2553–2558

Cohen JJ, Duke RC (1984) Glucocorticoid activation of a calcium-dependent endonuclease in thymocyte nuclei leads to cell death. J Immunol 132:38–42

Colston K, Colston MJ, Feldman D (1981) $1\alpha25$-Dihydroxy-vitamin D_3 and malignant melanoma: the presence of receptors and inhibition of cell growth in culture. Endocrinology 108:1083–1086

Coman DR (1944) Decrease of mutual adhesiveness. A property of cells from squamous cell carcinomas. Cancer Res 4:625–629

Conte N, Cecchettin M, Manente P et al. (1984) Calcitonin in hepatoma and cirrhosis. Acta Endocrinol 106:109–111

Criss WE, Kakiuchi S (1982) Calcium, calmodulin and cancer. Fed Proc 41: 2289–2291

Cunningham D, Gilchrist NL, Cowan RA et al. (1985) Alfacalcidol as a modulator of growth of low grade non-Hodgkin's lymphomas. Br Med J 291:1153–1156

Curtis GL, Stenback F, Ryan WL (1974) Enhancement of 7,12-dimethylbenzanthracene skin carcinogenesis by adenosine 3',5'-cyclic monophosphate. Cancer Res 34:2192–2195

Cuthbertson KSR, Cobbold PH (1985) Phorbol ester and sperm activate mouse oocyte by inducing sustained oscillations of cell Ca^{2+}. Nature 316:541–542

Dasdia T, DiMarco A, Goffredi M et al. (1979) Ion level and calcium fluxes in HeLa cells. Pharmacol Res Comm 11:19–29

Davies AA, Crumpton MJ (1985) Identification of calcium binding proteins associated with the lymphocyte plasma membrane. Biochem Biophys Res Comm 128:571–577

Deery WJ, Means AR, Brinkley BR (1984) Calmodulin–microtubule association in cultured mammalian cells. J Cell Biol 98:904–908

deLong RP, Coman DR, Zeidman I (1950) The significance of low calcium and high potassium in neoplastic tissue. Cancer 3:718–721

Dunham LJ, Nicholes S, Braunschweig A (1946) Potassium and calcium content of carcinomas and papillomas of the colon. Cancer Res 6:233–234

Durkin JP, Brewer LM, MacManus JP (1983) Occurrence of the tumor-specific calcium-binding protein, oncomodulin, in virally transformed normal rat kidney cells. Cancer Res 43:5390–5394

Eilam Y, Szydel N (1981) Calcium transport and cellular distribution in quiescent and serum-stimulated primary cultures of bone cells and skin fibroblasts. J Cell Physiol 106:225–243

Fearn JC, King AC (1985) EGF receptor affinity is regulated by intracellular calcium and protein kinase C. Cell 40:991–1000

Feuerstein N, Copper HL (1984) Studies of the differentiation of promyelocytic cells by phorbol ester. II. A methylation inhibitor, 3-deazaadenosine, inhibits the induction of specific differentiation proteins, lack of effect on early and late phosphorylation events. Biochim Biophys Acta 781:247–256

Fisher PB, Weinstein IB (1981) Enhancement of cell proliferation in low calcium medium by tumor promotors. Carcinogenesis 2:89–95

Frampton RJ, Omond SA, Eisman JA (1983) Inhibition of human cancer cell growth by 1α25-dihydroxy-vitamin D_3 metabolites. Cancer Res 43:4443–4447

Franks DJ, MacManus JP, Whitfield JF (1971) Effect of prostaglandins on cyclic AMP production and cell proliferation in thymic lymphocytes. Biochem Biophys Res Comm 44:1177–1183

Frantz CN (1985) Effects of platelet-derived growth factor on Ca^{2+} in 3T3 cells. Exp Cell Res 158:287–300

Freedman MH (1979) Early biochemical events in lymphocyte activation. Cell Immunol 44:290–313

Freedman MH, Raff MC (1975) Induction of increased calcium uptake in mouse T lymphocytes by concanavalin A and its modulation by cyclic nucleotides. Nature 155:378–382

Friedman DL (1976) Role of nucleotides in cell growth and differentiation. Physiol Rev 56:652–708

Froscio M, Guy GR, Murray AW (1981) Calmodulin inhibitors modify cell surface changes triggered by a tumor promotor. Biochem Biophys Res Comm 98:829–835

Fujita T (1985) Calcium and ageing. Calcif Tissue Int 37:1–2

Fujita T (1986) Ageing and calcium. Mineral Electrolyte Metab 12:149–156

Fujita T, Matsui T, Nakao T et al. (1984) T-lymphocyte subsets in osteoporosis. Mineral Electrolyte Metab 10:375–378

Fulton BP, Whittingham DG (1978) Activation of mammalian oocytes by intracellular injection of calcium. Nature 273:149–151

Ganapathi R, Grabowski D, Rouse W et al. (1984) Differential effect of the calmodulin inhibitor trifluorperazine on cellular accumulation, retention and cytotoxicity of anthrocyclines in doxorubicin (adriamycin)-resistant P388 mouse leukemia cells. Cancer Res 44:5056–5061

Garland CF, Garland FC (1980) Do sunlight and vitamin D reduce the likelihood of colon cancer? Int J Epidemiol 9:227–231

Garland C, Shekelle RB, Barett-Connor E et al. (1985) Dietary vitamin D and calcium and risk of colorectal cancer. A 19 year prospective study in men. Lancet I:307–309

Gately MK, Martz E (1979) Early steps in specific tumor lysis by sensitized mouse T lymphocytes. J Immunol 122:482–488

Gearing AJH, Wadhwa M, Perris AD (1985) Interleukin 2 stimulates T cell proliferation using a calcium flux. Immunol Lett 10:297–302

Gelfand EW, Cheung RK, Grinstein S (1984) Role of membrane potential in the regulation of lectin-induced calcium uptake. J Cell Physiol 121:533–539

Gelfand EW, Cheung PK, Mills GB et al. (1985) Mitogens trigger a calcium-dependent signal for proliferation in phorbol ester triggered lymphocytes. Nature 315:419–420

Goldberg ND, O'Dea BF, Haddox MK (1973) Cyclic GMP. Adv Cyclic Nucleotide Res 3:155–223

Gorbackhevaskaya LV, Borovkova PV, Rybin UO et al. (1983) Effect of exogenous calmodulin on lymphocyte proliferation in normal subjects. Bull Exp Med Biol 95:361–363

Grinstein S, Cohen S, Goetz JD et al. (1985) Characterization of the activation of Na^+/H^+ exchange in lymphocytes by phorbol esters. Changes in cytoplasmic pH dependence of the antiport. Proc Natl Acad Sci USA 82:1429–1433

Hazelton BJ, Tupper JT (1979) Calcium transport and exchange in mouse 3T3 and SV40-3T3 cells. J Cell Biol 81:538–542

Heidrik ML, Ryan WL (1970) Cyclic nucleotides on cell growth in vitro. Cancer Res 30:376–378

Heilbrunn LU (1937) An outline of general physiology. Saunders, Philadelphia

Hepler PK (1985) Calcium restriction prolongs metaphase in dividing *Tradescantia* stamen hair cells. J Cell Biol 100:1363–1368

Hesketh TR, Moor JP, Morris JDH et al. (1985) A common sequence of calcium and pH signals in the mitogenic stimulation of eukaryotic cell. Nature 313:481–484

Hickie A, Galant H (1967) Calcium and magnesium content of rat liver and Morris hepatoma 5123. Cancer Res 27:1053–1057

Hirayama T (1977) Changing patterns of cancer in Japan, with special reference to the decrease of stomach cancer mortality. In: Hiatt HH et al. (eds) Origins of human cancer. Cold Spring Harbor Laboratory, pp 55–75 (Cold Spring Harbor Conferences on Cell Proliferation, vol 4)

Homma Y, Hozumi M, Abe E et al. (1983) 1α25-Dihydroxy-vitamin D_3 and 1α-hydroxy-vitamin D_3

prolong survival time of mice inoculated with myeloid leukemia cells. Proc Natl Acad Sci USA 80:201–204

Horowitz M, Vignery A, Gershon RK et al. (1984) Thymus-derived lymphocytes and their interactions with macrophages are required for the production of osteoclast activating factor in the mouse. Proc Natl Acad Sci USA 81:2181–2185

Horton JE, Raisz LG, Simmons HA et al. (1972) Bone resorbing activity in supernatant fluid from cultured human peripheral blood leukocytes. Science 177:793–795

Horton JE, Oppenheim JJ, Mergenhagen SE et al. (1974) Macrophage–lymphocyte synergy in the production of osteoclast activating factor. J Immunol 113:1278–1287

Ito H, Hidaka H (1983) Antitumor effect of a calmodulin antagonist on the growth of solid sarcoma-180. Cancer Lett 19:215–220

Iwasaki Y, Iwasaki J, Freake HC (1983) Growth inhibition of human breast cancer cells induced by calcitonin. Biochem Biophys Res Comm 110:235–242

Izant JG (1983) The role of calcium ions during mitosis. Calcium participates in the anaphase trigger. Chromosoma 88:1–10

Jacobs C, Reach I, Degoulet P (1979) Cancer in patients on hemodialysis. N Engl J Med 300:1279–1280

Jobin J, Taylor CM, Caverzasio J et al. (1984) Calcium restriction and parathyroid hormone enhance renal compensatory growth. Am J Physiol 246:F685–690

Johnston GS, Friedman RM, Pastan I (1971) Restoration of several morphological characteristics of normal fibroblasts in sarcoma cells treated with adenosine-3',5'-cyclic monophosphate and its derivatives. Proc Natl Acad Sci 68:425–429

Jones HP, Lenz RW, Palevitz BA et al. (1980) Calmodulin localization in mammalian spermatozoa. Proc Natl Acad Sci 77:2772–2776

Kaiser N, Edelman IS (1977) Calcium dependence of glucocorticoid-induced lymphocytolysis. Proc Natl Acad Sci USA 74:638–642

Kaplan L, Katz AD, Ben-Isaac C et al. (1971) Malignant neoplasm and parathyroid adenoma. Cancer 28:401–407

Kikkawa U, Takai Y, Tanaka Y et al. (1983) Protein kinase C as a possible receptor protein of a tumor-promoting phorbol ester. J Biol Chem 258:11442–11445

Kikuchi Y, Inwano I, Kato K (1984) Effects of calmodulin antagonists on human ovarian cancer cell proliferation in vitro. Biochem Biophys Res Comm 123:385–392

Knowles AF, Salas-Prato M, Villela J (1985) Epidermal growth factor inhibits growth while increasing the expression of an ecto-Ca ATPase of a human hepatoma cell line. Biochem Biophys Res Comm 126:8–14

Koeffler HP (1983) Induction of differentiation of human acute myelogenous leukemia cells: therapeutic implications. Blood 62:709–721

Koizumi T, Nakao Y, Matsui T et al. (1985) Effects of corticosteroid and 1,24R-dihydroxy-vitamin D_3 administration on lymphoproliferation and autoimmune disease in MRL/MP-1pr/1pr mice. Int Arch Allergy Appl Immunol 77:396–404

Koretzky GA, Daniele RP, Greene WC et al. (1983) Evidence for an interleukin-independent pathway for human lymphocyte activation. Proc Natl Acad Sci USA 80:3444–3447

Kostellow AB, Morril GA (1980) Calcium dependence of steroid and guanine 3',5'-monophosphate induction of germinal vesicle breakdown in Rana pipiens oocytes. Endocrinology 106:1012–1019

Kram R, Manont P, Tomkins GM (1973) Pleiotypic control by adenosine 3',5'-cyclic monophosphate: a model for growth control in animal cells. Proc Natl Acad Sci 70:1432–1436

Kuribayashi T, Tanaka H, Abe E et al. (1983) Functional defect of variant clones of a human myeloid leukemia cell line (HL-60) resistant to 1α25-dihydroxy vitamin D_3. Endocrinology 113:1992–1998

Lansing AI (1947) Calcium and growth in ageing and cancer. Science 106:187–188

Lazo JS, Hait WN, Kennedy KA et al. (1984) Enhanced bleomycin-induced DNA damage and cytotoxicity with calmodulin antagonists. Molec Pharmacol 27:387–393

Lechner JF (1984) Interdependent regulation of epithelial cell replication by nutrients, hormones, growth factors and cell density. Fed Proc 43:116–120

Lechner JF, Kaighn ME (1969) Reduction of the calcium requirement of normal human epithelial cells by EGF. Exp Cell Res 121:432–435

Lichtman AH, Segal GB, Lichtman MA (1983) The role of calcium in lymphocyte proliferation. Blood 61:413–422

Lindahl-Kiessling KM (1976) Calcium dependency of the binding and mitogenicity of phytohemagglutinin. Differentiation between calcium-dependent and independent binding events. Exp Cell Res 103:151–157

Linnoila I, Becker KL, Silva OL et al. (1984) Calcitonin as a marker for diethylnitrosamine-induced pulmonary endocrine cell hyperplasia in hamsters. Lab Invest 51:39–45

Lipkin M, Newmark A (1985) Effect of added dietary calcium on colonic epithelial-cell proliferation in subjects at high risk for familial colonic cancer. N Engl J Med 313:1381–1383

Luben RA, Mundy GR, Trummel CL et al. (1974) Partial purification of osteoclast activating factor from phytohemagglutinin-stimulated human leukocytes. J Clin Invest 53:1473–1480

Macara IG, Marinetti GV, Balduzzi PC (1984) Transforming protein of avian sarcoma virus UR 2 is associated with phosphatidyl-inositol kinase activity. Possible role in tumorigenesis. Proc Natl Acad Sci USA 81:2728–2732

MacManus JP (1979) Occurrence of a low molecular calcium binding protein in neoplastic liver. Cancer Res 39:3000–3005

MacManus JP (1980) The purification of a unique binding protein from Morris hepatoma 5123. Biochim Biophys Acta 621:296–304

MacManus JP (1981) Development and use of a quantitative immunoassay for the calcium binding protein (molecular weight 11500) of Morris hepatoma 5123. Cancer Res 41:974–979

MacManus JP, Whitfield JF (1969a) Stimulation of desoxyribonucleic acid synthesis and mitotic activity of thymic lymphocytes by cyclic adenosine 3′,5′-monophosphate. Exp Cell Res 58:188–191

MacManus JP, Whitfield JF (1969b) Mediation of the mitogenic action of growth hormone by adenosine 3′5′-monophosphate C cyclic AMP. Proc Soc Exp Biol Med 132:409–412

MacManus JP, Whitfield JF (1971) Cyclic AMP-mediated stimulation by calcium of thymocyte proliferation. Exp Cell Res 69:281–288

MacManus JP, Whitfield JF, Youdale T (1971) Stimulation by epinephrine of adenyl cyclase activity, cyclic AMP formation, DNA synthesis and cell proliferation in a population of rat thymic lymphocytes. J Cell Physiol 77:103–116

MacManus JP, Boynton AL, Whitfield JF et al. (1975) Acetylcholine-induced initiation of thymic lymphoblast DNA synthesis and proliferation. J Cell Physiol 85:321–330

MacManus JP, Bruceland BM, Rixon RH et al. (1981) An increase in calmodulin during growth of normal and cancerous liver in vivo. FEBS Lett 133:99–102

Mangelesdorf DJ, Koeffler HP, Donaldson CA et al. (1984) 1,25-Dihydroxy vitamin D₃-induced differentiation in a human promyelocytic leukemia cell line (HL-60). Receptor-mediated maturation to macrophage-like cells. J Cell Biol 98:391–398

Mastro AM, Smith MC (1983) Calcium dependent activation of lymphocytes by ionophore A23187, and a phorbol ester tumor promoter. J Cell Physiol 116:51–56

Matas AJ, Simmons RL, Kjellstrand CM et al. (1975) Increased incidence of malignancy during chronic renal failure. Lancet I:883–886

Matsui T, Nakao Y, Kobayashi N et al. (1984) Phenotypic differentiation-linked growth inhibition in human leukemia cells by active vitamin D₃ analogues. Cancer 33:193–202

Matsui T, Nakao Y, Koizumi T et al. (1985a) 1α25-dihydroxy-vitamin D₃ regulates proliferation of activated T-lymphocyte subsets. Life Sci 37:95–101

Matsui T, Nakao Y, Kobayashi N et al. (1985b) Effects of calmodulin antagonists and cytochalasins on proliferation and differentiation of human promyelocytic leukemia cell line HL-60. Cancer Res 45:311–316

Matsui T, Takahashi R, Mihara K et al. (1985c) Cooperative regulation of c-myc expression in differentiation of human promyelocytic leukemia induced by recombinant γ-interferon and 1α25-dihydroxy-vitamin D₃. Cancer Res 45:4366–4371

May WS Jr, Sahyoun N, Wolf M et al. (1985) Role of intracellular calcium mobilization in the regulation of protein kinase C-mediated membrane processes. Nature 317:549–551

Mazia D (1937) The release of calcium in Arbacia eggs upon fertilization. J Cell Comp Physiol 10:291–304

Mazzetti G, Bagnave G, Monti MG et al. (1984) 1α25-Dihydroxycholecalciferol and human histocytic lymphoma cell line (U-937): the presence of receptor and inhibition of proliferation. Life Sci 34:2185–2191

McKeehan WL, McKeehan KA (1979) Epidermal growth factor modulates extracellular Ca^{2+} requirement for multiplication of normal human skin fibroblasts. Exp Cell Res 123:397–400

Means AR, Dedman JR (1980) Calmodulin – an intracellular calcium receptor. Nature 285:73–77

Medoff J, Parker N (1971) Stimulation of growth by cyclic AMP in cultured embryonic tissues. J Cell Biol 55:173

Michelangeli VP, Livesey SA, Martin TJ (1984) Effect of pertussis toxin on adenylate cyclase responses to prostaglandin E₂ and calcitonin in human breast cancer cells. Biochem J 224:371–377

Michell RH (1975) Inositol phospholipids and cell surface receptor function. Biochim Biophys Acta 421:81–147

Mills GB, Cragoe EJ Jr, Gelfand EW et al. (1985) Interleukin 2 induces a rapid increase in intracellular pH through activation of a Na^+/H^+ antiport. J Biol Chem 260:12500–12507

Mix LL, Dinerstein RL, Villereal ML (1984) Mitogens and melittin stimulate an increase in intracellular free calcium concentration in human fibroblasts. Biochem Biophys Res Comm 119:69–75

Miyaura C, Abe E, Kuribayashi T et al. (1981) $1\alpha25$-dihydroxy-vitamin D_3 induces differentiation of human myeloid leukemia cells. Biochem Biophys Res Comm 102:937–943

Miyaura C, Abe E, Nomura H et al. (1982) $1\alpha25$-dihydroxy-vitamin D_3 suppresses proliferation of murine granulo-cytomacrophage progenitor cells (CFU-C). Biochem Biophys Res Comm 108:1728–1733

Moens W, Vokaer AA, Kram R (1975) Cyclic AMP and cyclic GMP concentrations in serum and density-restricted fibroblast cultures. Proc Natl Acad Sci USA 72:1063–1067

Monvae JG, Niedel JE, Cambien JC (1984) B cell activation. IV. Induction of cell membrane depolarization and hyper-T-A expression by phorbol diesters suggests a role for protein kinase C in murine B lymphocyte activation. J Immunol 132:1472–1478

Moolenaar WH, Tertoolen LGT, deLaat SW (1984) Growth factors immediately raise cytoplasmic free calcium in human fibroblasts. J Biol Chem 259:8066–8069

Moon TD, Morley JE, Vessella RL et al. (1983) The role of calmodulin in human renal cell carcinoma. Biochem Biophys Res Comm 114:843–849

Morimoto S, Yoshikawa K, Kozuka T et al. (1986) Treatment of psoriasis vulgaris by oral administration of 1α-hydroxyvitamin D_3 – open design study. Calcif Tissue Int 39:209–212

Mundy GR, Salmon SE (1974) Evidence for the secretion of an osteoclast stimulating factor in myeloma. N Engl J Med 291:1041–1046

Mundy GR, Raisz LG, Shapiro JL et al. (1977) Big and little forms of osteoclast activating factor. J Clin Invest 60:122–128

Murao S, Gemmell MA, Callahan MF et al. (1983) Control of macrophage cell differentiation in human promyelocytic H-60 leukemia cells by 1,25-dihydroxyvitamin D_3 and phorbol-12-myristate-13-acetate. Cancer Res 43:4989–4996

Nakai T, Mita S, Yamamoto S et al. (1984) Inhibition by palmitoylcarnitine of adhesion and morphological changes in HL-60 cells induced by 12-O-tetradecanoyl-phorbol-13-acetate. Canc Res 44:1908–1912

Newmark HL, Wargovich MJ, Bruce WR (1985) Colon cancer and dietary fat, phosphate, and calcium: a hypothesis. J Natl Cancer Inst 72:1323–1325

Niedel JE, Kuhn LJ, Vanderbark GR (1983) Phorbol diester receptor copurifies with protein kinase C. Proc Natl Acad Sci USA 80:36–40

Nisbet-Brown E, Cheung RK, Lee JWW et al. (1985) Antigen-dependent increase in cytosolic free calcium in specific T-lymphocyte clone. Nature 316:545–547

Nishizuka Y (1984) The role of protein kinase C in cell surface signal transduction and tumor promotion. Nature 308:693–697

Ohta M, Okabe T, Ozawa K et al. (1985) $1\alpha25$-Dihydroxy-vitamin D_3 stimulates proliferation of human circulating monocytes in vitro. FEBS Lett 185:9–13

O'Keefe EJ, Payne RE (1983) Modulation of the epidermal growth factor receptor of human keratinocytes by calcium ion. J Invest Dermatol 81:231–235

Orellana SA, Solski PA, Brown JH (1985) Phorbol ester inhibits phosphoinositide hydrolysis and calcium mobilization in cultured astrocytoma cells. J Biol Chem 260:5236–5239

Owada NK, Hakura A, Iida K et al. (1984) Occurrence of caldesmon (a calmodulin binding protein) in cultured cells. Comparison of normal and transformed cells. Proc Natl Acad Sci USA 81:3133–3137

Owen M (1978) Histogenesis of bone cells. Calcif Tissue Int 25:205–207

Owen M (1980) The origin of bone cells in the postnatal organism. Arthritis Rheum 23:1073–1080

Paul D, Ristow HJ (1979) Cell cycle control by Ca ions in mouse 3T3 cells. J Cell Physiol 98:31–40;

Perris AD, Whitfield JF (1967) The effect of calcium on mitosis in the thymus of normal and irradiated rats. Nature 214:307–313

Perris AD, Whitfield JF (1969) The mitogenic action of bradykinin and its dependence on calcium. Proc Soc Exp Biol Med 130:1198–1201

Perris AD, Whitfield JF, Rixon RH (1967) Stimulation of mitosis in bone marrow and thymus of normal and irradiated rats by divalent cations and parathyroid extract. Radiat Res 32:550–553

Perris AD, Whitfield JF, Tölg PK (1968) Role of calcium in the control of growth and cell division. Nature 219:527–529

Perris AD, Edwards DJ, Atkinson MJ (1984) Some effects of parathyroidectomy on cell-mediated immune response in the rat. J Endocrinol 102:257–263

Pfyffer GE, Haemmerli G, Heizmann CW (1984) Calcium binding proteins in human carcinoma cell lines. Proc Natl Acad Sci 18:6632–6636

Poenie M, Aderton J, Tsien RY et al. (1985) Changes of free calcium levels with stages of the cell division cycle. Nature 315:147–149

Polla B, Rochemonteix B, de Junod AF et al. (1985) Effects of LTB4 and Ca^{++} ionophore A23187 on the release of human alveolar macrophages of factors controlling fibroblast functions. Biochem Biophys Res Comm 129:560–567

Raisz LG, Kream BE (1984) Regulation of bone formation. N Engl J Med 310:456–457

Ralph RK (1983) Cyclic AMP, calcium and control of cell growth. FEBS Lett 161:1–8

Rasmussen H (1981) Calcium and cAMP as synarchic messengers. John Wiley, New York

Reitsma PH, Rothberg PG, Astrin SM et al. (1983) Regulation of *myc* gene expression in HL-60 leukemia cells by a vitamin D metabolite. Nature 306:492–493

Rigby WFC, Shen L, Ball ED et al. (1984) Differentiation of a human monocyte cell line by 1,25-dihydroxy-vitamin D$_3$ (calcitriol). A morphologic, phenotypic and functional analysis. Blood 64:1110–1115

Rixon RH, Whitfield JF (1961) The radioprotective action of parathyroid extract. Int J Radiat Res 3:361–367

Rixon RH, Whitfield JF (1963) Effect of multiple injections of calcium compounds on the survival of X-irradiated rats. Nature 199:821–822

Rixon RH, Whitfield JF, Youdale T (1955) Increased survival of rats irradiated with X-rays and treated with parathyroid extract. Nature 182:1374

Rixon RH, Whitfield JF, MacManus JP (1970) Stimulation of mitotic activity in rat bone marrow and thymus by exogenous adenosine 3′,5′-monophosphate (cyclic AMP). Exp Cell Res 63:110–116

Rixon RH, MacManus JP, Whitfield JF (1979) The control of liver regeneration by calcitonin, para-thyroid hormone and 1α25-dihydroxy-cholecalciferol. Mol Cell Endocrinol 15:79–89

Roth SI, Schiller AL (1976) Comparative anatomy of parathyroid glands. In: Greep RO et al. (eds) American Physiological Society Handbook of Physiology, section 7, Endocrinology, vol 3, parathyroid gland. Williams and Wilkins, Baltimore, pp 281–312

Rudland PS, Gospodarowicz D, Seifert W (1974) Activation of guanyl cyclase and intracellular cyclic GMP by fibroblast growth factor. Nature 250:741–742

Sanfelippo PM, Sanfelippo MJ, Surak JG (1965) The radioprotective effects of parathyroid hormone and parenteral calcium in rats given whole body irradiation. Radiat Res 25:235–236

Sato T, Takusagawa K, Asoo M et al. (1982) Anti-tumor effect of 1α-hydroxyvitamin D$_3$. Tohoku J Exp Med 138:445–446

Sato T, Takusagawa K, Asoo N et al. (1984) Effect of 1α-hydroxyvitamin D$_3$ on metastasis of rat ascites hepatoma K-231. Br J Cancer 50:123–125

Schreiner GG, Unanue ER (1976) Calcium-sensitive modulation of Ig capping. Evidence supporting cytoplasmic control of ligand–receptor complexes. J Exp Med 143:15–31

Schuetz A (1972) Induction of nuclear breakdown and meiosis in *Spirula solidissima* oocyte by calcium ionophore. J Exp Zool 191:433–440

Seiff CA, Chessells JM, Levinsky RJ (1983) Allogenic bone marrow transplantation in infantile malignant osteopetrosis. Lancet I:439–441

Seltzer MH, Rosato FE, Fletcher MJ (1970) Serum and tissue calcium in human breast carcinoma. Cancer Res 30:615–616

Short J, Tsukada K, Rudent WA, Lieberman I (1975) Cyclic adenosine 3′,5′-monophosphate and the induction of DNA synthesis in the liver. J Biol Chem 250:3602–3606

Smith BM, Sturm RJ, Carchman RA (1983) Calcium modulation of phorbol-ester-induced alterations in murine macrophage morphology. Cancer Res 43:3385–3391

Sporn MB, Todaro GJ (1980) Autocrine secretion and malignant transformation of cells. N Engl J Med 303:878–880

Suda S, Enomoto S, Abe E et al. (1984) Inhibition by 1α25-dihydroxy vitamin D$_3$ of dimethyl sulfoxide-induced differentiation of Friend erythroleukemic cells. Biochem Biophys Res Comm 119:807–813

Suda T, Abe E, Miyaura C et al. (1985) Modulation of cell differentiation and tumor promotion by 1α25(OH)$_2$D$_3$. In: Norman AW et al. (eds) Vitamin D. A biochemical and clinical update. Walter de Gruyter, Berlin, pp 187–196

Sugimoto Y, Whitman M, Cantley LC et al. (1984) Evidence that the Rous sarcoma virus transforming gene product phosphorylates phosphatidylinositol and diacylglycerol. Proc Natl Acad Sci USA 81:2117–2121

Swierenga SHH, Whitfield JF, Gillan DJ (1976a) Alteration by malignant transformation of the calcium requirements for cell proliferation in vitro. J Natl Cancer Inst 57:125–129

Swierenga SHH, MacManus JP, Braceland BM et al. (1976b) Regulation of the primary immune response in vivo by parathyroid hormone. J Immunol 117:1608–1611

Swierenga SHH, Whitfield JF, Karasaki S (1978) The loss of proliferative calcium dependence. A simple in vitro indicator of tumorigenicity. Proc Natl Acad Sci USA 75:6069–6072

Swierenga SHH, Goyette R, Marceau N (1984) Differential effects of calcium deprivation on the cytoskeleton of non-tumorigenic and tumorigenic rat liver cells in culture. Exp Cell Res 153:39– 49

Takemoto D, Jilka C (1983) Increased content of calmodulin in human leukemia cells. Leukemia Res 7:97–100

Tanaka H, Abe E, Miyaura C et al. (1982) 1α-Dihydroxy-cholecalciferol and a human myeloid leukemia cell line (HL-60). Biochem J 204:713–719

Thomas EW, Murad F, Looney WB et al. (1973) Adenosine 3′,5′-monophosphate and guanosine 3′,5′-monophosphate concentrations in Morris hepatoma of different growth rates. Biochim Biophys Acta 297:564–567

Tisman G, Herbert V (1973) Studies of effects of cyclic adenosine 3′,5′-monophosphate in regulation of human hemopoiesis in vitro. In Vitro 9:86–91

Trummel CL, Mundy GR, Raisz LG (1975) Release of osteoclast activating factor by normal human peripheral blood leukocytes. J Lab Clin Med 85:1001–1007

Truneh A, Albert F, Golstein P et al. (1985) Early steps of lymphocyte activation bypassed by synergy between calcium ionophores and phorbol ester. Nature 313:318–320

Tsuruo T, Iida H, Tsukagoshi S et al. (1982) Increased accumulation of vincristine and adriamycin in drug-resistant P388 tumor cell following incubation with calcium antagonists and calmodulin inhibitors. Cancer Res 42:4730–4733

Tupper JT, Bodine PV (1983) Calcium effects on epidermal growth factor receptor mediated endocytosis in normal and SV40-transformed human fibroblasts. J Cell Physiol 115:159–166

Tupper JT, Zorgniotti F (1977) Calcium content and distribution as a function of growth and transformation in the mouse 3T3 cells. J Cell Biol 75:12–22

Tyler A (1941) Artificial parthenogenesis. Biol Rev 16:291–336

Veigl ML, Sedwick WD, Vanamean TC (1982) Calmodulin and Ca^{2+} in normal and transformed cells. Fed Proc 41:2283–2291

Vensshi K, Criss WE, Kakibuchi S (1980) Calcium-activable phosphodiesterase and calcium-dependent modulator protein in transplantable hepatoma tissues. J Biochem 87:601–607

Wei JW, Hickie RA (1981) Increased content of calmodulin in Morris hepatoma 5123. Biochem Biophys Res Comm 100:1562–1568

Wei JW, Morris HP, Hickie RA (1982) Positive correlation between calmodulin content and hepatoma growth rate. Cancer 42:2571–2574

Whitfield JF (1982) The role of calcium and magnesium in cell proliferation. An overview. In:Boynton AL (ed) Ions, cell proliferation and cancer. Academic Press, New York, pp 283–294

Whitfield JF, Rixon RH (1962) Prevention of postirradiation mitotic delay in cultures of mouse cells by calcium salt. Exp Cell Res 27:154–157

Whitfield JF, Youdale T (1966) Effect of calcium, agmatine and phosphate on mitosis in normal and irradiated population of rat thymocytes. Exp Cell Res 43:602–610

Whitfield JF, Brohee H, Youdale T (1966) Mitotic stimulation in normal and irradiated suspension cultures of rat bone marrow by an elevated salt concentration. Exp Cell Res 41:49–54

Whitfield JF, MacManus JP, Gillan DJ (1971) Inhibition by thyrocalcitonin (calcitonin) of the cyclic AMP-mediated stimulation of thymocyte proliferation by epinephrine. Horm Metab Res 3:348–351

Whitfield JF, MacManus JP, Braceland BM et al. (1972a) Inhibition by calcium of the cyclic AMP-mediated stimulation of thymic lymphoblast proliferation by prostaglandin E_1. Horm Metab Res 4:304–308

Whitfield JF, MacManus JP, Braceland BM et al. (1972b) The influence of calcium on the cyclic AMP-mediated stimulation of DNA-synthesis and cell proliferation by prostaglandin E_1. J Cell Physiol 79:353–362

Whitfield JF, Rixon RH, MacManus JP et al. (1973) Calcium, cyclic adenosine 3′,5′-monophosphate and the control of cell proliferation. A review. In Vitro 8:257–278

Whitfield JF, MacManus JP, Boynton AL et al. (1974) Concanavalin A and the initiation of thymic lymphoblast DNA synthesis and proliferation by a calcium-dependent increase in cyclic GMP level. J Cell Physiol 84:445–457

Whitney RB, Sutherland RM (1972a) Enhanced uptake of calcium by transforming lymphocytes. Cell Immunol 5:137–147

Whitney RB, Sutherland RM (1972b) Requirement for calcium ions in lymphocyte transformation stimulated by phytohemagglutinin. J Cell Physiol 80:329–333

Wood AW, Chang RL, Huang MT et al. (1983) 1α25-Dihydroxy-vitamin D_3 inhibits phorbol-ester dependent chemical carcinogenesis in mouse skin. Biochem Biophys Res Comm 116:605–611

Yalowich JC, Ross WE (1985) Potentiation of etoposide-induced DNA damage by calcium antagonists in L1210 cells in vitro. Cancer Res 44:3360–3365

Yanovich S, Preston L (1984) Effects of verapamil on daunomycin cellular retention and cytotoxicity in P388 leukemic cells. Cancer Res 44:1743–1747

Yates L, Vanaman TC (1980) Studies on the role of calmodulin in T-lymphocyte mitogenesis. Fed Proc 39:1626

Yoneda T, Mundy GR (1979) Monocytes regulate osteoclast-activating factor production by releasing prostaglandins. J Exp Med 150:338–350

Zajac JD, Levesey SA, Martin TJ (1984) Selective activation of cyclic AMP dependent protein kinase by calcitonin in a calcitonin secreting lung cancer cell line. Biochem Biophys Res Comm 122:1040–1046

Zavortink M, Welsh MJ, McIntosh JR (1983) The distribution of calmodulin in living mitotic cells. Exp Cell Res 149:375–385

Ziv Y, Rubin M, Lobrozo R et al. (1985) Primary hyperparathyroidism associated with pancytosis. N Engl J Med 313:187

Zwierzchowski L, Renca J, Grochowska I (1984) Role of calcium in the insulin-dependent stimulation of DNA synthesis in mouse mammary gland in vitro. Exp Cell Res 152:105–116

Chapter 17

Dietary Requirements for Calcium

B. E. C. Nordin and D. H. Marshall

Introduction

Normal, healthy adults must be assumed to be in a state of nutritional equilibrium – at least over a period of time. They must be taking in nitrogen, phosphorus, sodium and other elements at the same rate as they are excreting them. For some of these elements, the calculation of a body balance is very difficult, but in the case of calcium it is relatively simple. Discounting dermal losses [which may be quite significant (Charles et al. 1983)], the calcium balance is the difference between the dietary calcium on the one hand and the faecal and urinary calcium on the other.

Using this technique, the calcium requirement of normal adults can be defined as the amount of calcium required to maintain calcium balance. This in turn is determined by the rate of calcium absorption from the gastrointestinal tract and the rate of calcium excretion by the kidneys. The calculation of calcium requirement therefore involves an analysis of the absorption and excretion mechanisms and these will both be reviewed in this chapter before an attempt is made to estimate the requirement itself.

Calcium Absorption

Calcium is absorbed predominantly in the proximal small intestine, but before it can be absorbed it must be in solution. In normal individuals on normal diets, binding of calcium in the lumen does not significantly limit its availability for absorption. However, excessive dietary phosphate or oxalate and/or an abnormally high pH in the small intestine can limit the availability of calcium as can malabsorption of fat, which may result in the formation of calcium complexes in the intestinal lumen.

The absorptive mechanism itself has an active (saturable) component and a diffusion component (Ireland and Fordtran 1973; Heaney et al. 1975; Wilkinson

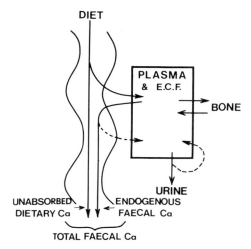

Fig. 17.1. Major calcium movements in the body.

1976). At low calcium intakes, calcium is mainly absorbed by active transport but with increasing intake this mechanism becomes saturated and additional calcium is absorbed by diffusion. As a consequence, fractional calcium absorption falls with intake as can be shown by intubation studies (Wilkinson 1976). This is not readily apparent when absorption is measured as the difference between dietary intake and faecal output because of the confounding effect of the digestive juice calcium. It is therefore necessary to distinguish between absorption measured in this way (net calcium absorption) and the actual fraction of dietary calcium that is absorbed (true calcium absorption).

When calcium is ingested, some is absorbed and the rest appears in the faeces as "unabsorbed dietary calcium". However, digestive juice calcium is also being secreted into the intestines and a proportion of dietary calcium absorbed is reabsorbed. The rest appears in the faeces as "endogenous faecal calcium". Thus the faeces contain unabsorbed dietary calcium and unabsorbed digestive juice calcium (Fig. 17.1). Since the maximum fractional true absorption is about 70%, and the digestive juice calcium is about 300 mg daily, the maximum reabsorption of the latter is 200 mg and the minimum obligatory endogenous faecal calcium at zero intake is measured at about 100 mg per day.

With increasing calcium intake, the *absolute* amount of calcium absorbed increases but the *fraction* absorbed falls and endogenous faecal calcium therefore rises. The cross-over point at which dietary calcium and faecal calcium are equal (zero net absorption) in normal subjects is about 200 mg daily (Fig. 17.2). At a typical daily calcium intake of 700 mg, and a digestive juice calcium of 300 mg, the total intestinal load is 1000 mg. About 500 mg of this load is absorbed and 500 mg appears in the faeces. The true fractional absorption in this situation is 50% (500/1000) but the net fractional absorption is only 29% (200/ 700).

It follows that the relationships between dietary calcium intake and net absorption on the one hand and dietary calcium and true absorption on the other are different. Fractional true absorption falls with intake but fractional net

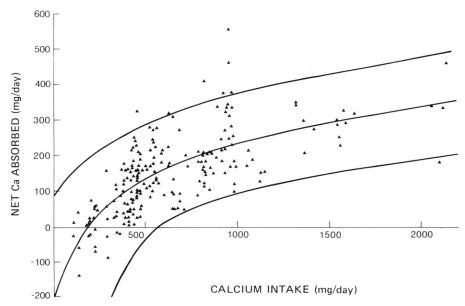

Fig. 17.2. The relation between dietary calcium and net calcium absorbed in 212 balances on 84 normal subjects. The equation is:

$$y = \frac{491x}{287 + x} + 0.06x - 206 \pm 77 \text{ mg/day}$$

where x is dietary calcium and y is net calcium absorbed. The figure shows the mean $\pm$ 2 SD and the equation is based on the assumption that there is an active (saturable) calcium transport mechanism and a diffusion component.

absorption rises from a negative value to a maximum of about 30% and then falls towards the diffusion fraction (Fig. 17.3). True absorbed calcium is always greater than net absorbed calcium by a quantity equivalent to the reabsorbed digestive juice calcium.

   The above considerations are applicable in the ideal situation. There are of course many factors which influence the availability of calcium for absorption, and the absorption mechanism itself. The former include molecules which form insoluble complexes with calcium. The most significant of these is the phosphate ion, but the dietary phosphate : calcium ratio has to be very high (probably over 3 : 1) to interfere significantly with calcium availability. (However, the relatively low phosphate content of human breast milk may be a factor in the higher absorption of calcium from breast milk than cow's milk – see below). Phytic acid, present in the husks of many cereals, was thought at one time to be potentially harmful in this respect but is probably hydrolysed by phytases in the bowel and therefore of little significance unless calcium intake is very low (Fourman and Royer 1968). Oxalic acid in excess can also precipitate calcium in the bowel but is not an important factor in normal diets (Wilkinson 1976). Conversely, certain sugars (notably lactose) enhance calcium absorption by a mechanism which is not fully understood but may relate to calcium availability in the lumen or to transit time (Cochet et al. 1983).

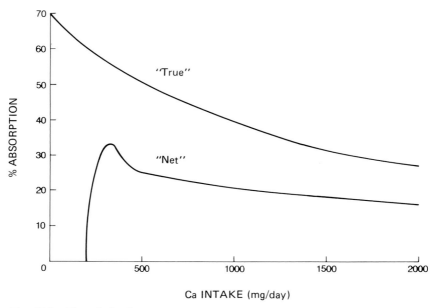

Fig. 17.3. The relation between calcium intake and percentage true and net calcium absorption.

As far as the actual absorptive process is concerned, the dominant influence is vitamin D, now known to operate mainly through its most active metabolite $1,25(OH)_2D_3$ (DeLuca 1973). Malabsorption of calcium is a feature of vitamin D deficiency and this abnormality was once thought to be the immediate cause of rickets and osteomalacia but this is no longer accepted (Nordin 1960). Conversely, high plasma levels of $1,25(OH)_2D_3$ are known to be associated with hyperabsorption of calcium, as in primary hyperparathyroidism and hypercalciuric stone disease (Nordin 1976a). It is possible that the less active metabolite $25(OH)D_3$ also regulates calcium absorption in its own right but this is more controversial.

Calcium Excretion

The urinary calcium is that part of the plasma water calcium [in the fasting state about 5 mg/100 ml (1.25 mmol/l)] which is not reabsorbed in the renal tubules. At a normal glomerular filtration rate of 100 ml/min, the filtered load of calcium is about 5 mg/min of which 4.9 mg is reabsorbed and 0.1 mg excreted. This corresponds to a 24-h calcium excretion of about 144 mg which is a representative urinary calcium output on a low-calcium diet. Calcium excretion is, however, extremely sensitive to changes in filtered load and a change in

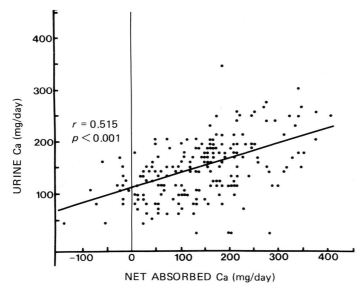

Fig. 17.4.   The relation between net calcium absorbed and urine calcium in the same studies as Fig. 17.2. The equation is:

$$y = 0.284x + 114 \pm 50 \text{ mg/day}.$$

plasma calcium of only 0.2 mg/100 ml, which is barely detectable, may be enough to double or halve the urinary calcium (Nordin 1976b). This very sensitive response tends to stabilize the plasma calcium concentration and preserves the equilibrium between calcium input and output from the plasma over a wide range of calcium intakes.

However, although plasma and urinary calcium rise in response to a calcium load, the converse is only partially true. The plasma calcium has to be maintained at a minimum value of 9.0 mg/100 ml (2.25 mmol/l) by the parathyroid/vitamin D system, and this sets a lower limit to urinary calcium whatever the calcium intake. In normal subjects, the relationship between absorbed and excreted calcium is such that equality between them is not reached until the net absorbed calcium is about 150 mg (Fig. 17.4). When absorbed calcium falls below this value, it is less than the obligatory urinary calcium and the subject goes into negative calcium balance.

It must be emphasized that obligatory calcium excretion varies with the sodium (Goulding 1983; Sabto et al. 1984) and protein intakes (Margen et al. 1974; Schuette et al. 1980; Linkswiler et al. 1981). The values quoted above for minimum urine calcium are derived from Western diets and are probably appreciably lower in cultures where the protein and/or sodium intakes are lower. Table 17.1 shows the dependence of urinary calcium on urinary sodium in both pre- and postmenopausal women; each mmol of sodium takes out of the body 0.01 mmol of calcium. Table 17.2 shows the effect of protein intake on urinary calcium. It will be seen that urinary calcium is almost more dependent on salt and protein intake than it is on calcium intake.

Table 17.1. Fasting urinary sodium/creatinine, calcium/creatinine and hydroxyproline/creatinine ratios in 515 normal postmenopausal women and 47 normal premenopausal women

Group	Mean values ($\pm$ SE)				
	Na/Cr	Ca/Cr		OHPr/Cr	
Pre-	12.1 (0.93)	0.20 (0.18)		0.014 (0.0006)	
Post-	12.0 (0.30)	0.25 (0.007)		0.020 (0.0003)	
"t"	—	2.2		6.5	
p	NS	<0.025		<0.001	
Group	Regression equations		t		p
Pre-	Ca/Cr = 0.010 $\times$ Na/Cr + 0.086		3.9		<0.001
Post-	Ca/Cr = 0.0083 $\times$ Na/Cr + 0.15		9.3		<0.001

Table 17.2. Mean urinary calcium at different levels of calcium and protein intake (Margen et al. 1974; Linkswiler et al. 1981)

Calcium intake (mg/day)	Protein intake (g/day)	Urinary calcium (mg/day)
100	6	51
	78	99
	150	161
900	6	105
	24	131
	78	155
1300	6	80
	78	163
	150	274
1600	6	46
	78	92
	387	318
2300	78	81
	300	176
	600	380

Effects of Calcium Deficiency

Osteoporosis

Experimental Data

Because of the large calcium reserve in the skeleton, and the PTH-mediated increase in bone resorption which occurs if the plasma calcium cannot be maintained by input from the gut, dietary calcium deficiency does not cause hypocalcaemia but rather causes bone breakdown. It appears that the organism is more concerned to maintain the plasma calcium concentration (presumably to protect the neuromuscular system and the cellular calcium) than it is to preserve the integrity of the skeleton. Parathyroid hormone does not remove calcium

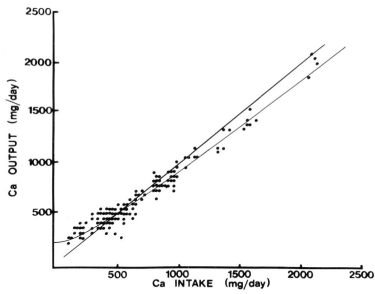

Fig. 17.5. The relation between calcium intake and calcium output in the same balance studies as Fig. 17.2. The equation is:

$$y = 0.957 \left(\frac{352x}{287 + x} \right) + 261 \pm 70 \text{ mg/day}$$

where x is calcium intake and y is calcium output.

selectively from bone but breaks down whole bone to provide the calcium required for the extracellular fluid. The system depends of course upon the existence of a well-mineralized mature skeleton. In the young infant and young animal, where the skeleton is only partially mineralized and where there is a great demand for calcium for bone growth, it is possible to produce hypocalcaemia by calcium deprivation.

The result of calcium mobilization from the skeleton and its associated bone breakdown is the decline in bone "density" which is known as osteoporosis. The evidence that this is the sequence of events in experimental animals is unequivocal (Nordin 1960). However, although it is clear that calcium deficiency causes osteoporosis, it does not follow that human osteoporosis is necessarily due to calcium deficiency and there has been considerable resistance in the nutrition world to any attempt to link the two. The FAO/WHO Committee (1961), Walker (1972) and Hegsted (1963) all point to the low calcium intakes consumed by many populations, particularly in the Third World, that apparently result in no ill effect, the failure to establish any relationship between calcium intake and metacarpal cortical width in population studies (Garn 1970) and the evidence that fracture rates are, if anything, lower in undeveloped than in developed countries and lower among Blacks than Whites, both in South Africa and the United States (Bollet et al. 1965; Solomon 1968). It is therefore necessary to look in more detail at the nature of osteoporosis, at its relationship to ageing and to fractures, and at its pathogenesis.

Osteoporosis in Humans

Age- and Menopause-Related Bone Loss. Loss of bone with age occurs in both sexes and in all races. In women, the process starts at or about the menopause and is due in the first instance to an increase in bone resorption resulting from oestrogen deficiency (Nordin et al. 1984). In men, the age of onset is less well defined and the process is probably more due to a reduction in bone formation (Nordin et al. 1984).

Population studies on the American continent have failed to reveal any relationship between calcium intake and metacarpal cortical width in ageing populations (Garn 1970). In a Yugoslav study, however, populations on low and high calcium intakes were directly compared and lower bone density and higher fracture rates were recorded in the population on the lower calcium intake (Matkovic et al. 1979). There have been several prospective controlled studies in which the effect of calcium supplements on postmenopausal bone loss was examined and in most of these some inhibition of peripheral (cortical) bone loss was observed (e.g. Horsman et al. 1977; Recker et al. 1977; Ettinger et al. 1987; Riis et al. 1987). However, in the two latter studies, calcium supplementation did not prevent loss of trabecular bone from the spine, although in one of them (Ettinger et al. 1987) the combination of calcium and oestrogen was effective at an oestrogen dose which was ineffective by itself.

That calcium administration should inhibit bone loss is hardly surprising when it lowers serum PTH (Goddard et al. 1986; Reid et al. 1986; Horowitz et al. 1987) and suppresses urinary hydroxyproline (Horowitz et al. 1984, 1985, 1987). Why it should not prevent trabecular bone loss is less clear but both the negative studies were performed on women very close to the menopause when trabecular bone loss is very rapid. In any event, prevention of cortical bone loss would be expected to reduce fracture rates since fracture risk and peripheral bone density are closely related (Jensen et al. 1983; Wasnich et al. 1985; Nordin et al. 1987).

The preventive action of oestrogens on postmenopausal bone loss is well documented (Horsman et al. 1977; Nachtigall et al. 1979; Christiansen et al. 1980; Lindsay et al. 1980; Genant et al. 1982; Horsman et al. 1983) although this effect is clearly dose-related and significant bone loss continues on the low oestrogen doses which are frequently prescribed because of the risk of endometrial cancer (Christiansen et al. 1982; Horsman et al. 1983; Lindsay et al. 1984). It may be that this can be prevented if the low dose of oestrogen is combined with a calcium supplement (see above).

The biochemical changes that occur at the menopause do not conclusively explain the pathophysiology of postmenopausal bone loss. There is a well-documented rise in the plasma calcium concentration and in urinary calcium and hydroxyproline (Gallagher et al. 1972; Nordin et al. 1975). Some workers believe that there is a simultaneous fall in calcium absorption (Gallagher et al. 1979) but others do not (Crilly et al. 1981). The plasma and urinary changes are reversible with oestrogen (Gallagher and Nordin 1975) but the urinary hydroxyproline can also be reduced by calcium administration in subjects with normal calcium absorption (Horowitz et al. 1984). The data are compatible with the concept that calcium requirement rises at the menopause as a result of the rise in urinary calcium, which is secondary to the rise in plasma calcium, but it leaves the cause of this mild hypercalcaemia unexplained. The traditional view is that it results from the loss of the protective action of oestrogen on bone (Jasani et al. 1965; Heaney 1965) but calcitonin deficiency has also been suggested as the cause (Stevenson et al. 1981).

The rate of obligatory calcium loss after the menopause varies with the urinary sodium, as it does before the menopause, but the theoretical urinary calcium at zero sodium excretion is higher after the menopause because of the higher plasma calcium concentration (Table 17.1). Moreover, this sodium-dependent variation in urinary calcium is associated with a corresponding variation in urinary hydroxyproline in postmenopausal women (Table 17.1).

Urinary calcium excretion varies directly with protein intake (Anand and Linkswiler 1974; Hegsted and Linkswiler 1981) because of the effect of the latter on tubular resorption of calcium (Schuette et al. 1981; Yuen et al. 1984) but it is not clear whether this is due to the sulphur content of the protein (causing sulphate–calcium complexing in the kidney) or to its acid component. Since this effect of protein on calcium excretion is not associated with any change in calcium absorption, it must be concluded that high protein intakes (like high sodium intakes) increase the obligatory calcium loss and therefore increase the calcium requirement, and that this effect would be more deleterious in post-than in premenopausal women.

However, phosphate feeding neutralizes the calciuretic effect of protein and it has been suggested that the high phosphorus content of meat may offset the calciuretic effect of its acid and sulphate content. Moreover, most studies in vegetarians have failed to show any difference in bone status between them and omnivorous populations (Ellis et al. 1974). An exception is a study of elderly women in which the vegetarians had a higher bone density than the meat eaters (Sanchez et al. 1980).

Osteoporotic Fractures. The principal fractures associated with osteoporosis are those of the wrist (predominantly in women), of the vertebrae and of the femoral neck. The age-specific incidence of these fractures is at least twice as high in women as in men and this effect, combined with the greater longevity of women, results in a great excess of fractures, particularly of the femoral neck, in women compared with men. Among Western women, the prevalence of osteoporotic fractures (distal radius, spine and proximal femur) reaches 25% by age 80 (Nordin et al. 1980) and is probably rising (Lewis 1981). If mortality from falls in the elderly is used as a measure of the hip fracture rate, then the variation across the world is very great, extending from about 40 cases per 10000 per annum in Western women over the age of 75 to about 1 per 10000 per annum in places like Sri Lanka and Singapore (WHO 1978). Any attempt to link calcium status with osteoporosis must take account of these facts.

Most evidence indicates that the mean bone "density" is significantly lower in fracture cases than in age-matched controls (Elsassar et al. 1980; Horsman et al. 1982), but in all studies there is a large overlap between fracture and non-fracture cases. For the wrist and hip fractures at least, trauma (generally a fall at ground level) is a major contributory factor and the rise in falls with age undoubtedly contributes to the high fracture rates in old people (Cook et al. 1982).

It is likely that the more severe osteoporosis associated with fractures represents an exaggeration of the normal bone-losing process due to the operation of additional risk factors. The most carefully studied group includes those with vertebral compression (spinal osteoporosis) who present in their early

60s and in whom the bone resorption rate is high (Nordin et al. 1981). The abnormality most frequently identified in these patients is malabsorption of calcium (Gallagher et al. 1979; Nordin et al. 1984). Some authorities believe that this is the *result* of the bone-resorbing process while others believe that it is a significant *pathogenetic factor* causing osteoporosis by secondary calcium deficiency. The urinary hydroxyproline [which is high in these cases (Nordin et al. 1984)] can be reduced by the administration of $1,25(OH)_2D_3$ (Gallagher et al. 1982; Need et al. 1985). This therapeutic response is compatible with the latter concept. However, a reduced bone formation rate is also a feature of these cases (Parfitt 1981) and this is more likely to be due to hormonal factors than to calcium deficiency or malabsorption of calcium.

In hip fracture cases yet another factor appears to be operating, namely vitamin D deficiency. Many studies from various parts of the world, including Australia, have shown significantly reduced plasma 25(OH)D levels in those patients with femoral neck fractures (Baker et al. 1979; Lips et al. 1982; Morris et al. 1984) but others have found no such reduction (Lund et al. 1975; Wootton et al. 1979). This abnormality (when present) is probably attributable to the relatively high proportion of such cases who were housebound *before* their fracture.

No causal relationship has been established between this vitamin D deficiency and the hip fracture but a high incidence of osteomalacia in these patients has been reported by some workers (Aaron et al. 1974; Faccini et al. 1976) and denied by others (Hodkinson 1974; Parfitt et al. 1982).

The above observations may help to explain why Third World populations, despite their lower calcium intakes, do not experience the high rates of fractures in the elderly which are a feature of Western peoples. If the reported facts are correct, it follows either that osteoporosis and calcium intake are unrelated (in which case there is really no such thing as a calcium deficiency disease) or that the calcium requirement of Third World peoples is lower than that of Western populations. Since calcium requirement depends upon the relationship between calcium absorption and calcium excretion, they must either have a higher absorption of calcium or a lower excretion or both.

As far as absorption of calcium is concerned, the main regulator of this process is vitamin D. Little is known about the vitamin D status of Third World populations but since it depends almost entirely on sunlight exposure (Parfitt et al. 1982) it is likely that the vitamin D status of rural populations living near the Equator would be better than that of urban populations at more extreme latitudes. As far as calcium excretion is concerned, the higher protein and sodium intakes of Western populations are well documented and it is therefore extremely probable that urinary calcium loss is higher in Western countries than in the Third World.

It is therefore unwise to assume that age-related bone loss, and in particular the excessive loss associated with fractures, cannot be attributed in part, at least, to "calcium deficiency" in the widest sense. The extent to which this deficiency arises from dietary inadequacy, calcium malabsorption or excessive urinary excretion of calcium (of hormonal or nutritional origin) is impossible to judge but the rise in urinary calcium is the major factor in women, in the same way as blood loss is a more important cause of anaemia than low iron intake. It is not possible to say whether calcium intake plays any role in determining bone status at maturity, although some suggestive evidence for this was presented at a recent conference (Kanders et al. 1984). In the final analysis, differences in rates of bone loss are probably less important than bone density at maturity in determining bone status in later life.

The calcium deficiency theory of osteoporosis is supported by a growing body of opinion (Draper and Scythes 1981; Marcus 1982) but it is not universally accepted (Paterson 1978). Nonetheless, it would be prudent to assume that calcium deficiency (relative or absolute) contributes to the high prevalence of osteoporosis in the Western world.

Growth

As far as children are concerned, the possible manifestation of calcium deficiency may be limitation of growth. In his classical experiments, Sherman (1952) showed that calcium deficiency limited the growth of rats and he subsequently reviewed balance studies on children which showed that calcium retention by children could be increased up to a point by increasing calcium intake. There are some old experimental and epidemiological studies which also suggest that calcium intake may affect growth. The largest of these was carried out in Lancashire in the 1920s and showed significantly faster growth in 10 000 schoolchildren given 0.75 pint of extra milk daily for 4 months than in 10 000 age-matched controls (Leighton and McKinlay 1930). In the United Kingdom the mean calcium intake rose from about 600 mg per person per day at the turn of the century to about 1150 mg in 1960 (Greaves and Hollingworth 1966). During the same period there was very little change in mean protein or calorie intake but the average height of 12-year-old urban schoolboys increased by about 2 inches. This increase in average height is widely attributed to the general improvement in nutritional status but the proportionate increase in calcium intake was actually much greater than the proportionate increase in other nutrients.

There are similar suggestive data from Japan. Until about 1950, the calcium intake of the Japanese was the lowest in the world – about 200 mg per person per day. The protein intake at that time was about 60 g per person per day and the calorie intake about 2000 per person per day. At this time the Japanese authorities became aware of the importance of calcium in nutrition and proceeded to fortify school bread with calcium and issue free milk to schoolchildren. At about the same time, milk production and consumption started to increase in Japan with the result that calcium intake has risen to about 600 mg per head per day (Japan Statistical Year Book 1970). These nutritional changes have been associated with a substantial increase in the height of Japanese schoolchildren. By 1970, 12-year-old Japanese boys were about 10 cm taller than boys of the same age 20 years before. Taken in conjunction with the anecdotal observation that the children of Japanese immigrants in the United States are substantially taller than their parents, these data suggest that the small stature of the Japanese is not wholly genetic but partly nutritional and attributable perhaps to the low calcium content of rice grain.

These observations are suggestive but no more. Walker (1972) has assembled conflicting data on the effect of calcium on growth and concluded that it has no beneficial effect.

Hypertension

Experimental Data

A growing body of data suggests a link between calcium deficiency and hypertension in experimental animals. Serum ionized calcium is said to be reduced and parathyroid hormone levels raised in spontaneous hypertensive rats (McCarron et al. 1981). Calcium supplementation is reported to reduce the blood pressure in these same animals and calcium restriction to raise it (McCarron 1982a). In Wistar–Kyoto rats, the blood pressure is reported to vary inversely with the calcium intake (Belizan et al. 1981; McCarron 1982b).

Observations in Humans

There was at one time thought to be a link between cardiovascular disease and soft water (Neri et al. 1972). This was almost certainly incorrect but it led on to the hypothesis that pre-eclamptic toxaemia might be due to calcium deficiency (Belizan and Villar 1980). From there it was a short step to the administration of calcium supplements to normal adults, with the reported outcome of a fall in blood pressure after 2–3 months on a 1 g/day calcium supplement (Belizan et al. 1983).

This subject is highly controversial and it is unfortunate that nearly all the reports linking calcium deficiency with high blood pressure come from only two laboratories. However, the issue is clearly important and the possibility that calcium intake plays a role in blood-pressure regulation requires further examination.

Calcium Requirements and Recommended Allowances

Normal Adults

The calcium requirement of an adult can only be defined as the amount of calcium required to preserve calcium balance. From what has been said above, it is clear that this value is determined on the one hand by the relationship between calcium intake and net absorption and on the other by the relationship between net absorption and urinary excretion. The equilibrium value has been repeatedly calculated from calcium balances on normal subjects and is generally found to fall within the range 400 to 800 mg daily or 6 to 12.5 mg/kg daily (Table 17.3). The lowest of these values (430 mg or 6.1 mg/kg) was obtained in "fully adapted" young male prisoners who had been on low-calcium diets for up to 1 year.

An exceptionally low estimate of calcium requirement at 200 mg (Hegsted et al. 1952) was derived from calcium balances at two intakes on 10 adult men in Peru. It may be in error for technical reasons or it may represent a racial or cultural difference between Peruvians and "Western" peoples, since all the

Table 17.3.   Calcium requirements of adults

Source	Subjects	Mean requirement (mg/day)	(mg/kg)
Mitchell and Curzon (1939)	107 men	—	9.75
Steggerda and Mitchell (1946)	19 young men	644	9.2
Steggerda and Mitchell (1951)	13 young adults	—	7.4
Outhouse et al. (1941)	7 young men and women	622	10.7
McKay et al. (1942)	124 college women	810	12.5
Patton and Sutton (1952)	9 young women	750	—
Hegsted et al. (1952)	10 adult men	200	3.3
Malm (1958)	23 men	430	6.1
Marshall et al. (1976)	212 balances on 84 adults	578	8.4
Heaney et al. (1978)	207 balances on premenopausal women	975	—

other quoted studies were performed on Europeans or Americans. The highest estimate (975 mg) is American and is due to consistently lower calcium absorption than is found in most other studies. It is difficult to reconcile this figure with the fact that bone status does not fall with age in normal premenopausal women whose mean calcium intake according to the same authors is much less than this (Heaney et al. 1978).

In the analysis of 212 balances on 84 subjects already referred to, the mean intake at zero balance was 578 mg (Fig. 17.5). A similar figure can be arrived at by calculation from the value at which absorbed and excreted calcium are equal

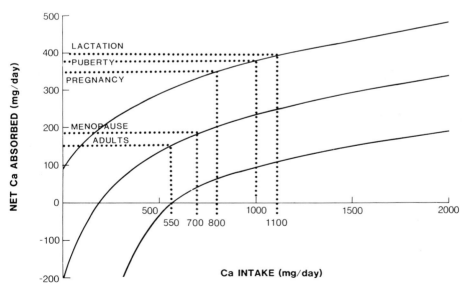

Fig. 17.6.   The estimated calcium requirement of five categories of normal subjects. On the vertical axis, the *interrupted lines* indicate the amount of calcium that needs to be absorbed. The corresponding values on the horizontal axis indicate the intake required to meet these needs. The regression lines (mean ± 2 SD) are taken from Fig. 17.2.

[about 150 mg (Fig. 17.4)] and the intake required to produce this absorption [about 550 mg (Fig. 17.2)]. This is shown in Fig. 17.6. On the basis of all these studies, a reasonable estimate of the mean calcium requirement of young western adults would appear to be 500–600 mg/day.

The range on this requirement is substantial. It is impossible to know how much of this variation is due to technical error and how much to true biological variation, but in the analysis of 212 balances referred to above it was calculated that a calcium intake of 900 mg was required to prevent negative calcium balance in 95% of normal subjects, and we recommend that the calcium allowance for adults should be at least 800 mg daily (Marshall et al. 1976).

Special Cases

Pregnancy

The calcium content of the newborn infant is about 25 g, most of which is laid down in the last trimester during which the foetus retains about 250 mg daily (American Academy of Pediatrics 1978). There is some evidence that pregnancy is associated with an increase in calcium absorption, due to a rise in the plasma $1,25(OH)_2D_3$ level (Whitehead et al. 1981) and a possible slight reduction in urinary calcium (Heaney and Skillman 1971). Assuming an obligatory maternal calcium excretion of 100 mg/day (Heaney and Skillman 1971), the required absorption to meet this and the needs of the foetus is 350 mg daily. Even if calcium absorption is enhanced, the required intake would need to be 800 mg (Fig. 17.6), and the recommended allowance about 1100 mg.

Lactation

The calcium content of human milk is about 35 mg per 100 ml. Since a lactating woman produces about 850 ml of milk daily, this represents about 300 mg of calcium. At a urine calcium of 100 mg, the required absorption is therefore 400 mg daily. Assuming maximal calcium absorption, i.e. two standard deviations above the normal mean [possibly due to the effect of prolactin on the production of $1,25(OH)_2D_3$ (Robinson et al. 1982)], the required intake is therefore 1100 mg (Fig. 17.6) and the recommended allowance 1300 mg.

Infancy

In the first 2 years of life, the daily calcium increment in the skeleton is about 100 mg (American Academy of Pediatrics 1978). The urinary calcium of newborn infants is low in absolute terms but comparable on a body weight basis to the obligatory loss in adults (about 2–3 mg/kg) (Table 17.3). Thus, the absolute excretion in the newborn is about 6 mg/day, and at the age of 1 year about 25 mg/day. When this is added to the daily skeletal increment of 100 mg, infants thus need to absorb about 125 mg/day.

Some calcium absorption studies in newborn infants are listed in Table 17.4 and illustrated in Fig. 17.7. The absorption of calcium from cow's milk

Table 17.4. Calcium balances in newborn infants (mg/day)[a]

Source	n	Milk	Calcium	Absorbed	Urine
Williams et al. (1970)	20	Cow	228	41	4
	5	Cow	240	47	3
	10	Cow	243	43	3
	10	Cow	302	115	3
	10	Human	174	44	5
Hanna et al. (1970)	11	Human	121	72	8
	15	Cow	251	72	6
	6	Cow	227	60	4
	6	Human	137	69	8
Widdowson (1965)	10	Human	131	67	8
	10	Cow	197	14	2
	10	Cow	296	16	2
Shaw (1976)	3	Cow	240	70	4
	3	Cow	251	85	4
	3	Cow	167	49	5
	2	Human	148	41	7
	2	Human	100	66	8
Widdowson et al. (1963)	11	Human	131	74	3

[a] Assumed body weight of 3.0 kg.
There are apparent differences in calcium absorption from different cow's milk preparations but further work is required to establish whether any of them are comparable to breast milk in this respect.

preparations is about 0.5 SD above the adult mean, and from human milk more than 1 SD above the adult mean. [The same inference can be drawn from the data of Fomon et al. (1963)]. Thus, different recommendations must be made for breastfed infants and those fed on cow's milk preparations. On human milk, the absorption of 125 mg requires a mean intake of about 225 mg (Fig. 17.8) and an allowance of about 300 mg which is very close to the amount provided in the average breast-milk production of 850 ml/day. On cow's milk preparations, the corresponding absorption requires a daily intake of about 350 mg (Fig. 17.7) or a daily allowance of about 500 mg. Allowing for the increase in body weight with age, this allowance should be increased to 600 mg/day in the 2nd year of life.

Childhood

From the age of 1 to 7 years, the daily skeletal calcium increment continues at about 100 mg daily (American Academy of Pediatrics 1978). The obligatory urinary calcium must rise with increasing body weight and a reasonable estimate at the end of this period for calcium requirement might be 75 mg/day. When this is added to the skeletal calcium increment of 100 mg/day, the net absorption needs to be 175 mg/day by the age of 8 years. Assuming that calcium absorption in children is comparable to that of infants on cow's milk, the required intake by age 8 is therefore 500 mg/day (Fig. 17.7) and the allowance should be 800 mg/day. Intermediate values of 600 and 700 mg respectively should be interpolated for children age 1–3 and 4–7 years.

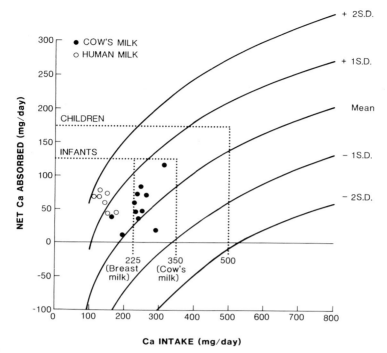

Fig. 17.7. The relation between calcium intake and net calcium absorbed in infants on cow's milk formulae (●) and human milk (○) (Table 17.4). The reference lines are those of normal adults (Fig. 17.2). The horizontal *interrupted lines* indicate the quantity of calcium that infants and children (aged 8) need to absorb and the vertical *interrupted lines* the intake required to provide this absorption. (Each point is the mean of several balances.)

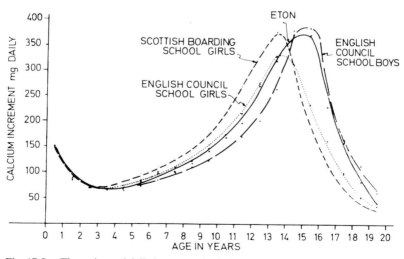

Fig. 17.8. The estimated daily increment in skeletal calcium of children at various ages. (Leitch and Aitken 1959)

Puberty

There is a striking increase in the skeletal calcium increment at puberty, spread over an age span from about 10 to 17 years of age. The peak increment in this period is about 350 mg daily (Leitch and Aitken 1959; Garn 1970; American Academy of Pediatrics 1978) but it occurs earlier and is less intense in girls than in boys (Fig. 17.8). Assuming a target figure of 300 mg for the skeleton and an obligatory urinary calcium of 75 mg, the net absorbed calcium in this period needs to be 375 mg daily. Even assuming enhanced calcium absorption, this requires an intake of 1000 mg daily and an allowance of perhaps 1200 mg daily during the peak growth phase, at least in boys (Fig. 17.6). Some increase in the allowance after this peak period provides for individual variation in age at puberty and for continuing growth in late adolescence. In girls, puberty tends to start earlier than in boys but the maximum growth rate is less. The peak requirement is therefore about 800 mg daily and the allowance 1000 mg, but the prepubertal allowance should be slightly greater than in boys because of their earlier growth spurt. It is clear that these recommended intakes are not achieved by many adolescent children (Marino and King 1980; Truswell and Darnton-Hill 1981) but the effect of this on their growth and bone density is unknown.

Ageing and Menopause

There is considerable evidence of an increase in calcium requirement in women at the menopause. There is a well-documented and significant rise of about 20% in the fasting urinary calcium (Table 17.1) and in the 24-h calcium (Table 17.5) which reflect a small but significant rise in plasma calcium. These changes are associated with a very significant rise in urinary hydroxyproline indicating increased bone resorption (Table 17.1). Balance studies from two centres indicate that the equilibrium between absorbed and excreted calcium in postmenopausal women is reached at a value of about 175–200 mg calcium per day (Table 17.6). Assuming normal calcium absorption, this represents an increase in calcium requirement to about 700 mg daily (Fig. 17.6) which would entail an allowance of about 1000 mg. This would of course have to be greater if there were a fall in calcium absorption at the menopause as Heaney et al. (1978) have suggested but we doubt whether this occurs in normal women (Table 17.5). The National Institutes of Health Consensus Conference on Osteoporosis (1984)

Table 17.5. Comparisons of pre- and postmenopausal women (mean±SE)

	Premenopause	Postmenopause	p
Radiocalcium absorption (fraction/h)[a]			
n	93	130	
	0.66±0.027	0.67±0.021	NS
24-h calcium excretion (mg)			
n	53[b]	178[c]	
	146±10	175±7	<0.05

[a] Marshall and Nordin (unpublished).
[b] Cochran and Fazid (unpublished).
[c] Nordin and Polley (1987) (less than 6 years since menopause).

Table 17.6.   Regression of urine calcium (y) on net absorption (x)

	n	Regression equation	Equilibrium value	r	p
Postmenopausal women					
Nordin					
All	105	$y = 0.28x + 124$ mg	170 mg	0.45	<0.001
Normal absorbers	43	$y = 0.41x + 112$ mg	190 mg	0.51	<0.001
Malabsorbers	62	$y = 0.17x + 124$ mg	149 mg	0.31	<0.02
Heaney	125	$y = 0.46x + 101$ mg	186 mg	0.58	<0.001
Premenopausal women					
Heaney	273	$y = 0.45x + 78$ mg	143 mg	0.14	<0.05
Young adults					
Nordin	212	$y = 0.28x + 113$ mg	158 mg	0.52	<0.001

recommended a daily calcium allowance of 1000–1500 mg after the menopause, based essentially on the Heaney et al. data.

Comparison with Current Intakes and Recommended Allowances Worldwide

Calcium intakes shown in FAO Food Balance Sheets in 1976 ranged from 242 mg per head per day in Thailand to 1288 mg per head per day in Iceland. The American figure was 969 mg per head per day and the British figure 921 mg per head per day. In all countries, the calcium intake is closely related to the intake of milk and milk products; the contribution of non-milk foods to calcium intake is relatively constant in different countries at about 200–300 mg daily (Fig 17.9).

Recommended calcium allowances vary greatly from country to country as shown in Table 17.7, which is based on the report by Committee 1/5 of the International Union of Nutritional Sciences (1982). The table shows the median value, the absolute range, the FAO recommendation, the American and British recommendations and the estimated requirements and recommended allowances based on the calculations and assumptions outlined in the earlier part of this chapter. It will be seen that existing recommendations for the various categories of subjects vary by a factor of up to 3 and that the American figures are close to the median in most categories and tend to be higher than the FAO recommendations, whereas the British figures tend to be lower.

The recommended allowances derived by our calculations differ from the median of the recommendations in different ways in the different groups.

For pregnancy and lactation, our recommended allowances are close to the current median values but slightly higher than the current FAO figure for lactation.

For infants, our recommended allowance is the same as the current median value when cow's milk is used but substantially lower when human breast milk is provided.

For young children, our estimated *requirement* is very close to current recommended *allowances* and it therefore follows that our recommended

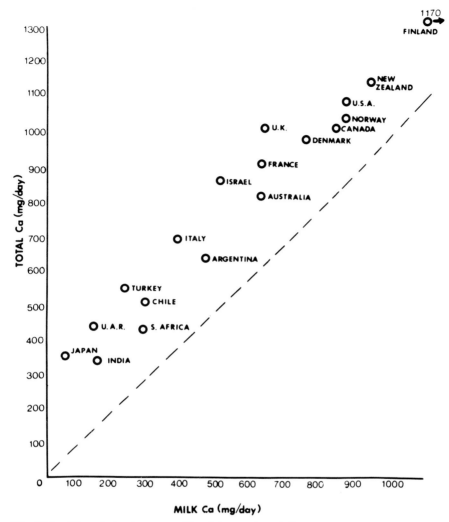

Fig. 17.9. The relation between milk calcium and total dietary calcium in different parts of the world. The line of equality is indicated.

allowance is substantially higher than most current ones though identical with the American recommendation.

Our recommendations for pubertal boys (1200 mg/day) and girls (1000 mg/day) are substantially higher than the median recommendation (700–800 mg/day) and near the top of the range, though very close to the American figures. We consider that the substantial increase in the skeletal calcium increment at puberty has not been adequately taken into account in current recommendations, particularly in the FAO and UK figures.

Our estimated requirement for adults (550 mg/day) is the most firmly based of our various calculations because it derives from large numbers of calcium

Table 17.7. Current diet calcium recommendations (mg/d) (40 countries)

	Median	Range	FAO	Australia	New Zealand	USA	UK	Estimated requirement	Recommended allowance
Pregnancy (late trimester)	1200	600–1200	1100	1100	1200	1200	1200	800	1100
Lactation	1200	400–2000	1100	1100	1200	1200	1200	1100	1300
Infants	500	360–1000	550	600	450	360	600	350 (cow's milk) 225 (breast milk)	500 300
Children	500	400–1000	450	800	400	800	600	500	800
Puberty									
Boys	800	500–1400	650	1200	600	1200	600	1000	1200
Girls	700	500–1500	650	1000	600	1200	700	800	1000
Adults									
Males	575	400–1000	450	800	500	800	500	550	800
Females	575	400–800	450	800	500	800	500		
Elderly (6 countries only)									
Males	600	500–800		—				—	—
Females	650	500–700		1000				700	1000

balances on normal subjects. It entails a recommended allowance well above the current median and FAO figures but the same as the current American figure.

There have been no previous allowances for postmenopausal women. The estimated requirement in this group is based on the rise in urine calcium but is of course critically dependent on the effect of the menopause on calcium absorption. We are not convinced that calcium absorption falls at the menopause in normal women, but this has been claimed by Heaney et al. (1978) and if correct could further raise the calcium requirement. There is of course a subset of women with severe malabsorption of calcium associated with severe spinal osteoporosis (Gallagher et al. 1973) but it is not possible to cater for this subset by means of general allowances, nor does an increase in calcium intake appear to be an appropriate response to this problem. Likewise, the needs of housebound individuals who may be at risk for vitamin D deficiency and consequent malabsorption of calcium fall outside the scope of these recommendations.

There is no reason whatever to suppose that the increased calcium intake which some of these recommendations involve could be injurious to health. Although hypercalcaemia is a dangerous condition, it cannot be produced in normal individuals by increasing the calcium intake owing to the inverse relation between calcium intake and absorption described above and to the compensatory rise in calcium excretion which follows any rise in the plasma calcium concentration. Moreoever, because of the response of the gastrointestinal tract, it is virtually impossible to produce significant hypercalciuria by calcium feeding, except in individuals with an abnormality of calcium absorption, such as occurs, for instance, in sarcoidosis and "idiopathic hypercalciuria". Nor is there any evidence that the calcification of atheromatous plaques or other dystrophic calcification is in any way related to calcium intake.

References

Aaron JE, Gallagher JC, Anderson J et al. (1974) Frequency of osteomalacia and osteoporosis in fractures of the proximal femur. Lancet I:229–233

American Academy of Pediatrics Committee on Nutrition (1978) Calcium requirements in infancy and childhood. Pediatrics 62:826–832

Anand CR, Linkswiler HM (1974) Effect of protein intake on calcium balance of young men given 500 mg calcium daily. J Nutr 104:695–700

Baker MR, McDonnell H, Peacock M, Nordin BEC (1979) Plasma 25-hydroxy vitamin D concentrations in patients with fractures of the femoral neck. Br Med J i:589

Belizan JM, Villar J (1980) The relationship between calcium intake and edema-proteinuria and hypertension gestosis: an hypothesis. Am J Clin Nutr 33:2202–2210

Belizan JM, Pineda O, Sainz E, Menendez LA, Villar J (1981) Rise of blood pressure in calcium-deprived pregnant rats. Am J Obstet Gynecol 141:163–169

Belizan JM, Villar J, Pineda O et al. (1983) Reduction of blood pressure with calcium supplementation in young adults. JAMA 249:1161–1165

Bollet AJ, Engh G, Parson W (1965) Sex and race incidence of hip fractures. Arch Intern Med 116:191–194

Charles P, Taagehoj F, Jensen L, Mosekilde L, Hansen HH (1983) Calcium metabolism evaluated by Ca^{47} kinetics: estimation of dermal calcium loss. Clin Sci 65:415–422

Christiansen C, Christensen MS, McNair P, Hagen C, Stocklund K, Transbol IB (1980) Prevention of early postmenopausal bone loss: controlled 2-year study in 315 normal females. Eur J Clin Invest 10:273–279

Christiansen C, Christensen MS, Larsen N-E, Transbol IB (1982) Pathophysiological mechanisms of estrogen effect on bone metabolism. Dose–response relationships in early postmenopausal women. J Clin Endocrinol Metab 55:1124–1130

Cochet B, Jung A, Griessen M, Bartholdi P, Schaller P, Donath A (1983) Effects of lactose on intestinal calcium absorption in normal and lactase-deficient subjects. Gastroenterology 84:935–940

Committee 1/5 of the International Union of Nutritional Sciences (1982) Recommended dietary intakes around the world. Nutr Abstr Rev 53:939–1119

Cook PJ, Exton-Smith AN, Brocklehurst JC, Lempert-Barber SM (1982) Fractured femurs, falls and bone disorders. J R Coll Physicians Lond 16:45–50

Crilly RG, Francis RM, Nordin BEC (1981) Steroid hormones, ageing and bone. J Clin Endocrinol Metab 10:115–139

DeLuca HF (1973) The kidney as an endocrine organ for the production of 1,25 dihydroxyvitamin D_3 a calcium-mobilising hormone. N Engl J Med 289:359–365

Draper HH, Scythes CA (1981) Calcium, phosphorus, and osteoporosis. Fed Proc 40:2434–2438

Ellis FR, Holesh S, Sanders TAB (1974) Osteoporosis in British vegetarians and omnivores. Am J Clin Nutrit 27:769–770

Elsassar U, Hesp R, Klenerman L, Wootton R (1980) Deficit of trabecular and cortical bone in elderly women with fracture of the femoral neck. Clin Sci 59:393–395

Ettinger B, Genant HK, Cann CE (1987) Postmenopausal bone loss is prevented by treatment with low-dosage estrogen with calcium. Ann Intern Med 106:40–45

Faccini JM, Exton-Smith AN, Boyde A (1976) Disorders of bone and fracture of the femoral neck. Evaluation of computer image analysis in diagnosis. Lancet I:1089–1092

Fomon SJ, Owen GM, Jensen RL, Thomas LN (1963) Calcium and phosphorus balance studies with normal full term infants fed pooled human milk or various formulas. Am J Clin Nutr 12:346–357

Food and Agriculture Organisation (1976) Food balance sheets. FAO, Rome

Food and Agriculture Organisation/World Health Organisation (1961) Expert group report on calcium requirements. FAO, Rome

Fourman P, Royer P (1968) Calcium metabolism and the bone. Blackwell, Oxford, pp 48–50

Gallagher JC, Nordin BEC (1975) Effects of oestrogen and progestogen therapy on calcium metabolism in post-menopausal women. In: van Keep PA, Lauritzen C (eds) Estrogens in the post-menopause. Karger, Basel, pp 150–176 (Frontiers of hormone research, vol 3)

Gallagher JC, Young MM, Nordin BEC (1972) Effects of artificial menopause on plasma and urine calcium and phosphate. Clin Endocrinol 1:57–64

Gallagher JC, Aaron J, Horsman A, Marshall DM, Wilkinson R, Nordin BEC (1973) The crush fracture syndrome in postmenopausal women. Clin Endocrinol Metab 2:293–315

Gallagher JC, Riggs BL, Eisman J, Hamstra A, Arnaud SB, DeLuca HF (1979) Intestinal calcium absorption and serum vitamin D metabolites in normal subjects and osteoporotic patients. J Clin Invest 64:729–736

Gallagher JC, Jerpbak CM, Jee WSS, Johnston KA, DeLuca HF, Riggs BL (1982) 1,25-dihydroxy-vitamin D_3: short and long-term effects on bone and calcium metabolism in patients with post-menopausal osteoporosis. Proc Natl Acad Sci USA 79:3325–3329

Garn SM (1970) The earlier gain and the later loss of cortical bone. CC Thomas, Springfield, Illinois

Genant HK, Cann CE, Ettinger B, Gordan GS (1982) Quantitative computed tomography of vertebral spongiosa: a sensitive method for detecting early bone loss after oophorectomy. Ann Intern Med 97:699–705

Goddard M, Young G, Marcus R (1986) Short-term effects of calcium carbonate, lactate, gluconate on the calcium–parathyroid axis in normal elderly men and women. Am J Clin Nutr 44:653–658

Goulding A (1983) Effects of varying dietary salt intake on the fasting urinary excretion of sodium, calcium and hydroxyproline in young women. NZ Med J 96:853–854

Greaves JP, Hollingworth DH (1966) Trends in food consumption in the United Kingdom. World Rev Nutr Diet 6:35–89

Hanna FM, Navarrete DA, Hsu FA (1970) Calcium–fatty acid absorption in term infants fed human milk and prepared formulas simulating human milk. Pediatrics 45:216–224

Heaney RP (1965) A unified concept of osteoporosis. Am J Med 39:877–880

Heaney RP, Skillman TG (1971) Calcium metabolism in normal human pregnancy. J Clin Endocrinol Metab 33:661–670

Heaney RP, Saville PD, Recker RR (1975) Calcium absorption as a function of calcium intake. J Lab Clin Med 85:881–890

Heaney RP, Recker RR, Saville PD (1978) Menopausal changes in calcium balance performance. J Lab Clin Med 92:953–963

Hegsted DM (1963) Symposium on human calcium requirements. JAMA 185:588–593

Hegsted JM, Moscoso I, Collazos CHC (1952) Study of minimum calcium requirements by adult men. J Nutr 46:181–201

Hegsted M, Linkswiler HM (1981) Long-term effects of level of protein intake on calcium metabolism in young adult women. J Nutr 111:244–251

Hodkinson HM (1974) Osteomalacia and femoral fractures. Lancet I:731

Horowitz M, Need AG, Philcox JC, Nordin BEC (1984) Effect of calcium supplementation on urinary hydroxyproline in osteoporotic postmenopausal women. Am J Clin Nutr 39:857–859

Horowitz M, Need AG, Philcox JC, Nordin BEC (1985) The effect of calcium supplements on plasma alkaline phosphatase and urinary hydroxyproline in postmenopausal women. Horm Metab Res 17:311–312

Horowitz M, Morris HA, Hartley TF et al. (1987) The effect of an oral calcium load on plasma ionized calcium and parathyroid hormone concentrations in osteoporotic postmenopausal women. Calcif Tissue Int 40:133–136

Horsman A, Gallagher JC, Simpson M, Nordin BEC (1977) Prospective trial of oestrogen and calcium in postmenopausal women. Br Med J ii:789–792

Horsman A, Nordin BEC, Simpson M, Speed R (1982) Cortical and trabecular bone status in elderly women with femoral neck fracture. Clin Orthop 166:143–151

Horsman A, Jones M, Francis R, Nordin BEC (1983) The effect of estrogen dose on postmenopausal bone loss. N Engl J Med 309:1405–1407

Ireland P, Fordtran JS (1973) Effect of dietary calcium and age on jejunal calcium absorption in humans studied by intestinal perfusion. J Clin Invest 52:2672–2681

Japan Statistical Year Book (1970) Bureau of Statistics, Office of the Prime Minister, Tokyo

Jasani C, Nordin BEC, Smith DA, Swanson I (1965) Spinal osteoporosis and the menopause. In: Proceedings of the Royal Society of Medicine 58:441–444

Jensen GF, Christiansen C, Boesen J, Hegedus V, Transbol I (1983) Relationship between bone mineral content and frequency of postmenopausal fractures. Acta Med Scand 213:61–63

Kanders B, Lindsay R, Dempster D, Markhard L, Valiquette G (1984) Determinants of bone mass in young healthy women. In: Christiansen C, Arnaud CD, Nordin BEC, Parfitt AM, Peck WA, Riggs BL (eds) Osteoporosis. Proceedings of the Copenhagen international symposium on osteoporosis. Glostrup Hospital, Denmark, pp 337–339

Leighton G, McKinlay PL (1930) Milk consumption and the growth of schoolchildren. Department of Health for Scotland, Her Majesty's Stationery Office, London

Leitch I, Aitken FC (1959) The estimation of calcium requirement: a re-examination. Nutr Abstr Rev 29:393–411

Lewis AF (1981) Fracture of neck of the femur: changing incidence. Br Med J 283:1217–1219

Lindsay R, Hart DM, Forrest C, Baird C (1980) Prevention of spinal osteoporosis in oophorectomised women. Lancet II:1151–1154

Lindsay R, Hart DM, Clark DM (1984) The minimum effective dose of estrogen for prevention of postmenopausal bone loss. Obstet Gynecol 63:759–763

Linkswiler HM, Zemel MB, Hegsted M, Schuette S (1981) Protein-induced hypercalciuria. Fed Proc 40:2429–2433

Lips P, Netelenbos JC, Jongen MJM et al. (1982) Histomorphometric profile and vitamin D status in patients with femoral neck fracture. Metab Bone Dis 4:85–93

Lund B, Sorensen OH, Christensen AB (1975) 25-Hydroxycholecalciferol and fractures of the proximal femur. Lancet II:300–302

Malm OJ (1958) Calcium requirment and adaptation in adult men. Scand J Clin Lab Invest 10 [Suppl 36]:1–289

Marcus R (1982) The relationship of dietary calcium to the maintenance of skeletal integrity in man – an interface of endocrinology and nutrition. Metabolism 31:93–102

Margen S, Chu J-Y, Kaufman NA, Calloway DH (1974) Studies in calcium metabolism. I. The calciuretic effect of dietary protein. Am J Clin Nutr 27:584–589

Marino DD, King JC (1980) Nutritional concerns during adolescence. Pediatr Clin North Am 27:125– 139

Marshall DH, Nordin BEC, Speed R (1976) Calcium, phosphorus and magnesium requirement. Proc Nutr Soc 35:163–173

Matkovic V, Kostial K, Simonovic I, Buzina R, Brodarec A, Nordin BEC (1979) Bone status and fracture rates in two regions of Yugoslavia. Am J Clin Nutr 32:540–549

McCarron DA (1982a) Calcium, magnesium and phosphorus balance in human and experimental hypertension. Hypertension 4:1127–1133

McCarron DA (1982b) Blood pressure and calcium balance in the Wistar-Kyoto rat. Life Sci

30:683–689

McCarron DA, Yung NN, Ugoretz BA, Krutzik S (1981) Disturbances of calcium metabolism in the spontaneously hypertensive rat. Hypertension 3:162–167

McKay H, Patton MB, Oblson MS et al. (1942) Calcium, phosphorus and nitrogen metabolism in young college women. J Nutr 24:367–384

Mitchell HH, Curzon EG (1939) The dietary requirement of calcium and its significance. Hermann, Paris, pp 36–101 (Actualités scientifiques et industrielles, no. 771)

Morris HA, Morrison GW, Burr M, Thomas DW, Nordin BEC (1984) Vitamin D and femoral neck fractures in elderly South Australian women. Med J Aust 140:519–521

Nachtigall LE, Nachtigall RH, Nachtigall RD, Beckman M (1979) Estrogen replacement therapy. I. A 10-year prospective study in the relationship of osteoporosis. Obstet Gynecol 53:277–281

National Institutes of Health Consensus Conference on Osteoporosis (1984) JAMA 252:799–802

Need AG, Horowitz M, Philcox JC, Nordin BEC (1985) 1,25-Dihydroxycalciferol and calcium therapy in osteoporosis with calcium malabsorption. Miner Electrolyte Metab 11:35–40

Need AG, Horowitz M, Philcox JC, Nordin BEC (1987) Biochemical effects of a calcium supplement in osteoporotic postmenopausal women with normal absorption and malabsorption of calcium. Miner Electrolyte Metab 13:112–116

Neri LC, Mandel JS, Hewitt D (1972) Relation between mortality and water hardness in Canada. Lancet I:931–934

Nordin BEC (1960) Osteomalacia, osteoporosis and calcium deficiency. Clin Orthop 17:235–258

Nordin BEC (1976a) Diagnostic procedures. In: Nordin BEC (ed) Calcium, phosphate and magnesium metabolism. Churchill Livingstone, Edinburgh, pp 469–524

Nordin BEC (1976b) Plasma calcium and plasma magnesium homeostasis. In: Nordin BEC (ed) Calcium, phosphate and magnesium metabolism. Churchill Livingstone, Edinburgh, pp 186–216

Nordin BEC (1976c) Nutritional considerations. In: Nordin BEC (ed) Calcium, phosphate and magnesium metabolism. Churchill Livingstone, Edinburgh, pp 1–35

Nordin BEC, Polley KJ (1987) Metabolic consequences of the menopause. Calcif Tissue Int 41(Suppl 1):1–59

Nordin BEC, Gallagher JC, Aaron JE, Horsman A (1975) Post-menopausal osteopenia and osteoporosis. In: van Keep PA, Lauritzen C (eds) Estrogens in the post-menopause. Karger, Basel, pp 31–149 (Frontiers of hormone research, vol 3)

Nordin BEC, Peacock M, Aaron J et al. (1980) Osteoporosis and osteomalacia. Clin Endocrinol Metab 9:177–205

Nordin BEC, Aaron J, Speed R, Crilly RG (1981) Bone formation and resorption as the determinants of trabecular bone volume in postmenopausal osteoporosis. Lancet II:277–280

Nordin BEC, Crilly RG, Smith DA (1984) Osteoporosis. In: Nordin BEC (ed) Metabolic bone and stone disease. Churchill Livingstone, Edinburgh, pp 1–70

Nordin BEC, Chatterton BE, Walker CJ, Wishart J (1987) The relation of forearm mineral density to peripheral fractures in postmenopausal women. Med J Aust 146:300–304

Outhouse J, Breiter H, Rutherford E, Dwight J, Mills R, Armstrong W (1941) Calcium requirement of man; balance studies on 7 adults. J Nutr 21:565–575

Parfitt AM (1981) The integration of skeletal and mineral homeostasis. In: DeLuca HF, Frost HM, Jee WSS, Johnston CC Jr, Parfitt AM (eds) Osteoporosis: recent advances in pathogenesis and treatment. University Park Press, Baltimore, pp 115–126

Parfitt AM, Gallagher JC, Heaney RP, Johnston CC, Neer R, Whedon GD (1982) Vitamin D and bone health in the elderly. Am J Clin Nutr 36:1014–1031

Paterson CR (1978) Calcium requirements in man: a critical review. Postgrad Med J 54:244–248

Patton MB, Sutton TS (1952) Utilisation of calcium from lactate, gluconate, sulphate and carbonate salts by young college women. J Nutr 48:443–452

Recker RR, Saville PD, Heaney RP (1977) Effect of estrogens and calcium carbonate on bone loss in postmenopausal women. Ann Intern Med 87:649–655

Reid IR, Schooler BA, Hannan SF, Ibbertson HK (1986) The acute biochemical effects of four proprietary calcium preparations. Aust NZ J Med 16:193–197

Riis B, Thomsen K, Christiansen C (1987) Does calcium supplementation prevent postmenopausal bone loss? N Engl J Med 316:173–177

Robinson CJ, Spanos E, James MF, Pike JW, Haussler MR, Makeen AM, Hillyard CJ, MacIntyre I (1982) Role of prolactin in vitamin D metabolism and calcium absorption during lactation in the rat. J Endocrinol 94:443–453

Sabto J, Powell MJ, Breidahl MJ, Gurr FW (1984) Influence of urinary calcium sodium on calcium excretion in normal individuals. Med J Aust 140:354–356

Sanchez TV, Mickelsen O, Marsh AG, Garn SM, Mayor GH (1980) Bone mineral mass in elderly vegetarian females and omnivorous females. In: Mazess RB (ed) Proceedings of the fourth international conference on bone measurement. NIAMDD, Bethesda, Maryland, pp 94–98 (NIH publication no. 801938)

Schuette SA, Zemel MB, Linkswiler HM (1980) Studies on the mechanism of protein-induced hypercalciuria in older men and women. J Nutr 110:305–315

Schuette SA, Hegsted M, Zemel MB, Linkswiler HM (1981) Renal acid, urinary cyclic AMP, and hydroxyproline excretion as affected by level of protein, sulfur amino acid, and phosphorus intake. J Nutr 111:2106–2116

Shaw JCL (1976) Evidence for defective skeletal mineralization in low birthweight infants: the absorption of calcium and fat. Pediatrics 57:16–25

Sherman HS (1952) Chemistry of food and nutrition. Macmillan, New York

Solomon L (1968) Osteoporosis and fracture of the femoral neck in the South African Bantu. J Bone Jt Surg [Br] 50–B 2:2–13

Steggerda FR, Mitchell HH (1946) Variability in calcium metabolism and calcium requirement of adult human subjects. J Nutr 31:407–422

Steggerda FR, Mitchell HH (1951) Calcium balance of adult human subjects on a high and low fat (butter) diet. J Nutr 45:201–211

Stevenson JC, Abeyasekera G, Hillyard CJ, Phang KG, MacIntyre I (1981) Calcitonin and the calcium-regulating hormones in postmenopausal women: effect of oestrogens. Lancet I:693–695

Truswell AS, Darnton-Hill I (1981) Food habits of adolescents. Nutr Rev 39:73–88

Walker ARP (1972) The human requirement of calcium: should low intakes be supplemented? Am J Clin Nutr 25:518–530

Wasnich RD, Ross PD, Heilbrun LK, Vogel JM (1985) Prediction of postmenopausal fracture risk with use of bone mineral measurements. Am J Obstet Gynecol 153:745–751

Whitehead M, Lane G, Young O et al. (1981) Interrelations of calcium-regulating hormones during normal pregnancy. Br Med J 283:10–12

Widdowson EM (1965) Absorption and excretion of fat, nitrogen, and minerals from "filled" milks by babies one week old. Lancet II:1099–1105

Widdowson EM, McCance RA, Harrison GE, Sutton A (1963) Effect of giving phosphate supplements to breast-fed babies on absorption and excretion of calcium, strontium, magnesium and phosphorus. Lancet II:1250–1251

Wilkinson R (1976) Absorption of calcium, phosphorus and magnesium. In: Nordin BEC (ed) Calcium, phosphate and magnesium metabolism. Churchill Livingstone, Edinburgh, pp 36–112

Williams ML, Rose CS, Morrow III G, Sloan SE, Barness LA (1970) Calcium and fat absorption in the neonatal period. Am J Clin Nutr 23:1322–1330

Wootton R, Brereton PJ, Clark MB et al. (1979) Fractured neck of femur in the elderly: an attempt to identify patients at risk. Clin Sci 57:93–101

World Health Organisation (1978) Statistical year book. WHO, Geneva

Yuen DE, Draper HH, Trilok G (1984) Effect of dietary protein on calcium metabolism in man. Nutr Abstr Rev 54:447–459

Subject Index